This book brings together our understanding of the structure and properties of cellular solids and the ways in which they can be exploited in engineering design. By unifying the modelling of many different types of cellular solid (for instance, engineering honeycombs, foams, wood, cancellous bone, cork), similarities in the behaviour in these diverse materials are explained. Case studies show how the models for foam behaviour can be used in the selection of the optimum foam for a particular engineering application.

In this new edition of their classic work on Cellular Solids, the authors have brought the book completely up to date, including: recent developments in processing of metallic and ceramic foams; new results for the post-yield behaviour, fracture toughness, creep and creep buckling, multiaxial failure criteria, and dynamic crushing of cellular solids; new models for the effect of variability and defects in the cell structure on mechanical properties; and a new description of the electrical and acoustic properties of cellular solids. Data for commercially available foams are presented on material property charts; two new case studies show how the charts are used for selection of foams in engineering design.

This book will be of interest to researchers working on cellular materials in materials science and engineering, structural mechanics and biomechanics.

Cambridge Solid State Science Series

EDITORS

Professor D. R. Clarke
Department of Materials,
University of California, Santa Barbara

Professor S. Suresh
Department of Materials Science and Engineering,
Massachusetts Institute of Technology

Professor I. M. Ward FRS
IRC in Polymer Science and Technology,
University of Leeds

Cellular solids
Structure and Properties

Second Edition

Cellular solids

Structure and properties

Second edition

Lorna J. Gibson
Department of Materials Science and Engineering,
Massachusetts Institute of Technology,
Cambridge, MA 02139, USA

Michael F. Ashby
Cambridge University Engineering Department,
Cambridge, UK

PUBLISHED BY THE PRESS SYNDICATE OF THE UNIVERSITY OF CAMBRIDGE
The Pitt Building, Trumpington Street, Cambridge, United Kingdom

CAMBRIDGE UNIVERSITY PRESS
The Edinburgh Building, Cambridge CB2 2RU, UK   http://www.cup.cam.ac.uk
40 West 20th Street, New York, NY 10011-4211, USA   http://www.cup.org
10 Stamford Road, Oakleigh, Melbourne 3166, Australia

First edition © Lorna J. Gibson and Michael F. Ashby, 1988
Second edition © Lorna J. Gibson and Michael F. Ashby, 1997

First published by Pergamon Press Ltd., 1988
Second edition published by Cambridge University Press, 1997
First paperback edition (with corrections), 1999

Typeface   10¼/13½pt Times   *System*   3B2   [KW]

A catalogue record for this book is available from the British Library

Library of Congress Cataloguing in Publication data

Gibson, Lorna J.
    Cellular solids : structure & properties / Lorna J. Gibson,
Michael F. Ashby. – 2nd ed.
        p.   cm. – (Cambridge solid state science series)
    Includes bibliographical references and index.
    ISBN 0-521-49560-1 (hc)
    1. Foamed materials.   2. Porous materials.   I. Ashby, M. F.
II. Title.   III. Series.
    TA418.9.F6G53   1997
    620.1′1–dc20   96-31571 CIP

ISBN 0 521 49560 1 hardback
ISBN 0 521 49911 9 paperback

Transferred to digital printing 2001

Contents

Preface to second edition

The ten years since the first edition of this book appeared have seen a remarkable increase in interest in cellular solids. New techniques for making ceramic and metallic foams have widened the range of man-made materials and the diversity of their applications; and the continued interest in wood, cork, and cancellous bone has stimulated new experiments and models to characterise natural cellular structures. This second edition was stimulated by these developments and by the very diverse community of enthusiastic scientists and engineers who have contacted us to discuss aspects of the structure and properties of cellular solids. Each chapter of the first edition has been extensively revised and brought up to date, but this, we found, was not enough. Techniques for making foams are more fully described; recent work on the mechanical response of solid foams to multiaxial loading, to deformation under creep conditions, and to impact loading have required new sections; acoustic properties are included for the first time; and there is a new chapter on the selection of foams to meet specified design criteria and on data sources and databases for foam properties.

As with the first edition, we have been greatly helped by many generous colleagues and friends. In addition to those listed already in the Preface to the 1st edition, we would particularly like to acknowledge the help of Dr. D. Cebon and Mr. C. Seymour of the Engineering Department, Cambridge; Professor Yves Brechet and Mr. D. Bassetti of the Department of Physical Chemistry, University of Grenoble; Professor J.-S. Huang of the National Cheng Kung University, Taiwan; Dr. M. Silva, Orthopaedic Biomechanics Laboratory, Beth Israel Hospital, Harvard Medical School, Boston; and Professor W. C. Hayes, Orthopaedic Biomechanics Laboratory, Beth Israel Hospital, Harvard Medical School,

Boston. The production of the manuscript and figures would not have been possible without the word-processing and drawing skills (and patience) of Mrs. Jo Ladbrooke and Mrs. Sheila Mason, to whom we are very grateful.

Permission to reproduce figures in the text was generously given by Cambridge University Press, Crown, Elsevier Science Limited, John Wiley and Sons, Krieger Publishing Corporation, Munksgaard International Publishers Limited, Plenum Publishing Corporation, Pulp and Paper Research Institute of Canada, Springer-Verlag and Taylor and Francis.

Finally we would like to thank the Royal Society of London; the NATO Programme for International Collaborative Research and the National Science Foundation Faculty Award for Women for providing the financial support which has made this new edition possible.

L. J. Gibson and M. F. Ashby, January 1997

Preface to first edition

Low-density, cellular solids appear widely in nature and are manufactured on a large scale by man. Aspects of their structure, and of their mechanical, thermal and other properties, have been studied in some detail by mathematicians, by physicists, by engineers and even by food technologists, each interested in one particular aspect of their geometry, or behaviour. This has led to a literature which is perhaps more scattered and diverse than that relating to any other class of engineering material: and there is not, at present, any source from which the interested reader can derive a reasonably broad and comprehensive picture. We were led to write this book in an attempt to bring together, in one text, and using a common nomenclature, a broad survey of the understanding of cellular solids.

We have been greatly helped in doing this by the advice, comments and critical readings of sections of the text, by a large number of most helpful colleagues and friends. We would particularly like to acknowledge the help of Professor C. Calladine, Dr. J. Woodhouse, Mrs. T. Shercliff, and Mr. J. Zhang of the Engineering Department, Cambridge: Dr. P. Echlin of the Botany Department, Cambridge; Ms. L. A. Demsetz, Ms. A. T. Huber, and Mr. T. C. Triantafillou of the Department of Civil Engineering, MIT; Professor L. Glicksman of the Department of Architecture, MIT; Professors B. Budiansky, J. Hutchinson, and T. McMahon of the Division of Applied Sciences, Harvard University; Professor K. E. Easterling of the University of New South Wales; Professor S. K. Maiti of the Indian Institute of Technology, Bombay; Professor D. Weaire of Trinity College, Dublin; Dr. N. Rivier of the Blackett Laboratory, Imperial College, London; and Professor M. Fortes of the Department of Metallurgy and Materials,

xiii

University of Lisbon. The drawings and micrographs, respectively, were prepared by Ms. S. Mason and Mr. J. Godlonton of the Engineering Department, Cambridge; their assistance is sincerely appreciated. We are also grateful for the financial support provided at various stages of this project by the Bernard M. Gordon (1948) Engineering Curriculum Development Fund and the Sloan Fund at MIT and by the NATO programme for international collaborative research.

Permission to reproduce figures in the text was generously given by The Royal Society of London, Elsevier Sequoia, Taylor and Francis Inc., Plenum Publishing Corp., Cambridge University Press, Van Nostrand Reinhold Inc., The Building Research Establishment, Princes Risborough, U.K., Springer-Verlag, the Pulp and Paper Institute of Canada, and Munksgaard International Publishers Ltd., Copenhagen, Denmark.

L. J. Gibson and M. F. Ashby, August 1987

Units and conversion tables

Physical constants in SI units

Absolute zero of temperature	$-273.2°C$
Acceleration due to gravity, g	$9.807 \, \text{m/s}^2$
Avogadro's number, N_A	$6.022 \times 10^{23} \, \text{mol}$
Base of natural logarithms, e	2.718
Boltzmann's constant, k	$1.381 \times 10^{-23} \, \text{J/K}$
Faraday's constant, F	$9.648 \times 10^4 \, \text{C/mol}$
Gas constant, $\bar{R}$	$8.314 \, \text{J/mol/K}$
Permeability of vacuum, μ_0	$1.257 \times 10^{-6} \, \text{H/m}$
Permittivity of vacuum, ε_0	$8.854 \times 10^{-12} \, \text{F/m}$
Planck's constant, h	$6.626 \times 10^{-34} \, \text{J/s}$
Velocity of light in vacuum, c	$2.998 \times 10^8 \, \text{m/s}$
Volume of perfect gas at STP	$22.41 \times 10^{-3} \, \text{m}^3/\text{mol}$

Conversion of units

Angle, θ	1 rad	$57.30°$
Density, ρ	1 lb/ft^3	$16.03 \, \text{kg/m}^3$
Energy, U	See next page	
Diffusion coefficient, D	1 cm^3/s	$1.0 \times 10^{-4} \, \text{m}^2/\text{s}$
Force, F	1 kgf	$9.807 \, \text{N}$
	1 lbf	$4.448 \, \text{N}$
	1 dyne	$1.0 \times 10^{-5} \, \text{N}$
Length, l	1 ft	$304.8 \, \text{mm}$
	1 inch	$25.40 \, \text{mm}$
	1 Å	$0.1 \, \text{nm}$
Mass, M	1 tonne	$1.000 \, \text{Mg}$
	1 short ton	$0.908 \, \text{Mg}$
	1 long ton	$1.107 \, \text{Mg}$
	1 lb mass	$0.454 \, \text{kg}$
Power, P	See next page	
Stress, σ	See next page	
Specific Heat, C_p	1 cal/g°C	$4.188 \, \text{kJ/kg°C}$
	1 Btu/lb°F	$4.187 \, \text{kJ/kg°C}$
Stress Intensity, K_{IC}	1 ksi$\sqrt{\text{in}}$	$1.10 \, \text{MN/m}^{3/2}$
Surface Energy, γ	1 erg/cm^2	$1 \, \text{mJ/m}^2$
Temperature, T	1°F	$0.556° \, \text{K}$
Thermal Conductivity, λ	1 cal/s.cm.°C	$4.188 \, \text{W/m.°C}$
	1 Btu/h.ft.°F	$1.731 \, \text{W/m.°C}$
Volume, V	1 Imperial gall	$4.546 \times 10^{-3} \, \text{m}^3$
	1 US gall	$3.785 \times 10^{-3} \, \text{m}^3$
Viscosity, η	1 poise	$0.1 \, \text{N.s/m}^2$
	1 lb ft.s	$0.1517 \, \text{N.s/m}^2$

Conversion of units – stress and pressure[†]

	MN/m^2	dyn/cm^2	lb/in^2	kgf/mm^2	bar	long ton/in^2
MN/m^2	1	10^7	1.45×10^2	0.102	10	6.48×10^{-2}
dyn/cm^2	10^{-7}	1	1.45×10^{-5}	1.02×10^{-8}	10^{-6}	6.48×10^{-9}
lb/in^2	6.89×10^{-3}	6.89×10^4	1	7.03×10^{-4}	6.89×10^{-2}	4.46×10^{-4}
kgf/mm^2	9.81	9.81×10^7	1.42×10^3	1	98.1	63.5×10^{-2}
bar	0.10	10^6	14.48	1.02×10^{-2}	1	6.48×10^{-3}
long ton/in^2	15.44	1.54×10^8	2.24×10^3	1.54	1.54×10^2	1

[†]To convert row unit to column unit, multiply by the number at the column–row intersection, thus $1 \, MN/m^3 = 10$ bar.

Conversion of units – energy[†]

	J	erg	cal	eV	Btu	ft lbf
J	1	10^7	0.239	6.24×10^{18}	9.48×10^{-4}	0.738
erg	10^{-7}	1	2.39×10^{-8}	6.24×10^{11}	9.48×10^{-11}	7.38×10^{-8}
cal	4.19	4.19×10^7	1	2.61×10^{19}	3.97×10^{-3}	3.09
eV	1.60×10^{-19}	1.60×10^{-12}	3.83×10^{-20}	1	1.52×10^{-22}	1.18×10^{-19}
Btu	1.06×10^3	1.06×10^{10}	2.52×10^2	6.59×10^{21}	1	7.78×10^2
ft lbf	1.36	1.36×10^7	0.324	8.46×10^{18}	1.29×10^{-3}	1

[†]To convert row unit to column unit, multiply by the number at the column–row intersection, thus $1 \, J = 10^7$ erg.

Conversion of units – power†

	kW	erg/s	hp	ft lbf/s
kW(kJ/s)	1	10^{10}	1.34	7.38×10^2
erg/s	10^{-10}	1	1.34×10^{-10}	7.38×10^{-8}
hp	7.46×10^{-1}	7.46×10^9	1	5.50×10^2
ft lbf/s	1.36×10^{-3}	1.36×10^7	1.82×10^{-3}	1

†To convert row unit to column unit, multiply by the number at the column–row intersection, thus 1 kW = 1.34 hp.

Chapter 1

Introduction

1.1 Introduction and synopsis

The word 'cell' derives from the Latin *cella*: a small compartment, an enclosed space. Our interest is in clusters of cells – to the Romans, *cellarium*, to us (less elegantly) *cellular solids*. By this we mean an assembly of cells with solid edges or faces, packed together so that they fill space. Such materials are common in nature: wood, cork, sponge and coral are examples (cellulose is from the Latin diminutive *cellula*: full of little cells).

Man has made use of these natural cellular materials for centuries: the pyramids of Egypt have yielded wooden artefacts at least 5000 years old, and cork was used for bungs in wine bottles in Roman times (Horace, 27 BC). More recently man has made his own cellular solids. At the simplest level there are the honeycomb-like materials, made up of parallel, prismatic cells, which are used for lightweight structural components. More familiar are the polymeric foams used in everything from disposable coffee cups to the crash padding of an aircraft cockpit. Techniques now exist for foaming not only polymers, but metals, ceramics and glasses as well. These newer foams are increasingly used structurally – for insulation, as cushioning, and in systems for absorbing the kinetic energy from impacts. Their uses exploit the unique combination of properties offered by cellular solids, properties which, ultimately, derive from the cellular structure.

This book brings together the understanding of the structure and properties of cellular solids, and of the ways their properties can be exploited in engineering design. They are an important class of engineering material, yet a curiously neglected one. Economically speaking, they are far more important than are fibre-composites, for example, but the literature on them is, by comparison, tiny – most undergraduate texts do not even mention them. They are produced and used on an enormous scale; if we include wood, the economics of the business is comparable with that of the aluminium or glass industry, yet they are less researched, less well understood and less adequately documented than almost any other class of material.

In this chapter we introduce briefly the structure of cellular materials, ways of making them, their properties and the applications for which they are used. We conclude the chapter with an outline of the remainder of the book and a brief description of the sources of literature on cellular materials.

1.2   What is a cellular solid?

A cellular solid is one made up of an interconnected network of solid struts or plates which form the edges and faces of cells. Three typical structures are shown in Fig. 1.1. The simplest (Fig. 1.1(a)) is a two-dimensional array of polygons which pack to fill a plane area like the hexagonal cells of the bee; and for this reason we call such two-dimensional cellular materials *honeycombs*. More commonly, the cells are polyhedra which pack in three dimensions to fill space; we can such three-dimensional cellular materials *foams*. If the solid of which the foam is made is contained in the cell edges only (so that the cells connect through open faces), the foam is said to be open-celled (Fig. 1.1(b)). If the faces are solid too, so that each cell is sealed off from its neighbours, it is said to be closed-celled (Fig. 1.1(c)); and of course, some foams are partly open and partly closed. The geometry and characterization of cells is an interesting subject in its own right, and one which has led to ingenious analysis; we deal with it in more depth in Chapter 2.

The single most important feature of a cellular solid is its *relative density*, ρ^*/ρ_s; that is, the density of the cellular material, ρ^*, divided by that of the solid from which the cell walls are made, ρ_s. Special ultra-low-density foams can be made with a relative density as low as 0.001. Polymeric foams used for cushioning, packaging and insulation have relative densities which are usually between 0.05 and 0.2; cork is about 0.14; and most softwoods are between 0.15 and 0.40. As the relative density increases, the cell walls thicken and the pore space shrinks; above about 0.3 there is a transition from a cellular structure to one which is better thought of as a solid containing isolated pores (Fig. 1.2). In this book we are concerned with the true cellular solids, and thus with relative densities of less than 0.3.

1.3   **Making cellular solids**

Almost any material can be foamed. Polymers, of course, are the most common. But metals, ceramics, glasses, and even composites, can be fabricated into cells. Pictures of a wide range of cellular materials are shown in Chapter 2. Here we briefly summarize the ways of making them.

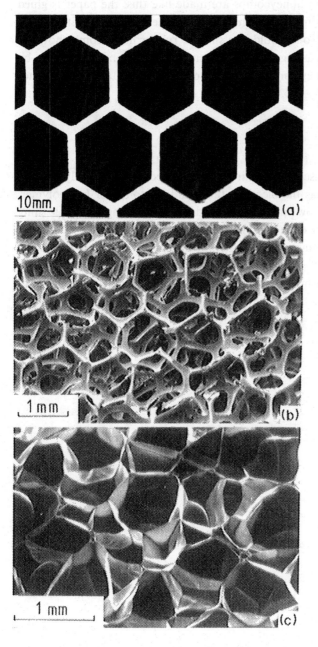

Figure 1.1 Examples of cellular solids: (a) a two-dimensional honeycomb: (b) a three-dimensional foam with open cells: (c) a three-dimensional foam with closed cells.

(a) Honeycombs

Structures like the honeycomb shown in Fig. 1.1(a) can be made in at least four ways. The most obvious is to press sheet material into a half-hexagonal profile and glue the corrugated sheets together. More commonly, glue is laid in parallel strips on flat sheets, and the sheets are stacked so that the glue bonds them together along the strips. The stack of sheets is pulled apart ('expanded') to give a honeycomb. Paper–resin honeycombs are made like this; the paper is glued and expanded, and then dipped into the resin to protect and stiffen it. Honeycombs can also be cast into a mould; the silicone rubber honeycomb shown in the figure was made by casting. And, increasingly, honeycombs are made by extrusion; the ceramic honeycombs used to support exhaust catalysts in automobiles are made in this way.

(b) Foams

Different techniques are used for foaming different types of solids. Polymers are foamed by introducing gas bubbles into the liquid monomer or hot polymer, allowing the bubbles to grow and stabilize, and then solidifying the whole thing by cross-linking or cooling (Suh and Skochdopole, 1980). The gas is introduced either by mechanical stirring or by mixing a blowing agent into the polymer. *Physical blowing agents* are inert gases such as carbon dioxide or nitrogen; they are forced into solution in the hot polymer at high pressure and expanded into bub-

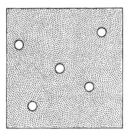

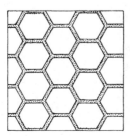

Figure 1.2 Comparison between a cellular solid and a solid with isolated pores.

bles by reducing the pressure. Alternatively, low melting point liquids such as chlorofluoro-carbons or methylene chloride are mixed into the polymer and volatilize on heating to form vapour bubbles. Microcellular foams, with cell sizes on the order of 10μ, can be made by saturating, under pressure and at room temperature, a polymer with an inert gas and then relieving the pressure and heating the supersaturated polymer to the glass transition temperature, causing cell nucleation and growth to occur. *Chemical blowing agents* are additives which either decompose on heating, or which combine together when mixed to release gas; sodicarbonamide is an example. Each process can produce open- or closed-cell foams; the final structure depends on the rheology and surface tension of the fluids in the melt. Closed-cell foams then sometimes undergo a further process called reticulation, in which the faces of the cells are ruptured to give an open-cell foam. Finally, low-density microcellular polymer foams and aerogels with relative densities as low as 0.002 and cell sizes as small as 0.1μ can be made by a variety of phase separation methods: one is to precipitate the polymer as a low-density gel in a fluid and then remove the fluid by evaporation (LeMay *et al.*, 1990).

Metallic foams can be made using either *liquid or solid state processing* (Shapovalov, 1994 and Davies and Zhen, 1983). Powdered metal and powdered titanium hydride or zirconium hydride can be mixed, compacted and then heated to the melting point of the metal to evolve hydrogen as a gas and form the foam. Mechanical agitation of a mixture of liquid aluminium and silicon carbide particles forms a froth which can be cooled to give aluminium foam. Liquid metals can also be infiltrated around granules which are then removed: for instance, carbon beads can be burned off or salt granules can be leached out. Metals can be coated onto an open-cell polymer foam substrate using electroless deposition, electrochemical deposition or chemical vapor deposition. Metal foams can also be made by a eutectic transformation: the metal is melted in an atmosphere of hydrogen and then cooled through the eutectic point, yielding the gas as a separate phase within the metal. Solid state processes usually use powder metallurgy. In the powder sintering method, the powdered metal is mixed with a spacing agent which decomposes or evaporates during sintering. Alternatively, a slurry of metal powder mixed with a foaming agent in an organic vehicle can be mechanically agitated to form a foam which is then heated to give the porous metal. Metal foams can also be formed by coating an organic sponge with a slurry of powdered metal, drying the slurry and firing to remove the organic sponge. In one of the most remarkable processes, single crystal silicon can be made porous by anodization: a silicon wafer is immersed in a solution of hydrofluoric acid, ethanol and water and subjected to a current for a brief time (Bellet and Dolino, 1994). The anodizing process tunnels, giving an interconnected network of pores with a cell size of 10 nm and a relative density as low as 0.1; yet the material remains a single crystal.

Glass foams are made by methods which parallel those for polymers; principally, by the use of blowing agents (often, H_2S). Carbon foams can be made by graphitizing polymeric foams in a carefully controlled environment. Ceramic foams are made by infiltrating a polymer foam with a slip (a fine slurry of ceramic in water or some other fluid); when the aggregate is fired the slip bonds to give an image of the original foam which, of course, burns off. Ceramic foams can also be made by chemical vapour deposition onto a substrate of reticulated carbon foam. Cement foams can be made by mixing a slurry of cement with a pre-formed aqueous foam made by mixing compressed air with an aqueous solution of suitable foaming agents. Microcellular silica foams, with cell sizes less than 100 nm and densities as low as 4 kg/m³, have been made by the sol–gel polymerization of tetraalkoxy silanes.

Cellular solids can also be made by bonding together previously expanded spheres or granules. Polystyrene is sometimes moulded in this way. Glass and metal foams can be made by sintering hollow spheres together. Syntactic foams are made by mixing hollow spheres, usually made of glass, with a binder such as an epoxy resin. And fibres can be bonded to give low-density mats (like felt) which have much in common with other types of cellular solids.

Finally, it is worth mentioning that many foodstuffs are foams. Some, like meringue, are made by mechanical beating. Others, like breads, use a biological blowing agent (yeast). Still others, like cornflakes, rely on a physical blowing agent (steam). And nature has devised her own ways of making foams either as part of the growth process of a single organism (as in bone, woods, cork and leaves) or as the product of a community of organisms (like coral, or sponge, or the nests of certain insects).

1.4   Properties of cellular solids

Foaming dramatically extends the range of properties available to the engineer. Cellular solids have physical, mechanical and thermal properties which are measured by the same methods as those used for fully dense solids. Figure 1.3 shows the range of four of these properties: the density, the thermal conductivity, the Young's modulus, and the compressive strength. The bar with dotted shading shows the range of the property spanned by conventional solids; the solid bar shows the extension of this range made possible by foaming. This enormous extension of properties creates applications for foams which cannot easily be filled by fully dense solids, and offers potential for engineering ingenuity. The low densities permit the design of light, stiff components such as sandwich panels and large portable structures, and of flotation of all sorts. The low thermal conductivity allows cheap, reliable thermal insulation that can be bettered only by expensive vacuum-based methods. The low stiffness makes foams ideal for a wide range of cushioning applications; elastomeric foams, for instance, are the standard materials for seating. The low strengths and large compressive strains make foams attractive for energy-absorbing applications; there is an immense market for cellular solids for the protection of everything from computers to canisters of hazardous wastes. The next section discusses applications of foams in a little more detail. The reader might bear Fig. 1.3 in mind while reading it.

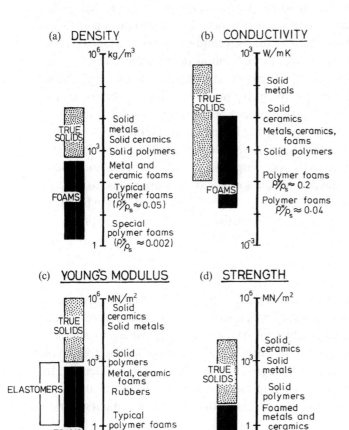

Figure 1.3 The range of properties available to the engineer through foaming: (a) density; (b) thermal conductivity (c) Young's modulus; (d) compressive strength.

1.5   **Applications of cellular solids**

The four diagrams of Fig. 1.3 relate directly to four major areas of application of cellular materials: thermal insulation, packaging, structural use, and buoyancy. Between them they account for most of the foams produced today, so we shall start with those. But there are other, smaller areas of application which are important, and growing; we will return to some of these at the end of the section.

(a) Thermal insulation

The largest single application for polymeric and glass foams is as thermal insulation. Products as humble as disposable coffee cups, and as elaborate as the insulation of the booster rockets for the space shuttle, exploit the low thermal conductivity of foams. Modern buildings, transport systems (refrigerated trucks and railway cars), and even ships (particularly those designed to carry liquid natural gas) all take advantage of the low thermal conductivity of expanded plastic foams. When fire hazard is a major consideration (as in some buildings), or when a very long life is envisaged (as in pipes and roofs) glass foams can be used instead. A particular advantage of foams for ultra-low-temperature research is their low thermal mass, reducing the amount of refrigerant needed to cool the insulation itself. The same is true, at higher temperatures, in the design of kilns and furnaces: a large part of the energy dissipated in the kiln is used to raise the temperature of the structure to its operating level; the lower the thermal mass, the greater the efficiency. The thermal mass of a foam is proportional to its relative density, so it is only a small percentage of that of the solid from which it is made. Thermal properties of foams are discussed in more detail in Chapter 7.

(b) Packaging

The second major use of man-made cellular solids is in packaging (Kiessling, 1961). An effective package must absorb the energy of impacts or of forces generated by deceleration without subjecting the contents to damaging stresses. Foams are particularly well suited for this. Figure 1.3 shows that the strength of a foam can be adjusted over a wide range by controlling its relative density. In addition, foams can undergo large compressive strains (0.7 or more) at almost constant stress, so that large amounts of energy can be absorbed without generating high stresses.

We leave the details of package design and energy absorption to Chapter 8. Here we note that foams also offer a number of secondary advantages as packaging materials. The low density means that the package is light, reducing handling and shipping costs. The low cost per unit volume and ease of moulding means that an article of curious shape can be completely embedded in a foam package, protecting it cheaply. Currently, the foams most widely used in packaging are polystyrene, polyurethane and polyethylene (Suh and Skochdopole, 1980).

(c) Structural

Many natural structural materials are cellular solids: wood, cancellous bone, and coral all support large static and cyclic loads, for long periods of time. The structural use of natural cellular materials by man is as old as history itself. Wood is still the world's most widely used structural material. The understanding of the way in which its properties depend on the density and on the direction of loading (Chapter 10) can lead to improved design with wood. Interest in the mechanics of cancellous bone (Chapter 11) stems from the need to understand bone diseases and attempts to devise materials to replace damaged bone. And, increasingly, man-made foams and honeycombs are used in applications in which they perform a truly structural function.

The most obvious example is their use in sandwich panels. The innovative design of the deHavilland Mosquito (a World War II bomber) used panels made from thin plywood skins bonded to balsa wood cores (Hoff, 1951); in later designs the balsa wood was replaced by a cellulose acetate foam. Today, sandwich panels in modern aircraft use glass or carbon-fibre composite skins separated by aluminium or paper–resin honeycombs, or by rigid polymer foams, giving a panel with enormous specific bending stiffness and strength. The same technology has spread to other applications where weight is critical: space vehicles, skis, racing yachts and portable buildings. Sandwich panels are found in nature, too: the skull is made up of two layers of dense, compact bone separated by a lightweight core of spongy, cancellous bone; some types of leaves are structured around the sandwich principle; and the cuttle bone of the cuttle fish is an elaborate multi-layer sandwich panel. The mechanics and design of sandwich panels are discussed in Chapter 9.

(d) Buoyancy

Cellular materials found one of their earliest markets in marine buoyancy. Pliny (AD 77) describes the use of cork for fishing floats. Today, closed-cell plastic foams are extensively used as supports for floating structures and as flotation in boats. Foams are much more damage-tolerant than flotation bags or chambers; because of their closed cells they retain their buoyancy even when extensively damaged; they are unaffected by extended immersion in water; and they do not rust or corrode. Flotation is commonly made from foamed polystyrene, polyethylene, polyvinyl chloride or silicones, all of which can be foamed easily to give closed cells and have outstanding resistance to water and to common pollutants. Buoyancy foams are conveniently characterized by the buoyancy factor, B, which is used to calculate the volume of foam required in a given application. It is defined by

$$B = \frac{\rho_{\text{water}} - \rho_{\text{foam}}}{\rho_{\text{water}}}.$$

Taking the density of water as $1000 \, \text{kg/m}^3$ and that of a typical flotation foam as $40 \, \text{kg/m}^3$ gives a typical buoyancy factor of 0.96. In modern sailboat design, cellular materials have been used as the core of a sandwich structure forming the deck and hull of the boat, providing structural rigidity as well as buoyancy.

(e) Other applications

Foams and honeycombs are used as *filters* at many different levels. High-quality metal castings must be free of inclusions: pouring the metal through an open-cell ceramic foam is the best way to filter them out. Foam pads can be used as cheap, disposable air filters. Most exciting is the development of the molecular filters used in membrane technology for separation of one sort of molecule from others in solution. The membrane itself is a rather special open-cell foam.

Foam sheets can be used as *carriers* for inks, dyes and lubricants, and even for enzymes for chemical processing. The cells are saturated with the dye or chemical, which either leaches slowly out (giving a controlled rate of release) or is expelled when the sheet is pressed or struck. Carriers of catalysts are important, too. Ceramic foams or honeycombs, lightly coated with platinum, are currently used as car exhaust catalysts; and the same ceramic foams can be used to carry the nickel and other catalysts used in hydrogenation and other energy-related applications.

Foams have special advantages as *water-repellent membranes* which still allow free passage of air. Open-cell polytetrafluoroethylene (PTFE) is used as a microporous, hydrophobic barrier in high-quality sporting and leisurewear, providing a fabric which breathes, yet excludes water. A similar material is used as *artificial skin*, providing protection to burn victims, while still allowing free access of air.

The special mechanical properties of foams, mentioned already, lead to a number of special uses. Their compressibility makes them unsurpassed as stoppers for bottles (Chapter 12), and the same property is exploited in bulletin boards and other functions requiring insertion and removal of pins or nails. Their slightly rough surface gives them a *high coefficient of friction*, so they are used as non-slip surfaces for trays, tables or floors. Their *high damping capacity* means that they absorb sound well, and they are used to line ceilings and walls of lecture theatres and halls. And one should not forget that many foods are foams; the cellular structure being easier to bite, chew and digest.

Finally, foams have useful *electrical properties*. The attenuation of electromagnetic waves, for instance, depends on the dielectric loss in the medium through which they travel. The low density of polymer foams gives them an exceptionally low loss factor per unit volume, recommending them for the skins of radomes and radio transmitters. And with appropriate fillers, polymer foams can be made conducting, allowing them to be used as antistatic shields and as cheap sensors.

1.6   **Outline of the book**

This book is concerned with the structure of cellular solids, with their properties, and with the way these properties are exploited in engineering design and in nature. The structure of honeycombs and foams is described in Chapter 2. Chapter 3 summarizes the properties of the solid materials from which they are made. Chapters 4, 5 and 6 describe the mechanics of honeycombs and of foams; in them the mechanisms of deformation are identified and analysed to give relationships between the properties of the cellular material and its structure. Thermal, electrical and acoustic properties are summarized in Chapter 7. Chapters 8 and 9 concern engineering design with cellular materials; here, the methods of selecting foams for the absorption of energy and for the design of sandwich panels are developed and illustrated by case studies. In Chapters 10, 11 and 12 the origins of the mechanical behaviour of natural cellular materials are described, with particular applications to wood, cancellous bone and cork. Finally, in Chapter 13, a compilation of the properties of commonly available commercial foams is provided along with the names and addresses of materials suppliers.

1.7   **The literature**

The literature on cellular solids is a large one. A *journal* – the *Journal of Cellular Plastics* – is devoted to the subject, and several others publish papers on foams. The more important of these are *Plastics and Polymers, Polymer Engineering Science*, the *International Journal of Mechanical Science*, the *Journal of Applied Polymer Science, Packaging Engineering, Packaging*, and *Composite Materials*. Others carry occasional, but often important, studies – among them, the *Proceedings of the Royal Society A*; *Metallurgical Transactions, Acta Metallurgica*, the *Journal of the American Ceramic Society* and the *Journal of Materials Science*. Papers on natural cellular materials such as wood, bone and leaf have, naturally enough, tended to appear in biological and botanical journals: those containing articles on wood are: *Wood Science, Wood Science and Technology*, and *Wood Fibre*; those with papers on bone are: the *Journal of Biomechanics*, the *Journal of Biomechanical Engineering*, the *Journal of Bone and Joint Surgery, Clinical Orthopaedics and Related Research*, and *Acta Orthopaedica Scandinavica*; for leaf, see the *Journal of Experimental Botany*, the *Annals of Botany, Plant Physiology*, and the *Journal of Cell Science*. The *Journal of Microscopy* contains occasional articles on the preparation of cellular materials for microscopic observation.

A number of *reviews* of the properties of cellular materials are available. General articles include those by Ashby (1983), Gibson (1989) and Weaire and Fortes (1994) who describe liquid as well as solid foams. Polymer foams have been reviewed by Suh and Skochdopole (1980) and Shutov (1983) while metal foams have been described by Davies and Zhen (1983). The *Materials Research Society Bulletin* publishes occasional issues on cellular materials: the first, edited by LeMay, Hopper, Hrubesh and Pekala (1990), focussed on low density microcellular materials while the second, edited by Shaefer (1994), gave a broader overview of engineered porous materials.

Proceedings of conferences on foams have recently been published by the Materials Research Society (ed. Sieradzki, *et al.*, 1991), by the American Society of Mechanical Engineers (ed. Kumar and Advani, 1992, and Kumar, 1994) and by Rapra Technology Ltd. (1991, 1993, 1995).

Several *books* on aspects of cellular solids are in print. That by D. W. Thompson (1961) is remarkable for its breadth and insight. The books edited by Hilyard (1982) and by Hilyard and Cunningham (1994) contain a series of articles on the mechanics of cellular plastics, but do not touch on metal, ceramic, glass or natural foams. That edited by Wendle (1976), and those by Semerdjiev (1982) and Shutov (1986) are primarily concerned with the selection and use of 'structural' foams: sandwich-like mouldings in which a low-density core is sandwiched between two high-density skins, of the same polymer. The book by Mustin (1968) concentrates on cushion and package design, which it covers in considerable detail. The mechanics of structural sandwich panels are described by Allen (1969). Natural cellular materials are covered by a number of books: wood by Mark (1967), Dinwoodie (1981), and Bodig and Jayne (1982); and bone by Currey (1984). Wainwright *et al.* (1976), Vogel (1988) and Vincent (1990) survey the mechanical behaviour of a wide range of biological materials including wood, bone, leaves, and stalks.

Nowhere in the literature is there an attempt to unify the ideas about cellular solids. That is the goal we set ourselves here.

References

Allen, H. G. (1969) *Analysis and Design of Structural Sandwich Panels*. Pergamon Press, Oxford.

Ashby, M. F. (1983) The mechanical properties of cellular solids. *Met. Trans.* **14A**, 1755–69.

Bellet, D. and Dolino, G. (1994) X-ray observation of porous-silicon wetting. *Phys. Rev.* **B50**, 17–162.

Bodig, J. and Jayne, B. A. (1982) *Mechanics of Wood and Wood Composites*. Van Nostrand Reinhold, New York.

Cellular Polymers: An International Conference. 20–22 March 1991 London, UK. RAPRA Technology Ltd., Shrewsbury, UK.

Cellular Polymers: 2nd International Conference. 23–25 March 1993 Edinburgh, UK. RAPRA Technology Ltd., Shrewsbury, UK.

Cellular Polymers: 3rd International Conference. 27–28 April 1995 Coventry, UK. RAPRA Technology Ltd., Shrewsbury, UK.

Currey, J. D. (1984) *The Mechanical Adaptations of Bones*. Princeton University Press, Princeton, NJ.

Davies, G. J. and Zhen, S. (1983) Metallic foams: their production, properties and applications. *J. Mat. Sci.* **18**, 1899–1911.

Dinwoodie, J. M. (1981) *Timber, Its Nature and Behaviour*. Van Nostrand Reinhold, New York.

Gibson, L. J. (1989) Modelling the mechanical behaviour of cellular materials. *Mat. Sci. Eng.* **A110**, 1–36.

Griffin, J. D. and Skochdopole. R. E. (1964) in *Engineering Design for Plastics*, ed. E. Baer. Van Nostrand Reinhold, New York.

Hilyard, N. C. (1982) *Mechanics of Cellular Plastics*. Applied Science, UK.

Hilyard, N. C. and Cunningham, A. (1994) *Low Density Cellular Plastics*, Chapman and Hall, London.

Hoff, N. J. (1951) *Structural Problems of Future Aircraft*. Third Anglo-American Aeronautical Conference. The Royal Aeronautical Society, London, p. 77.

Horace, Q. (27 BC) *Odes*. Book III, Ode 8 line 10.

Kiessling, G. C. (1961) *Modern Packaging*, **35**(3A), 287.

Kumar, V. (1994) *Cellular and Microcellular Materials*, Winter Annual Meeting of the American Society of Mechanical Engineers Chicago, IL November 6–11, 1994. ASME Symposium Volume MD-54.

Kumar, V. and Advani, S. G. (eds) (1992) *Cellular Polymers*. Winter Annual Meeting of the American Society of Mechanical Engineers Anaheim, CA Nov 8–13, 1992. ASME Symposium Volume MD-38.

LeMay, J. D., Hopper, R. W., Hrubesh, L. W. and Pekala, R. W. (1990) Low Density Microcellular Materials. *Mat. Res. Soc. Bull.* **15** (12), 19–45.

Mark, R. E. (1967) *Cell Wall Mechanics of Tracheids*. Yale University Press, New Haven, CT.

Mustin, G. S. (1968) *Theory and Practice of Cushion Design*. The Shock and Vibration Center, US Department of Defence.

Pliny, C. (AD 77) *Natural History*, Volume 16, section 34.

Semerdjiev, S. (1982) *Introduction of Structural Foams*. Society of Plastics Engineers, Brookfield Center, CT.

Shaefer, D. W. (ed.) (1994) Engineered Porous Materials. *Mat. Res. Soc. Bull.* **19** (4), 14–49.

Shapovalov, V. (1994) Porous metals. *Mat. Res. Soc. Bull.* **19** (4), 24–28.

Shutov, F. A. (1983) Foamed polymers, cellular structures and properties, *Advances in Polymer Science*, **51**, 155.

Shutov, F. A. (1986) *Integral/Structural Polymer Foams: Technology, Properties and Applications*. Springer-Verlag, Berlin.

Sieradzki, K., Green, D. J. and Gibson, L. J. (eds.) (1991) *Mechanical Properties of Porous and Cellular Materials. Mat. Res. Soc. Sym. Proc.* Vol 207.

Skochdopole, R. E. (1965) in *Encyclopedia of Polymer Science and Technology*, ed. N. M. Bikales, Wiley, Vol. 3, p. 80.

Suh, K. W. and Skochdopole, R. E. (1980) *Encyclopedia of Chemical Technology*, 3rd edn. ed. Kirk-Othmer. Vol. II, p. 82.

Thompson, D. W. (1961) *On Growth and Form*, abridged edition, ed. J. T. Bonner. Cambridge University Press, Cambridge.

Vincent, J. F. V. (1990) *Structural Biomechanics*, 2nd edn., Princeton University Press, Princeton, NJ.

Vogel, S. (1988) *Life's Devices: The Physical World of Animals and Plants*, Princeton University Press, Princeton, NJ.

Wainwright, S. A., Biggs, W. D., Curry, J. D. and Gosline, J. M. (1976) *Mechanical Design in Organisms*, Princeton University Press, Princeton, NJ.

Weaire, D. and Fortes, M. A. (1994) Stress and strain in liquid and solid foams. *Adv. Phys.* **43**, 685–738.

Wendle, B. C. (1976) *Engineering Guide to Structural Foams*. Technomic Publishing Co., Westport, CT.

Chapter 2

The structure of cellular solids

2.1   **Introduction and synopsis**

The structure of cells has fascinated natural philosophers for at least 300 years. Hooke examined their shape, Kelvin analysed their packing, and Darwin speculated on their origin and function. The subject is important to us here because the properties of cellular solids depend directly on the shape and structure of the cells. Our aim is to characterize their size, shape and topology: that is, the connectivity of the cell walls and of the pore space, and the geometric classes into which these fall.

The single most important structural characteristic of a cellular solid is its *relative density*, ρ^*/ρ_s (the density, ρ^*, of the foam divided by that of the solid of which it is made, ρ_s).[†] The fraction of pore space in the foam is its *porosity*; it is simply $(1 - \rho^*/\rho_s)$. Generally speaking, cellular solids have relative densities which are less than about 0.3; most are much less – as low as 0.003. At first sight one might suppose that the *cell size*, too, should be an important parameter, and sometimes it is; but (as later chapters show) most mechanical and thermal properties depend only weakly on cell size. *Cell shape* matters much more; when the cells are equiaxed the properties are isotropic, but when the cells are even slightly elongated or flattened the properties depend on direction, often strongly so. And there are important topological distinctions, too. The first is that between two-dimensional cells (in which the cell walls have a common generator, as in a *honey-*

[†]   Throughout this book symbols with a subscript 's' refer to properties of the solid while those with a superscripted asterisk '*' refer to the properties of the cellular material.

comb) and three-dimensional ones, in which cell walls have random orientations in space (as in a *foam*). The distinction is a useful one: the modelling of properties in two dimensions (detailed in Chapter 4) is much simpler than in three, yet much of the simpler analysis can be extended, by using approximate arguments, to the more complex geometry of foams (as we do in Chapter 5). In three dimensions there is also the distinction between cells which are *closed* (that is, each cell is sealed off from its neighbours by membrane-like faces) and those which are *open* (so that the cells interconnect). Subtler topological details – the *connectivity*[†] of the cell edges and faces, for instance – may appear less important (and have largely been neglected); but they, too, can have a profound effect on properties.

So it is helpful to develop a familiarity with cell structure, and it is rewarding too – many are aesthetically pleasing. The following sections show examples of natural and man-made cellular solids. Geometric idealizations for the cell shape allow the relative density to be calculated from the dimensions of the cell edges and faces. Topological laws govern connectivity of edges and faces, and impose constraints on the dispersion of cell sizes which help in understanding the more complex aspects of cell geometry. Anisotropy, a neglected topic, can be quantified approximately. From all this we draw a procedure for classifying and describing cellular structures.

2.2   Cell structure

Honeycombs and foams

The structure of cellular solids ranges from the near-perfect order of the bee's honeycomb to the disordered, three-dimensional networks of sponges and foams. When, around 1660, Robert Hooke perfected his microscope, one of the first materials he examined was cork (Chapter 12). What he saw led him to identify the basic unit of plant and biological structure; it was he who called it 'the cell'. His book *Micrographia* (Hooke, 1664) records it thus:

> I no sooner descern'd these (which were indeed the first microscopical pores I ever saw, and perhaps, that were ever seen, for I had not met with any Writer or Person that had made any mention of them before this) but me thought I

[†] The connectivity of cell faces is the number of faces which meet at an edge; it is usually three but can be as high as six. The connectivity of cell edges is the number of edges that meet at a node or vertex: usually three in honeycombs and four in foams, though it can be much higher. Examples are given later in this chapter. In graph theory, connectivity is used in a different sense ('a graph is *n*-connected if it is disconnected by removal of *n*, and not less than *n*, edges') which should not be confused with our simpler meaning.

had in the discovery of them, presently hinted to me the true and intelligible reason of all the *Phenomena* of Cork.

Hooke's careful drawings of cork cells show their roughly hexagonal shape in one plane and their box-like shape in the plane normal to this (Fig. 2.1). The cells were stacked in long rows, with very thin walls 'as those thin films of Wax in a Honey-comb'.

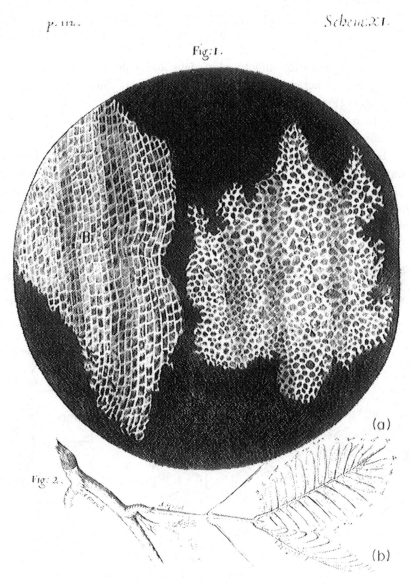

Figure 2.1   (a) Radical and (b) tangential sections of cork as observed by Robert Hooke (1664).

And if ever there was a structure that fascinated mathematicians, physicists and biologists, it is the bee's honeycomb (Fig. 2.2). It is certainly the most studied of all cellular conformations, and one of the most beautiful. Euclid (3rd century BC) admired its regularity, and Pliny (AD 77) speaks of men who devoted a lifetime to its study. The early literature is vast – Thompson's (1961) list of more than 30 learned treatises on the subject published before 1860 includes works by men of the stature of Colin MacLaurin (1742), Georges Louis Leclere Buffon (1753) and Charles Darwin (1859). The regularity of the bee's cell is remarkable (though it is not as perfect as is generally thought; see Wyman, 1865), and it will serve to epitomize two-dimensional cellular solids.

Honeycombs are manufactured of metal and paper for the cores of sandwich panels, and of ceramic for supports for catalysts and components of heat exchangers (Fig. 2.3). Most have cells which, like those of the bee, are *hexagons*; then the edge connectivity is three. Honeycombs with *square* or *triangular* cells, with an edge connectivity of four or of six (Fig. 2.3) can be made, but they are less efficient in the sense that they use more solid to enclose the same pore volume. A *random* honeycomb (a soap froth between glass plates, for instance) has small cells with as few as three sides, and large cells with, commonly, as many as nine sides (there is no upper limit); but if the connectivity

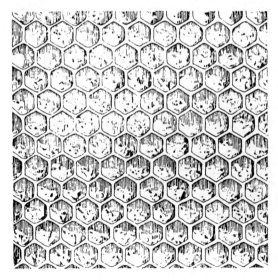 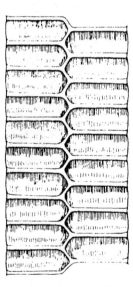

Figure 2.2   The honeycomb of a bee. (After T. Rayment, A cluster of bees. *The Bulletin,* Sydney; cited by Thompson, 1961 (Fig. 42) courtesy of Cambridge University Press.)

is fixed (it is three in the soap froth) the average number of edges per cell is determined by topological laws that we shall come to in the next section.

The study of the geometry of three-dimensional cellular solids (which we shall call *foams*) has a pedigree almost as distinguished as that of the honeycomb. Plateau (1873), in his treatise on solid geometry, identified the cell shape as a rhombic dodecahedron (a 12-faced, garnet shaped figure), and it is certainly possible to partition space into an array of cells with this shape. But it is not the most efficient way to do so. For over a century, it was thought that the space-filling cell which minimizes surface area per unit volume was Kelvin's tetrakaidecahedron with slightly curved faces (Kelvin, 1887) (Fig. 2.4(a)). Recently, using computer software for minimization of surface area (Brakke, 1992), Weaire and Phelan (1994) have identified a unit cell of even lower surface area per unit volume (by about 0.3%) (Fig. 2.4(b)). The unit cell is made up of six 14-sided cells (with 12 pentagonal and 2 hexagonal faces) and two pentagonal dodecahedra, all of equal volume. The 14-sided cells are arranged in three orthogonal axes with the 12-sided cells lying in the interstices between them, giving an overall simple cubic lattice structure. Only the hexagonal faces are planar: all of the pentagonal faces are curved. But efficient space-filling is not the only factor influencing cell shape, and when other factors dominate, the cell shape can be quite different from either of these two figures.

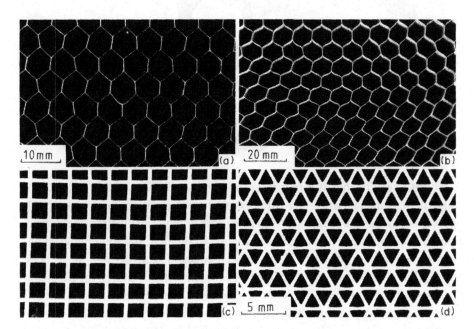

Figure 2.3 Two-dimensional cellular materials: (a) aluminium honeycomb, (b) paper-phenolic resin honeycomb, (c) ceramic honeycomb with square cells, (d) ceramic honeycomb with triangular cells.

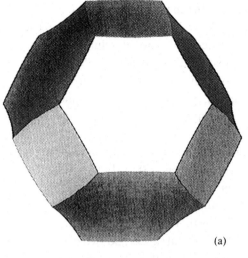

Figure 2.4 (a) Kelvin's tetrakaidecahedral cell (b) Weaire and Phelan's unit cell, consisting of six 14-sided polyhedra and two 12-sided polyhedra (courtesy of Professor Denis Weaire).

(a)

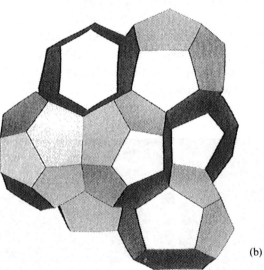

(b)

To get a feel for this, it is best to examine some real foams. Fig. 2.5 shows a range of man-made foams: three are polymeric, two are metallic, two are made of ceramics and one is a glass. The top pair (both polymeric) illustrate the distinction between open-cell and closed-cell foams. In the first, the solid material has been drawn into struts which form the cell edges. The struts join at vertices; usually, but not always, four cell edges meet at each vertex – that is, the edge-connectivity is four. In the second, solid membranes close off the cell faces, but it is rare that the solid is uniformly distributed between the edges and faces. During foaming, surface tension draws solid into the cell edges, leaving a thin skin framed by thicker edges; the volume fraction ϕ of the solid in the edges is an important parameter in characterizing foams and in understanding their properties. The

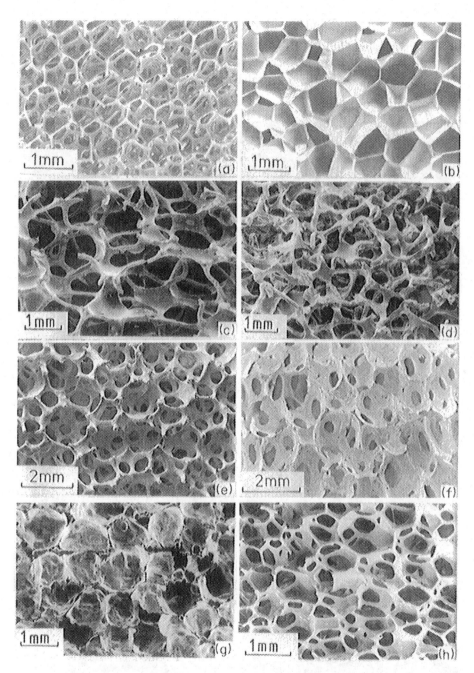

Figure 2.5 Three-dimensional cellular materials: (a) open-cell polyurethane, (b) closed-cell polyethylene, (c) nickel, (d) copper, (e) zirconia, (f) mullite, (g) glass, (h) a polyether foam with both open and closed cells.

connectivity of the edges and faces of closed cells is harder to establish. If surface
tension is the dominant force which shapes the structure, then four edges meet at
109.4° at each vertex, and three faces meet at 120° at each edge: the metal, ceramic
and glass foams shown in Fig. 2.5 are all like this.

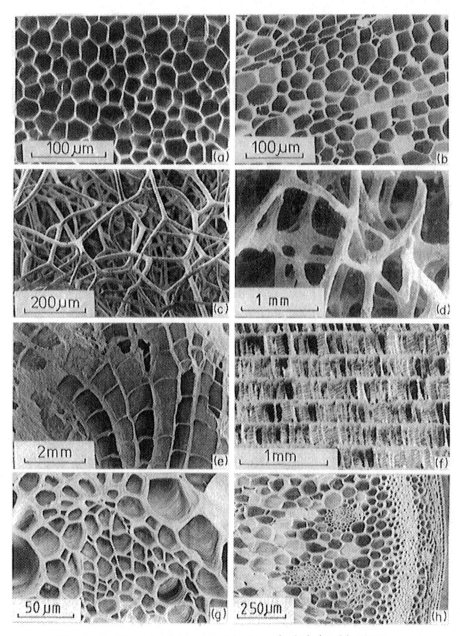

Figure 2.6   Natural cellular materials: (a) cork, (b) balsa, (c) sponge,
(d) cancellous bone, (e) coral, (f) cuttlefish bone, (g) iris leaf, (h) stalk of a plant.

Natural foams show much more variation (Fig. 2.6). Some, like cork or balsa (top pair), have closed cells which are almost as regular as a honeycomb. Others, like sponge or cancellous bone (second pair), are an open network of struts with a connectivity of three, four, five or even six. Still others, like coral or cuttle bone (third pair), are obviously anisotropic: the cells are elongated or aligned in particular directions, and this gives them properties which depend on the direction in which they are measured. Almost all natural cellular solids are like this: the enormous anisotropy of wood, and of leaf and stalk (the last pair of micrographs in Fig. 2.6) is largely caused by the elongated shape of their cells.

Many foods are foams, too (Fig. 2.7). Bread (first picture) usually has closed cells, expanded by the fermentation of yeast or by CO_2 from bicarbonate of

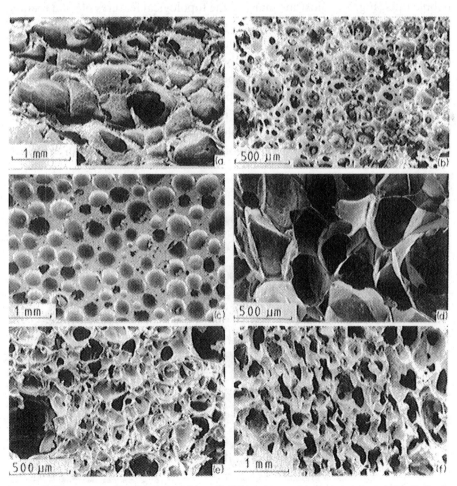

Figure 2.7   Food foams: (a) bread, (b) meringue, (c) chocolate bar, (d) junk food crisp, (e) Malteser, (f) Jaffa cake.

soda. Meringue (second picture) is foamed egg white and sugar. Chocolate (third picture) and other hard, brittle candies are often expanded to make them more attractive to eat or cheaper, per unit volume, to make. And some of the most profitable sectors of the food industry – breakfast cereals and snack-foods are examples – rely on steam-foaming to produce their texture and crunchiness.

Low-density solids can be assembled in other ways. One is to create a random mat of fibres, the contact points of which may be bonded together. Paper, felt and cotton wool are like this; so, too, are the thermal tiles which protect the American space shuttle on re-entry (Fig. 2.8). Their mechanical and thermal properties have much in common with those of cellular materials, and although they are not the main subject of this book, the methods developed here can be applied to them also.

Bubbles, too, can be bonded to give low-density structures. Soap foams and bubble rafts (Fig. 2.9) illustrate many of the topological features of solid foams. Solid bubbles can be sintered together to give stiff, low-density structures. In two dimensions bubbles pack with great regularity. In three dimensions this is unusual (though not impossible); more commonly, they pack in a dense-random arrangement so that they occupy a fraction of about 0.64 of the available space, not 0.72 as in close packing; but because each bubble is hollow, the relative density of the entire structure is much lower than this. Figure 2.9, bottom picture, shows a structure made by aggregating aluminium spheres (Akutagawa, 1985); similar structures can be made of other metals and of glass. Each particle is a thin spherical

Figure 2.8   Fibrous materials: (a) felt, (b) paper, (c) cotton wool, (d) space shuttle tile (courtesy of Nancy Shaw, NASA Lewis Research Center, Cleveland, OH).

shell and is bonded to its neighbours at its contact points. Like the more conventional cellular solids it has a low density, and is a good thermal insulator. Its properties can be calculated by adaptations of the methods developed in later chapters.

The properties of any of these cellular solids depend on the way the solid is distributed in the cell faces and edges. So the first step in understanding them is to quantify, as far as possible, their structures. We first describe the topological laws governing the shape and size of the cells, and then develop equations relating relative density to cell wall thickness and length. Finally, we suggest a method for characterizing the structure of honeycombs and foams.

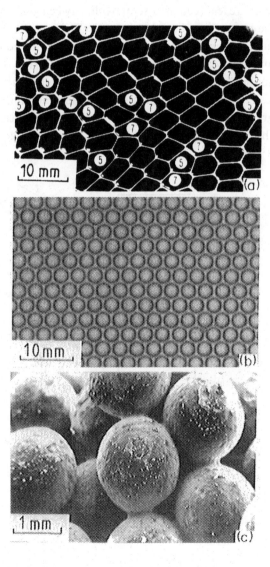

Figure 2.9 Bubble mat materials: (a) two-dimensional soap honeycomb, (b) bubble raft, (c) hollow sintered aluminium spheres (courtesy of Wes Akutagawa, Jet Propulsion Lab, Pasadena CA).

2.3   Shape, size and topology

(a) Cell shape

The unit cells which pack to fill a plane in two dimensions are sketched in Fig. 2.10, which shows the shapes available for both isotropic and anisotropic cells. Even when the cell shape is fixed, the cells can be stacked in more than one way (Fig. 2.11) giving structures which have differing edge connectivity, and different properties. Man-made honeycombs use these regular shapes. In most, three cell edges meet at a vertex (they are *three-connected*), forming an array of six-sided cells as in Fig. 2.3(a) or (b); but some are four-connected, giving the four-sided cells of Fig. 2.3(c) or six-connected, giving triangular cells shown in Fig. 2.3(d). Natural two-dimensional cells are less regular: a soap froth between glass slides, or the cells of the retina of the eye, even the bee's honeycomb (Wyman, 1865) contains elements of randomness which are visible as four, five, seven and even eight-sided cells (labelled in Fig. 2.9a), and with this goes a dispersion of cell sizes. But even the most random of honeycombs obey certain topological laws, developed below, which mean that precise statements can be made about them.

In three dimensions a greater variety of cell shapes is possible. Figure 2.12 shows most of the shapes that can be packed together to fill space; their geome-

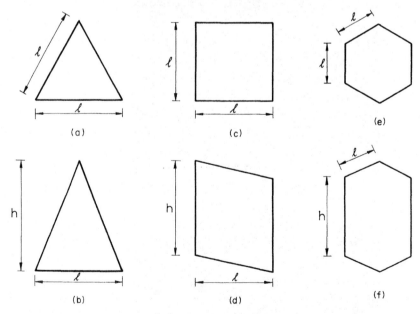

Figure 2.10   Polygons found in two-dimensional cellular materials: (a) equilateral triangle, (b) isosceles triangle, (c) square, (d) parallelogram, (e) regular hexagon, (f) irregular hexagon. Note that any triangle, quadrilateral or hexagon with a centre of symmetry will fill the plane.

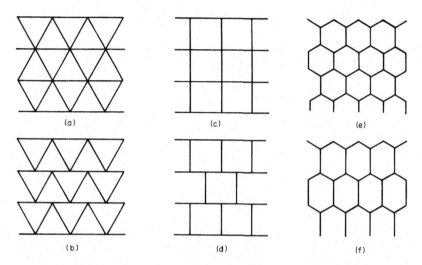

Figure 2.11 Packing of two-dimensional cells to fill a plane: (a, b) Two packings of equilateral triangles with $Z_e = 6$ and $Z_e = 4$, respectively. When $Z_e = 4$, $n = 4$, topologically. (c, d) Two packings of squares with $Z_e = 4$ and $Z_e = 3$, respectively. When $Z_e = 3$, $n = 6$ topologically. (e) Packing of regular hexagons. (f) Packing of irregular hexagons.

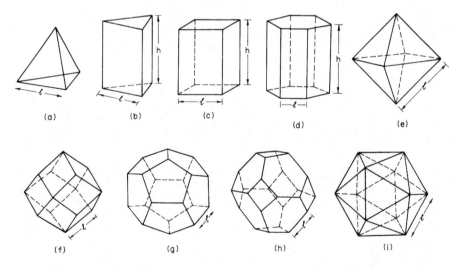

Figure 2.12 Three-dimensional polyhedral cells: (a) tetrahedron, (b) triangular prism, (c) rectangular prism, (d) hexagonal prism, (e) octahedron, (f) rhombic dodecahedron, (g) pentagonal dodecahedron, (h) tetrakaidecahedron, (i) icosahedron.

Table 2.1 Geometric properties of isolated cells

Cell shape	Number of faces, f (a)	Number of edges, n (b)	Number of vertices, v (c)	Cell volume (d) (e)	Surface area (a) (d) (e)	Edge length (b) (e)	Comments (f)
Tetrahedron	4	6	4	$0.118l^3$	$\sqrt{3}l^2$	$6l$	Regular
Triangular prism	5	9	6	$\frac{\sqrt{3}}{4}l^3 A_r$	$\frac{\sqrt{3}}{2}l^2(1+2\sqrt{3}A_r)$	$6(1+A_r/2)$	Packs to fill space
Square prism	6	12	8	$l^3 A_r$	$2l^2(1+2A_r)$	$8(1+A_r/2)$	Packs to fill space (cube is regular)
Hexagonal prism	8	18	12	$\frac{3\sqrt{3}}{2}l^3 A_r$	$3\sqrt{3}l^2(1+2A_r/\sqrt{3})$	$12(1+A_r/2)$	Packs to fill space
Octahedron	8	12	6	$0.471l^3$	$3.46l^2$	$12l$	Regular
Rhombic Dodecahedron	12	24	14	$2.79l^3$	$10.58l^2$	$24l$	Packs to fill space
Pentagonal Dodecahedron	12	30	20	$7.663l^3$	$20.646l^2$	$30l$	Regular
Tetrakaidecahedron	14	36	24	$11.31l^3$	$26.80l^2$	$36l$	Packs to fill space
Icosahedron	20	30	12	$2.182l^3$	$8.660l^2$	$30l$	Regular

Notes

(a) In an infinite, packed array every face is shared between two cells; the number of faces per cell and the surface area per cell in the array are one-half of these values.

(b) In an infinite, packed array every edge is shared between Z_f faces (usually three); the number of edges per cell and the edge length per cell in the array are $1/Z_f$ of these values.

(c) In an infinite, packed array every vertex is shared between Z_e edges (usually four); the number of vertices per cell in the array is $1/Z_e$ of this value.

(d) See, for instance, De Hoff and Rhines (1968), and the *Handbook of Chemistry and Physics* (1972).

(e) A_r is the aspect ratio: $A_r = h/l$ where h and l are defined in Fig. 2.12.

(f) Regular polyhedra have faces and edges which are all identical. Most do not pack to fill space.

tries are characterized in Table 2.1. Like the two-dimensional cells, they must pack to fill space. In the undistorted forms sketched in Fig. 2.12 only some of them do: the triangular, rhombic and hexagonal prisms, the rhombic dodecahedron (a body with 12 diamond-shaped faces)[†] and the tetrakaidecahedron (a body with six square and eight hexagonal faces) are true space-filling bodies (Plateau, 1873; Kelvin, 1887; Smith, 1952; Thompson, 1961; Ko, 1965). The resulting cellular arrays are sketched in Fig. 2.13. All of these have been suggested, at one time or another, as idealizations for the cells in foams, together with others which, by themselves, do not pack properly unless distorted: the tetrahedron (four faces), the icosahedron (20 faces); and the pentagonal dodecahedron (12 identical five-sided faces – Jones and Fesman, 1965; Harding, 1967; Chan and Nakamura, 1969; Menges and Knipschild, 1975; Barma *et al.*, 1978). Most foams, of course, are not regular packings of identical units but contain cells of different sizes and shapes with differing numbers of faces and edges. But even the most random of foams obey topological rules like those that govern honeycombs, and again this means that precise and useful statements can be made about them.

The topology of cells is a field which has fascinated physicists, biologists and metallurgists for centuries: the considerable literature contains contributions from the likes of Robert Hooke, Lord Kelvin, and, of course, Leonhard Euler; and it continues to challenge scientists with a bent for this sort of geometry (see, for example, Rivier, 1983, 1985, 1986; Weaire, 1983; Weaire and Rivier, 1984; Ferro and Fortes, 1985; Rosa and Fortes, 1986, Fortes, 1986a,b). From this body of work we extract three simple results which help in the characterization of honeycombs and foams: Euler's law, the Aboav–Weaire law and Lewis's rule; and we examine the deductions that can be made from them. We then examine ways of characterizing anisotropy, and the implications of connectivity in cellular solids.

(b) Euler's law: faces, edges and vertices

From a geometric point of view it is helpful to think of a cellular structure as *vertices*, joined by *edges*, which surround *faces*, which enclose *cells*. (In two dimensions we lose one dimension and think of vertices joined by edges which enclose faces or cells.) The number of edges which meet at a vertex is the edge-connectivity, Z_e (it is usually three in a honeycomb and four in a foam but it can have other values). The number of faces which meet at an edge is the face-connectivity, Z_f (usually three for a foam but it, too, can have other values). The number of

---

[†] The trapezo-rhombic dodecahedron is closely related to the rhombic dodecahedron, and also packs to fill space (Ko, 1965).

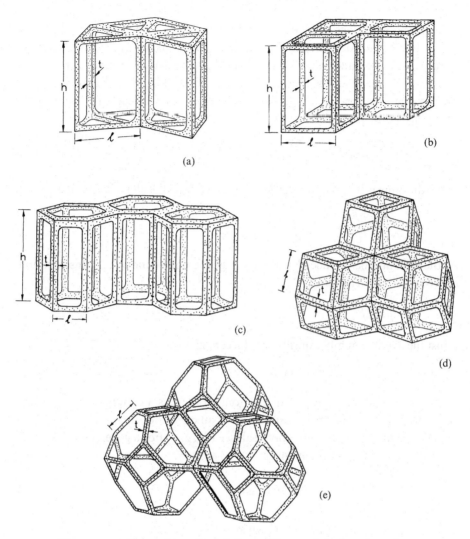

Figure 2.13 The packing of polyhedra to fill space: (a) triangular prisms, (b) rectangular prisms, (c) hexagonal prisms, (d) rhombic dodecahedra, (e) tetrakaidecahedra.

vertices V, of edges E, of faces F and of cells C are related by *Euler's law* (Euler, 1746; Lakatos, 1976) which, for a large aggregate of cells, states that:

$$F - E + V = 1 \quad \text{(two dimensions)} \tag{2.1}$$

$$-C + F - E + V = 1 \quad \text{(three dimensions).} \tag{2.2}$$

A honeycomb with regular hexagonal cells obviously has six edges surrounding each face. An immediate consequence of Euler's law, shown below, is that an *irregular* three-connected honeycomb also has, on average, six sides per face; this means that a five-sided face can only be introduced if, somewhere, a compensating seven-sided face is created also; a four-sided face requires an eight-sided face or two seven-sided ones; and so on. In practice this usually means that most cells have six sides, and those which do not are paired, as in the soap honeycomb of Fig. 2.9a. The result is obtained by noting that, if $Z_e = 3$, then $E/V = 3/2$ (the 2 appears because each edge is shared between two vertices). If F_n is the number of faces with n sides, we have that:

$$\sum \frac{nF_n}{2} = E \tag{2.3}$$

(the 2 this time is because an edge separates two faces). Euler's law then gives:

$$\left(6 - \frac{\sum nF_n}{F}\right) = \frac{6}{F}.$$

As the total number of faces F becomes large the right-hand side of this equation approaches zero. The term

$$\sum nF_n/F$$

is just the average number of sides per face, $\bar{n}$, giving:

$$\bar{n} = 6 \quad \text{(two dimensions)} \tag{2.4}$$

so that a face with five edges can exist only if there is a complementary seven-edged face, and so on. These results are for an edge-coordination, Z_e, of 3; this value, though common, is not universal (see Fig. 2.3). For a general edge-coordination Z_e, the same argument gives:

$$\bar{n} = \frac{2Z_e}{Z_e - 2} \quad \text{(two dimensions)} \tag{2.5}$$

which, of course, reduces to $\bar{n} = 6$, when $Z_e = 3$.

The simplicity of this result encourages the hope that analogous equations might hold in three dimensions. It is not quite so easy. The problem is less constrained (Eqn. (2.2) has one more variable in it), so the deductions one can make are less general. One, however, is of interest. For an isolated cell ($C = 1$), the average number of edges per face $\bar{n}$ is related to the number of faces f (for an edge-connectivity of three on the surface of the cell) by:

$$\bar{n} = 6\left(1 - \frac{2}{f}\right) \quad \text{(three dimensions)} \tag{2.6a}$$

or, more generally,

$$\bar{n} = \frac{Z_e Z_f}{Z_e - 2}\left(1 - \frac{2}{f}\right). \tag{2.6b}$$

An important consequence follows: in most foams, most cells have faces with five edges, no matter what the shape of the cell. If, for example, the cells are on average dodecahedra ($f = 12$), the average face, according to Eqn. (2.6) has $\bar{n} = 5.0$ edges exactly. If they are tetrakaidecahedra ($f = 14$), the average is 5.14; and even if they are icosahedra ($f = 20$), the average is only 5.4. So frequent sightings of pentagonal faces in foams do *not* mean that the cells are pentagonal dodecahedra (as is often claimed), though a proper statistical study of the average number of edges per face would, via Eqn. (2.6), give information about cell shape. Many foam-like structures (grain boundaries in metals and ceramics for instance) have $f \approx 14$ and $\bar{n} \approx 5.1$ (see, for example, Smith, 1952), but, as Weaire and Rivier (1984) and Fortes (1986b) emphasize, there is nothing that requires the cells to have an average of 14 faces: both f and $\bar{n}$ depend on how the foam is made, and on the forces which shape its cells.

(c) Origins and consequences of dispersion in cell size

Many honeycombs and all foams have cells with a range of sizes. The size distribution can be very narrow, as in the bee's honeycomb, or it can be so broad that the largest cells are hundreds of times bigger than the smallest. Dispersion in cell size does not imply anisotropy – that is a different problem, which we discuss later. But both dispersion and anisotropy relate ultimately to the way the cellular material was made.

In many foaming processes a supersaturated gas is made to separate from a liquid. Gas bubbles nucleate throughout the liquid and grow, initially as spheres but later (as they start to interact) as polyhedral cells. If the bubbles all nucleate randomly in space but at the same time, and all grow with the same linear growth rate, then the initial structure is a *random Voronoi honeycomb* (two dimensions) or a *Voronoi foam* (three dimensions). The characteristic polyhedron is the cell, centred on the point of nucleation, which contains all points which are closer to this nucleation point than to any other (Voronoi, 1908[†]). The cells obviously fill space, and are random. A honeycomb made in this way (Fig. 2.14(a)) looks very different from the neat hexagons of Figs. 2.2 and 2.3, but it still has $\bar{n} = 6$ as Euler's law requires. A Voronoi foam, too, has curiously angular cells, and cells with, on average, 15.54, not 14, faces (Meijering, 1953; Rivier, 1982). The well-defined construction of these structures leads to other results; for the Voronoi honeycomb the average area of an n-sided face is roughly proportional to n (Crain, 1972, 1978) and the distribution functions for face areas and edge lengths can be calculated exactly (Gilbert, 1962; Meijering, 1953).

[†] In practice it is constructed by joining the random nucleation points by straight lines and bisecting these with plane surfaces. The envelope of the surfaces which surrounds a point is its Voronoi cell.

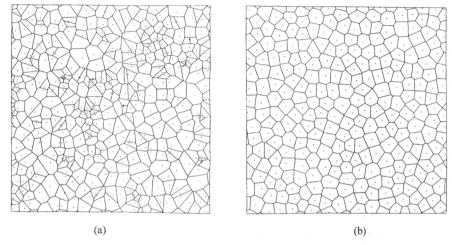

(a)   (b)

Figure 2.14   (a) A Voronoi honeycomb for a set of random points (marked). (This is the structure that would form if cells nucleate at random points, all appearing at the same instant, and grow with the same linear growth rate.) (b) A Voronoi honeycomb for a set of points, initially random, from which all points closer than a critical spacing were removed. (This is the structure that would form if cells cannot nucleate closer together than a critical spacing and grow with the same linear growth rate.)

The essential simplicity of the Voronoi foam has made it alluring to theoreticians, who devise generalizations to include randomness in time as well as in space (the bubbles do not all appear at once but nucleate progressively). The nearest thing to Voronoi structures in nature are the cellular solids created by the competitive building of sea creatures and of insects: coral and some sponges; the nests of wasps and ants; and, of course, the bee's honeycomb. Their great regularity looks quite different from the Voronoi honeycomb of Fig. 2.14(a). That is because the creature has a finite size, excluding the nucleation of two cells from points which are closer than this. This, too, can be modelled (Weaire and Rivier, 1984) and the result, as one might expect, is a foam or honeycomb of much greater regularity. An example is shown in Fig. 2.14(b), in which an array of points are random with the constraint that no two can be closer than a chosen 'exclusion distance'. The result is not unlike some natural structures (particularly the nest of the wasp). But it still has a slightly angular look that most familiar honeycombs lack.

That is because competitive growth is not the only factor which shapes foams. There are several others. The most obvious of these is surface tension. When this is the dominant shaping force, and when it is isotropic (independent of orientation) the structure is one which minimizes the surface area at constant cell volume. Then the cell edges in a honeycomb, and the cell faces in a foam, meet at 120°. The faces (in general) have a curvature which is related to the pressure difference Δp between the pair of cells which meet at that face. This pressure difference in a honeycomb is related to the radius of curvature of the face, r, by

$$\Delta p \propto \frac{T}{r} \qquad (2.7a)$$

or, in foams, to the principal radii, r_1 and r_2, by

$$\Delta p \propto T \left(\frac{1}{r_1} + \frac{1}{r_2} \right) \qquad (2.7b)$$

where T is the surface tension. When cells are of identical size and shape (so $\Delta p = 0$) these requirements are satisfied in two dimensions by an ordered packing of regular hexagons ($\bar{n} = 6$) and in three dimensions by an ordered packing of tetrakaidecahedra ($f = 14, \bar{n} = 5.14$). When cells are random, and of different size, less can be said about the structure, though it must, of course, satisfy Euler's law. Surface-tension foams are of particular interest because a mechanism can be imagined for the way in which they coarsen – and it leads to an interesting topological consequence. Cells coarsen if the gas or fluid in one cell diffuses through its walls into the surrounding cells (as it is known to do in soap foams, and may also do during the foaming of polymers). The rate of diffusion is proportional to the pressure difference, Δp (Eqn. 2.7) times the area of the shared face. From this may be derived[†] von Neumann's growth law (von Neumann, 1952; Rivier, 1986; Fortes, 1986a) which states that the area growth-rate of a two-dimensional cell is proportional to its number of sides minus six:

$$\frac{\mathrm{d}A}{\mathrm{d}t} = C_1(n - 6) \qquad (2.8a)$$

where A is the cell area and C_1 is a constant. The equivalent three-dimensional result for describing the volume-growth rate of a cell is given by Rivier (1983):

$$\frac{\mathrm{d}V}{\mathrm{d}t} = C_2(f - \bar{f}) \qquad (2.8b)$$

where f is the number of faces on the cell, $\bar{f}$ is the average number of faces per cell (often 14) and C_2 is another constant. Thus many-sided cells grow, few-sided cells shrink; and six-sided cells (in two dimensions) and 14-faced cells (in three dimensions) remain unchanged. The consequence is a trend to a more and more inhomogeneous structure, reaching, at long times, a dispersion which can be very broad (Weaire and Kermode, 1983). Such a structure, shown in Figs. 2.15 and 2.16, is dramatically different both from a Voronoi foam and from the initial surface-tension foam. In the foam of Fig. 2.16, the smaller cells make up the cell walls of the larger cells, giving a hierarchical cellular material with cells at two

---

[†] If all edges are of length l, and all meet at 120°, the edges of a cell with n edges must bend through the angle $2\alpha = 2 (60 - 360/n)$, and its radius of curvature, R, is given by $1/R = 1/l (60 - 360/n)$. The rate of change of area of the cell, $\mathrm{d}A/\mathrm{d}t$, is proportional to $n (60 - 360/n)$, or to $(n - 6)$.

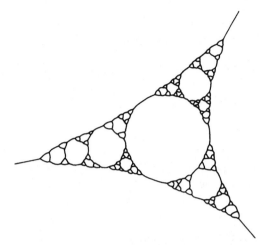

Figure 2.15   A schematic of a foam with a very wide distribution of cell sizes (Weaire and Rivier, 1984, Fig. 17 courtesy of Taylor and Francis).

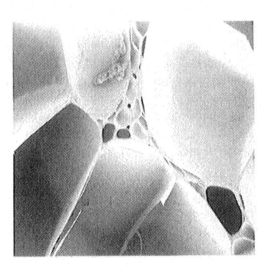

Figure 2.16
A micrograph of a polymeric foam showing a structure which resembles the schematic of Fig. 2.15.

levels of scale (Lakes, 1993). Although higher order hierarchical cellular materials, with cells at even more levels of scale, might be envisioned theoretically, processes for making such materials have yet to be developed.

We will return to this point in a moment to introduce two further topological results. But it would be misleading to imply, by omission, that cell shape is always controlled either by competitive growth or by surface tension. In a liquid, viscous forces can also have a profound effect on topology (Harvey, 1954). As cells grow in a liquid, the liquid drains out of the cell walls which simultaneously stretch. A high viscosity limits this drainage, and can generate forces which are larger than those of surface tension giving structures which are not at equilibrium (in the sense of having cell walls which meet at 120° and so on). Most polymer foams are made by foaming the liquid polymer. When this is done in a mould, so that the volume expansion causes it to rise in one direction, the cells are generally elongated in the rise direction because of viscous forces. Not only are such foams anisotropic but the average number of edges and faces per cell can be changed too. In natural cellular materials, such as bone and wood, there are further considerations: the shapes of the cells are clearly influenced by the loads that the material has to carry (Currey, 1984). The cell walls in trabecular bone (Chapter 11), for instance, follow the trajectories of the principal stresses to which the bone as a whole is subjected. Mechanical considerations determine the orientation of cells in wood (Dinwoodie, 1981) and may be responsible for the cell patterns in the leaves of plants and the wings of insects. Figure 2.17, a drawing of cells in the wing of a dragonfly, shows how ribs of equal thickness form a hexagonal net, meeting at 120°, but ribs of unequal thickness meet in a way that reduces the angle of bending of the thicker rib, probably because it maximizes the stiffness-to-weight ratio of the wing. In these examples it is mechanical efficiency, not minimum length or surface area, that shapes the structure.

(d) The Aboav–Weaire law and Lewis's rule

We saw earlier that a seven-edged cell in a honeycomb had a five-edged partner, often as a neighbour; in foams, a 16-faced cell is likely to have a 12-faced mate. It is generally true that a cell with more sides than average has neighbours which, taken together, have less than the average number. This correlation was noted by Aboav (1970, 1980) in Smith's (1952, 1964) pictures of soap honeycombs, and can be seen in Fig. 2.9(a) of this chapter. The observation is described for honeycombs by the rule (Aboav, 1970, given a formal derivation by Weaire, 1974):

$$\bar{m} = 5 + \frac{6}{n} \tag{2.9}$$

where n is the number of edges of the candidate cell and $\bar{m}$ is the average number of edges of its n neighbours (for a fuller discussion, see Weaire and Rivier, 1984). One might speculate that a similar result holds in three dimensions, such that:

$$\bar{g} = 13 + \frac{14}{f} \tag{2.10}$$

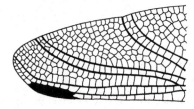

Figure 2.17 The wing of a dragonfly (redrawn from Thompson, 1961 Fig. 36, courtesy of Cambridge University Press). Cell edges of equal thickness meet at 120°, but thin cell edges meet thick ones at about 90°.

where f is the number of faces on a cell and $\bar{g}$ the average number of faces of its neighbours; calculations by Fortes (private communication) show Eqn. (2.10) to be a reasonable approximation when $\bar{f} = 14$. Rivier (1985) gives a more formal derivation of an alternative three-dimensional generalization. All express the same topological rule: that the more sides one cell acquires, the fewer, on average, have those which immediately surround it.

The study of cell topology (biological cells this time) has turned up another remarkable result. Lewis (1923, 1928, 1943), examining a variety of two-dimensional cellular patterns, found that the area of a cell varied linearly with the number of its edges:

$$\frac{A(n)}{A(\bar{n})} = \frac{n - n_0}{\bar{n} - n_0} \tag{2.11}$$

where $A(n)$ is the area of a cell with n sides, $A(\bar{n})$ is that of the cell with the average number of sides, $\bar{n}$, and n_0 is a constant (Lewis finds $n_0 = 2$). We have already mentioned that this rule holds for Voronoi cells; Lewis finds that it holds for most other two-dimensional cells as well. It is obviously plausible, but it was more than 50 years after Lewis first published his observations that a proof was found (Rivier and Lessowski, 1982). The argument can be generalized to three dimensions (Rivier, 1982), for which

$$\frac{V(f)}{V(\bar{f})} = \frac{f - f_0}{\bar{f} - f_0} \tag{2.12}$$

where $V(f)$ is the volume of a polyhedral cell with f faces, $V(\bar{f})$ that for a cell with the average number of faces $\bar{f}$, and f_0 is a constant about equal to 3.

These results, taken together, give some insight into the topological consequences of cell coarsening. They quantify the fact that a bigger-than-average cell is always surrounded by smaller-than-average neighbours. The pressure difference between cells is related to the size difference, and it is often this pressure difference which drives coarsening (by forcing the cell fluid – usually a gas – to diffuse through the cell walls). When this is so the local coarsening rate accelerates with time until the smaller, shrinking, cells disappear completely, consumed by the bigger ones. The process clearly leads to an increasing dispersion of cell sizes, with the result shown schematically in Fig. 2.18, which is based in part on the measurements taken by Aboav (1980) from Smith's (1952) photographs of soap froths.

2.4   Calculating relative density

The properties of a honeycomb or a foam depend, above all else, on its relative density. Models for these properties given in later chapters concern themselves

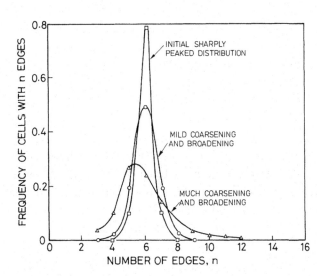

Figure 2.18 The distribution of cell sizes as cells in a soap froth grow by diffusion of gas from small, high-pressure cells to large, low-pressure cells. As growth proceeds there is an increasing dispersion of cell sizes. (Based partly on Aboav's (1980) analysis of Smith's (1952) data.)

with the microscopic struts and plates that make up the cell edges and faces, and the way these respond to load, or transmit heat, or dissipate energy. But models which contain microscopic parameters are not useful – the engineer cannot be expected to check on the cell-wall thickness of every foam he or she uses. All he or she knows is its density. So we will need equations which relate the cell dimensions and shape to the density.

The proper choice of equation depends on the *dimensionality* of the structure (honeycomb versus foam) and on whether (if a foam) it has *open* or *closed* cells. If the cell edge-length is l and the cell-wall thickness is t, and $t \ll l$ – that is, the relative density is low – then for all honeycombs:

$$\frac{\rho^*}{\rho_s} = C_1 \frac{t}{l} \qquad (2.13a)$$

for all open-cell foams with edges of length l and thickness t:

$$\frac{\rho^*}{\rho_s} = C_2 \left(\frac{t}{l}\right)^2 \qquad (2.13b)$$

and for all closed-cell foams with faces of side l and uniform thickness t:

$$\frac{\rho^*}{\rho_s} = C_3 \frac{t}{l} \qquad (2.13c)$$

where the Cs are numerical constants, near unity, that depend on the details of the cell shape.

For most purposes this is enough. The properties are calculated in terms of t and l; these are then converted to relative density via Eqns. (2.13), introducing a single numerical constant C. The constant is then determined by making a single experimental measurement, thereby 'calibrating' the result for all densities. But when the relative density is large these simple expressions overestimate the density because of *double-counting*: the corners of open cells and the edges and corners of closed cells are counted twice. Then the next order of approximation may be useful. Trivial geometry shows that, for honeycombs, it has the form:

$$\frac{\rho^*}{\rho_s} = C_1 \frac{t}{l}\left(1 - D_1 \frac{t}{l}\right) \tag{2.14a}$$

where D_1 is another constant. The introduction of this second constant is a nuisance (it makes the scaling procedure used in Chapter 5 much more difficult); it is removed by noting that the correction term in brackets is important only when t/l is large, and that it should make ρ^*/ρ_s go to the value unity when the cell thickness has the value that just causes the cell to fill completely. For regular hexagonal honeycombs this happens when $t = \sqrt{3}l$, giving:

$$\frac{\rho^*}{\rho_s} = \frac{2}{\sqrt{3}} \frac{t}{l}\left(1 - \frac{1}{2\sqrt{3}} \frac{t}{l}\right). \tag{2.14b}$$

The analogous results for square and triangular honeycombs are given in Table 2.2.

The geometry for foams is more difficult. On dimensional grounds the vertex-corrected density of open-cell foams must have the form

$$\frac{\rho^*}{\rho_s} = C_2\left(\frac{t}{l}\right)^2\left(1 - D_2\left(\frac{t}{l}\right)\right). \tag{2.14c}$$

For closed-cell foams the simple expression Eqn. (2.13c) double-counts both edges and corners, and, again for dimensional reasons,

$$\frac{\rho^*}{\rho_s} = C_3 \frac{t}{l}\left(1 - D_3 \frac{t}{l}\left(1 - D_4 \frac{t}{l}\right)\right). \tag{2.14d}$$

For any given cell shape the correction factors D_2, D_3 and D_4 can be evaluated. But the correction is significant only when the relative density is large: 0.2 or more. Usually, experimental scatter, and corrections of other sorts, mask the small differences that double-counting introduces. One of these is the *distribution of solid* between edges and faces. In many foams which have closed cells, solid is drawn preferentially into the cell edges, which are thicker than the faces (it can be seen most easily in Fig. 2.16, but is true of most closed-cell foams). Let ϕ be the volume fraction of the solid contained in the cell edges; the remaining fraction $(1 - \phi)$ is in the faces. Then, if the cell edge thickness is t_e and the face thickness is t_f,

$$\phi = \frac{\dfrac{\bar{n}f}{2Z_f}lt_e^2}{\dfrac{\bar{n}f}{2Z_f}lt_e^2 + \dfrac{f}{2}l^2 t_f} = \frac{t_e^2}{t_e^2 + \dfrac{Z_f}{\bar{n}}t_f l} \tag{2.15}$$

where (as before) $\bar{n}$ is the average number of edges per face on a single cell, f is the number of faces on a single cell, and Z_f is the number of faces that meet at an edge. The relative density of the foam is

$$\frac{\rho^*}{\rho_s} = \frac{\dfrac{\bar{n}f}{2Z_f}lt_e^2 + \dfrac{f}{2}l^2 t_f}{C_4 l^3} = \frac{f}{C_4}\left\{\frac{\bar{n}}{2Z_f}\frac{t_e^2}{l^2} + \frac{1}{2}\frac{t_f}{l}\right\} \tag{2.16}$$

where C_4 is the constant relating cell volume to l^3 (Table 2.1). From these we obtain equations for t_f/l and t_e/l which are used later:

$$\frac{t_f}{l} = \frac{2C_4}{f}(1 - \phi)\frac{\rho^*}{\rho_s} \tag{2.17a}$$

and

$$\frac{t_e}{l} = \left(\frac{2Z_f}{\bar{n}}\frac{C_4}{f}\phi\frac{\rho^*}{\rho_s}\right)^{1/2}. \tag{2.17b}$$

For most foams, $Z_f = 3$, $\bar{n} \approx 5$, $f \approx 14$ and $C_4 \approx 10$ so that, to a good approximation,

$$\frac{\rho^*}{\rho_s} = 1.2\left(\frac{t_e^2}{l^2} + 0.7\frac{t_f}{l}\right) \tag{2.18}$$

$$\frac{t_f}{l} = 1.4(1 - \phi)\frac{\rho^*}{\rho_s} \tag{2.19a}$$

and

$$\frac{t_e}{l} = 0.93\phi^{1/2}\left(\frac{\rho^*}{\rho_s}\right)^{1/2}. \tag{2.19b}$$

Sometimes it is important to know the constants C_1, C_2 and C_3, and it is instructive to examine their magnitudes, if only to be convinced that they are close to unity. They can only be determined for regular, space-filling structures: a hexagonal honeycomb, for instance, or a tetrakaidecahedral foam. Cell shapes like a pentagon in two dimensions or icosahedron in three do not pack to fill space; nor (because of topological requirements) can they be distorted to do so, and this means that there is no sensible way of calculating the relative density of a hypothetical cellular material made of them. They *can* be made to fill space by mixing them with other figures (heptagons for the pentagons; tetrahedra for the icosahedra) but even then distortion is needed. Calculating the relative density of these mixed structures is difficult – and there are too many alternative combinations to give the results much meaning. But the polyhedra that fill space are another matter: for these, the relative densities can be calculated (De Hoff and Rhines, 1968). They are listed in Table 2.2. The honeycombs are straightforward; note only that the constant C_1 of Eqns. (2.13a) and (2.14a) has values in the range 1.15 to 3.46 (the best choice is the value for hexagons: 1.15). Five foam

Table 2.2   The relative densities of cell aggregates

Honeycombs

Equilateral triangles
$(Z_e = 6, n = 3 \text{ or } Z_e = 4, \text{ or } n = 4)^{\dagger}$

$$\frac{\rho^*}{\rho_s} = 2\sqrt{3}\,\frac{t}{l}\left(1 - \frac{\sqrt{3}}{2}\frac{t}{l}\right)$$

Squares
$(Z_e = 4, n = 4 \text{ or } Z_e = 3, n = 6)^{\dagger}$

$$\frac{\rho^*}{\rho_s} = 2\,\frac{t}{l}\left(1 - \frac{1}{2}\frac{t}{l}\right)$$

Regular hexagons
$(Z_e = 3, n = 6)$

$$\frac{\rho^*}{\rho_s} = \frac{2}{\sqrt{3}}\,\frac{t}{l}\left(1 - \frac{1}{2\sqrt{3}}\frac{t}{l}\right)$$

Three dimensions: open cells (aspect ratio $A_r = h/l$)

Triangular prisms
$(Z_e = 8, Z_f = 4.5, \bar{n} = 3.6, \bar{f} = 5)$

$$\frac{\rho^*}{\rho_s} = \frac{2}{\sqrt{3}}\,\frac{t^2}{l^2}\left\{1 + \frac{3}{A_r}\right\}$$

Square prisms
$(Z_e = 6, Z_f = 4, \bar{n} = 4, \bar{f} = 6)$

$$\frac{\rho^*}{\rho_s} = \frac{t^2}{l^2}\left\{1 + \frac{2}{A_r}\right\}$$

Hexagonal prisms
$(Z_e = 5, Z_f = 3.6, \bar{n} = 4.5, \bar{f} = 8)$

$$\frac{\rho^*}{\rho_s} = \frac{4}{3\sqrt{3}}\,\frac{t^2}{l^2}\left\{1 + \frac{3}{2A_r}\right\}$$

Rhombic dodecahedra
$(Z_e = 5.33, Z_f = 3, \bar{n} = 4, \bar{f} = 12)$

$$\frac{\rho^*}{\rho_s} = 2.87\,\frac{t^2}{l^2}$$

Tetrakaidecahedra
$(Z_e = 4, Z_f = 3, \bar{n} = 5.14, \bar{f} = 14)$

$$\frac{\rho^*}{\rho_s} = 1.06\,\frac{t^2}{l^2}$$

Three dimensions: closed cells (aspect ratio $A_r = h/l$)

Triangular prisms
$(Z_e = 8, Z_f = 4.5, \bar{n} = 3.6, \bar{f} = 5)$

$$\frac{\rho^*}{\rho_s} = 2\sqrt{3}\,\frac{t}{l}\left\{1 + \frac{1}{2\sqrt{3}A_r}\right\}$$

Square prisms
$(Z_e = 6, Z_f = 4, \bar{n} = 4, \bar{f} = 6)$

$$\frac{\rho^*}{\rho_s} = 2\,\frac{t}{l}\left\{1 + \frac{1}{2A_r}\right\}$$

Hexagonal prisms
$(Z_e = 5, Z_f = 3.6, \bar{n} = 4.5, \bar{f} = 8)$

$$\frac{\rho^*}{\rho_s} = \frac{2}{\sqrt{3}}\,\frac{t}{l}\left\{1 + \frac{\sqrt{3}}{2A_r}\right\}$$

Rhombic dodecahedra
$(Z_e = 5.33, Z_f = 3, \bar{n} = 4, \bar{f} = 12)$

$$\frac{\rho^*}{\rho_s} = 1.90\,\frac{t}{l}$$

Tetrakaidecahedra
$(Z_e = 4, Z_f = 3, \bar{n} = 5.14, \bar{f} = 14)$

$$\frac{\rho^*}{\rho_s} = 1.18\,\frac{t}{l}$$

---

†See Fig. 2.11.

packings are listed, three based on prisms and two on equiaxed bodies. The prismatic structures have cell edges of two sorts (the base l and the height h of the prism), so for these the equation contains the aspect ratio, $A_r = h/l$. When the aspect ratio is 1 the constant C_2 for open cells (Eqns. 2.13b and 2.14b) ranges from 1.06 to 4.61 (the best choice is the tetrakaidecahedral value of 1.06). For closed cells the constant C_3 (eqns. 2.13c and 2.14c) ranges from 1.18 to 4.46 (the best choice is again the tetrakaidecahedral value 1.18). The derivation of the equations in Table 2.2 assumes that $t \ll l$, an approximation that starts to break down, as we have said, when $\rho^*/\rho_s > 0.2$. Then the correction term shown in Eqns. (2.14) can be added, but it is worthwhile only if the property data are exceptionally precise: usually, experimental scatter is too large to make the sophistication worthwhile.

The point, then, is that the relative density of a honeycomb and of a closed-cell foam always scales as t/l; that of an open-celled foam as $(t/l)^2$, with a constant of proportionality near unity. Extensive use is made of this throughout the rest of the book.

2.5 Characterizing cellular materials

What are the parameters which characterize the structure of a cellular material? Much of the information needed to analyse its properties is contained in a record of the *material* of which the foam is made; its *relative density* (or porosity); a statement of whether its cells are *open or closed*; and the *mean cell diameter*. This sort of information is obtained from optical or scanning-electron micrographs. Without much extra effort the same micrographs can give a more complete characterization. Tables 2.3 and 2.4 list the information which is useful in characterizing a cellular solid. It may not be possible to assign values or answers to all the headings, but by responding to as many as possible the maximum amount of information is recorded.

In some instances, such as in the study of cancellous bone (Chapter 11), more detailed information may be available from techniques such as X-ray quantitative computed tomography and micro-magnetic resonance imaging. Quantitative analysis of the digitized images allows stereologic examination of any plane section of a specimen. Detailed descriptions of these techniques are given by Feldkamp *et al.* (1989), Chung *et al.* (1995) and Hipp *et al.* (1996).

(a) Honeycombs

True honeycombs, though much studied, are relatively rare; they appear mostly as the hexagonal aluminium or paper–resin structures which are used as the cores of sandwich panels and as the ceramic grids used as supports in catalytic

Table 2.3 Characterization chart for honeycombs

Material	Alumina ceramic (Fig. 2.3(c))
Density, ρ^* (kg/m^3)	1400
Edge connectivity, Z_e	4
Mean edges/cell, $\bar{n}$	4
Cell shape (and angles)	Square
Symmetry of structure	Square (mm)
Largest principal cell dimension, $\bar{L}_1$ (mm)	2.42
Smallest principal cell dimension, $\bar{L}_2$ (mm)	2.42
Shape anisotropy ratio, $R = \bar{L}_1/\bar{L}_2$	1
Standard deviation of cell size	0
Cell wall thickness, t (mm)	0.48
Relative density, ρ^*/ρ_s	0.36
Other specific features (periodic variations in density, cell size, etc.)	Highly regular

Table 2.4 Characterization chart for foams

Material	Rigid polyurethane (Fig. 2.19)
Density, ρ^* (kg/m^3)	32
Open or closed cells	Closed
Edge connectivity, Z_e	4
Face connectivity, Z_f	3
Mean edges/face, $\bar{n}^*$	—
Mean faces/cell, $\bar{f}^*$	—
Cell shape*	—
Symmetry of structure	Axisymmetric
Cell edge thickness, t_e (μm)	30
Cell face thickness, t_f (μm)	3
Fraction of material in cell edges, ϕ	0.70
Largest principal cell dimension, $\bar{L}_1$ (mm)	0.53
Smallest principal cell dimension, $\bar{L}_3$ (mm)	0.44
Intermediate principal cell dimension, $\bar{L}_2$ (mm)	0.53
Shape anisotropy ratios, $R_{12} = \bar{L}_1/\bar{L}_2$ and $R_{13} = \bar{L}_1/\bar{L}_3$	$R_{12} = 1.0$; $R_{13} = 1.2$
Standard deviation of cell size (mm)	0.075
Other specific features (periodic variations in density, cell size, etc.)	None

*From these micrographs it is not possible to measure all of the parameters, but in open–cell foams it is usually possible to do so.

converters (Fig. 2.3). But many natural materials – particularly woods – have cells so elongated that, in the plane transverse to the direction of elongation, the cells look and behave like a honeycomb. Then the structural features listed in Table 2.3 are the important ones. The data in the table give an example of its use; they describe the honeycomb of Fig. 2.3(c).

The *material description* should be as full as possible, to allow the solid material properties (Chapter 3) to be identified. The *density* is straightforward: it is usually adequate to cut a block of foam, measure its dimensions, and weigh it. But in seeking correlations between properties and structure it is important to measure the density of each sample, since densities can vary by 10% or more in a single batch of material. The *edge-connectivity* Z_e (the average number of edges which meet at a vertex) is easily measured from micrographs, the mean number of *edges/cell*, $\bar{n}$, then follows from Euler's law (Eqn. 2.5). This begins to define the *cell shape*. If the honeycomb is a uniform one the shape will be obvious: hexagons ($Z_e = 3$, $\bar{n} = 6$), or parallelograms ($Z_e = 4$, $\bar{n} = 4$), or triangles ($Z_e = 6$, $\bar{n} = 3$). Then it is only necessary to add a *characterizing angle* θ between a pair of sides to fix the shape completely. It is helpful to record the *symmetry of the structure* because the properties have this symmetry too. Ordered, regular hexagons have hexagonal symmetry (six mirror planes) and this is enough to make the properties isotropic in the plane. If the hexagons are not regular, but are randomly oriented (that is, the structure is disordered), the macroscopic properties are again isotropic. Macroscopic anisotropy can arise from anything that, when averaged over a large number of cells, destroys hexagonal symmetry. If the cell walls tend to align in a preferred direction, or tend to greater thickness in a preferred direction, or the cells on average have one axis which is longer than the other two, then properties are anisotropic. Cell elongation is the commonest cause of anisotropy; it turns out (Chapter 4) that a slight elongation can produce large differences in properties. So it makes sense to record the average values of the largest and smallest *principal cell dimensions*, $\bar{L}_1$ and $\bar{L}_2$. This is done by counting the number of cells, N_c, per unit length of a straight line lying parallel to each of the two principal directions (which are usually obvious); if the measurements are made on a plane section the mean cell diameter is given (De Hoff and Rhines, 1968) by:

$$\bar{L}_1 = \frac{1.5}{N_c} \tag{2.20}$$

and so on. The *shape anisotropy-ratio*, R, is then defined as $R = \bar{L}_1/\bar{L}_2$ where $\bar{L}_1$ is the larger principal dimension. Finally certain properties depend on the *dispersion of cell size*, that is the range of cell diameters within the honeycomb. Dispersed structures are characterized in terms of the distribution function of a local topological or geometric property, such as the number of edges per cell (Fig. 2.18) or the area of a cell. The area-distribution is perhaps the one which most easily relates to the properties we examine later. It is measured by plotting the frequency of occurrence, $P(A)$, of the cell of area A, against A, normalized so that:

$$\int_0^\infty P(A)\mathrm{d}A = 1.$$

Then the mean cell area is:

$$\bar{A} = \int_0^\infty AP(A)\mathrm{d}A. \tag{2.21}$$

The mean cell diameter is:

$$\bar{L} = \frac{2}{\sqrt{\pi}}\bar{A}^{1/2} \tag{2.22}$$

and the standard deviation of cell size (the measure of dispersion) is:

$$\mu_A = \int_0^\infty (A - \bar{A})^2 P(A)\,\mathrm{d}A. \tag{2.23}$$

This completes the formal description. But there may be other features that should be recorded. Many natural honeycombs, and some which are man-made, contain periodic variations in density or cell size (the growth rings in woods, for example). There may be occasional exceptionally large cells, cracks, broken walls, or other defects. These, too, should be recorded; they may later turn out to have a critical influence on the properties.

(b) Foams

The procedure for foams (Table 2.4) parallels that for honeycombs but is slightly more elaborate. The first new distinction is between *open* and *closed* cells – something which is usually obvious from the micrographs, but can also be determined from the permeability of the foam to a gas or a liquid. Both *edge-connectivity* (Z_e, usually 4) and *face-connectivity* (Z_f, usually 3) should be recorded, together with the mean number of edges per face, $\bar{n}$. The corresponding number of faces per cell, $\bar{f}$, is then given by Euler's law (Eqn. 2.6). The cell shape can be inferred from the connectivity and the number of faces per cell, if these can be established. Table 2.2 lists the characteristics of the simpler shapes. Most foams, of course, are not regular: then the cells vary in shape but – depending on the mean values of Z_e, Z_f, etc., may still have a 'typical' shape. Most have high symmetry: asymmetry is introduced only if the cells are elongated or flattened. The *cell edge thickness*, t_e, and *face thickness*, t_f, can be measured from micrographs (allowing for, or avoiding, distortion caused by the angle of sectioning). The *fraction of solid in the cell edges*, ϕ, can be obtained from the relative areas of edge and face on a plane section (the fraction in the faces is, of course, $1 - \phi$). The *principal cell dimensions* are obtained (as before) by counting the number, N_c, of cells per unit length of straight lines drawn parallel to the principal directions (usually obvious): if the micrographs are plane sections, the mean cell diameter (De Hoff and Rhines, 1968) is:

$$\bar{L}_1 = \frac{1.5}{N_c} \qquad\qquad (2.24)$$

and so on for $\bar{L}_2$ and $\bar{L}_3$. *The shape anisotropy ratios* $R_{12} = \bar{L}_1/\bar{L}_2$ and $R_{13} = \bar{L}_1/\bar{L}_3$ (where $\bar{L}_1$ is the largest principal dimension) are calculated from these. Anisotropy can also be characterized by the fabric tensor, H, which, in the principal coordinate system, has the form, in matrix notation,

$$H = \begin{bmatrix} H_1 & 0 & 0 \\ 0 & H_2 & 0 \\ 0 & 0 & H_3 \end{bmatrix}$$

where H_i represents the mean intercept length in the x_i direction (Harrigan and Mann, 1984). The *dispersion of cell sizes*, in foams, is more difficult to measure accurately; though crude, it is often adequate to measure the projected areas of cell on scanning micrographs, and then follow the method described for honeycombs. Finally (as before) any exceptional structural features or defects should be recorded.

The table contains data for the polymer foam shown, in three mutually perpendicular sections, in Fig. 2.19. The cells are elongated so that the mean cell diameter in the rise direction, L_1 is larger by the factor $R_{13} = 1.2$, than those in the other two directions at right angles to it. This may not seem much, but it is enough to make the foam almost twice as stiff in the rise direction as in the other two.

2.6   Conclusions

Cells come in many shapes and sizes. Two-dimensional arrays (honeycombs) if regular, are assemblies of triangles, squares or hexagons. Many man-made honeycombs, and some naturally occurring ones, are regular; but in nature there are frequent deviations from regularity caused by the way in which the individual cells nucleate and grow, and the rearrangements which take place when they impinge. The random Voronoi honeycomb (the product of cells which grow from random points until they impinge) is an example; and in it the cells have a dispersion of size, and vary in the number of edges per cell, from three to nine or more.

Three-dimensional arrays of cells (which we call foams), similarly, can be equiaxed – then they are made up of an array of one of the shapes shown in Fig. 2.12. Almost all man-made and natural foams are anisotropic, but the structure can still be thought of as typified by one of the regular units: the tetrakaidecahe-

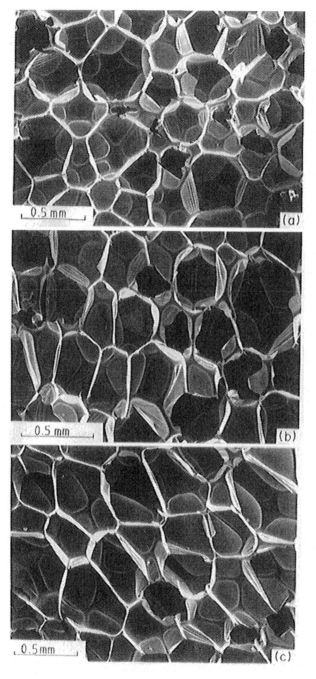

Figure 2.19 Three orthogonal sections through an anisotropic polyurethane foam. (a) A section normal to the rise direction. (b,c) Two perpendicular planes in the rise direction. The section shown in (c) has been rotated roughly 45° in the microscope stage. Data derived from this figure are given in Table 2.4.

dron is a popular choice because it has an average number of edges per face, and of faces per cell, which seems to match observations of grains in metals and cell shapes in some biological tissues; but the matter has not been adequately investigated.

Certain geometric aspects of the cell structure are crucial to the understanding of the mechanical properties, developed in later chapters. First among these is the relative density; the relations which link this to the dimensions of the cell are listed in Table 2.2. Second in importance is anisotropy: the tendency for cells to be elongated, or flattened, or have walls of unequal thickness. Ways of characterizing this are discussed in Section 2.5. Finally, there are the other topological aspects of cell shape: edge and face connectivity, number of contact neighbours and so forth. These vary from one structure to another, and can influence properties in important ways. Those for simple cell shapes are listed in Table 2.1, but in structures with a dispersion of cell sizes the problem is more difficult. Help can be derived from a number of topological rules, among them, Euler's law, the Aboav–Weaire law and Lewis's rule. Their application is described in Section 2.3.

It is appropriate to conclude this chapter, as it began with one of Hooke's drawings, since it was he who coined the word 'cell'. It is shown in Fig. 2.20, and is of sponge; a micrograph of the same material is shown in Fig. 2.6(c). At first sight it is just another open-cell foam. But it is unusual in having edges with a connectivity of three (the usual number is four), and this starts, as it were, a chain reaction, influencing the number of faces per cell, and thus the typical cell shape. Hooke's drawing records all this with remarkable accuracy. If, without seeming presumptuous, one might draw a lesson from his work which is relevant to all studies of the properties of cellular solids, it is this: first characterize your cells.

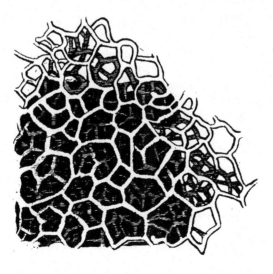

Figure 2.20 Robert Hooke's sketch of sponge, which should be compared with Fig. 2.6c. Note the unusual edge connectivity.

References

Aboav, D. A. (1970) *Metallography*, **3**, 383.

Aboav, D. A. (1980) *Metallography*, **13**, 43.

Akutagawa, W. (1985) Private communication.

Barma, P., Rhodes, M. B. and Salovey, R. (1978), *J. Appl. Phys.*, **49**, 4985.

Brakke, K. (1992) *Exp. Math.* **1**, 141.

Buffon, G. L. L. (1753) *Histoire naturelle*, Paris, Ch. 4, p. 211.

Chan, R. and Nakamura, M. (1969) *J. Cell. Plastics*, **4**, 112.

Chung, H., Wehrli, F., Williams, J., Kugelmass, S and Wehrli, S. (1995) Quantitative analysis of trabecular bone microstructure by 400 MHz nuclear magnetic resonance imaging. *J. Bone & Min. Res.* **10**, 803–11.

Crain, I. K. (1972) *Search*, **3**, 220.

Crain, I. K. (1978) *Comput. Geosci.*, **4**, 383.

Currey, J. D. (1984) *The Mechanical Adaptations of Bones*. Princeton University Press, Princeton, NJ.

Darwin, C. (1859) *Origin of Species*, 6th edn, Ch. 8, p. 99.

De Hoff, R. T. and Rhines, F. N. (1968) *Quantitative Microscopy*, McGraw-Hill, New York, p. 93.

Dinwoodie, J. M. (1981) *Timber, its Nature and Behaviour*, Van Nostrand Reinhold, New York.

Euclid (3rd century BC; exact dates unknown) *Elements*; first English edition, *The Elements of Geometry of the Most Ancient Philosopher Euclid*, translated and published by H. Billingsley London (1570).

Euler, L. (1746) Thoughts on the elements of bodies, *Memoirs of the Prussian Academy of Sciences*, Berlin.

Feldkamp, L., Goldstein, S. A., Parfitt, A. M., Jesion, G. and Kleerekoper, M. (1989) The direct examination of three-dimensional bone architecture by computed tomography. *J. Bone & Min. Res.* **4**, 3–11.

Ferro, A. C. and Fortes, M. A. (1985) *Z. Kristallographie*, **173**, 41.

Fortes, M. A. (1986a) *J. Mater. Sci.* **21**, 2509.

Fortes, M. A. (1986b) *Acta Metall.*, **34**, 1617.

Gilbert, E. N. (1962) *Ann. Math. Stat.*, **13**, 958.

Handbook of Chemistry and Physics (1972) 52nd edition, ed. R. C. Weast. Chemical Rubber Co., Cleveland, Ohio.

Harding, R. H. (1967) in *Resinography of Cellular Plastics*, ASTM Publication STP 414. American Society for Testing Materials, p. 3.

Harrigan, T. P. and Mann, R. W. (1984) *J. Mater. Sci.*, **19**, 761.

Harvey, E. N. (1954) *Photoplasmatologia*, **2**, 1.

Hipp, J. A., Jansujwicz, A., Simmons, C. A. and Snyder, B. D. (1996) Trabecular bone morphology from micro-magnetic resonance imaging. *J. Bone & Min. Res.*, **11**, 286–92.

Hooke, R. (1664) *Micrographia*. The Royal Society, London, p. 112.

Jones, R. E. and Fesman, A. (1965) *J. Cell. Plast*, **1**, 200.

Kelvin, Lord (Sir W. Thompson) (1887) *Phil. Mag.*, **24**, 503.

Ko, W. L. (1965) *J. Cell. Plast.*, **1**, 45.

Lakatos, I. (1976) *Proofs and Refutations, the Logic of Mathematical Discovery*, ed. J. Worall and E. Zahar, Cambridge University Press, Cambridge.

Lakes, R. (1993) Materials with structural hierarchy. *Nature* **361**, 511.

Lewis, F. T. (1923) *Proc. Amer. Acad. Arts Sci.*, **58**, 537.

Lewis, F. T. (1928) *Anat. Rec.*, **38**, 341.

Lewis, F. T. (1943) *Amer. J. Bot.*, **30**, 766.

MacLaurin, C. (1742) *Phil. Trans. Roy. Soc.*, **42**, 56.

Meijering, J. L. (1953) *Philips Research Report*, **8**, 270.

Menges, G. and Knipschild, F. (1975) *Polymer Eng. Sci.*, **15**, 623.

Plateau, J. A. F. (1873) *Statique Experimentale et Theorique des Liquides Soumis aux Seules Force Moleculaires*. Ghent.

Pliny, C. (AD 77) *Natural History*, vol. 16.

Rivier, N. (1982) *J. Physique*, **43**, C9–91.

Rivier, N. (1983) *Phil. Mag.*, **47**, L45.

Rivier, N. (1985) *Phil. Mag.*, **52**, 795.

Rivier, N. (1986) *Physica, D* **23**, 129–37.

Rivier, N. and Lessowski, A. (1982) *J. Phys. A.*, **15**, L143.

Rosa, M. E. and Fortes, M. A. (1986) *Acta Cryst.*, **A42**, 282.

Shaw, N. (1985) Private communication.

Smith, C. S. (1952) in *Metal Interfaces*, ed. C. Herring. American Society of Metals, New York, p. 108.

Smith, C. S. (1964) *Metal Rev.*, **9**, 1.

Thompson, D. W. (1961) *On Growth and Form*, abridged edition, ed. J. T. Bonner, Cambridge University Press, Cambridge.

von Neumann, J. (1952) in *Metal Interfaces*, ed. C. Herring. American Society of Metals, Cleveland, p. 108.

Voronoi, G. F. (1908) *J. Reine, Angew. Math.*, **134**, 198.

Weaire, D. (1974) *Metallography*, **7**, 157.

Weaire, D. (1983), in *Topological Disorder in Condensed Matter*, edited by T. Ninomiya and F. Yonezawa. Springer-Verlag, Berlin.

Weaire, D. and Kermode, J. P. (1983) *Phil. Mag.*, **48**, 245.

Weaire, D. and Phelan, R. (1994) A counter-example to Kelvin's conjecture on minimal surfaces. *Phil. Mag. Lett.* **69**, 107–10.

Weaire, D. and Rivier, N. (1984) *Contemp. Phys.*, **25**, 59.

Wyman, J. (1865) *Proc. Amer. Acad. Arts and Sci.*, **7**, 68.

Material properties

3.1 Introduction and synopsis

Foams can be made out of almost anything: metals, plastics, ceramics, glasses, and even composites. Their properties depend on two separate sets of parameters. There are those which describe the geometric structure of the foam – the size and shape of the cells, the way in which matter is distributed between the cell edges and faces, and the relative density or porosity; these are described in Chapter 2. And there are the parameters which describe the intrinsic properties of the material of which the cell walls are made; those we describe here.

The cell wall properties which appear most commonly in this book are the density, ρ_s, the Young's modulus, E_s, the plastic yield strength, σ_{ys}, the fracture strength, σ_{fs}, the thermal conductivity, λ_s, the thermal expansion coefficient, α_s, and the specific heat C_{ps}. Throughout, the subscript 's' indicates a property of the solid cell wall material while a superscripted '*' refers to a property of the foam itself. Thus, E^*/E_s means 'the foam modulus divided by that of the cell wall material'; this is also referred to as 'the relative modulus'.

It is helpful to start with an overview of the properties of solid materials, which have values which lie in certain characteristic ranges. A perspective on these is given by *material property charts* (Ashby, 1992), of which Figs. 3.1 and 3.2 are examples. The first shows Young's modulus, E_s, plotted against the density, ρ_s. Metals lie near the top right: they have high moduli and high density. Fine ceramics like alumina or silicon carbide are stiffer still, but, on average, a little less dense than the average metal. Glasses based on silica are less dense than most ceramics; they occupy a position just to the left of the 'metals' envelope. Polymers and elastomers, because they are predominantly made of light atoms, have the lowest densities of all solids and are much less stiff than the other classes because of the soft van der Waals bonding between the molecules. Polymer–matrix composites, not surprisingly, lie somewhere between polymers and the glasses or carbons with which they are reinforced. A general envelope for 'polymer foams' is also shown for comparison (though here, of course, we show E^*, not E_s). They lie at the extreme lower left.

The second chart – Fig. 3.2 – shows the thermal expansion coefficient, α_s, plotted against the thermal conductivity, λ_s. It reveals that polymers, and in particular, elastomers, expand a great deal, and conduct heat very poorly – only foams (for which α^* and λ^* are plotted) have more extreme values of these properties. Metals and certain ceramics conduct heat well and have relatively low values of α_s. Glasses and composites lie in between. Note the considerable range displayed by solid properties – typically a factor of between 10^2 and 10^3 separates

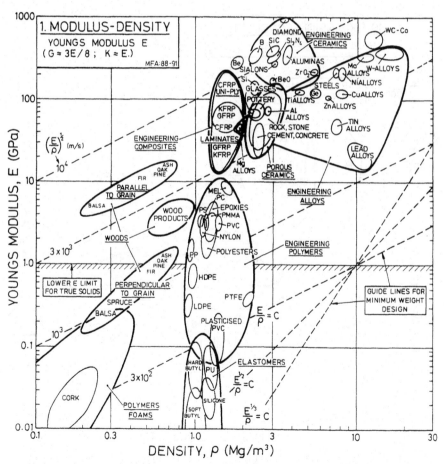

Figure 3.1 A chart showing Young's modulus, E_s, and density, ρ_s, for materials. Each material class occupies a characteristic field on the chart.

the lowest from the highest. But even this is small when compared with the range which can be achieved by foaming, which, for properties such as modulus and strength extend the range by a further factor of up to 10^4.

The groupings revealed by these charts are the reason that we elect to describe material properties under three broad headings: *polymers* and *elastomers*, *metals*, and *ceramics* and *glasses*; this is done in the next three sections. In each, the constitutive equations used in analysing foam properties are developed. But to tackle any practical problem, like that of selecting a foam to package a delicate instrument or to serve as the core of a sandwich panel – one also needs data for the cell wall properties. Selected data for mechanical and thermal properties for solids are presented in Tables 3.1, 3.4 and 3.5; data for electrical and acoustic properties are given, when relevant, in the text of this chapter and in Chapter 7. Often (particularly for polymers) precise data for cell-wall properties may not

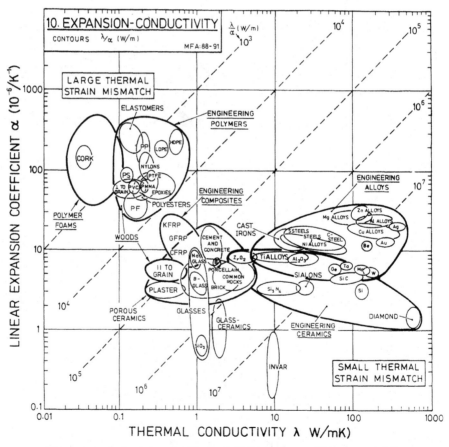

Figure 3.2   A chart showing the thermal expansion coefficient, α_s, and the thermal conductivity, λ_s, for materials. As in Fig. 3.1, each material class occupies a characteristic field.

be available; then one has to estimate them. Methods for doing this are given in each section.

There is an underlying difficulty in attempting to present a balanced overview of material properties in a single chapter – each of its sections condenses information which, adequately treated, requires a book to itself. As with pocket histories of Europe, or quick guides to World architecture, superficiality is a problem. It is minimized by focusing on specific goals. Ours are these: to introduce the cell-wall properties which are essential for the later development of models of cellular solids; to present typical data for these properties; and to provide sufficient physical understanding that sensible estimates can be made for them when no data are available. But the reader who needs a comprehensive understanding of his or her material will need more than this: additional information will be found in the books listed in the reference section of this chapter under the special heading 'General references for compilations of material properties'.

3.2   Polymers and elastomers

Engineering polymers have structures which, at first sight, seem impossibly complicated. Some are amorphous; others partly crystalline. In some the molecules are linear and tangled like lengths of string; in others, they are cross-linked like a fishnet. Some have a narrow spread of molecular weights; others a broad one. We will, as far as possible, avoid this complexity, calling only on those aspects of structure which directly relate to the mechanical and thermal properties. Properties for typical polymers are given in Table 3.1. Polymer names and their abbreviations are listed in Table 3.2.

(a) Structure and density of polymers

Roughly speaking, three sorts of polymer are used to make foams: thermoplastics, thermosets and elastomers. *Linear thermoplastics* like polyethylene (PE), polystyrene (PS) and polymethylmethacrylate (PMMA) have long, snake-like molecules. If the molecules have awkward side groups (like PMMA) which prevent neat packing, they solidify in an *amorphous* structure; if (like PE) they do not, they may coil together like fan-folded computer paper to form *chain-folded crystals*, though with amorphous material in between. The density of such linear polymers is always near $1 \, \text{Mg/m}^3$.[†] That for low-density (largely amorphous)

[†] For the following reason. The unit of the polymer chain or side-chain is, typically, $—CH_2—$, with a molecular weight of 14, or a weight per unit of $14/N_A$ where N_A is Avogadro's number: the result is $2.3 \times 10^{-26} \, \text{kg/unit}$. The volume occupied by the unit is a constant equal to $4 \times l^2_{C-H} \times l_{C-C}$, where l_{C-H} is the carbon–hydrogen bond length and where l_{C-C} is the carbon–carbon bond length. Its value is roughly $2.5 \times 10^{-29} \, \text{m}^3$. The weight divided by this volume is just about $1 \, \text{Mg/m}^3$.

Table 3.1 Properties of solid polymers

Material	Density ρ_s (Mg/m^3)	Glass temp. T_g (K)	Thermal expansion $\alpha_s \times 10^{-6}$ (K^{-1})	Thermal conductivity λ_s (W/m K)	Specific heat C_{ps} (J/kg K)	Young's modulus at 20°C E_s (GN/m^2)	Yield strength σ_{ys} (MN/m^2)	Fracture strength σ_{fs} (MN/m^2)	Fracture toughness K_{ICs} (MN/m$^{3/2}$)
Cellulose	1.5	—	—	—	—	25	350	—	—
Epoxies	1.25–1.7	400	55–90	0.2–0.5	1700–2000	5–10	40–80	40–85	0.6–1.0
Latex rubber	0.9	—	—	—	—	0.0026	—	—	—
Lignin	1.4	—	—	—	—	2.0	—	—	—
Nylon 66	1.15	340	80–90	0.2–0.25	1200–1900	2–3.5	50–110	55–120	1.5–2.2
Polybutadiene	1.05	203	130–150	0.14	1800–2500	0.001–0.05	5–15	5–15	0.07–0.1
Polychloroprene	0.94	200	500–600	0.1–0.12	2100	0.002–0.1	3.5–20	3.5–20	0.1–0.2
Polyester	1.25–1.4	340	70–100	0.2–0.24	800–1500	1.3–4.5	45–60	45–85	0.5
Polyethylene, PE (low D)	0.91–0.94	270	160–200	0.35	2250	0.15–0.24	6–10	7–17	1
Polyethylene, PE (high D)	0.95–0.97	300	120–160	0.52	2200	0.55–1.0	20–28	20–37	2
Polyisoprene	0.91	220	150–450	0.14	1800–2500	0.002–0.1	10–20	10–20	0.07–0.1
Polymethylmethacrylate, PMMA	1.2	378	54–72	0.2	1500	3.3	81	95	0.9–1.7

Polypropylene, PP	0.91	253	70–110	0.2	1900	1.2–1.7	30–35	30–40	3–5.8
Polystyrene, PS	1.05	373	70–100	0.1–0.15	1350	1.4–3.0	30–35	30–40	2
Polytetrafluorethylene, PTFE	2.2	360	100	0.25	1050	0.35	17–21	17–28	1.0–1.8
Polyurethane, PU (rigid)	1.2	200–250	150–165	0.28–0.3	1700	1.6	127	130	0.3–0.4
Polyurethane, PU (flexible)	1.2	290–300	70–100	0.28–0.3	1650	0.045	25–50	25–50	0.1–0.2
Polyvinyl chloride, PVC	1.4	290–300	50–60	0.12–0.18	1400	2.4–3.0	40–59	45–65	2.45
Polyvinylidene fluoride, PVDF	1.78	235	80–120	0.18–0.22	1150	1.8–2.4	20–25	35–40	0.7–1.4
Protein	1.2–1.4	—	—	—	—	—	—	—	—
Suberin	0.90	375	—	—	—	9.0	—	—	—

Table 3.2   Common polymers and their short names

Thermoplastics	PE	Polyethylene
	LDPE	Low-density polyethylene
	HDPE	High-density polyethylene
	PS	Polystyrene
	PP	Polypropylene
	PMMA	Polymethylmethacrylate (acrylic)
	PVC	Polyvinylchloride
	PC	Polycarbonate
	PA	Polyamide (nylon)
	PVDF	Polyvinylidene fluoride
	PTFE	Polytetrafluroethylene
	PET	Polyethyleneterephthalate
Thermosets	EP	Epoxy
	UP	Unsaturated polyester
	PUR	Polyurethane
	PF	Phenolic
Elastomers	NR	Natural rubber
	BR	Polybutadiene
	IR	Synthetic polyisoprene
	CR	Polychloroprene
	BUTYL	Butyl rubber
	PIB	Polyisobutylene

polyethylene, for instance, is $0.92\,\mathrm{Mg/m^3}$; that for polystyrene is $1.05\,\mathrm{Mg/m^3}$. Crystallization gives slightly better packing; for polyethylene it increases the density to $0.97\,\mathrm{Mg/m^3}$ (HDPE). Polymers with heavier atoms (such as cellulose which contains oxygen, or PTFE which contains fluorine) are a little denser – their density can be as high as $2.2\,\mathrm{Mg/m^3}$.

The second class of polymers are the *cross-linked thermosets*. In these, strong primary C—C bonds link the molecules to form a three-dimensional network with hydrogen or other side groups where necessary to satisfy valence requirements. The density of thermosets (epoxies, for instance) is usually larger than that of thermoplastics, partly because most contain heavy oxygen or nitrogen atoms, and partly because the cross-linking itself increases the density by pulling the chains a little closer together. The result is that most thermosets have densities between 1.2 and $1.5\,\mathrm{Mg/m^3}$.

The final class is that of *elastomers*, or rubbers. Structurally, they can be thought of as linear-chain polymers with a few widely spaced cross-links attaching each molecule to its neighbours. The difference lies in the glass temperature, a material property which is as significant for amorphous solids as the melting point is for crystals. Unlike the first two classes of polymer, the glass temperature of an elastomer is well below room temperature. This means that at room temperature the secondary bonds have already melted, and the molecules can slide relative to each other with ease. Were it not for the cross-links, the material would be a viscous liquid, but the cross-links give it a degree of mechanical stability. Elastomers can undergo enormous deformations (500% is not uncommon); but on unloading the thermal motion of the molecules, coupled with the memory built into the arrangement of the cross-links, causes the material to return to its original shape. The restoring force is a small one, so the moduli of elastomers are less, by a factor of 1000 or more, than those of thermoplastics or thermosets. The packing of the loosely tangled molecules is similar to that of non-crystalline linear, polymers, so the density of most elastomers is, like that of polyethylene, about $1 \, Mg/m^3$ (Table 3.1).

When the cell-wall density of a polymeric foam is not known, it can be estimated, with an error which is unlikely to exceed $\pm 15\%$, by selecting the appropriate line in Table 3.3. Its basic features have already been explained. Amorphous polymers containing only carbon and hydrogen always have a density near $1 \, Mg/m^3$. Heavier atoms (the commonest are oxygen, chlorine and fluorine) increase the density: teflon —$(CF_2)_n$— with two fluorines per carbon, is considerably denser than polyethylene, —(CH_2)—. As a rule of thumb the density is roughly $(1 + 0.7 \, m) \, Mg/m^3$, where m is the number of heavy atoms (meaning O, N, F or Cl) per carbon in the polymer chain. Cross-linked polymers tend to be denser than simple linear polymers for two reasons given earlier: they almost always contain heavy atoms, and the cross-linking draws the atoms closer together. The polymers on which plant-life is based, containing a large fraction of cellulose, are denser still, largely because the molecules in crystalline cellulose

Table 3.3 Estimates of polymer density

Type of polymer	Approximate density (Mg/m^3)
Linear polymers not containing heavy atoms	1.0
Simple elastomers (latex)	1.0
Linear polymers containing m heavy atoms (F, Cl, O) per carbon	$(1 + 0.7 \, m)$
Simple cross-linked polymers (epoxies, polyesters)	1.4
Plant polymers based on cellulose and lignin	1.5

contain oxygen, and pack efficiently; the average density of the cell wall of both trees and smaller plants is close to $1.5\,\text{Mg}/\text{m}^3$.

(b) Thermal properties of polymers

It is common experience that solids are solid until they melt, when they change discontinuously to liquids. Amorphous and semicrystalline linear polymers do not have a sharp *melting point*; instead, they transform from a viscoelastic solid to a viscous liquid over a well-defined range of temperature above which they can be moulded or foamed. Cross-linked polymers do not melt properly at all (though they may show rubbery-viscous flow). If the temperature is raised too high, they simply decompose.

Important from our point of view is the *glass transition temperature*, T_g, which characterizes the transition from the solid to rubber or viscous liquid. Below T_g the amorphous component of the polymer is glassy. When it is loaded, it distorts elastically and then yields or fractures; a foam made of it would be called 'rigid'. Above T_g the polymer is viscoelastic or elastomeric and can sustain large, often recoverable, deformations without true yield or fracture; a foam made of it would be called 'flexible'. Most commercial polymers have a glass temperature in the range 0–100°C, as shown in Table 3.1. The glass transition temperature increases with molecular weight and with the degree of cross-linking.

When a sheet of polymer separates a volume at temperature T_1 from one at a lower temperature T_2, heat flows from the hotter side to the colder one. If this arrangement is left long enough, *steady state* is established. Then the heat flux, q (measured in W/m^2) is related to the sheet thickness t and the temperature difference $T_1 - T_2$ by

$$q = \lambda_s \frac{(T_1 - T_2)}{t}$$

where λ_s is the *thermal conductivity*. But when heat is applied suddenly to one surface of the sheet, the conditions are *transient*; the heat diffuses inwards at a rate which is determined by the *thermal diffusivity*, a_s. It is related to the thermal conductivity by

$$a_s = \frac{\lambda_s}{\rho_s C_{ps}} \tag{3.1}$$

where ρ_s is the density and C_{ps} is the specific heat. Heat increases the local amplitude of vibration of the atoms which are strung together in the polymer chain. Heat conduction, which involves the transport of energy, relies on the propagation of phonons – travelling lattice vibrations; their efficiency in carrying heat depends on how strongly the vibrations of adjacent atoms are coupled (and thus on the modulus of the material) and on the phonon mean-free-path – the distance a phonon moves before it is scattered or reflected. Covalent bonds in a crystallographically ordered array give high moduli and strong coupling; and in very perfect crystals phonons travel a long way before they are scattered (so diamond, for instance, is a very good thermal conductor). Weak, van der Waals bonds give low moduli and weak coupling, and the disordered structure characteristic of virtually all polymers, gives a mean-free-path which is only three or four atom-diameters in length. That is why polymers have thermal conductivities which are lower than those of any other class of dense solid (Fig. 3.2).

As a general rule, the thermal conductivities of polymers lie in the surprisingly narrow range 0.3 ± 0.1 W/m.K; if no data are available, this value can be used. Polymers with some crystallinity (such as PE, PP, PTFE) lie at the high end of the range; amorphous polymers like PMMA and the elastomers polyisoprene and polybutadiene, at the lower end. The thermal conductivity changes with temperature by less than 10% between 0 and 100°C, so the room temperature value is usually adequate.

The *specific heat* C_{ps}, is the energy required to raise the temperature of the unit mass by 1°C. The value of C_{ps} depends on the vibration and rotational motions excited in the solid, and this in turn is mainly dictated by the chemical structure of the polymer. The specific heats of all rigid polymers are large; they lie in the range 1.2–1.8 kJ/kgK; a value in the middle of this range is the best estimate if no data are available. Elastomers have still larger specific heats; the range, typically, is 1.8–2.5 kJ/kgK.

As the material property chart of Fig. 3.2 showed, polymers as a class show higher *coefficients of thermal expansion*, α_s, than metals or ceramics. The differences between polymers are not well understood. Rigid polymers have values in the range $50\text{–}150 \times 10^{-6}$/K; the middle of this range is a usable estimate if no data are available. Elastomers can have much larger values: up to 500×10^{-6}/K.

(c) Elastic properties of polymers

Polymers show three distinct regimes of elastic behaviour: the glassy regime, the glass transition and the rubber regime (Fig. 3.3). Each has certain well-defined characteristics illustrated, for a thermoplastic, a thermoset and an elastomer, in Figs. 3.4, 3.5 and 3.6. The axes of these figures are the modulus E_s, and the normalized temperature T/T_g, where T_g is the glass temperature. Each figure is divided into fields corresponding to the main regimes of behaviour. Contours show the value of the relaxation modulus at various loading times.[†] The modulus varies widely: between absolute zero and the temperature at which the polymer decomposes or melts (typically, 150–300°C) the modulus of a thermoplastic

[†]The relaxation modulus at a time t is the stress σ at t divided by the strain ϵ in a stress relaxation test.

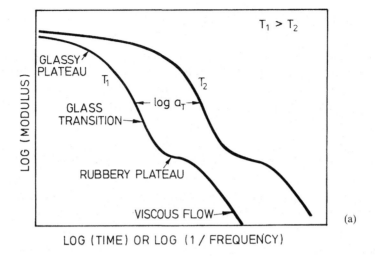

LOG (MODULUS)

$T_1 > T_2$

GLASSY PLATEAU

T_1 T_2

GLASS TRANSITION→   ←log a_T→

RUBBERY PLATEAU

VISCOUS FLOW→   (a)

LOG (TIME) OR LOG (1 / FREQUENCY)

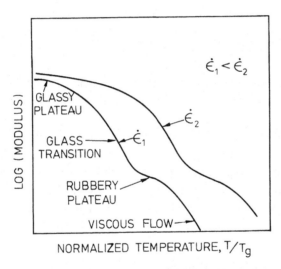

LOG (MODULUS)

$\dot{\epsilon}_1 < \dot{\epsilon}_2$

GLASSY PLATEAU

$\dot{\epsilon}_2$

GLASS TRANSITION $\dot{\epsilon}_1$

RUBBERY PLATEAU

VISCOUS FLOW→

NORMALIZED TEMPERATURE, T/T_g (b)

Figure 3.3 (a) Schematic showing the way in which the modulus of a polymer varies with loading time or frequency, at two fixed temperatures, illustrating the shift factor, a_T. (b) Schematic showing the way in which the modulus of a polymer varies with temperature, at two fixed strain-rates.

drops by a factor of 1000 or more, that of a thermoset by a factor of about 10. The mechanisms which underlie this behaviour are described below.

Within the low-temperature, *glassy, regime* (0–0.9 T_g), amorphous polymers all have a modulus of about 3 GN/m². That value – 3 GN/m² – directly reflects the stiffness of the van der Waals bonds which glue the molecules together. It is these, not the stiff covalent bonds between carbon atoms, which stretch and deflect when the polymer is loaded. The spring constant of a single van der Waals bond is known (it is about 1 N/m at room temperature). The number of such bonds per unit volume depends on the packing density in the polymer, but that does not vary much either. That is why the glassy modulus, which is roughly the bond stiffness divided by the atom spacing, is almost the same for all amorphous polymers. Foams made of glassy polymers are 'rigid', that is, they are relatively stiff, and their elastic range is truncated by plastic yielding or brittle fracture (next section).

Increasing the temperature within the glassy regime has two effects. First, thermal expansion increases the molecular separation and lowers the van der Waals forces. Second, the thermal energy of the molecules permits small relative movements of side groups, or of short segments of chain, at 'loose sites' in the structure. This gives time-dependent strain and a further drop in modulus. Most polymers show several of these sub-T_g molecular relaxation processes, which are known, in order of decreasing temperature, as the 'β', 'γ' and 'δ' relaxations. If a description is needed of the temperature-dependence of the modulus in the glassy regime, it can be approximated by the linear equation:

$$E_s = E_s^0\left(1 - \alpha_m \frac{T}{T_g}\right) \qquad (3.2)$$

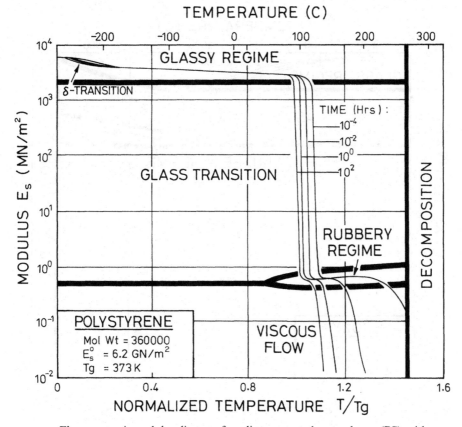

Figure 3.4 A modulus diagram for a linear-amorphous polymer (PS), with a glass temperature above room temperature ($T_g = 100°C$). It shows three important regimes–the glassy regime, the rubbery regime and the viscoelastic transition. The contours show the relaxation modulus for times between 0.3 s and 100 hours.

with $E_s^0 \approx 5\,\mathrm{GN/m^2}$ and $\alpha_m \approx 0.5$; this value of α_m then includes, in an approximate way, both effects. The dependence of E_s on strain-rate in the glassy regime is so small that it can usually be neglected.

Cross-linked polymers have moduli which are greater than those which are amorphous, but the difference, at the same fraction of T_g, is not great. The glass temperature itself is higher for cross-linked polymers, and this gives them a somewhat higher modulus at a given absolute temperature (epoxies, for instance, have moduli in the range 5–$10\,\mathrm{GN/m^2}$). In the glassy regime it is usually adequate to assume that E_s is described by Eqn. (3.2) with $E_s^0 \approx 10\,\mathrm{GN/m^2}$, and is independent of strain-rate. Only if all the weak bonds are replaced by strong ones does the modulus really increase; and this can be achieved (in a fibre or a sheet) by orienting the chains along the fibre or in the plane of the sheet by drawing or by crystallization. Cellulose is like this; and

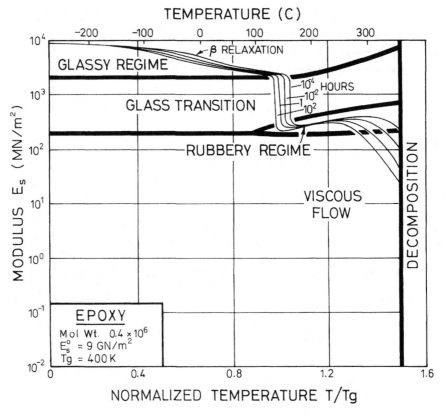

Figure 3.5 A modulus diagram for a cross-linked polymer (an epoxy), with a glass temperature above room temperature ($T_g = 127°C$). The drop in modulus from the glassy regime is less than that for linear polymers, and decreases with increasing cross-link density.

its modulus (measured along the cellulose fibre, of course) is not $3\,GN/m^2$, but almost $30\,GN/m^2$.

Between 1.1 and about 1.4 T_g most polymers exhibit a *rubbery* plateau (identified on Figs. 3.4, 3.5 and 3.6). For *elastomers* – linear polymers with a few, deliberately introduced, cross-links – the rubbery regime is particularly wide (Fig. 3.6), and spans room temperature. These materials show enormous elastic strains: a good elastomer can be repeatedly stretched to five times its original length, completely reversibly. The rubbery modulus is low – roughly 1000 times less than the glassy modulus – and, because the strains are large, the linear approximation on whch Hookean elasticity is based breaks down, and has to be replaced by a more elaborate, non-linear, elasticity theory (see, for example, McClintock and Argon, 1966).

In the rubbery state the weak intermolecular bonds have melted while the strong covalent bonds linking units in the polymer chain and forming the occasional cross-links are still very much intact. The long chain-like molecules of the

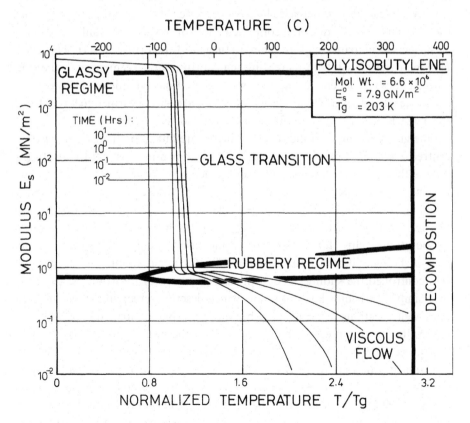

Figure 3.6 A modulus diagram for a typical elastomer (PIB). The glass temperature is now below room temperature ($T_g = -70°C$) and the rubbery plateau is greatly extended.

elastomer can assume a variety of disordered configurations in response to the thermal vibration, which causes a Brownian wriggling of the units of the chain. The most probable configurations are those which maximize the entropy of the system. But there is a constraint: the cross-links tie the chains together at widely spaced points along their lengths. When strained, the chains tend to order (by lining up) and the entropy decreases; but the cross-links provide a memory of the particular set of disordered configurations which corresponded to the original shape, and when the load is removed the molecules relax back to this state. Standard texts (e.g. Treloar, 1958) summarize the molecular theory of rubber elasticity. For small strains the modulus, E_s, is related to the density, ρ_s, and the average molecular weight between cross-links, M_e, by

$$E_s = \frac{3\rho_s RT}{M_e}.$$

(3.3)

For many elastomers, M_e is around $10\,\mathrm{kg/mol}$, giving a rubbery modulus of about $1\,\mathrm{MN/m^2}$ at room temperature; but it is a sensitive function of cross-link density. Vulcanization increases the cross-link density, reducing M_e and raising the modulus; so, too, does the combined effect of oxygen and sunlight, causing elastomers (and foams made of them) to turn hard and brittle. The rubbery modulus increases slowly with increasing temperature, in agreement with the Eqn. (3.3). It does not depend strongly on strain-rate. Foams made of elastomers are called 'flexible'; they are relatively compliant and easily deformed to large strains from which they recover completely when unloaded.

The glassy regime and the rubbery regime are linked by the *glass transition*, centred around the glass temperature, T_g, as shown in Figs. 3.3 to 3.6. In this regime the modulus drops steeply, and it depends strongly on time for static loading or on frequency for cyclic loading. This is because, at and above T_g, segments of the molecular chains are able to slide relative to each other. The relative motions are thermally activated (meaning that the thermal vibration of the atoms helps sliding) so that an increase of temperature increases the rate of sliding and makes the modulus fall. If a single, unique, molecular process controlled the sliding, its dependence on temperature would be described by a simple Arrhenius law of the sort which describes creep in metals and ceramics well (Sections 3.3 and 3.4); but the amorphous structure of polymers makes the process more complicated. Where the molecular packing is loose, sliding is easy; where molecules pack more tightly, it is more difficult. This complication means that there is no simple equation (like Eqns. 3.2 and 3.3) to describe the modulus E_s as a function of temperature T and time t for the viscoelastic regime. But it is found that, if the modulus is known at one temperature, its value at another is given by applying a simple scaling law to the line axis (Williams *et al.*, 1955; well described in many standard texts such as Hall, 1981, or Ward, 1983). Thus

$$E_s(t, T_0) = E_s\left(\frac{t}{a_T}, T_1\right) \tag{3.4}$$

where a_T, the shift factor, is a function of T_0 and T_1 (see Fig. 3.3). This *time–temperature correspondence* is a useful one because it allows the modulus in very short-time or extremely long-time tests at one temperature to be obtained from data at more reasonable times at other temperatures. This is aided by the discovery (Williams *et al.*, 1955) of a simple empirical description for a_T when $T_0 = T_g$:

$$\log a_T = \frac{-C_1(T_1 - T_g)}{C_2 + T_1 - T_g} \tag{3.5}$$

where C_1 and C_2 are constants ($C_1 = 17.4$; $C_2 = 143°$K).

Foams made of polymers which are in the viscoelastic transition at room temperature (like PE) are themselves viscoelastic. Their moduli (Chapter 5) are proportional to E_s, and they are expected to follow the same time – temperature correspondence as the polymer of which they are made. It is not possible to estimate the modulus of a polymer in the viscoelastic transition with any useful precision – it varies too widely. The important thing to remember, particularly in selecting foams for packaging or safety padding, is that a small change in temperature (from 20°C to 0°C for instance) or a change of loading rate (from laboratory rates of 10^{-3}/s to impact rates of 10^3/s for instance) may cause E_s to change by a factor of 2 or more when the polymer of which it is made is in its viscoelastic transition.

We have focussed so far on Young's modulus, E_s. Sometimes it is the shear modulus G_s or the bulk modulus K_s which are needed. They are related to Young's modulus, E_s, and Poisson's ratio, v_s, for an isotropic, linear-elastic solid, by

$$G_s = \frac{E_s}{2(1 + v_s)} \tag{3.6}$$

and

$$K_s = \frac{E_s}{3(1 - 2v_s)}. \tag{3.7}$$

For crystalline solids, and for amorphous polymers below T_g (in the glassy regime), $v_s \approx 1/3$, giving $G_s \approx 3E_s/8$ and $K_s \approx E_s$. This approximation breaks down completely for elastomers, for which $v_s \approx 0.49$; then $G_s \approx E_s/3$ and $K_s \gg E_s$.

(d) Plasticity and fracture of polymers

Polymers are remarkable for the large elastic strains they can undergo before they yield or fracture. Except for elastomers this elastic deformation is limited to about 5%, but this is 50 or more times greater than that of most metals or ceramics. At low temperatures $(T \ll T_g)$ polymers are linear-elastic to fracture (Fig. 3.7). At higher temperatures $(T \approx 0.8 T_g)$ the mode of failure changes from brittle to ductile, characterized by a yield point followed by rapid drop in load, with small overall extension. A further increase in temperature leads to necking and inhomogeneous cold drawing, with extensions which are now large. Finally, at higher temperatures still $(T > T_g)$, the polymer deforms homogeneously by viscous processes, giving large extension to failure. As with the modulus, a number of mechanisms are involved, each with a characteristic range of temperature and stress over which it is dominant. But there is an added complication: plastic flow and (particularly) fracture depend on the stress state. Mechanisms such as brittle fracture and cold drawing which appear in tensile modes of loading are largely suppressed in compression, others appearing to take their place.

One way of dealing with this is to summarize the plastic and fracture response on *failure-mechanism diagrams* (Ahmad and Ashby, 1988) with separate diagrams for tension (Fig. 3.8) and compression (Fig. 3.9). The axes are strength normalized by modulus, σ_s/E_s^0, and reduced temperature T/T_g. Each diagram is divided into fields in which a given mechanism (brittle fracture, for instance) is dominant. Superimposed on these are contours of constant strain-rate which show the strength at that rate of extension. Cellular solids made from polymers have plastic and fracture properties which are related (in ways analysed in Chapters 4, 5 and 6) to those of the parent polymer, so an understanding of these mechanisms is helpful later.

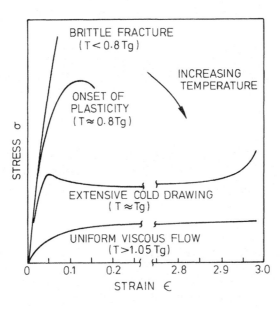

Figure 3.7 Schematic illustrating the way in which the tensile stress–strain curve of a polymer changes with temperature. The polymer is completely brittle at low temperatures but becomes increasingly ductile above $0.8\,T_g$.

At low temperatures (below about $0.8\ T_g$) polymers fail in tension by *brittle fracture*, triggered by crazing or shear-band formation (Kinloch and Young, 1983). The fracture strength σ_{fs} is related to the fracture toughness K_{ICs} of the polymer (Table 3.1) and to the length $2c$ of the largest crack or flaw it contains by

$$\frac{\sigma_{fs}}{E_s} = \frac{K_{ICs}}{E_s\sqrt{\pi c}}.\qquad(3.8)$$

Tests in which cracks of various lengths $2c$ are deliberately introduced into polymer samples follow Eqn. (3.8) when $2c$ is large, but when it is smaller than a value in the range 0.05 to 0.5 mm the strength becomes independent of the length of the deliberate flaws. This is because internal flaws in the form of crazes or shear bands are nucleated by the stress, and these replace the deliberate flaws as the source of brittle fracture. A craze is a small lens-shaped zone of dilation, like a crack with faces linked by fibrils of drawn polymer, embedded in the otherwise

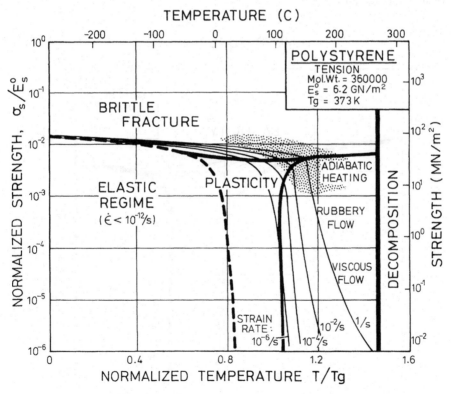

Figure 3.8 A failure-mechanism diagram for a linear-amorphous polymer (PS) loaded in tension. It shows a field in which deformation is elastic (defined as the regime in which the plastic strain-rate is less than 10^{-12}/s), one in which brittle fracture is the dominant failure mechanism, and fields of plasticity and viscous flow. Adiabatic heating becomes important in the shaded area.

undeformed polymer. A shear band is a lens-like region in which a large plastic shear has taken place, embedded in unsheared matrix. Both concentrate stress at their sharp periphery in very much the same way that a true crack does, and both can act as a nucleus for crack growth. The *intrinsic brittle strength* is then given by Eqn. (3.8) with $2c$ set equal to the natural craze or shear-band length. It is useful to have an estimate of this strength. Most polymers have a fracture toughness K_{ICs} close to $1\,\mathrm{MN/m^{3/2}}$ (Table 3.1); taking the craze length $2c = 0.2\,\mathrm{mm}$ and $E_s = 3\,\mathrm{GN/m^2}$ gives the *intrinsic brittle strength*:

$$\frac{\sigma_{\mathrm{fs}}}{E_{\mathrm{s}}} \approx \frac{1}{50}$$

in broad agreement with the observed tensile fracture strength (Fig. 3.8 is typical). It is almost independent of temperature and strain-rate. Brittle fracture is possible in compression, too, but the stress required to cause it is larger by a factor of about 10 than that for tension and, almost always, some form of plasticity replaces it (Fig. 3.9).

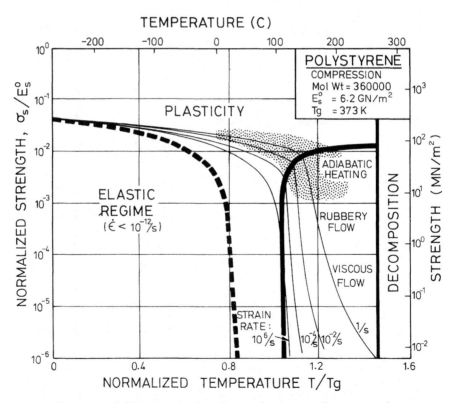

Figure 3.9 A failure-mechanism diagram for the same linear-amorphous polymer of Fig. 3.8 (PS), loaded in compression. Brittle fracture is suppressed and the stresses required for plastic flow are about 1.4 times larger than in tension.

At low temperatures, the *yield strength*, σ_{ys}, is greater than the intrinsic brittle strength, so the polymer is brittle in tension. But as the temperature is raised, σ_{ys} falls faster than σ_{fs}. The two intersect at the ductile-to-brittle transition temperature, roughly $0.9T_g$ (but really a range of temperatures, depending on strain-rate). In this regime the polymer shows yielding and cold drawing in tension, and shear banding in compression. The understanding of yield in polymers is still incomplete; reviews of models and their drawbacks can be found in texts such as Ward (1983) and Kinloch and Young (1983). For our purposes it is sufficient to note that most models simplify to the following relationship between strength, σ_{ys}, temperature, T, and strain-rate, $\dot{\epsilon}$ which adequately describes experimental data:

$$\frac{\sigma_{ys}}{E_s} = \frac{\sigma_{ys}^0}{E_s^0}\left(1 - \frac{AT}{T_g}\ln\frac{\dot{\epsilon}_0}{\dot{\epsilon}}\right). \tag{3.9}$$

Here σ_{ys}^0 is the yield strength at $0°K$ (typically, $E_s^0/30$), A is a dimensionless constant, typically 0.02, and $\dot{\epsilon}_0$ is a kinetic factor, typically $10^{14}/s$. Because strain-rate appears in Eqn. (3.9), the contours showing the strength at a given strain-rate in Figs. 3.8 and 3.9 fan out in the plastic field. The yield strength, too, depends on stress state, but much less strongly than the fracture strength: the yield strength in compression is, typically, 1.4 times greater than that in tension.

Above about $1.05T_g$ large viscoelastic or viscous deformations become possible (Figs. 3.8 and 3.9). Well above T_g the strength corresponding to a strain-rate $\dot{\epsilon}$ can be approximated by a linear-viscous flow equation:

$$\frac{\sigma_{ys}}{E_s} = \frac{3\dot{\epsilon}\eta}{E_s} \tag{3.10}$$

where η is a viscosity. Nearer T_g the temperature-dependence of η is complicated, following a WLF time–temperature law (Eqn. (3.5)), and the material shows time-dependent recovery on unloading (the regime of *rubbery flow*). Here the polymer rheology is best described by a viscoelastic model: a combination of springs (describing the elastic, recoverable part of the deformation) and dashpots (describing the contribution of viscous molecular sliding). Details are beyond our scope; they can be found in Ward (1983).

High-rate deformation of polymers is complicated by *adiabatic heating*. The work per unit volume, $\sigma_{ys}\epsilon_{pl}$, in a plastic strain ϵ_{pl} is almost entirely converted to heat. If the specific heat is C_{ps} and the density is ρ_s, this heat is sufficient to raise the temperature by

$$\Delta T = \frac{\sigma_{ys}\epsilon_{pl}}{C_{ps}\rho_s}$$

provided, of course, the heat is not lost by conduction. At low strain-rates the heat generated locally by plastic work in a shear band is conducted harmlessly away into the bulk of the polymer, and ultimately to its surroundings. But polymers have low thermal conductivities; at higher rates (typically, at $\dot{\epsilon} > 10^{-3}/s$) conduction is not fast enough to prevent the temperature rising, softening the shear band and the matrix on either side of it, and causing non-uniform deformation. The region in which adiabatic heating is significant is shown as a shaded band on Figs. 3.8 and 3.9.

Elastomers, in the regime in which they are rubbery, show large recoverable deformations without yield or fracture. But below T_g (which, for many elastomers is around $-70°C$) they, too, become stiff and brittle like any other glassy polymer; and above about $2T_g$ they melt. Figure 3.10 shows a failure-mechanism diagram for a typical elastomer. Note the glass temperature at $-70°C$; room temperature is in the middle of the rubbery regime.

A foam made of a polymer has mechanical properties which reflect the elastic, plastic and fracture properties of the polymer – though the relationship is often a complex one (Chapters 5 and 6). The diagrams shown here give an idea of how the parent polymer behaves, and are helpful in guiding the choice of the values for E_s, σ_{ys} and σ_{fs} which are needed later.

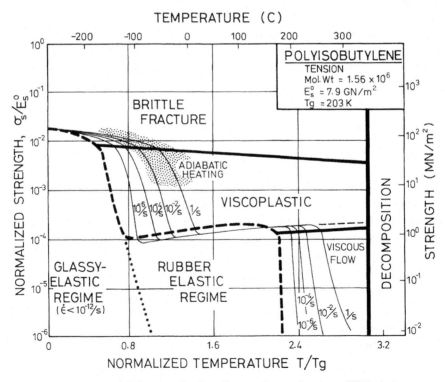

Figure 3.10 A failure-mechanism diagram for an elastomer (PIB). At low absolute temperatures it shows the same failure mechanisms as PS (note that the normalized temperatures, T/T_g, are the same). Above T_g there is a large rubbery regime in which enormous elastic extensions are possible; ultimate failure in this regime is by tearing. At still higher temperatures the rubber melts and decomposes.

(e) Electrical properties of polymers

When a potential gradient is applied to all but perfectly insulating materials, an electric current flows. Most polymers are near-perfect insulators with enormous *resistivities* in the range $10^{13}\,\Omega\text{m}$ – a reasonable estimate if no data are available. It is because of their high resistivity that polymers are so widely used as insulators, for switch gear and in electrical isolation. Doping of certain rather special polymers (such as polyacetylene) with metal ions, however, gives materials with useful conductivity. Development in conducting and semi-conducting polymers is rapid.

Polymers, and foams made from them, can be made conducting in a much simpler way: by blending them with a conducting filler such as graphite, or metallized ceramic flakes. Antistatic mats and casings generally use this route to achieve sufficient conductivity to bleed away static charge.

In an electric field, insulators become polarised: charges appear at the surfaces which screen the interior from the external field. This requires motion of electric charge, but over a very much smaller length scale than that for conduction: it is sufficient for molecules which carry a dipole moment to rotate such that the moment is favourably oriented in the field. The *dielectric constant* ϵ measures the magnitude of the surface charge: no dipole moment, no charge. Thus polyethylene (almost no dipole moment) has a low dielectric constant of roughly 2; nylon, with a polar side group (giving a dipole moment), has a dielectric constant of 3.5, and PVDF, which is strongly polar, has a value of 8. The dielectric constant is of importance in choosing insulators for electrical condensers (choose a high ϵ) and for radar and microwave-transparent structures (choose a low ϵ). Certain polymer foams have exceptionally low ϵ and are, for this reason, used as sandwich cores in radar-transparent structures.

When subjected to an alternating field, dielectrics dissipate energy and heat up; the effect is known as the *dielectric loss*. Dielectric loss can be useful – it's the basis of microwave heating – but in structures the aim is normally to avoid it. The heating arises from the energy dissipated when the dipoles which give a high dielectric constant reverse their orientation in response to the oscillating field. Polymers with a low dielectric constant generally have a low dielectric loss as well. Foams have a lower dielectric loss than the solid polymer from which they are made simply because a fraction of the space they fill is gas, not solid.

Even a modest electric field applied across a very thin polymeric film generates a high-field gradient (field strength divided by the film thickness, and measured in volts/m). All insulators have a *breakdown potential gradient*, a critical gradient at which one electron, breaking free and starting to 'conduct' (though in an insulator) provokes, by collisions, a cascade of further electrons which – like a lightning strike – terminally damages the material. The breakdown potential gradient governs the choice of polymers or of foams for high-voltage, thin-film applications.

3.3   Metals

The recent development of metal forms for catalysis, for filters, and for absorbing energy, has stimulated interest in their properties. These are analysed in later chapters. Here we summarize the properties of metals which are required for this analysis. Data for them are listed in Table 3.4. It is seldom necessary to estimate properties for metals; most are known with adequate precision.

(a) Structure and density of metals

The metallic bond is a relatively strong one, giving melting points and moduli which are much larger than those of polymers (Figure 3.1). The bond itself derives from the electrostatic attraction between the ionized metal atoms and the cloud of free electrons which surrounds them; and, unlike the covalent bond of polymers and many ceramics, it is non-directional. This has an important consequence: the energy of the metal crystal is minimized by arranging the ions in a way that gives the greatest packing density (the face-centred cubic or the close-packed hexagonal structures), or one that maximizes the number of nearest and next-nearest neighbours (the body-centred cubic structure). Such arrangements have a high *density*: those for common engineering metals range from $2.7\,Mg/m^3$, for aluminium, to $8.9\,Mg/m^3$, for copper, compared with $1\,Mg/m^3$ for polymers. To a first approximation the density of an alloy or a composite is given by a 'rule-of-mixtures', that is, by a linear interpolation between the densities of the components it contains.

(b) Thermal properties of metals

Pure metals have sharp *melting points*, T_m, at which all their properties change discontinuously. Alloys have a melting range, but their mechanical properties still change discontinuously at the onset of melting. The strong metallic bond gives melting points in the range $1000\text{--}3000°K$ – much higher than those of polymers.

As a solid is heated, energy is absorbed by thermally induced vibrations of its atoms. The high moduli and great crystallographic order of a metal crystal give strong coupling of the vibrations of neighbouring atoms. This alone would cause metals to transmit heat well. But the *thermal conductivity* is in fact higher than this would suggest, because of the even larger contribution of the free electrons, which, because of their mobility, transmit both heat and electricity rapidly through the crystal (the Wiedemann–Franz Law expresses the fact that the two conductivities are proportional). Table 3.4 shows that the mean thermal conductivity of metals (very roughly, 100 W/m K) is about 500 times greater than that of polymers.

At and above room temperature, the thermal energy of a simple solid like a metal is $3kT$ per atom (where k is Boltzmann's constant). Expressed per mole, it is $3RT$, or per kg, $3RT/M$ where M is the atomic weight (in kg/mol). The *specific heat* is the change in this energy per kg per $^\circ$K, giving

$$C_{ps} \approx \frac{3R}{M}. \tag{3.11}$$

The specific heats for metals are listed in Table 3.4. Metals with a low atomic weight (like aluminium) have the highest specific heats, and the values of the specific heat itself are very close to that given by Eqn. (3.11). Polymers, because of their complicated structures, cannot be treated in this simple way, but it is still true that, because they are made up of light atoms, they have large specific heats. Note, however, that this specific heat is the energy to raise a unit *mass* of the material by 1°K; often it is more relevant to ask for the energy required to increase the temperature of a unit *volume* of the material ($C_{ps}\rho_s$) and then the difference between metals and polymers almost disappears.

The atomic vibrations which absorb thermal energy also increase the average atom spacing, giving *thermal expansion*. The thermal energy tending to cause expansion is about $3kT$ per atom; and the binding energy per atom tending to oppose it is proportional to $E_s\Omega$ where Ω is the atomic volume. We might therefore expect the total expansion at a temperature T to be proportional to $kT/E_s\Omega$, giving a thermal expansion coefficient α_s of

$$\alpha_s \approx \frac{Ck}{E_s\Omega} \tag{3.12}$$

where C is a constant. It is not a bad estimate: the expansion coefficients for metals do vary inversely as the moduli (Table 3.4); because of their high moduli, metals have expansion coefficients which are much smaller than those for polymers.

(c) Elastic properties of metals

The metallic bond is a strong one. At small strains ($<10^{-3}$), metals are linear-elastic solids. An applied stress simply stretches the atomic bonds, and Young's modulus, which measures their stiffness, is large (Table 3.4) and independent of the rate of loading. The melting point, too, reflects the strength of the interatomic bonds, so it is not surprising that it and the modulus are proportional: E_s is well described by

$$E_s \approx \frac{100\,kT_m}{\Omega} \tag{3.13}$$

Table 3.4 Properties of solid metals

Material	Density ρ_s (Mg/m³)	Melting point T_m (K)	Thermal expansion $\alpha_s \times 10^{-6}$ (K⁻¹)	Thermal conductivity λ_s (W/m K)	Specific heat C_{ps} (J/kg K)	Young's modulus at 20°C E_s (GN/m²)	Yield strength σ_{ys} (MN/m²)	Fracture strength σ_{fs} (MN/m²)	Fracture toughness K_{ICs} (MN/m^{3/2})
Aluminium	2.7	933	24	230	1080	69	40	200	50
Aluminium alloys	2.6–2.9	≤933	20–24	88–160	920–960	69–79	100–627	300–700	23–45
Beryllium	1.85	1277	12.4	158	1883	296	34	380	4
Brasses and bronzes	7.2–8.9	≤1200	14–20	110–230	380	103–124	70–640	230–890	50–100
Chromium	7.2	2148	6.2	67	936	289	350–430	400–690	20–30
Cobalt and alloys	8.1–9.1	<1768	12–13	40–69	400–600	200–248	180–2000	500–2500	25–40
Copper	8.9	1356	16.5	384	493	124	60	400	>100
Copper alloys	7.5–9.0	≤1356	14–17	50–230	320–400	120–150	60–960	250–1000	50–100
Gold	19.3	1336	14.2	297	130	82	40	220	>100
Iron	7.9	1809	11.7	75	460	196	50	200	5–100
Lead and alloys	10.7–11.3	≤600	28–30	27–35	130–140	14	11–55	14–60	>50
Magnesium alloys	1.74–2.0	923	26	42–140	1020	41–45	80–300	125–380	15–40

Material								
Molybdenum and alloys	10–13.7	≤2880	4–5	116–146	250–270	320–365	560–1450	665–1650	20–40
Nickel	8.9	1726	13.3	92	730	214	70	400	>100
Nickel alloys	7.8–9.2	≤1726	12–14	15–65	380–460	130–234	200–1600	400–2000	50–100
Niobium and alloys	7.9–10.5	≤2740	7–8	30–58	200–340	80–110	240–550	330–600	20–40
Platinum	21.4	2042	8.9	71	130	172	25–140	150–280	>100
Silver	10.5	1234	19.7	420	234	76	55	300	>100
Steels	7.6–8.1	≤1809	11–12	24–66	420–500	190–210	200–1500	500–2300	50–200
Tantalum and alloys	16.6–16.9	≤3250	6–7	30–54	100–168	150–186	330–1090	400–1100	20–40
Titanium and alloys	4.3–5.1	1943	8–9	15–27	500–550	80–130	180–1320	300–1400	55–115
Tungsten and alloys	13.4–19.6	≤3680	4–5	160–190	140–145	350–406	1000–1400	1500–1800	20–40
Uranium	18.9	1405	17	26	126	172	180–250	385–530	20–40
Zinc and alloys	5.2–7.2	≤505	20–40	100–112	400–420	43–96	169–421	200–500	10

where, as before, k is Boltzmann's constant and Ω is the atomic volume (Frost and Ashby, 1982). An increase in temperature expands the crystal lattice, reducing the stiffness of the bonds a little and causing a drop in the modulus. Its temperature dependence is well approximated by

$$E_s = E_s^0 \left(1 - \alpha_m \frac{T}{T_m} \right) \tag{3.14}$$

where E_s^0 is the value of the modulus at absolute zero and α_m is a constant. Its value, almost always, is close to 0.5, meaning that the modulus drops by a factor of about 2 in the temperature interval between absolute zero and the melting point.

Poisson's ratio for most metals is close to 0.3. The bulk modulus and shear modulus can be estimated from Eqns. (3.6) and (3.7). The moduli of metals do not show the enormous changes with temperature and time which characterize polymers, so modulus diagrams are unnecessary.

(d) Plastic and fracture properties

As the load is increased, the stress reaches the yield strength, σ_{ys}, at which the metal yields plastically and further deformation is irrecoverable (Fig. 3.11). On a microscopic scale the carriers of deformation are crystal dislocations; the yield strength is the stress required to drive the dislocations large distances through the grains of the metal, overcoming the lattice friction and the pinning effects of impurities, precipitates and other defects. But in moving through a grain the dislocations multiply, interact, and form tangles; and the tangles, by making movement more difficult, cause work-hardening (Fig. 3.11). In many metals this work-hardening is well described by a power law:

$$\sigma_s = \sigma_{ys} + B\epsilon_{pl}^m \tag{3.15}$$

where σ_{ys} is the initial yield strength, ϵ_{pl} is the plastic strain and B is a constant with the dimensions of stress. An annealed metal work-hardens strongly; then the power, m, has a value of about 0.5. A work-hardened metal is already close to its saturation strength; then the power, m, is small – typically 0.05 – and the material behaves as an elastic–perfectly-plastic solid, as shown in Fig. 3.11.

At and near room temperature the yield strength depends so weakly on temperature and strain-rate that it is sometimes treated as a constant. But this is not strictly true: most metals show a yield strength which decreases slowly with increasing temperature, and depends slightly on the strain-rate $\dot{\epsilon}$ at which the testing is done. On a microscopic scale again, thermal energy activates dislocations which are held up at obstacles, causing them to move a little more easily, and this imparts a temperature dependence to the yield strength. Kinetic models

(Kocks *et al.*, 1975) lead to equations for yield by *low-temperature plasticity* which simplify to a form very like that for yield in polymers (Eqn. 3.9.):

$$\frac{\sigma_{ys}}{E_s} = \frac{\sigma_{ys}^0}{E_s^0}\left(1 - \frac{AT}{T_m}\ln\frac{\dot{\epsilon}_{0s}}{\dot{\epsilon}}\right) \tag{3.16}$$

where σ_{ys}^0 is the yield strength at $0°K$, A is a constant of general order 0.04 and $\dot{\epsilon}_{0s}$ is a kinetic constant of about $10^6/s$ (data for individual metals and alloys are tabulated by Frost and Ashby, 1982). The equation summarizes the experimental observations that, near room temperature, the yield strength decreases linearly with increasing temperature and logarithmically with decreasing strain-rate.

At temperatures above $0.3\,T_m$ metals show slow, time-dependent deformation at stresses well below the yield strength given by Eqn. (3.16). In this regime two new classes of mechanism operate: *power-law creep* and *diffusional flow*. The first, like low-temperature plasticity, involves dislocations, but now the thermal energy is sufficient to permit them to climb as well as glide, by-passing obstacles which hold them up. The details of the mechanism are still not entirely clear, but it is well characterized by a relation between strain-rate $\dot{\epsilon}$, stress σ and temperature T of the form

$$\dot{\epsilon} = A_s\left(\frac{\sigma}{\sigma_{0s}}\right)^{n_s}\exp-\frac{Q_s}{RT} \tag{3.17}$$

or, on inversion

$$\frac{\sigma}{E_s} = \frac{\sigma_{0s}}{E_s}\left(\frac{\dot{\epsilon}}{A_s}\right)^{1/n_s}\exp\frac{Q_s}{n_s RT} \tag{3.18}$$

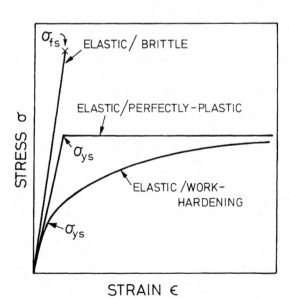

Figure 3.11 Schematic illustrating the stress–strain curves of an elastic–perfectly-plastic solid, and elastic–work-hardening solid and an elastic–brittle solid. Metals, in general, are elastic–plastic; ceramics are elastic–brittle.

where A_s, and n_s and σ_{0s} are creep constants and Q_s is an activation energy. The power n_s, typically about 5, gives the mechanism its name 'power-law creep'. The other mechanism – diffusional flow – is encountered less often. Above about $0.5\,T_m$, diffusion becomes rapid enough that a polycrystalline metal can change its shape by diffusion alone, leading to slow viscous flow. This diffusional flow, too, can be described by Eqn. (3.18) but the power n is now close to 1 and the constant A depends on grain size. The data needed to evaluate Eqn. (3.18) for both mechanisms are given by Frost and Ashby (1982).

The plastic response of a metal is conveniently summarized in a deformation-mechanism map, now available for a wide range of metals and alloys (Frost and Ashby, 1982). Fig. 3.12 is an example. The axes are the strength normalized by Young's modulus, σ_s/E_s^0, and the normalized temperature, T/T_m. On it are plotted the fields of dominance of each mechanism described above. Superim-

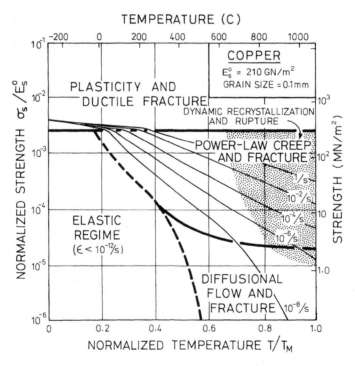

Figure 3.12 A failure-mechanism diagram for a ductile metal (copper). It shows fields in which deformation is elastic (meaning that the plastic strain-rate, by any mechanism, is less than $10^{-12}/\text{s}$), one for plasticity, and one for each of the two sorts of creep. Dynamic recrystallization is important in the shaded zone. The solid ultimately fractures by the fracture mechanism indicated on each field, but only after substantial plastic strain. For the analysis of metallic foams, 'failure' usually means plastic collapse or extensive creep, not fracture. Plasticity and creep are almost independent of stress state, so one diagram adequately describes both tension and compression.

posed on these are contours of constant strain-rate: the strength, σ_s, corresponding to any given temperature and strain-rate, can be read from the diagram. At high temperatures metals recrystallize during deformation, and this can cause further loss of strength; the regime in which this happens is shaded on the diagram.

When a polycrystalline metal is loaded in tension or bending (as it is in the cell walls of a foam) it may deform in the ways described already, or it may fracture. Some metals (those with b.c.c. or h.c.p. structure) fracture in a brittle way if the temperature is sufficiently low. *Brittle fracture* is caused by the propagation of small cracks, either along the boundaries of the grains or across the cleavage planes of the individual crystals. The cracks sometimes pre-exist in the metal (a result of growth defects when the metal was cast or formed) or, more usually, they are nucleated by slip when the metal starts to deform. In either case the initial cracks have a size $2c$ which is roughly that of the grains themselves, d, and they propagate when the Griffith criterion (Eqn. 3.8) is met, giving the fracture strength

$$\frac{\sigma_{fs}}{E_s} \approx \frac{K_{ICs}}{E_s \sqrt{\pi d/2}}. \tag{3.19}$$

Here K_{ICs} is the fracture toughness of the metal. For b.c.c. metals (like iron and steel) and for h.c.p. metals (like zinc) K_{ICs} is small at low temperatures; it can be as low as $1\,\mathrm{MN/m^{3/2}}$. Then, if the grain size is $200\,\mu m$, the metal fractures at around $50\,\mathrm{MN/m^2}$. Face-centred cubic metals (like copper) are much tougher. Even at $0°K$ the fracture toughness is around $50\,\mathrm{MN/m^{3/2}}$, making it defect-tolerant: the stress required to propagate a grain-size crack from Eqn. (3.19) is nearly $3\,\mathrm{GN/m^2}$.

As the temperature of b.c.c. and h.c.p. metals is raised above the ductile-to-brittle transition temperature, they, too, become tougher and more ductile. When the toughness is large, materials fail by a new mechanism. Deformation causes holes to nucleate at inclusions; further plastic strain makes them grow; and when they are large enough, or when the sample itself becomes mechanically unstable, they coalesce to give a *ductile fracture*.

Further increase in the temperature can lead to other fracture mechanisms. Within the creep regime (that is, above about $0.3\,T_m$) holes nucleate and grow by creep, either within the grains to give a *transgranular creep fracture*, or in the grain boundaries to give an *intergranular creep fracture*. And at the highest temperatures, in the range in which *dynamic recrystallization*[†] takes place (Fig.

[†]'Dynamic recrystallization' is continuous recrystallization which takes place during deformation. The work-hardening (which both stabilizes deformation against necking and drives the stress up to the level at which a fracture mechanism can operate) is continuously removed, and the result is the extensive plastic deformation with the formation of a neck in tension. The neck grows in severity until the material parts because the section has shrunk to zero.

3.1), fracture occurs by *rupture*: necking so severe that the sample section is reduced locally to zero.

Most metal foams available today are made from ductile f.c.c. metals like aluminium, copper and nickel. Then the failure mechanisms which are important in analysing foam properties are those associated with plasticity and creep; the ductility of the metal is so great that fracture only occurs after very large strains. It is possible to map fracture mechanisms in a way which parallels Fig. 3.12 for deformation (Ashby *et al.*, 1979; Gandhi and Ashby, 1979), but the information is not relevant to us here. Instead, the fields which quantitatively describe plasticity and creep in Fig. 3.12 have also been labelled with the approximate fracture mechanism.

The normalized axes of Fig. 3.12 makes it broadly typical of all f.c.c. metals (see Frost and Ashby, 1982, for diagrams for other classes of solid). Diagrams like this, although approximate, are useful in giving an immediate idea of the dominant mechanism in any given application, and will guide us later in the selection of the appropriate constitutive equations for the analysis of metal foams.

(e) Electrical properties of metals

All metals contain electrons which are free to move, making them good electrical conductors. For this reason, they cannot function as dielectrics, in which charge can reorient, but it cannot flow over large distances. The dielectric constant, the dissipation factor and the breakdown potential gradient have no meaning for metals. The one property which is of importance is the resistivity.

It is controlled by the electron density in the metal and the mean-free-path of electrons between scattering events. Anything which disrupts the perfection of the crystal lattice of a metal reduces the mean-free-path and increases the resistance: plastic deformation (slightly); precipitates (rather more); dissolved impurities and alloying elements (strongly). Good conductors, exemplified by pure copper and pure silver, have resistivities of about $10^{-8}\Omega$.m. Poor conductors (titanium, heavily alloyed steels) have resistivities up to 100 times larger: about $10^{-6}\Omega$.m.

3.4   Ceramics and glasses

Ceramic and glass honeycombs and foams occur in nature as pumice, coral and bone, and increasingly are manufactured from standard refractory materials such as zirconia, mullite, alumina and clays for use as high-temperature insulation and for filters. Their properties are analysed in later chapters. Here we summarize the material properties that are essential to this analysis. Data are listed in Table 3.5. Estimates are sometimes necessary, so ways of making them are given below, along with the likely errors associated with doing so.

(a) Structure and density of ceramics and glasses

Ceramics are crystalline, inorganic compounds. Some, like the refractory magnesia (MgO) or hydroxyapatite (a major component of bone) are ionically bonded. Others, like the ultra-hard silicon carbide (SiC) or the silicates which form the basis of clays, have covalent bonds. Both types of bond are strong, giving the materials higher melting points and moduli than any other class of solid. *Glasses* are amorphous, inorganic compounds. (If they partially crystallize, they become glass–ceramic composites or pyrocerams.) Most glasses are based on silica (SiO_2) and are held together by a network of strong covalent bonds which, again, give softening temperatures which are high and moduli which are large compared with those of polymers.

Broadly speaking, engineering ceramics have one of two types of structures (Kingery *et al.*, 1976). The first is characterized by a close-packing of oxygen with metal ions contained in the interstices between the oxygens. If the oxygen is in an f.c.c. packing, simply geometry shows that the density is simply

$$\rho_s = \frac{M_w}{8\sqrt{2}N_A b r_O^3}$$

where M_w is the molecular weight, N_A is Avogadro's number, b is the number of oxygen ions in the molecule (one in MgO, three in Al_2O_3 and so on) and r_O is the radius of an oxygen atom. This radius does vary somewhat from oxide to oxide, but a good approximate description is given by taking it to be 1.14×10^{-10} m, giving the rule of thumb:

$$\rho_s(Mg/m^3) = \frac{M_w(kg/kmol)}{10b} \tag{3.20}$$

(the maximum error is about 15%). As an example: the molecular weight of Al_2O_3 is 102 kg/kmol, and each molecule contains three oxygen ions, giving $\rho_s \approx 3.4$ Mg/m^3; the measured density is 3.9 Mg/m^3. The second type of ceramic, and all inorganic glasses, are characterized by covalent, directional bonds which give open network structures with low coordination (often four) and a lower density. Crystalline silica, for instance, has a density, in Mg/m^3, of only 2.65 compared with the 3.9 of alumina, even though silicon and aluminium have similar atomic weights. And the amorphous, even more open, structure of silica-based glasses gives them a slightly lower density still (see Table 3.5).

Table 3.5 Properties of solid ceramics and glasses

Material	Density ρ_s (Mg/m³)	Melting point T_m (K)	Thermal expansion $\alpha_s \times 10^{-6}$ (K⁻¹)	Thermal conductivity λ_s (W/mK)	Specific heat C_{ps} (J/kgK)	Young's modulus at 20°C E_s (GN/m²)	Compression strength σ_{cs} (MN/m²)	Fracture strength σ_{fs} (MN/m²)	Fracture toughness K_{ICs} (MN/m$^{3/2}$)
Alumina Al₂O₃	3.9	2323	8.8	25	795	380	1750–3000	250–300	3–5
Beryllia BeO	3.0	2700	9.0	200	1250	380	1550–1850	100–140	2–3
Calcite (limestone, coral)	2.7	—	8.0	7	—	63	—	30–80	0.9
Cement	2.4–2.5	1200–1500(s)†	1.2	10–14	670	45–50	20–60	1–4	0.2
Glass, silica	2.6	1100(s)†	0.5–1	2.0	750	94	2000–2200	350–360	0.6–0.7
Glass, soda	2.48	720(s)	8.5	1.6	990	74	2000–3000	35–55	0.7
Glass, borosilicate	2.23	820(s)	4.5	1.2	800	65	750–1000	55	0.8
Graphite	1.82	4000	2.5	120–200	120	27	45–350	10–100	1–2
Ice, H₂O	0.92	273	55	2–2.1	2100	9.1	610	6–7	0.12
Magnesia, MgO	3.5	3073	13.5	3.0	950	250	1400–2800	90–190	3.0
Mullite	3.2	2120	5.3	2–6	770	145	2600–3000	150–220	2–3
Porcelain	2.3–2.5	800(s)	6.0	1.5	—	70	—	45	1.0
Sialon	3.2	2750	3.2	20–25	710	300	3800–4500	400–800	5
Silicon carbide, SiC	3.2	3110	4.3	50–84	1420	410	2000–3500	200–500	3
Silicon nitride, Si₃N₄	3.2	2173(d)†	3.2	17	630	310	1100–2500	300–850	4
Titanium carbide, TiC	7.2	3500	7.4	17	550	370	3500–4500	260–330	2–3
Tungsten carbide, WC	14–17	3100	5.8	30–80	190	450–650	5000–8000	370–530	2–3.8
Zirconia, cubic, ZrO₂	5.6	2843	10.0	2.0	670	200	1650–3600	200–500	2–6

† (s) = softening; (d) = decomposition.

(b) Thermal properties of ceramics and glasses

The strong bonding in ceramics gives them high *melting points* (Table 3.5). Typically, T_m for crystalline ceramics is greater than 2000°K. Glasses soften at a lower temperature than this (typically, 1000°K); and since the crystalline phases in pottery and porcelain are held together by glass, they, too, soften at around the same temperature.

The strong bonding has another consequence: it gives strong elastic coupling between atoms. This gives ceramics *thermal conductivities* which are larger, by a factor of about 100, than those of polymers (Table 3.5). But ceramics do not contain the free electrons which give metals their exceptional ability to transmit heat and electricity, so their thermal conductivities are less, by a factor of about 4, than those of metals. Although the bonding in glasses is strong, their amorphous structure reduces considerably the vibrational coupling of neighbouring atoms, and their thermal conductivity is therefore lower than that of crystalline ceramics.

The *specific heat* of ceramics, like that of metals, tends to the constant value of $3R/M$ J/kg K well above the Debye temperature[†] (Eqn. 3.11). But for many ceramics the Debye temperature is near room temperature so that the full heat capacity is not realized. Because of this, ceramics have specific heats, at room temperature, which, per mole, are a little lower than metals, although (because of their low molecular weight) the specific heat in the usual units of J/kg K is still a little higher (Table 3.5).

As explained in Section 3.3(b), *thermal expansion coefficients* vary inversely as the moduli (Eqn. 3.12). Ceramics with dense-packed oxygen structures have large moduli, larger, in general, than those of metals, and the thermal expansion coefficients are correspondingly smaller (Table 3.5). Silica-based ceramics and glasses, which have network structures, show still lower expansion coefficients because vibrational energy can be absorbed in transverse modes which do not expand the structure.

(c) Elastic properties of ceramics and glasses

The strong ionic and covalent bonds give ceramics and glasses a relatively high modulus: as high, or higher than those of metals (Table 3.5). As with metals, there is a proportionality between modulus and melting point (Eqn. 3.13) which

[†]The Debye temperature is a material property which characterizes the temperature at which the atoms of the solid can absorb $3kT$ of thermal energy per atom. Above the Debye temperature the specific heat quickly approaches the value calculated here. But below the Debye temperature the vibrational modes of the atoms are progressively suppressed; the solid has a reduced capacity to absorb thermal energy and the specific heat falls.

can be useful in estimating moduli. The modulus varies slightly with temperature, in a way which is well described by Eqn. (3.14), with a temperature coefficient, α_m, which is about the same as that for metals: roughly 0.5. The moduli of glasses are generally a little lower than those of crystalline ceramics because of their amorphous structure and lower density. Poisson's ratios for ceramics and glasses are roughly 0.3. The bulk and shear moduli can be estimated by using this value in Eqns. (3.6) and (3.7). The dependence of the moduli on temperature and time is so simple that no modulus diagram is needed to describe it.

(d) Plastic and fracture properties of ceramics and glasses

If fracture is suppressed (for instance, by testing under a large hydrostatic pressure) then it is found that ceramics exhibit *plasticity* in much the same way that metals do. Dislocations move through the crystals of the ceramic when the shear stress overcomes the lattice friction: in metals it is generally small, $E_s/100$ or less; in ceramics it is large, rising to about $E_s/10$ at absolute zero. It is this which gives ceramics their great hardness and wear resistance; but it is also one of the factors which makes them brittle. When deformed, ceramics work-harden in the same way that metals do, following Eqns. (3.15) and (3.16). But it is rare that this kind of plasticity is encountered in practice. Brittle fracture (described later) almost always intervenes.

At temperatures above about $0.5\,T_m$, ceramics creep in much the same way that metals do. The mechanisms are the same as those in metals: *power-law creep* and *diffusional flow*. The creep response is approximately described by Eqn. (3.18) with a power which has a value of about 3 in the power-law creep regime, and close to 1 in the diffusional flow regime. Glasses are a little different. Below the glass temperature, T_g, the viscosity of the glass is so large that flow can, under almost all circumstances, be ignored. Above T_g there is sufficient thermal energy to allow relative motion of the molecules making up the glass, which then responds to stress by flowing in a viscous manner. The viscosity falls steeply with temperature, following a W.L.F. time–temperature scaling (Eqn. 3.5) near T_g and a simple Arrhenius law well above it (Kingery *et al.*, 1976).

The most marked mechanical characteristic of ceramics and glasses, however, is their brittleness in tension. As with metals, *brittle fracture* is caused by the propagation of small cracks, either along the boundaries of the grains or across their cleavage planes. The difference is that cracks always pre-exist in ceramics and in glasses, either as fabrication defects or because of corrosion or abrasion. If the length of these pre-existing cracks is $2c$, then the fracture strength of the ceramic in tension is given by Eqn. (3.8):

$$\frac{\sigma_{fs}}{E_s} = \frac{K_{ICs}}{E_s\sqrt{\pi c}}$$

where K_{ICs} is the fracture toughness of the ceramic or glass. Here lies the second major difference in mechanical properties between ceramics and metals: K_{ICs} for ceramics and glasses are low, typically $2\,MN/m^{3/2}$. The secret of making strong ceramics is to fabricate them in such a way that the intrinsic cracks they contain are short. By using very fine powders, and consolidating them with great care, it is possible to limit the crack length $2c$ to $1\,\mu m$ or less, and then the ceramic has a high strength. But in fabricating ceramic foams this level of process control is difficult to achieve; and most of them (firebrick, for instance) have cell walls which contain large flaws and have a very low fracture strength.

At higher temperatures ($>0.5\,T_m$), in the regime in which ceramics creep, they show other modes of fracture, very like those in metals: intergranular creep fracture, transgranular creep fracture and rupture (Section 3.3d). The overall fracture behaviour can, as before, be summarized in a failure-mechanism diagram. Figures 3.13 and 3.14 show a pair of such diagrams for alumina, one for tension and one for compression. At low temperatures the material is elastic at low stresses and fails by brittle fracture at high stresses. Brittle fracture depends on stress-state, so that the fracture stress in tension is lower than that in compression by a factor of about 12. As the temperature is raised, thermal activation makes dislocations mobile, and at the ductile-to-brittle transition the material starts to show limited creep plasticity. This temperature for all ceramics except ice is well

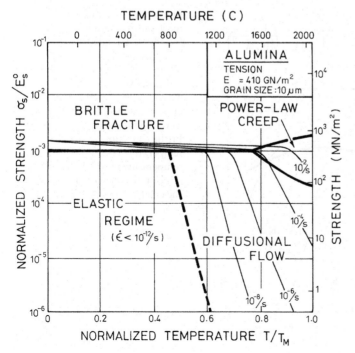

Figure 3.13 A failure-mechanism diagram for a ceramic (alumina) loaded in tension. It shows a field on which deformation is elastic (defined as the regime in which the plastic strain-rate is less than $10^{-12}/s$), one in which brittle fracture is the dominant failure mechanism, and fields of creep plasticity and fracture.

above room temperature, so that creep (and the associated creep fracture) is relevant only for the analysis of ceramic foams for use at elevated temperatures.

Most ceramics have diagrams which look roughly like Figs. 3.13 and 3.14 (Frost and Ashby, 1982; Gandhi and Ashby, 1979). They must, of course, be regarded as an approximate description only. Changes in processing, or in chemistry, can influence the placing of the boundaries which separate the regimes and the positions of the contours. But they give useful guidance in the selection of constitutive equations for the modelling of foam behaviour.

(e) Electrical properties of ceramics and glasses

Ceramics and glasses exhibit greater diversity in their electrical properties than any other class of solids. Many are good insulators, like alumina, Al_2O_3 (used for spark-plug insulators) and show dielectric behaviour. Some are semi-conductors – notably germanium, silicon, diamond and a range of compounds with rather similar structure; so, too, are chalcogenide glasses. Still others are conductors, among these metal carbides such as TiC and oxides such as ZnO.

The dielectric constants (defined in Section 3.2(e)) of ceramics span a wider range than those of polymers. Oxides, nitrides and silicates all contain two or more types of atom with different charges, which polarise in an electric field. The result is that none have as low a dielectric constant as polyethylene. Silica,

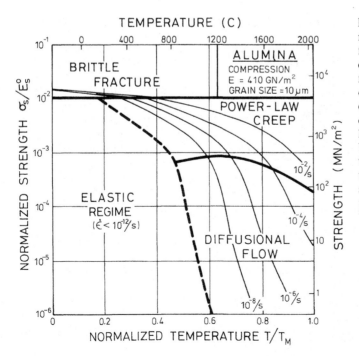

Figure 3.14 A failure-mechanism diagram for alumina, loaded in compression. Brittle fracture is made more difficult, expanding the elastic regime and the creep-plasticity field, and lowering the ductile-to-brittle transition temperature (the temperature at which the strain-rate contours enter the creep field).

SiO_2, has one of the lowest – about 3.7. Most have a dielectric constant of about 10, though some can be much higher. The dielectric loss in ceramics derives – as in polymers – from the localised charge movement in an oscillating electric field. It is low, comparable with that in polymers, so delectric ceramics are transparent to microwaves. Ice, happily, is an exception – it has a very large dielectric loss, allowing frozen food to be defrosted in a microwave oven.

Dielectric ceramics, like polymers, have a breakdown potential gradient. Generally speaking, it is lower than that of polymers. Alumina, Al_2O_3, has a high value – hence its use for power-line insulators. Cement, concrete and plaster are poor, largely because they are hydrated, and contain absorbed water.

As already mentioned, the resistivity of ceramics and glasses spans an enormous range. Silica, SiO_2, alumina, Al_2O_3, and many glasses have almost as high a resistivity as the best polymers. At the other extreme the metal carbides TaC, ZrC, WC and NbC have resistivities near $10^{-6}\Omega.m$, about the same as that of stainless steel and titanium alloys.

3.5   Summary

Cellular solids are made from very diverse materials. They include examples drawn from all known classes of solid: polymers, metals, ceramics, glasses and composites. In modelling their properties it is important to know the characteristics of the material from which the honeycomb or foam is made, and the regime of stress and temperature to which it is subjected. Then, using deformation and fracture diagrams, it is possible to identify the dominant mechanisms of deformation and failure, and to select a constitutive equation which is appropriate for the modelling of the foam behaviour.

Polymers at low temperatures $(T \ll T_g)$ behave in an elastic–brittle manner. A little below the glass temperature $(T \approx 0.9\,T_g)$ they become elastic–plastic solids, with extensive ductility. At $T \approx T_g$ they become viscoelastic, showing recoverable viscous deformation. At still higher temperatures $(T \approx 1.2\,T_g)$ they become rubbery, and, if they contain a few widely spaced cross-links, they show the characteristics of true elastomers and can accommodate very large, reversible, elastic distortions. At the highest temperatures $(T > 1.4\,T_g)$, polymers melt to viscous fluids or decompose.

Metals at room temperature generally behave as elastic–plastic solids, with large ductilities; then the most important properties are the modulus, E_s, and yield strength, σ_{ys}. Ceramics and glasses, by contrast, behave at room temperature in an elastic–brittle manner; their most important properties are the modulus, E_s, and the brittle fracture strength, σ_{fs}.

The thermal properties, like the mechanical properties, vary widely. Polymers have poor thermal conductivities, large specific heats and large coefficients of thermal expansion. Metals are almost the opposite: good conductors, but with low specific heats and fairly low expansion coefficients. Ceramics show the least thermal expansion, and have conductivities and specific heats intermediate between those of metals and those of polymers.

Electrical properties, too, show great diversity. Polymers and many ceramics and glasses are dielectrics – they conduct hardly at all, but respond to an electric field by polarising; the dielectric constant measures the extent of polarisation. Extremely large fields cause breakdown – catastrophic conduction, like that of a lightning strike. Oscillating fields are associated with losses, and thus with heating; in most polymers and ceramics this is low, making them transparent to microwaves. All metals and certain ceramics and glasses conduct well, but – in the greater world of materials – conduction is the exception, not the rule.

We now proceed to the analysis of honeycomb and foam properties, using for the cell wall the properties described above. The reader seeking a more detailed discussion of the properties of solids might consult Ward (1983) or Hall (1981) for polymers, Cottrell (1967) or Frost and Ashby (1982) for metals, and Kingery *et al.* (1976) or Creyke *et al.* (1982) for ceramics and glasses. Composites, which we have not discussed here because foams made from them are comparatively rare, are well introduced in the book by Hull (1981).

References

Ahmad, Z. B. and Ashby, M. F. (1988) *J. Mat. Sci.,* **23,** 2037.

Ashby, M. F. (1992) *Materials Selection in Mechanical Design.* Pergamon Press, Oxford UK

Ashby, M. F. and Hallam, S. D. (1986) *Acta Met.,* **34,** 497.

Ashby, M. F., Gandhi, C. and Taplin, D. M. R. (1979) *Acta Met.,* **27,** 699.

Cottrell, A. H. (1967) *An Introduction to Metallurgy,* Arnold, London.

Creyke, W. E. C., Sainsbury, I. E. J. and Morrell, R. (1982) *Design with Non Ductile Materials,* Applied Science, Barking.

Frost, H. J. and Ashby, M. F. (1982) *Deformation Mechanism Maps.* Pergamon Press, Oxford.

Gandhi, C. and Ashby, M. F. (1979) *Acta Met.,* **27,** 1565.

Gilbert, D G., Ashby, M. F. and Beaumont, P. W. R. (1986) *J. Mat. Sci.* **21,** 3194.

Hall, C. (1981) *Polymeric Materials.* Macmillan, London.

Hull, D. (1981) *An Introduction to Composite Materials.* Cambridge University Press, Cambridge.

Kingery, W. D., Bowen, H. K. and Uhlmann, D. R. (1976) *Introduction to Ceramics.* Wiley-Interscience, New York.

Kinloch, A. J. and Young, R. J. (1983) *Fracture Behaviour of Polymers,* Applied Science Publishers, London.

Kocks, U. F., Argon, A. S. and Ashby, M. F. (1975) *Thermodynamics and Kinetics of Slip,* Progress in Materials Science, vol. 19, p. 1. Pergamon Press, Oxford.

McClintock, F. A. and Argon, A. S. (1966) *Mechanical Behaviour of Materials.* Addison Wesley, Reading, Mass.

Treloar, L. R. G. (1958) *The Physics of Rubber Elasticity.* Clarendon Press, Oxford.

Ward, I. M. (1983) *Mechanical Properties of Solid Polymers,* 2nd edn. Wiley-Interscience, New York.

Williams, M. L., Landel, R. L. and Ferry, J. D. (1955) *J. Amer. Chem. Soc.,* **77,** 3701.

General references

(a) All materials

Few hard-copy data sources span the full spectrum of materials and properties. Eight which, in different ways, attempt to do so are listed below.

Materials Selector (1994), Materials Engineering, Special Issue. Penton Publishing, Cleveland, Ohio, U.S.A. *Tabular data for a broad range of metals, ceramics, polymers and composites, updated annually. Basic reference work.*

The Elsevier Materials Selector (1991), edited by N. A. Waterman and M. F. Ashby. Elsevier Science Publishers Ltd, Crown House, Linton Road, Essex IG11 8JU. U.K. and the CRC Press Inc, 2000 Corporate Blvd N.W., Boca Raton, Florida 33431, U.S.A. *A 3-volume compilation of data for all materials, with selection and design guide. Basic reference work.*

ASM Engineered Materials Reference Book 2nd Edition (1994), editor: Bauccio, M. L., ASM International, Metals Park, Ohio 44073, U.S.A. *Compact compilation of numeric data for metals, polymers, ceramics and composites.*

Materials Selector and Design Guide (1974), Design Engineering, Morgan-Grampian Ltd, London. *Resembles the Materials Engineering "Materials Selector", but less detailed and now rather dated.*

Handbook of Industrial Materials (1992) 2nd Edition, Elsevier, Oxford, UK. *A compilation of data remarkable for its breadth: metals, ceramics, polymers, composites, fibres, sandwich structures, leather ...*

Materials Handbook (1986) 2nd Edition, Editors: Brady, G. S. and Clauser, H. R., McGraw-Hill, New York, NY, USA. *A broad survey, covering metals, ceramics, polymers, composites, fibres, sandwich structures, leather ...*

Handbook of Thermophysical Properties of Solid Materials (1961), Goldsmith, A., Waterman, T. E. and Hirschhorn, J. J. Macmillan, NY, USA. *Thermophysical and thermochemical data for elements and compounds.*

Guide to Engineering Materials Producers (1994), editor: Bittence, J. C., ASM International, Metals Park, Ohio 44037, U.S.A. *A comprehensive catalog of addresses for material suppliers.*

(b) Polymers and elastomers

Polymers are not subject to the same strict specification as metals. Data tend to be producer-specific. Sources, consequently, are scattered, incomplete and poorly presented. Saechtling is the best; although no single hard-copy source is completely adequate, all those listed here are worth consulting. See also Data-bases as Software (Section 5; some (Plascams, CMS) are good on polymers.

Saechtling: International Plastics Handbook, editor: Dr. Hansjurgen Saechtling, Macmillan Publishing Co (English edition), London, UK (1983), *The most comprehensive of the hard-copy data-sources for polymers.*

Polymers for Engineering Applications, R. B. Seymour. ASM International, Metals Park, Ohio 44037, U.S.A. (1987). *Property data for common polymers. A starting point, but insufficient detail for accurate design or process selection.*

New Horizons in Plastics, a Handbook for Design Engineers, editor: J. Murphy, WEKA Publishing, London, UK. (1992).

ASM Engineered Materials Handbook, Vol. 2. Engineering Plastics (1989). ASM International, Metals Park, Ohio 44037, U.S.A.

Handbook of Plastics and Elastomers, Editor C. A. Harper, McGraw-Hill, New York, U.S.A. (1975).

International Plastics Selector, Plastics, 9th Edition. Int. Plastics Selector, San Diego, Calif, U.S.A. (1987).

Die Kunststoffe und Ihre Eigenschaften, editor: Hans Domininghaus, VDI Verlag, Dusseldorf, Germany (1992).

Properties of Polymers, 3rd edition, D. W. van Krevelen, Elsevier, Amsterdam, Holland, (1990). *Correlation of properties with structure; estimation from molecular architecture.*

Handbook of Elastomers, A. K. Bhowmick and H. L. Stephens. Marcel Dekker, New York, U.S.A. (1988).

ICI Technical Service Notes, ICI Plastics Division, Engineering Plastics Group, Welwyn Garden City, Herts, U.K. (1981).

Technical Data Sheets, Malaysian Rubber Producers Research Association, Tun Abdul Razak Laboratory, Brickendonbury, Herts.

SG13 8NL (1995). *Data sheets for numerous blends of natural rubber.*

(c) Metals

Metals and alloys conform to national and (sometimes) international standards. One consequence is the high quality of data. Hard copy sources for metals data are generally comprehensive, well-structured and easy to use.

ASM Metals Handbook (1986) 9th Edition, and (1990) 10th Edition. ASM International, Metals Park, Ohio, 44073 U.S.A. The 10th Edition contains Vol 1: Irons and Steels; Vol 2: Non-ferrous Alloys; Vol 3: Heat Treatment; Vol 4: Friction, Lubrication and Wear; Vol 5: Surface Finishing and Coating; Vol 6: Welding and Brazing; Vol 7: Microstructural Analysis; more volumes are planned. Vol 8: Friction, Lubrication and Wear Technology; *Basic reference work, continuously upgraded and expanded.*

ASM Metals Reference Book 3rd edition, (1993), ed. M. L. Bauccio, ASM International, Metals Park, Ohio 44073, U.S.A. *Consolidates data for metals from a number of ASM publications. Basic reference work.*

Smithells, C. J. (1992), *Metals Reference Book*, 7th Edition (Editor: E. A. Brandes and G. B. Brook). Butterworths, London, U.K. *A comprehensive compilation of data for metals and alloys. Basic reference work.*

Metals Databook (1990), Colin Robb. The Institute of Metals, 1 Carlton House Terrace, London SW1Y 5DB, U.K. *A concise collection of data on metallic materials covered by the U.K. specifications only.*

ASM *Guide to Materials Engineering Data and Information*, (1986). ASM International, Metals Park, Ohio 44073, U.S.A. *A directory of suppliers, trade organisations and publications on metals.*

The Metals Black Book, Volume 1, Steels, ed.: J. E. Bringas, Casti Publishing Inc. 14820–29 Street, Edmonton, Alberta T5Y 2B1, Canada, (1992), *A compact book of data for steels.*

The Metals Red Book, Volume 2, Nonferrous Metals, ed. J. E. Bringas, Casti Publishing Inc. 14820–29 Street, Edmonton, Alberta T5Y 2B1, Canada, (1993).

(d) Ceramics and glasses

The new engineering or 'fine' ceramics are still too new for adequate specification and characterisation. The hard-copy sources listed below try hard, but using them remains a frustrating experience.

Handbook of Structural Ceramics, editor: M. M. Schwartz, McGraw-Hill, New York, USA (1992). *Lots of data, information on processing and applications.*

ASM Engineered Materials Handbook, Vol. 4, Ceramics and Glasses. ASM International, Metals Park, Ohio 44073, U.S.A. (1991).

Concise Encyclopedia of Advanced Ceramic Materials editor: R. J. Brook, Pargamon Press, Oxford, UK (1991).

Handbook of Ceramics and Composites, 3 Vols, ed. N. P. Cheremisinoff, Marcel Dekker Inc., New York, USA (1990).

Richerson, D. W., *Modern Ceramic Engineering*, 2nd Edition, Marcel Dekker, New York, USA (1992).

Handbook of Properties of Technical and Engineering Ceramics Parts 1 and 2, Morrell, R., National Physical Laboratory, Teddington, U.K. (1985).

Creyke, W. E. C., Sainsbury, I. E. J. and Morrell, R., *Design with Non Ductile Materials*. App. Sci., London, U.K. (1982).

Handbook of Glass Properties, Bansal, N. P. and Doremus, R. H., Academic Press, New York, U.S.A. (1966).

Engineering Design Guide 05: The Use of Glass in Engineering, Oliver, D. S. Oxford University Press, Oxford, UK (1975).

What Every Engineer Should Know about Ceramics, S. Musikant, Marcel Dekker, London UK (1991). *Good on data.*

Chapter 4

The mechanics of honeycombs

4.1   Introduction and synopsis

The honeycomb of the bee, with its regular array of prismatic hexagonal cells, epitomizes a cellular solid in two dimensions. Here we use the word 'honeycomb' in a broader sense to describe any array of identical prismatic cells which nest together to fill a plane. The cells are usually hexagonal in section, as they are in the bee's honeycomb, but they can also be triangular, or square, or rhombic. Examples were shown in Fig. 2.3.

Man-made polymer, metal and ceramic honeycombs are now available as standard products. They are used in a variety of applications: polymer and metal ones for the cores of sandwich panels in everything from cheap doors to advanced aerospace components; metal ones for energy-absorbing applications (the feet of the Apollo 11 landing module used crushable aluminium honeycombs as shock absorbers); and ceramic ones for high-temperature processing (as catalyst carriers and heat exchangers, for example). And many natural materials – wood is one – can be idealized and analysed as honeycombs (as we do in Chapter 10). If such materials are to be used in load-bearing structures an understanding of their mechanics is important.

There is a second good reason for studying honeycombs: it is that the results shed light on the mechanics of the much more complex three-dimensional foams. Analysing foams is a difficult business: the cell walls form an intricate three-dimensional network which distorts during deformation in ways which are hard to identify. Honeycombs are much easier. Large-scale models can be

93

made of rubber, metal or ceramic, and their deformation modes observed and classified. And because honeycombs have a regular geometry their deformations can be analysed more or less exactly to give equations which describe their properties. Finally, the results of the analysis can be checked by experiments on the tailor-made models.

When this is done the following picture emerges. If a honeycomb is compressed *in-plane* (that is, in the plane of the photographs shown in Fig. 2.3), the cell walls at first *bend*, giving linear elastic deformation. Beyond a critical strain the cells *collapse* by elastic buckling, plastic yielding, creep or brittle fracture, depending on the nature of the cell wall material. Cell collapse ends once the opposing cell walls begin to touch each other; as the cells close up the structure *densifies* and its stiffness increases rapidly. In tension the cell walls at first bend (as in compression) but elastic buckling is not possible. If the cell wall material yields plastically the honeycomb itself shows extensive plasticity; if the cells are brittle it fractures. On loading *out-of-plane* (such that a component of stress acts parallel to the axis of the prismatic cells) the cell walls suffer *extension or compression* and the moduli and collapse stresses are much larger.

In this chapter we analyse the in-plane and the out-of-plane deformation and failure of honeycombs, and demonstrate the accuracy of the analysis by comparing the results with experimental data. The analysis is kept as simple as possible while still retaining a general hexagonal cell shape. Deformation mechanisms are discussed in Section 4.2; the results of analysing these for hexagonal cells, of particular relevance to the later analysis of foams, are mostly to be found in Sections 4.3 and 4.4. Results for triangular and square cells (which can have mechanical properties which differ greatly from those of hexagons) are given in Appendix 4A.

4.2   Deformation mechanisms in honeycombs

Figure 4.1 shows a hexagonal honeycomb, by far the commonest kind. The *in-plane* stiffnesses and strengths (that is, those in the X_1–X_2 plane) are the lowest because stresses in this plane make the cell walls bend. The *out-of-plane* stiffnesses and strengths (those in the X_3 direction) are much larger because they require the axial extension or compression of the cell walls. Throughout this chapter we distinguish between in-plane and out-of-plane properties, and analyse them with slightly different goals in mind. The study of in-plane properties highlights the mechanisms by which cellular solids deform and fail, and these ideas will be carried forward to develop the understanding of three-dimensional foams (Chapter 5) and of natural honeycomb-like materials such as wood, cancellous bone and cork (Chapters 10–12). The out-of-plane analysis gives the additional stiffnesses which are needed for the design of honeycomb cores in

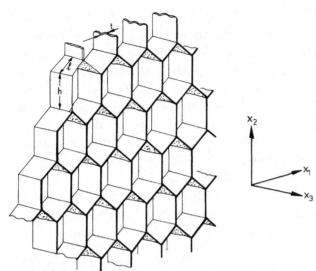

Figure 4.1 A honeycomb with hexagonal cells. The in-plane properties are those relating to loads applied in the X_1–X_2 plane. Responses to loads applied to the faces normal to X_3 are referred to as the out-of-plane properties.

sandwich panels and for the complete description of the behaviour of natural honeycomb-like materials, such as wood.

(a) In-plane deformation

Figure 4.2 shows compressive and tensile stress–strain curves for an elastomeric honeycomb (a rubber), an elastic–plastic honeycomb (a metal) and one which is elastic–brittle (a ceramic). They have broadly similar shapes, but for different reasons. In compression, all show a linear-elastic regime followed by a plateau of roughly constant stress, leading into a final regime of steeply rising stress. Each regime is associated with a *mechanism of deformation* which can be identi-fied by loading and photographing model honeycombs (Figs 4.3 and 4.4). On first loading, the cell walls *bend*, giving linear elasticity (provided, of course, that the cell wall material is itself linear-elastic). But when a critical stress is reached the cells begin to collapse: in elastomeric materials collapse is by the *elastic buckling* of the cell walls and so is recoverable; in materials with a plastic yield point it is by the formation of *plastic hinges* at the section of maximum moment in the bent members; and in brittle materials it is by *brittle fracture* of the cell walls; the last two, of course, are not recoverable. Eventually, at high strains, the cells collapse sufficiently that opposing cell walls touch (or their broken fragments pack together) and further deformation compresses the cell wall material itself. This gives the final, steeply rising portion of the stress–strain curve labelled *densification*.

An increase in the relative density of a honeycomb increases the relative thick-ness of the cell walls. Then the resistance to cell wall bending and cell collapse goes up, giving a higher modulus and plateau stress; and the cell walls touch sooner, reducing the strain at which densification begins. Figure 4.5 summarizes the mechanisms for compressive deformation of honeycombs, and shows how the stress–strain curve changes with increasing relative density or t/l.

Tensile deformation can be different (Fig. 4.2). Initially the cell walls bend, giving linear-elastic deformation with the same slope (and so the same modulus) as in compression. But in tension an elastomeric honeycomb does not buckle; instead, the cell walls rotate towards the tensile axis, and the stiffness rises. Plastic honeycombs behave in almost the same way as they do in compression: plastic hinges form, allowing large deformations at a nearly constant 'plateau' stress; only the geometry change introduces a difference usually pushing the tensile curve above the compressive one. Brittle honeycombs fail abruptly in tension, at a stress which is usually lower than the true crushing strength. As with any brittle

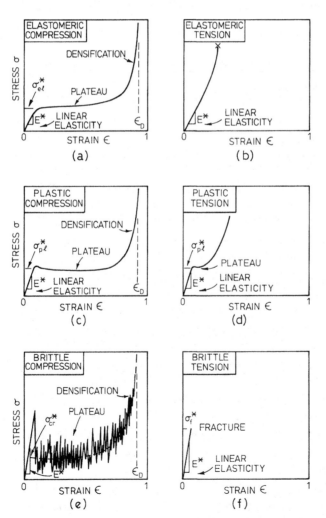

Figure 4.2 Compressive and tensile stress–strain curves for honeycombs: (a) and (b) an elastomeric honeycomb; (c) and (d) an elastic-plastic honeycomb; (e) and (f) an elastic-brittle honeycomb.

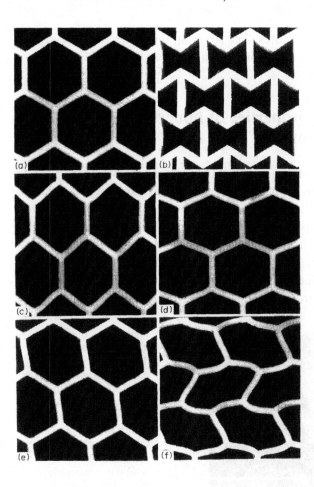

Figure 4.3 In-plane mechanisms of deformation in elastomeric honeycombs: (a) an undeformed rubber honeycomb with regular hexagonal cells ($h/l = 1; \theta = 30°$); (b) an undeformed rubber honeycomb with inverted cells ($h/l = 2; \theta = -30°$); (c) the cell wall bending caused by uniaxial compression in the X_1 direction; (d) the cell-wall bending caused by uniaxial compression in the X_2 direction; (e) bending of the cell walls in shear; (f) elastic buckling of the cell walls from compression in the X_2 direction.

solid, fracture in tension is controlled by the largest *defect* (a crack, or notch, or cluster of damaged cells) which propagates in a way which is calculated by the methods of fracture mechanics. Increasing relative density has a similar effect to that in compression: the elastic moduli, plastic yield stress and brittle fracture stress all increase.

(b) Out-of-plane deformation

Honeycombs are much stiffer and stronger when loaded along the cell axis – the X_3 direction in Fig. 4.1. The same is true for honeycombs loaded in out-of-plane shear (as they are in sandwich panels loaded in bending). In these cases the initial linear-elastic deformation involves significant axial or shear deformations of the cell walls themselves. In compression the linear-elastic regime is truncated by buckling (elastic for an elastomer, plastic for a metal or rigid polymer) and final failure is by tearing or crushing, giving a stress–strain curve like that shown in Fig. 4.6. In tension the honeycomb is elastic until it tears, yields plastically or fractures. The stress–strain curves for honeycombs with a range of relative densities again form a family as shown in Fig. 4.7.

4.3   The in-plane properties of honeycombs: uniaxial loading

Figure 4.8(a) shows a unit cell of a hexagonal honeycomb. We wish to analyse its response to loads applied in the X_1–X_2 plane. If the hexagon is regular (that is, the sides are equal and the angles are all 120°) and the cell walls are all of the same thickness, then the in-plane properties are isotropic: they do not depend on direction. Such a structure has two independent elastic moduli (a Young's modulus E^* and a shear modulus G^*, for instance) and a single value of the plateau stress, σ^*. But when the hexagon is irregular (like that in the figure) or the cell walls in one direction are thicker than those in the others, the properties are

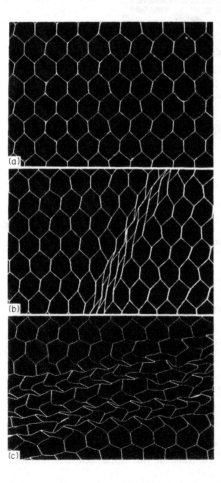

(a)

(b)

(c)

Figure 4.4 In-plane mechanisms of deformation in elastic-plastic honeycombs: (a) an undeformed aluminium honeycomb; (b) plastic yielding of the cell walls loaded in compression in the X_1 direction; (c) plastic yielding of the cell walls loaded in compression in the X_2 direction.

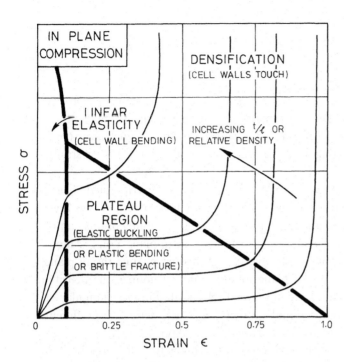

Figure 4.5 A schematic diagram for a honeycomb loaded in compression in the X_1–X_2 plane, showing the linear-elastic, collapse and densification regimes, and the way the stress strain curve changes with t/l.

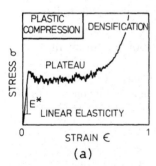

(a)

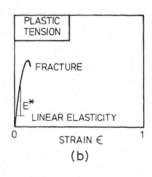

(b)

Figure 4.6 Stress–strain curves for the axial (X_3) loading of a honeycomb, in (a) compression and (b) tension.

anisotropic and a complete description of the in-plane properties (Appendix, page 496) requires four moduli (E_1^*, E_2^*, G_{12}^* and v_{12}^* where v_{12}^* is a Poisson's ratio) and two values for the plateau stress (σ_1^* and σ_2^*).

We now evaluate the moduli and collapse stresses for a general honeycomb for which h and l differ and for arbitrary cell wall angle, θ. The isotropic results are easily obtained from these. We assume that the honeycomb has a low relative density, ρ^*/ρ_s, so that t/l is small; simple geometry gives the relation between the two as:

$$\frac{\rho^*}{\rho_s} = \frac{t/l(h/l+2)}{2\cos\theta(h/l+\sin\theta)} \tag{4.1a}$$

which reduces to:

$$\rho^*/\rho_s = \frac{2}{\sqrt{3}} \frac{t}{l}$$

(4.1b)

when the cells are regular ($h = l; \theta = 30°$). We also assume that deformations are sufficiently small that changes in geometry can be neglected. Refinements to deal with these can be added (Gibson, 1981), but the results are much more complicated and the additional agony is necessary only when $t/l > 1/4$ or strains are greater than 20%.

Many commercial honeycombs are made by expanding strip-glued sheets. Then each cell has four walls of thickness t and two (those of height h in Figs. 4.1 or 4.8) which are doubled and have thickness $2t$. The doubling of this pair of cell walls does not (surprisingly) change the values of the in-plane Young's moduli and Poisson's ratios calculated below, though it does change the in-plane shear modulus and all the out-of-plane moduli, and, of course, the density. Some few honeycombs do not have hexagonal cells at all, but are triangular or square. Results for these exceptional geometries, which deform by different mechanisms than the hexagonal cells, are presented in Appendix 4A.

Properties of the solid cell wall materials are often taken from handbooks; reliable data for both the Young's modulus and yield strength of the cell wall of common metal honeycombs can be obtained this way. Ceramic and polymer honeycombs are more difficult. The cell walls of ceramic honeycombs are often themselves porous; then, their properties must be found from measurements on

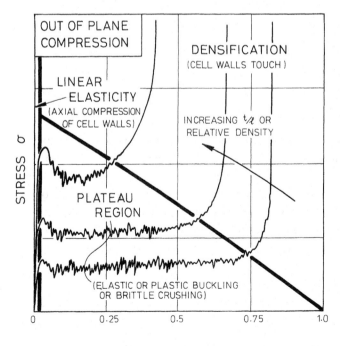

Figure 4.7 A schematic diagram for honeycombs loaded in the axial (X_3) direction, showing regimes of linear elasticity, collapse and densification, and the way in which the stress–strain curve changes with t/l.

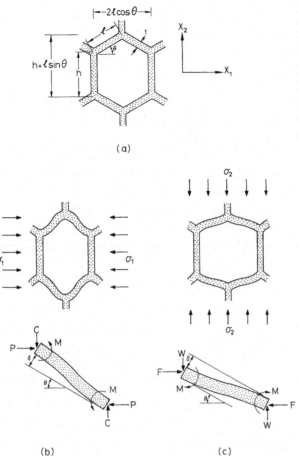

Figure 4.8 Cell deformation by cell wall bending, giving linear-elastic extension or compression of the honeycomb: (a) the undeformed honeycomb; (b) and (c) the bending caused by loads in the X_1 and X_2 directions.

individual cell wall members cut from the honeycomb. Young's modulus is conveniently estimated by exciting the cut member into vibration and measuring the frequencies using a microphone attached to a spectrum analyser; the flexural modulus is then calculated from the frequency and the density which are measured separately (Hunt, 1993). The modulus of rupture of the cell wall, too, can be measured on either a single strut, using an electronic balance to measure the fracture load (Hunt, 1993), or on a square unit cell loaded on the cell diagonal in a deformation stage in a scanning electron microscope (Huang and Gibson, 1991).

(a) Linear-elastic deformation

When a honeycomb, loaded in the X_1 or the X_2 direction, deforms in a linear-elastic way, the cell walls bend (Patel and Finnie, 1970; Abd El-Sayed, 1976; Abd El-Sayed *et al.*, 1979; Gibson *et al.*, 1982; Warren and Kraynik, 1987). The response is conveniently described by five moduli[†]: two Young's moduli E_1^* and E_2^*, a shear modulus G_{12}^*, and two Poisson's ratios, v_{12}^* and v_{21}^*. But the five are not independent. The reciprocal relation, derived in the Appendix (Eqn. (A6), page 498)

$$E_1^* v_{21}^* = E_2^* v_{12}^* \tag{4.2}$$

reduces the number of independent moduli to four. They are the same in tension as in compression, and are calculated as follows.

The two Young's moduli (Abd El-Sayed *et al.*, 1979; Gibson, 1981; Gibson *et al.*, 1982) are calculated by the method illustrated in Fig. 4.8(b) and (c). A stress σ_1 parallel to X_1 causes one set of cell walls – those of length l – to bend; one cell wall is shown in detail beneath the cell. Equilibrium requires that the component of force C parallel to X_2 be zero. The moment M tending to bend the cell wall (which we treat as a beam of length, l, thickness, t, depth, b, and Young's modulus, E_s) is

$$M = \frac{Pl\sin\theta}{2} \tag{4.3}$$

where

$$P = \sigma_1(h + l\sin\theta)b. \tag{4.4}$$

From standard beam theory (e.g. Roark and Young, 1976), the wall deflects by:

$$\delta = \frac{Pl^3\sin\theta}{12E_s I} \tag{4.5}$$

where I is the second moment of inertia of the cell wall ($I = bt^3/12$ for a wall of uniform thickness t). Of this, a component $\delta\sin\theta$ is parallel to the X_1 axis, giving a strain:

$$\epsilon_1 = \frac{\delta\sin\theta}{l\cos\theta} = \frac{\sigma_1(h + l\sin\theta)bl^2\sin^2\theta}{12E_s I\cos\theta}. \tag{4.6}$$

The *Young's modulus parallel to X_1* is just $E_1^* = \sigma_1/\epsilon_1$, giving

$$\boxed{\frac{E_1^*}{E_s} = \left(\frac{t}{l}\right)^3 \frac{\cos\theta}{(h/l + \sin\theta)\sin^2\theta}.} \tag{4.7}$$

Loading in the X_2 direction is shown in Fig. 4.8(c); the forces acting on the cell wall of length, l, and depth, b, are shown in the bottom part. By equilibrium $F = 0$ and $W = \sigma_2 lb\cos\theta$, giving:

[†] E_1^* is Young's modulus for tension or compression applied in the X_1 direction; G_{12}^* is the shear modulus for shear in the X_1–X_2 plane; and v_{ij} is the negative ratio of the strain in the X_j direction to that in the X_i direction, for normal loading in the X_i direction.

$$M = \frac{Wl\cos\theta}{2}.$$ (4.8)

The wall deflects by:

$$\delta = \frac{Wl^3\cos\theta}{12E_sI}.$$ (4.9)

Of this, a component $\delta\cos\theta$ is parallel to the X_2 axis, giving a strain:

$$\epsilon_2 = \frac{\delta\cos\theta}{h + l\sin\theta} = \frac{\sigma_2 bl^4\cos^3\theta}{12E_sI(h + l\sin\theta)}$$ (4.10)

from which the Young's modulus parallel to X_2 is simply $E_2^* = \sigma_2/\epsilon_2$, giving:

$$\boxed{\frac{E_2^*}{E_s} = \left(\frac{t}{l}\right)^3 \frac{(h/l + \sin\theta)}{\cos^3\theta}.}$$ (4.11)

For regular hexagons with walls of uniform thickness, both Young's moduli, E_1^* and E_2^*, reduce to the same value:

$$\frac{E_1^*}{E_s} = \frac{E_2^*}{E_s} = 2.3\left(\frac{t}{l}\right)^3$$ (4.12)

as they should: honeycombs made up of regular hexagons are isotropic.

When loaded in either the X_1 or the X_2 direction the inclined members of a hexagonal cell carry axial and shear loads in addition to the bending components considered here. This has two effects. First, it produces axial and shear deformations of the member, which, for small t/l, are negligible relative to the bending deflection; the exact expressions for the elastic moduli, including bending, shear and axial deformations of the members, are given in Appendix 4B. Second, it creates a beam–column effect: there is an additional moment in the member caused by the fact that the axial loads are no longer co-linear. This magnifies the moment and bending deflections by a factor of

$$\frac{1}{1 - P_a/P_{crit}}$$

producing a non-linearity in the stress–strain curve which becomes significant as the axial component of load, P_a, approaches the Euler load, P_{crit}. Here, we have calculated the elastic moduli in the limit of small deflections, allowing the beam–column effect to be neglected.

The Poisson's ratios are calculated by taking the negative ratio of the strains normal to, and parallel to, the loading direction. We find for loading in the X_1 direction:

$$\boxed{\nu_{12}^* = -\frac{\epsilon_2}{\epsilon_1} = \frac{\cos^2\theta}{(h/l + \sin\theta)\sin\theta}}$$ (4.13)

or, for regular hexagons, $\nu_{12}^* = 1$. For loading in the X_2 direction:

$$v_{21}^* = -\frac{\epsilon_1}{\epsilon_2} = \frac{(h/l + \sin\theta)\sin\theta}{\cos^2\theta} \tag{4.14}$$

again giving $v_{21}^* = 1$ for regular hexagons. There is an unusual feature to these results. When θ is less than zero (that is, the cells are inverted) Poisson's ratio is negative, implying that a compressive stress in one direction produces a contraction in the normal in-plane direction, rather than the usual expansion. A rubber honeycomb with negative Poisson's ratios is shown in Fig. 4.3(b).

Structures made up of rigid rods connected by elastic hinges, or of cylinders attached to elastic strips, are capable of showing negative Poisson's ratios, too (Evans, 1989; Lakes, 1991). Hypothetical two-dimensional microstructures giving a negative Poisson's ratio have also been described by Wojciechowski (1989), Wojciechowski and Branka (1989), Rothenburg et al. (1991) and Milton (1992).

Note also that, for the honeycomb, the reciprocal theorem (Eqn. (4.2)) applies:

$$E_1^* v_{21}^* = E_2^* v_{12}^* = E_s \left(\frac{t}{l}\right)^3 \frac{1}{\sin\theta\cos\theta}$$

so that only three of the four moduli E_1^*, E_2^*, v_{12}^* and v_{21}^* are independent.

But a general honeycomb does have four independent in-plane constants. The fourth is the shear modulus, G_{12}^*, calculated as shown in Fig. 4.9. Because of symmetry there is no relative motion of the points A, B and C when the honeycomb is sheared; the shear deflection u_s is entirely due to the bending of beam BD and its rotation (through the angle ϕ) about the point B. The forces are shown in the figure. Summing moments at B gives the moment applied to the members AB and BC:

$$M = \frac{Fh}{4}. \tag{4.15}$$

Then, using the standard result $\delta = Ml^2/6E_sI$, the angle of rotation is:

$$\phi = \frac{Fhl}{24E_sI}.$$

The shearing deflection u_s of the point D with respect to B is:

$$u_s = \tfrac{1}{2}\phi h + \frac{F}{3E_sI}\left(\frac{h}{2}\right)^3 = \frac{Fh^2}{48E_sI}(l + 2h).$$

The shear strain, γ, is given by:

$$\gamma = \frac{2u_s}{(h + l\sin\theta)} = \frac{Fh^2}{24E_sI}\frac{(l + 2h)}{(h + l\sin\theta)}. \tag{4.16}$$

The remote shear stress, τ, is $F/2lb\cos\theta$, giving the shear modulus $G_{12}^* = \tau/\gamma$, or:

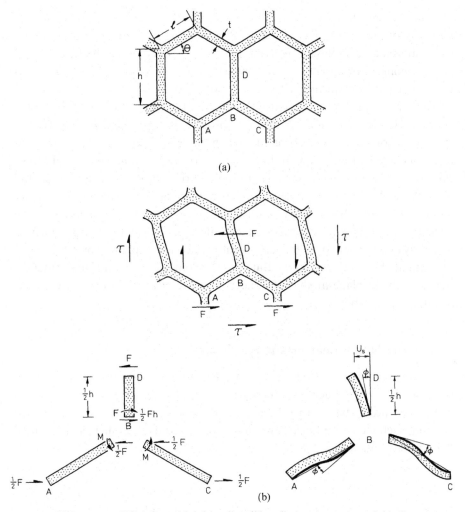

Figure 4.9 Cell deformation by cell wall bending and rotation, giving linear-elastic shear of the honeycomb: (a) the undeformed honeycomb; (b) the loads, moments, displacements and rotations caused by a shear stress.

$$\frac{G_{12}^*}{E_s} = \left(\frac{t}{l}\right)^3 \frac{(h/l + \sin\theta)}{(h/l)^2 (1 + 2h/l)\cos\theta}. \tag{4.17}$$

This reduces, for regular, uniform, hexagons, to:

$$\frac{G_{12}^*}{E_s} = 0.57\left(\frac{t}{l}\right)^3 = \frac{1}{4}\frac{E^*}{E_s} \tag{4.18}$$

which correctly obeys the relation $G = E/(2(1 + v))$ (Eqn. (3.6)) for isotropic solids.

Similar results have been found by Warren and Kraynik (1987), who include the effect of axial as well as bending stiffness of the cell wall in calculating the four independent in-plane elastic moduli. They also consider the effect of cell walls of non-uniform thickness: for low density honeycombs ($\rho^*/\rho_s < 0.2$), this can change the in-plane moduli by a factor of up to 2.

These results have been extensively checked by experiments on elastomeric and metal honeycombs (Gibson, 1981; Gibson *et al.*, 1982). Measured and calculated values of E^*, v^* and G^* are compared in Figs 4.10, 4.11 and 4.12. The experiments included structures with differing density (t/l), cell angle (θ) and aspect ratio (h/l). The moduli are sensitive to slight variations in wall thickness, t, and angle θ, which, in any real honeycomb, are always there because of poor quality control. The error bars with a width of one standard deviation show how the measured variations in these parameters affect the calculations. Agreement is generally good; where discrepancies occur they can be explained by refinements of the modelling (Gibson, 1981). But this is seldom necessary; for most practical purposes the in-plane moduli of honeycombs are adequately described by the equations given above.

(b) Non-linear elasticity; elastic buckling

The plateau of the compressive stress–strain curve, for elastomeric honeycombs, is caused by elastic buckling. Photographs like Fig. 4.3(f) show that the cell walls most nearly parallel to the loading direction behave like an end-loaded column; such a column buckles when the load exceeds the Euler buckling load (Timoshenko and Gere, 1961):

$$P_{\text{crit}} = \frac{n^2\pi^2 E_s I}{h^2}. \tag{4.19}$$

The factor, n, describes the *rotational stiffness* of the node where three cell walls meet. Figure 4.13 shows the buckling mode observed when rubber honeycombs are compressed in the X_2 direction (loads parallel to X_1 simply cause bending). The load per column (column EB in Fig. 4.13(b), for example) is related to the remote stress, as before, by:

$$P = 2\sigma_2 lb \cos\theta.$$

Elastic collapse occurs when $P = P_{\text{crit}}$, giving the elastic collapse stress, $(\sigma^*_{\text{el}})_2$, as:

$$\boxed{\frac{(\sigma^*_{\text{el}})_2}{E_s} = \frac{n^2\pi^2}{24}\frac{t^3}{lh^2}\frac{1}{\cos\theta}} \tag{4.20}$$

with $I = bt^3/12$. But a problem remains: that of calculating the end constraint factor, n. It depends on the degree of constraint to rotation at the node B caused by the walls AB and BC; if rotation is freely allowed, $n = 0.5$; if no

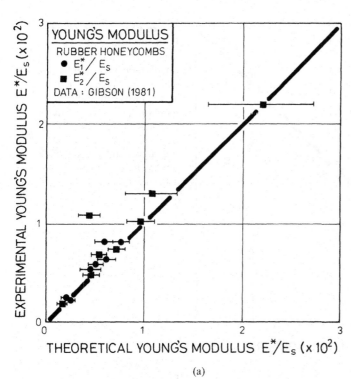

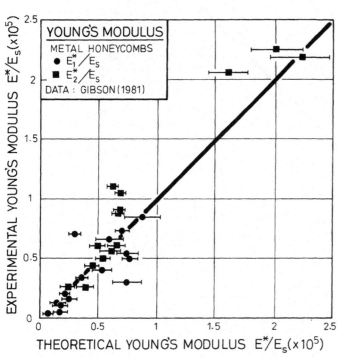

Figure 4.10 Young's moduli in the X_1 and X_2 directions for: (a) elastomeric honeycombs; (b) metal honeycombs. The honeycombs were specially fabricated with a wide range of t/l, h/l and θ. The generally good agreement supports the use of Eqns. (4.7) and (4.11) to describe the moduli in terms of the cell geometry.

(a)

(b)

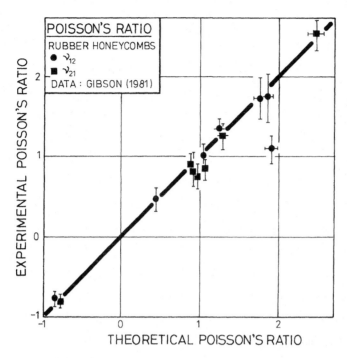

Figure 4.11 The in-plane Poisson's ratio in the X_1 and X_2 directions for elastomeric honeycombs with a wide range of t/l, h/l and θ. Two of the data points are for a honeycomb with a negative Poisson's ratio like the one shown in Fig. 4.3b. Agreement with theory (Eqns. 4.13 and 4.14) is excellent.

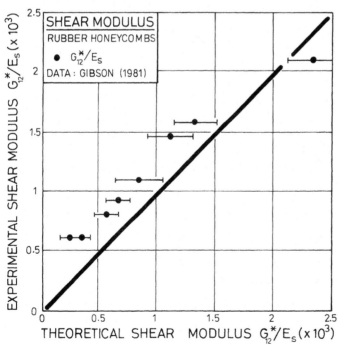

Figure 4.12 The in-plane shear moduli of elastomeric honeycombs with a wide range of t/l, h/l and θ. The measured shear moduli are systematically a little higher than those predicted by the simple model developed in the text, although the result, Eqn. (4.17), remains an adequate description of the shear modulus.

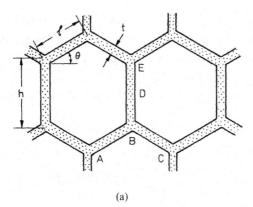

(a)

Figure 4.13   Cell deformation by elastic buckling: (a) the undeformed honeycomb; (b) the buckling mode in uniaxial loading, and the associated forces, moments, displacement and rotations.

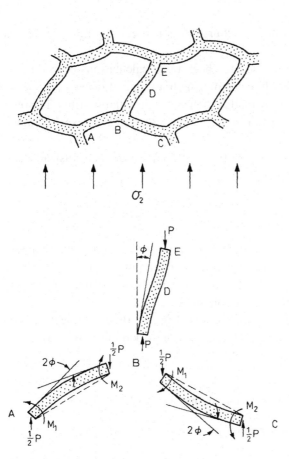

(b)

rotation is possible, $n = 2$. The constraint on the vertical wall caused by the walls to which it is connected lies between these limits; then n is greater than 0.5 and less than 2. Values of n as a function of h/l, derived in Appendix 4C, are listed in Table 4.1; for regular hexagons, $n = 0.69$.

We can now evaluate the collapse stress for regular hexagons (for which $\theta = 30°$, $h = l$ and t is constant). It is:

$$\frac{(\sigma_{el}^*)^2}{E_s} = 0.22 \left(\frac{t}{l}\right)^3. \tag{4.21}$$

The modulus for regular hexagons is $E^*/E_s = 2.3t^3/l^3$, giving an elastic collapse strain of:

$$(\epsilon_{el}^*)_2 = \frac{(\sigma_{el}^*)_2}{E^*} = \frac{1}{10} \tag{4.22}$$

regardless of density, provided, of course, it is low enough that simple beam theory is valid ($t/l < 1/4$).

Experiments have been devised to check these calculations (Gibson, 1981; Gibson et al., 1982). Rubber honeycombs with a range of density t/l, cell wall angle θ and aspect ratio h/l were loaded to their elastic collapse stress. The results are compared with Eqn. (4.20) in Fig. 4.14. The agreement gives every confidence that the analysis is adequate.

In tension, of course, buckling of this sort does not happen. Instead the cell walls continue to distort by bending.

(c) Plastic collapse

Metals, and many polymers, are elastic–plastic solids. Honeycombs made of them collapse plastically when the bending moment in the cell walls reaches the fully plastic moment. This gives a stress–strain curve with a plateau both in compression and in tension at the plastic collapse stress σ_{pl}^*.

Consider loading in the X_1 direction first (Fig. 4.15(b)). An upper bound on the plastic collapse stress is given by equating the work done by the force:

$$P = \sigma_1 (h + l \sin \theta) b$$

Table 4.1 End constraint factors for the elastic buckling of honeycombs with hexagonal cells

h/l	n
1	0.686
1.5	0.760
2	0.806

during a plastic rotation ϕ of the four plastic hinges A, B, C, and D to the plastic work done at the hinges giving:

$$4M_{\mathrm{p}}\phi \geq 2\sigma_1 b(h + l\sin\theta)\phi l\sin\theta$$

where M_{p} is the fully plastic moment of the cell wall in bending:

$$M_{\mathrm{p}} = \tfrac{1}{4}\sigma_{\mathrm{ys}}bt^2 \tag{4.23}$$

and where σ_{ys} is the yield stress of the cell-wall material. It follows that:

$$\boxed{\frac{(\sigma^*_{\mathrm{pl}})_1}{\sigma_{\mathrm{ys}}} = \left(\frac{t}{l}\right)^2 \frac{1}{2(h/l + \sin\theta)\sin\theta}\cdot} \tag{4.24}$$

A lower bound is given by equating the maximum moment in the beam to M_{p}. This maximum moment is

$$(M_{\mathrm{max}})_1 = \tfrac{1}{2}\sigma_1(h + l\sin\theta)bl\sin\theta \tag{4.25}$$

from which:

$$\boxed{\frac{(\sigma^*_{\mathrm{pl}})_1}{\sigma_{\mathrm{ys}}} = \left(\frac{t}{l}\right)^2 \frac{1}{2(h/l + \sin\theta)\sin\theta}\cdot} \tag{4.26a}$$

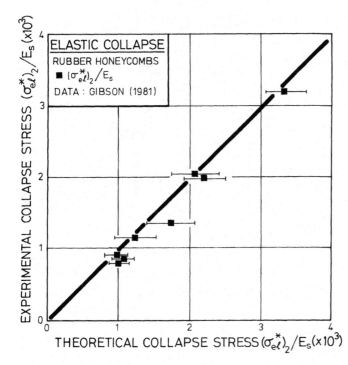

Figure 4.14 The elastic collapse stress of elastomeric honeycombs with a wide range of t/l, h/l and θ, compared with theory (Eqn. 4.20). Agreement is good.

ELASTIC COLLAPSE
RUBBER HONEYCOMBS
■ $(\sigma^*_{el})_2/E_s$
DATA : GIBSON (1981)

EXPERIMENTAL COLLAPSE STRESS $(\sigma^*_{el})_2/E_s$ (×10^3)

THEORETICAL COLLAPSE STRESS $(\sigma^*_{el})_2/E_s$ (×10^3)

The lower and upper bounds are identical, and thus define the exact solution to the problem. For regular uniform hexagons it reduces to:

$$\frac{\sigma^*_{pl}}{\sigma_{ys}} = \frac{2}{3}\left(\frac{t}{l}\right)^2. \tag{4.26b}$$

Plastic collapse in the X_2 direction is treated in the same way. An upper bound is given by equating the plastic work done in a postulated compatible deformation to the work done by the stress (Fig. 4.15(c)). A lower bound is given by equating the maximum moment in the beam (Fig. 4.15(c)):

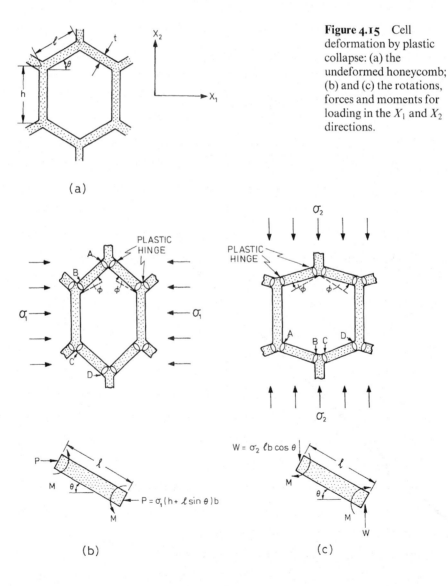

Figure 4.15 Cell deformation by plastic collapse: (a) the undeformed honeycomb; (b) and (c) the rotations, forces and moments for loading in the X_1 and X_2 directions.

$$(M_{\max})_2 = \frac{\sigma_2 l^2 b \cos^2 \theta}{2} \tag{4.27}$$

to its fully plastic moment. As before the two results are identical, and thus equal to the exact solution:

$$\boxed{\frac{(\sigma^*_{\text{pl}})_2}{\sigma_{\text{ys}}} = \left(\frac{t}{l}\right)^2 \frac{1}{2 \cos^2 \theta}} \tag{4.28}$$

which reduces to Eqn. (4.26b) when $\theta = 30°$: regular hexagons have the same plastic collapse stress in the X_1 and X_2 directions.

Plastic shear is dealt with in a similar way. If a honeycomb is loaded in simple shear with a shear stress τ_{12} acting in the X_1–X_2 plane, plastic hinges form in the vertical walls. The shear force on each vertical wall is $F = 2\tau_{12} bl \cos \theta$. If the moment $Fh/2$ exerted by this force exceeds the fully plastic moment of Eqn. (4.23), the honeycomb shears plastically, giving a lower bound for the in-plane shear strength of:

$$\boxed{\frac{(\tau^*_{\text{pl}})_{12}}{\sigma_{\text{ys}}} = \frac{1}{4} \left(\frac{t}{l}\right)^2 \frac{1}{h/l \cos \theta}} \tag{4.29a}$$

which reduces to

$$\frac{1}{2\sqrt{3}} \left(\frac{t}{l}\right)^2 \tag{4.29b}$$

for regular hexagons. The upper bound for the in-plane shear strength is found by equating the plastic work done by the force F on the vertical wall with the plastic work done by the plastic moment at the hinge. The result is the same as the lower bound (Eqn. (4.29a)) which therefore gives the exact solution.

Under some conditions elastic buckling can precede plastic collapse, though once the displacements become large, plasticity causes permanent deformation. For this to happen the density (roughly, t/l) must be lower than a critical value which is found by setting:

$$(\sigma^*_{\text{el}})_2 = (\sigma^*_{\text{pl}})_2$$

giving (from Eqns. (4.20) and (4.28)):

$$\left(\frac{t}{l}\right)_{\text{crit}} = \frac{12}{n^2 \pi^2 \cos \theta} \left(\frac{h}{l}\right)^2 \frac{\sigma_{\text{ys}}}{E_{\text{s}}}$$

or, for regular hexagons:

$$\left(\frac{t}{l}\right)_{\text{crit}} = 3 \frac{\sigma_{\text{ys}}}{E_{\text{s}}}. \tag{4.30}$$

For many metals $\sigma_{\text{ys}}/E_{\text{s}}$ is of order 10^{-3}; for these, elastic buckling precedes plastic collapse only when the relative density $(2t/\sqrt{3}l)$ is very low – less than 0.3%.

But many polymers yield at a stress of around $E_s/100$ and then elastic collapse occurs first for all densities below about 3%.

Data for the plastic collapse of metal honeycombs, loaded uniaxially in the X_1 and X_2 directions, are compared with Eqns. (4.26a) and (4.28) in Fig. 4.16. The plastic collapse stress is sensitive to small defects, and that may be why the experimental data are systematically lower than theory predicts. Despite this, it is clear that the models give a good general description of the way in which the plateau stress of plastic honeycombs depends on cell geometry.

The post-yield behaviour of aluminium honeycombs loaded in the X_2 direction has been described by Papka and Kyriakides (1994) who photographed the honeycomb deformation at increasing strain levels during displacement-controlled uniaxial compression tests (Fig. 4.17). In the linear-elastic regime, the deformation is homogeneous throughout the specimen. Beyond the first maximum in stress, or limit stress, the deformation in the specimen begins to localize. Within the localized band, the cells collapse in an asymmetric shearing mode of deformation while away from the band, deformation remains symmetric and homogeneous. With increasing strain, the deformation in the rows of cells adjacent to the collapsed band also becomes asymmetric. Eventually, the walls of the first row of cells to collapse come into contact: the increase in load required for further deformation is reflected in an upturn in the overall stress–strain response. This sequence of events repeats itself for each adjacent row of cells

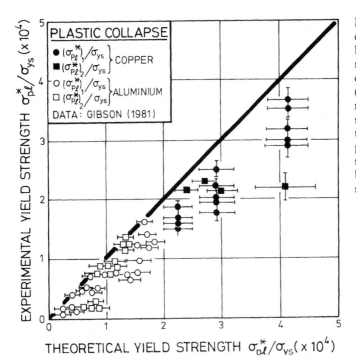

Figure 4.16 The plastic collapse stresses of metal honeycombs with a wide range of t/l, h/l and θ, compared with theory (Eqns. 4.26a and 4.28). The metal honeycombs are systematically weaker than the theory predicts, perhaps because of residual stress in the cell walls, or because of defects in the structure.

until all rows have collapsed, at which point the stress rises sharply corresponding to densification. Similar behaviour has been reported for arrays of cylindrical tubes confined within surrounding walls loaded normal to the axis of the cylinders (Stronge and Shim, 1988; Poirier *et al.*, 1992).

The post-yield behaviour can be analysed using the finite-element technique (Papka and Kyriakides, 1994). The solid cell wall material is modelled as bilinear, with a Young's modulus of 69 GPa, a yield stress of 292 GPa and a post-yield slope of $E/\alpha = 0.69$ GPa corresponding to $\alpha = 100$; the solid was assumed to follow the J_2 flow theory of plasticity with isotropic hardening. Both the local response of a unit cell of the honeycomb and the global response of the entire honeycomb were analysed; the plastic instability was induced by including a misalignment of $0.2°$ in one or two of the vertical members. The local response was qualitatively similar to that of the entire honeycomb. The finite-element results reproduced the deformation and stress–strain response of the honeycomb well (Fig. 4.18). The effect of relative density, solid cell wall yield stress and strain

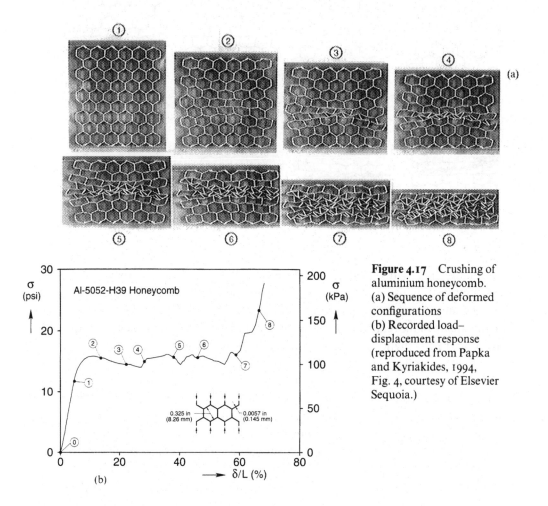

Figure 4.17 Crushing of aluminium honeycomb. (a) Sequence of deformed configurations (b) Recorded load–displacement response (reproduced from Papka and Kyriakides, 1994, Fig. 4, courtesy of Elsevier Sequoia.)

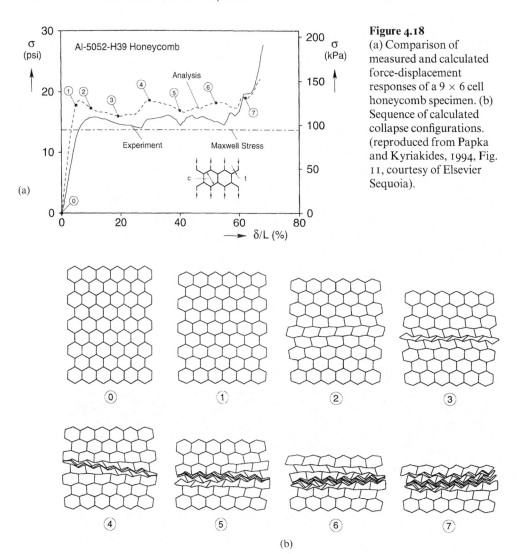

Figure 4.18
(a) Comparison of
measured and calculated
force-displacement
responses of a 9 × 6 cell
honeycomb specimen. (b)
Sequence of calculated
collapse configurations.
(reproduced from Papka
and Kyriakides, 1994, Fig.
11, courtesy of Elsevier
Sequoia).

hardening coefficient on the post-yield behaviour were then investigated in a
parametric study using the unit cell analysis. The experimental measurements
and the finite-element analysis both confirm the dependence of the limit stress
on the square of the relative density (Eqn. (4.28)); both the finite-element analysis
and Eqn. (4.28) slightly overestimate the limit stress. Increasing the solid cell
wall yield stress increases the difference between the limit stress and the local
minimum corresponding to contact of the cell walls in the unit cell. Low values
of the strain hardening coefficient ($\alpha < 6$) were found to eliminate the local maximum in the stress–strain curve, giving a monotonically increasing response.
Analysis of a nine by six cell honeycomb indicated that as a result of the monotonically increasing load, the deformation was uniform over the entire honeycomb,
eliminating localization.

(d) Brittle failure

Honeycombs made of ceramic or glass, or of brittle plastics, fail in a brittle mode. In compression the cells suffer progressive crushing; in tension the honeycomb fails by fast brittle fracture.

Consider compressive crushing first (Fig. 4.19). As before, a remote stress σ_1 or σ_2 applies bending moments to the cell walls. If the tensile stress on the surface

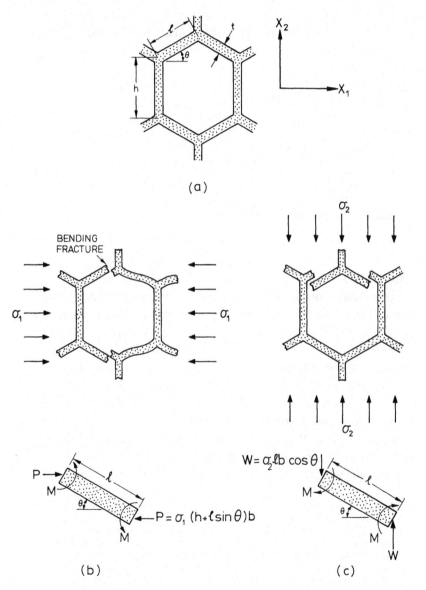

(a)

(b)   (c)

Figure 4.19   The compressive crushing of a brittle honeycomb: (a) the undeformed honeycomb; (b) and (c) the forces and moments which lead to fracture for loading in the X_1 and X_2 directions.

of a cell wall exceeds its modulus of rupture[†], σ_{fs}, the cell wall fractures. The surface tensile stress caused by a moment M_{max} is:

$$\sigma_{max} = \frac{6M_{max}}{bt^2}.$$

Setting $\sigma_{max} = \sigma_{fs}$ defines the moment which will cause crushing:

$$M_f = \tfrac{1}{6}\sigma_{fs}bt^2. \tag{4.31}$$

The moments caused by remote stresses σ_1 and σ_2 were calculated earlier (Eqns. (4.25) and (4.27)). Using these, in turn, in the last equation, gives the crushing strengths:

$$\frac{(\sigma^*_{cr})_1}{\sigma_{fs}} = \frac{1}{3(h/l + \sin\theta)\sin\theta}\left(\frac{t}{l}\right)^2 \tag{4.32}$$

and

$$\frac{(\sigma^*_{cr})_2}{\sigma_{fs}} = \frac{1}{3\cos^2\theta}\left(\frac{t}{l}\right)^2. \tag{4.33}$$

When $\theta = 30°$ and $h/l = 1$, both reduce to:

$$\frac{\sigma^*_{cr}}{\sigma_{fs}} = \frac{4}{9}\left(\frac{t}{l}\right)^2 \tag{4.34}$$

so that the crushing strength of regular, uniform hexagons is isotropic.

Tensile fracture is different. If, in a brittle honeycomb loaded to near its fracture stress, one cell wall fails, the stress on the neighbouring walls increases and they will fail too. The failed cluster is like a crack; the stress concentration at its periphery causes further walls to fail, and fracture propagates across the section. (The same thing does not usually happen in compression because the broken fragments of cell wall pack together, transmitting and redistributing the stress in a uniform way.) The problem is best approached by the methods of fracture mechanics.

Consider the brittle honeycomb containing a crack – a cluster of broken cells – as shown in Fig. 4.20. When it is loaded in tension the cell walls at first bend elastically. The load is transmitted through the honeycomb as a set of discrete forces and moments acting on each of the cell walls. But since the honey-

---

[†]The 'modulus of rupture' is the maximum surface stress in a bent beam at the instant at which it fractures. If the beam is made of a brittle solid, like the cell wall discussed here, the fracture initiates at a microcrack (usually a surface microcrack) in the wall and propagates catastrophically. On average, the modulus of rupture is a little larger than the tensile strength because, in bending, only one surface of the beam sees the maximum tensile stress; in simple tension the entire beam is stressed uniformly, so a given microcrack is less likely to be stressed in bending than in simple tension. The statistics of the problem (see, for example, Davidge, 1979) show that the modulus of rupture is typically about 1.1 times larger than the tensile strength.

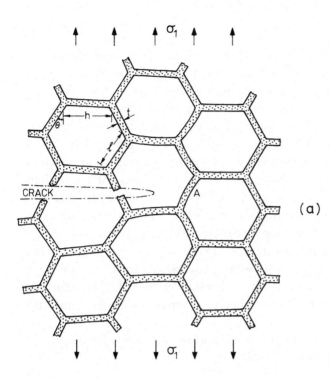

(a)

Figure 4.20 Crack propagation leading to brittle tensile failure in a honeycomb. The stress concentration at the crack tip causes the cell wall just beyond the tip to be loaded more heavily than any other: (a) the geometry and deflections for loading in the X_1 direction; (b) those for loading in the X_2 direction.

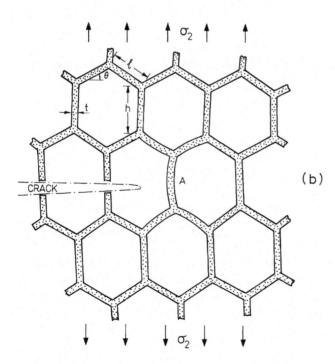

(b)

comb is linear-elastic the average force and moment on a given cell wall can be calculated from the stress field of the equivalent linear-elastic continuum. The discrete problem can be solved by taking the solution to the continuum problem (just as is done at the atomistic level by replacing discrete interatomic forces by continuous elastic properties), and using it to calculate the forces and moments on the discrete cell walls (Ashby, 1983). When the combination of forces is sufficient to fracture the cell wall just ahead of the crack tip, the crack advances. This condition defines K_{IC}^*, the fracture toughness of the honeycomb, which we now calculate. The method involves a number of assumptions. First, if the honeycomb is to be treated as a continuum, the crack length must be large relative to the cell size. Second, the contribution of axial forces in the cell walls to the internal stress ahead of the crack tip is neglected. Third, we assume that the cell wall material has a constant modulus of rupture, σ_{fs}. The calculation is best regarded as a dimensional argument leading to the proper dependence of K_{IC}^* on t, h, θ, etc., since the geometrical constants are not determined. This means that one experimental measurement is required to calibrate the model, which then has predictive capability.

A crack length $2c$ in an elastic solid lying normal to a remote tensile stress σ_1 creates a singular local stress field, σ_l of:

$$\sigma_l = \frac{\sigma_1 \sqrt{\pi c}}{\sqrt{2\pi r}} \tag{4.35}$$

at a distance r from its tip. Consider the first unbroken cell wall (labelled A in Fig. 4.20(a)), which we take to be half the cell width, $(h + l\sin\theta)/2$, beyond the crack tip. The force on it is:

$$P = \sigma_l(h + l\sin\theta)b$$

where, using Eqn. (4.35),

$$\sigma_l = \sigma_1 \sqrt{\frac{c}{h + l\sin\theta}}.$$

This force exerts a bending moment, M_1, which is proportional to P:

$$M_1 \propto Pl\sin\theta$$

on the wall marked A. We shall take the constant of proportionality to be 1, but this is an approximation, and it is for this reason that the calculation is not exact. When M_1 exceeds the fracture moment (Eqn. (4.31)) the wall fails and the crack advances. Assembling these results gives the tensile fracture strength for loading in the X_1 direction:

$$\frac{(\sigma_f^*)_1}{\sigma_{fs}} \approx \frac{1}{6(h/l + \sin\theta)^{1/2}\sin\theta} \sqrt{\frac{l}{c}}\left(\frac{t}{l}\right)^2. \tag{4.36}$$

Loading in the X_2 direction (Fig. 4.20(b)) gives, instead

$$P = 2\sigma_l lb \cos\theta$$

where

$$\sigma_l = \sigma_2 \sqrt{\frac{c}{2l\cos\theta}}$$

and the bending moment on a cell wall just ahead of the crack tip is proportional to:

$$M_2 \propto \frac{Pl\cos\theta}{2}$$

where, once again, we take the constant of proportionality to be 1. Assembling these gives the stress for tensile fracture when loading is in the X_2 direction:

$$\frac{(\sigma_f^*)_2}{\sigma_{fs}} = \frac{1}{3\sqrt{2}\cos^{3/2}\theta} \sqrt{\frac{l}{c}}\left(\frac{t}{l}\right)^2. \qquad (4.37)$$

If no cell walls have broken the largest 'flaw', $2c$, is equal to the cell size itself and the tensile strengths become identical with the compressive crushing strengths derived earlier (Eqns (4.32) and (4.33)) except for a small difference in the numerical constant arising from the approximations mentioned above. But when a true flaw, caused by broken cell walls, exists, the tensile strength is less than the compressive strength. For regular hexagons, both equations for tensile fracture reduce to essentially the same result, which, with sufficient accuracy for our purposes can be written:

$$\frac{\sigma_f^*}{\sigma_{fs}} = 0.3\left(\frac{l}{c}\right)^{1/2}\left(\frac{t}{l}\right)^2. \qquad (4.38)$$

It is helpful to rephrase this result in the terminology of fracture mechanics. Tensile fracture (we have shown) will occur when the fracture toughness is reached:

$$\sigma_f^* \sqrt{\pi c} = K_{IC}^* \qquad (4.39)$$

where the K_{IC}^*, the fracture toughness of the honeycomb, is given by:

$$K_{IC}^* = 0.3\sigma_{fs}\sqrt{\pi l}\left(\frac{t}{l}\right)^2. \qquad (4.40)$$

The result shows that the fracture toughness decreases rapidly with relative density (since $\rho^*/\rho_s \propto t/l$) and more slowly as the cell size decreases. So brittle honeycombs tend to be fragile; for realistic values of t/l and l, the fracture toughness is less, by a factor of 100 or so, than that of the solid of which it is made.

Calculation of the fracture toughness of a centrally cracked honeycomb plate using the finite-element method has shown that the continuum assumption is valid for semi-crack lengths over 7 times the cell size ($c/l > 7$) and that axial stresses in the members, neglected in our derivation, do not become significant until the relative density is increased above 0.2 (Huang and Gibson, 1991). Limited comparisons between the fracture toughness measured in notched three point bend specimens of cordierite honeycombs, Eqn. (4.40) and the finite element analysis suggest that the model gives a good description of the measured values (Fig. 4.21).

In reality, the modulus of rupture of the brittle cell walls is not constant: variations in the size of inherent cracks within the cell walls lead to variations in the modulus of rupture. The Weibull distribution, often used to describe the variability in the tensile strength of brittle materials, gives the probability of failure, P_f, in a brittle material of volume, V, loaded with a uniform tensile stress, σ, as (Weibull, 1951):

$$P_f = 1 - \exp\left[-V\left(\frac{\sigma - \sigma_u}{\sigma_0}\right)^{m_w}\right] \qquad \text{for } \sigma > \sigma_u \qquad (4.41)$$

and

$$P_f = 0 \qquad \text{for } \sigma < \sigma_u$$

where σ_u, σ_0 and m_w are material properties characterizing the inherent flaw size population in the material. In practice, it is common, for ceramics, to find that $\sigma_u = 0$. The Weibull modulus, m_w, an empirical constant, is related to the flaw

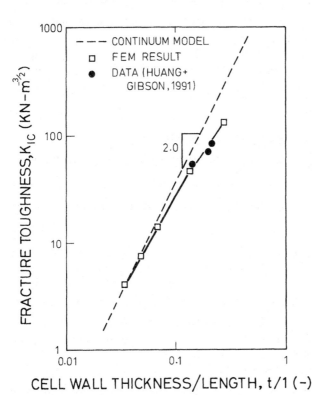

Figure 4.21 Dependence of fracture toughness on the cell-wall thickness/length ratio ($c = 15.2$ mm, $l = 1.8$ mm).

size distribution of the brittle material. One consequence of the Weibull distribution of failure strengths is that brittle materials exhibit a *size effect*: two specimens of different volumes, V_1 and V_2, loaded in uniform tension, will have different failure strengths, σ_1 and σ_2, according to:

$$\frac{\sigma_1}{\sigma_2} = \left(\frac{V_2}{V_1}\right)^{1/m_{\mathrm{w}}}. \tag{4.42}$$

The cell walls of the brittle honeycomb are loaded primarily in bending rather than uniform tension; it can be shown that the size effect is identical for members loaded in the same configuration, i.e. for honeycombs with the same cell shape. We then find that the variation in the modulus of rupture of the cell wall leads to a cell size effect for the fracture toughness. For two honeycombs of different cell sizes, the ratio of the mean moduli of rupture of the cell walls is:

$$\frac{\bar{\sigma}_{\mathrm{fs},1}}{\bar{\sigma}_{\mathrm{fs},2}} - \left(\frac{V_2}{V_1}\right)^{1/m_{\mathrm{w}}} = \left(\frac{bl_2 t_2}{bl_1 t_1}\right)^{1/m_{\mathrm{w}}}$$

or:

$$\frac{\bar{\sigma}_{\mathrm{fs},1}}{\bar{\sigma}_{\mathrm{fs},2}} = \left(\frac{l_2}{l_1}\right)^{2/m_{\mathrm{w}}} \left[\frac{\rho_2^*}{\rho_1^*}\right]^{1/m_{\mathrm{w}}}. \tag{4.43}$$

The ratio of the fracture toughnesses of two specimens of honeycombs of equal total volume and identical cell shape but with different cell sizes is then (using Eqn. (4.40)):

$$\frac{K_{\mathrm{IC},1}^*}{K_{\mathrm{IC},2}^*} = \left(\frac{l_1}{l_2}\right)^{\frac{1}{2} - \frac{2}{m_{\mathrm{w}}}} \left[\frac{\rho_1^*}{\rho_2^*}\right]^{2 - \frac{1}{m_{\mathrm{w}}}}. \tag{4.44}$$

If the densities of the two honeycombs are equal then:

$$\frac{K_{\mathrm{IC},1}^*}{K_{\mathrm{IC},2}^*} = \left(\frac{l_1}{l_2}\right)^{\frac{1}{2} - \frac{2}{m_{\mathrm{w}}}}. \tag{4.45}$$

The fracture toughness can either increase or decrease with cell size depending on the value of the Weibull modulus, m_{w}: for $m_{\mathrm{w}} > 4$ it increases with increasing cell size while if $m_{\mathrm{w}} < 4$ it decreases.

(e) Viscoelastic deformation

At room temperature many polymers are near their glass temperature, T_{g} (Chapter 3). When this is so, the polymer – and any honeycomb or foam made from it – has a modulus which is not constant, but depends on the time of loading. The cell walls no longer behave like linear-elastic beams loaded in bending; instead, they must be treated as *viscoelastic*.

Consider the case of a hexagonal honeycomb, loaded in the X_1 direction. The cell edges (as before) are loaded in bending, and respond in a way which can be described, at its simplest, by a *Voigt element*: a linear spring of stiffness E_{s2} in parallel with a dashpot with a viscosity (for uniaxial loading) of η_s. The force causing bending is now related to the deflection δ and deflections rate $\dot{\delta}$ of the beam by

$$P = \frac{12E_{s2}I}{l^3 \sin\theta}\delta_{el} + \frac{12\eta_s I}{l^3 \sin\theta}\dot{\delta}_v. \tag{4.46}$$

Here the first term is the force required to cause an elastic deflection δ_{el} and the second term, containing the viscosity, η_s, is the force which will cause a viscous deflection rate $\dot{\delta}_v$ (it is the limit of the creep equation (4.59) derived below, with $n_s = 1$). For a Voigt solid,

$$\dot{\delta} = \dot{\delta}_{el} = \dot{\delta}_v \tag{4.47}$$

where $\dot{\delta}$ is the overall deflection rate of the cell wall. Substituting into the equation for load gives the differential equation for viscoelastic bending of a cell edge:

$$\dot{\delta} + \frac{E_{s2}}{\eta_s}\delta - \frac{Pl^3 \sin\theta}{12\eta_s I} = 0. \tag{4.48}$$

But (as before) the stress σ_1 on the honeycomb is related to the force P by Eqn. (4.4), and the strain is related to δ by the first part of Eqn. (4.6), giving:

$$\dot{\epsilon}_1 + \frac{E_{s2}}{\eta_s}\epsilon_1 - \frac{C\sigma_1 bl^3}{\eta_s I} = 0 \tag{4.49}$$

with

$$C = \frac{(h/l + \sin\theta)\sin^2\theta}{12\cos\theta}.$$

The model is made more realistic by adding a second spring of stiffness E_{s1} in series with the Voigt element. Then, repeating the procedure above, the differential equation for the honeycomb is found as:

$$\dot{\epsilon}_1 + \frac{E_{s2}\epsilon_1}{\eta_s} = \frac{Cbl^3}{I}\left\{\frac{\dot{\sigma}_1}{E_{s1}} + \frac{\sigma_1}{\eta_s}\left(\frac{E_{s1} + E_{s2}}{E_{s1}}\right)\right\}. \tag{4.50}$$

The equation for loading in the X_2 direction follows in a similar way. For regular hexagons both reduce to:

$$\dot{\epsilon} + \frac{E_{s2}}{\eta_s}\epsilon = \frac{\sqrt{3}l^3}{4t^3}\left\{\frac{\dot{\sigma}}{E_{s1}} + \frac{\sigma}{\eta_s}\left(\frac{E_{s1} + E_{s2}}{E_{s1}}\right)\right\}. \tag{4.51}$$

This is the differential equation describing the viscoelastic honeycomb. It can be integrated, subject to appropriate boundary conditions, to describe its response to time-dependent loads. It is informative to examine a simple case, that for the application of a uniaxial stress, σ. When loading is very rapid the first term on each side of the equation dominates and the equation as a whole integrates to:

$$\epsilon = \frac{\sqrt{3}l^3}{4t^3}\frac{\sigma}{E_{s1}} \tag{4.52}$$

where E_{s1} is the *unrelaxed modulus* of the solid. The honeycomb is linear elastic with a modulus $E^* = \sigma/\epsilon$ which is identical with Eqn. (4.12). When, instead, loading is very slow, the first terms on either side of the equation become negligible, and the honeycomb is again perfectly elastic but with a lower modulus

$$\epsilon = \frac{\sqrt{3}l^3\sigma}{4t^3}\left(\frac{E_{s1} + E_{s2}}{E_{s1}E_{s2}}\right) \tag{4.53}$$

where $E_{s1}E_{s2}/(E_{s1} + E_{s2})$ is known as the *relaxed modulus* of the solid.

When the load is held constant the honeycomb shows time-dependent strain. Setting $\dot{\sigma} = 0$ and integrating gives:

$$\epsilon = \frac{\sqrt{3}l^3}{4t^3}\sigma\left(\frac{E_{s1} + E_{s2}}{E_{s1}E_{s2}}\right)\left(1 - \exp-\frac{E_{s2}}{\eta_s}\tau\right) \tag{4.54}$$

where τ is the time. The time constant for this relaxation is η_s/E_{s2}, and is a property of the cell wall material only; it is the same for all relative densities. That is the general characteristic of the viscoelastic response: the magnitude of the response is a characteristic of the honeycomb, and thus depends on the relative density; but the time constant for the relaxation is a property of the cell wall material and is independent of the cell dimensions.

(f) Creep and creep buckling

At and above their glass temperature, T_g, polymers show slow, permanent, time-dependent deformations, or *creep*. Metals and ceramics creep, too, though the rate of creep is significant only when the temperature is greater than about 0.3 of the melting temperature, T_m. Creep is a problem when cellular solids carry loads for long periods of time – refractory brick in furnace walls and polymer foams used in structural applications are examples. Here we calculate the creep-rate of two-dimensional honeycombs loaded in the X_2 direction from both the creep deflection rate of the inclined members and creep buckling of the vertical members. The analysis for loading in the X_1 direction is analogous, except that buckling does not occur. The analysis is extended to foams in Chapter 6.

Consider the creep of a horizontal beam loaded as in Fig. 4.22. Let the creep-rate of the cell wall material be described by Eqn. (3.17):

$$\dot{\epsilon} = \dot{\epsilon}_{0s}\left(\frac{\sigma}{\sigma_{0s}}\right)^{n_s} \tag{4.55}$$

where $\dot{\epsilon}_{0s}$, σ_{0s} and n_s are creep constants. We wish to calculate the deflection rate of the inclined wall, $\dot{\delta}$, and the stress at which the vertical one will buckle. To achieve the first, we treat the cell wall as a beam loaded in bending, and make three standard assumptions: that the bending displacements are small; that the neutral axis lies in the plane of symmetry of the beam section; and that plane sec-

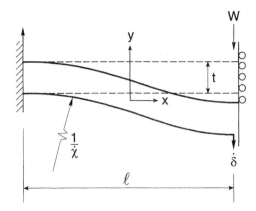

Figure 4.22   The rate of deflection, $\dot{\delta}$, of a creeping-beam.

tions remain plane. Equilibrium requires that the bending moment M carried by the beam be related to the stress within it, $\sigma(y)$ by:

$$M = 2b \int_0^{t/2} y\sigma(y) \, \mathrm{d}y. \tag{4.56}$$

Compatibility requires that the strain rate $\dot{\epsilon}(y)$ at any point in the section is related to the rate of change of curvature, $\dot{\kappa}$, by

$$\dot{\epsilon}(y) = y\dot{\kappa}.$$

Substituting this into the creep law of equation (4.55) gives the stress distribution

$$\sigma(y) = \sigma_{0s} \left(\frac{y\dot{\kappa}}{\dot{\epsilon}_{0s}} \right)^{1/n_s}.$$

Inserting this into equation (4.56), integrating and inverting gives the key equation for any creeping-beam problem:

$$\dot{\kappa} = \dot{\epsilon}_{0s} \left(\left(\frac{2n_s + 1}{2n_s} \right) \frac{M}{b\sigma_{0s}} \right)^{n_s} \left(\frac{2}{t} \right)^{2n_s+1}$$
$$= \dot{\kappa}_0 \left(\frac{M}{M_0} \right)^{n_s} \tag{4.57a}$$

with

$$\dot{\kappa}_0 = \frac{2\dot{\epsilon}_{0s}}{t} \tag{4.57b}$$

and

$$M_0 = \left(\frac{2n_s}{2n_s + 1} \right) \cdot \frac{\sigma_{0s} b t^2}{4}. \tag{4.57c}$$

The moment at any section of the horizontal beam of Fig. 4.22 is

$$M(x) = -Wx.$$

where x is measured from the mid-point of the beam segment shown in Fig. 4.22.

The curvature rate, $\dot{\kappa}(x)$, at the point x contributes to the deflection rate $\dot{\delta}$ of the beam

$$\frac{d^2\dot{\delta}}{dx^2} = \dot{\kappa}(x)$$

to the deflection rate. Substituting these two results into equation (4.57a) and integrating from $x = 0$ to $x = -l/2$ allows the deflection rate $\dot{\delta}$ to be calculated:

$$\dot{\delta} = \frac{1}{(n_s + 2)} \frac{\dot{\kappa}_o W^{n_s}}{M_o^{n_s}} \frac{l^{n_s+2}}{2^{n_s+1}}$$

from which

$$\dot{\delta} = \frac{1}{(n_s + 2)} \dot{\epsilon}_{os} \frac{l^2}{t} \left(\left(\frac{2n_s + 1}{n_s} \right) \cdot \frac{Wl}{\sigma_{os}bt^2} \right)^{n_s} \tag{4.58}$$

which correctly reduces to the result for plastic failure in the limit of $n \to \infty$. As before, the load $W = \sigma_2 bl \cos^2 \theta$ and the strain-rate $\dot{\epsilon}_2^* = \dfrac{\dot{\delta} \cos\theta}{h + l \sin\theta}$. This then gives

$$\frac{\dot{\epsilon}_2^*}{\dot{\epsilon}_{os}} = \frac{1}{(n_s + 2)} \left(\frac{l}{t} \right)^{2n_s+1} \left(\frac{2n_s + 1}{n_s} \frac{\sigma_2}{\sigma_{os}} \cos^2 \theta \right)^{n_s} \frac{\cos\theta}{h/l + \sin\theta}. \tag{4.59}$$

Consider now the buckling of the vertical cell wall shown in Fig. 4.23(a). It has length h, and depth b and thickness t, with $t \le b$ so that buckling occurs in the plane of the figure. In order to capture both elastic and buckling response, we must use a constitutive law for flexural response which includes terms describing both stiffness and creep (Abel, Ashby and Cocks, 1995). Combining the elastic curvature-rate with that for creep (Eqn. 4.57a) gives:

$$\dot{\kappa} = \frac{\dot{M}}{E_s I} + \dot{\kappa}_o \left(\frac{M}{M_o} \right)^{n_s} \tag{4.60}$$

where κ is the curvature, M is the moment, $E_s I$ is the flexural stiffness and $\dot{\kappa}_o$ and M_o are related to the constants $\dot{\epsilon}_o$ and σ_{os} of Eqn. (4.55) through the relationships given as Eqns. (4.57b) and 4.57c).

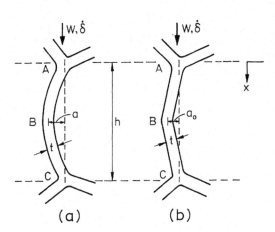

(a)   (b)

Figure 4.23 The creep-buckling of a cell wall. Because creep is non-linear, the sinusoidal shape of (a) is quickly replaced by one in which most of the curvature is at the mid-point, as shown in (b).

We assume that the strut has some imperfections, such as misalignment of the cell walls, inducing bending which can be represented by an initial mid-point deflection, a_o, transverse to the line of the strut. We will assume that $a_o > t/2$ where t is the cell wall thickness. In the elastic buckling analysis it was assumed that the deflected shape is a half sine-wave, and the strut adopts a similar shape throughout the buckling process (Fig. 4.23(a)). But if the body creeps in a non-linear way ($n \gg 1$) a much higher curvature develops in the vicinity of the mid-point B than along the rest of the strut. As a result the beam profile becomes more like that shown in Fig. 4.23(b), with only a slight curvature along most of its length, and a sharp change of slope at B. For the purpose of determining the distribution of moments we assume that the lengths AB and BC of the strut remain straight and there is a relative rotation of these two sections at B.

If the strut carries a constant load F, and x is the distance from the top of the strut, then over the length AB

$$M = 2F\frac{x}{h}a \qquad (4.61a)$$

and

$$\dot{M} = 2F\frac{x}{h}\dot{a}. \qquad (4.61b)$$

At a given instant during the buckling process the work-rate, $\dot{W}$ is

$$\dot{W} = 2\int_0^{h/2} M\dot{\kappa}\,\mathrm{d}x. \qquad (4.62)$$

Setting the rate of work done to the force, F, times the displacement rate, $\dot{\delta}$, and substituting for $\dot{\kappa}$ using Eqn. (4.60) gives

$$F\dot{\delta} = 2\int_0^{h/2}\left[M\frac{\dot{M}}{E_sI} + M\dot{\kappa}_o\left(\frac{M}{M_o}\right)^{n_s}\right]\mathrm{d}x$$

or, substituting equations (4.61):

$$F\dot{\delta} = 2\int_0^{h/2}4F^2\left(\frac{x}{h}\right)^2\frac{a\dot{a}}{E_sI}\,\mathrm{d}x + 2\int_0^{h/2}\dot{\kappa}_oM_o\left(\frac{2Fxa}{hM_o}\right)^{n_s+1}\mathrm{d}x.$$

Integrating and dividing through by F, gives

$$\dot{\delta} = \frac{1}{3}\frac{Fh}{E_sI}a\dot{a} + \frac{1}{(n_s+2)}ha\dot{\kappa}_o\left(\frac{Fa}{M_o}\right)^{n_s}. \qquad (4.63)$$

The strut essentially deforms as two straight beam elements hinged at B, giving a relationship between $\dot{\delta}$ and $\dot{a}$

$$\dot{\delta} = \frac{4\dot{a}a}{h}.$$

Substituting this into (4.63) gives

$$\frac{4\dot{a}}{h}\left(1 - \frac{Fh^2}{12E_sI}\right) = \frac{1}{(n_s + 2)}h\dot{\kappa}_o\left(\frac{Fa}{M_o}\right)^{n_s}.$$ (4.64)

According to this equation, the displacement rate, $\dot{a}$, goes to infinity when

$$F = 12\frac{E_sI}{h^2}.$$ (4.65)

This corresponds to the elastic-buckling load for the strut. If we had assumed a sinusoidal shape we would have obtained the Euler buckling load – close to this load elastic deformation would dominate and a sinusoidal shape would be more appropriate than that considered here. We by-pass this consequence of the simplifications by identifying the load of Eqn. (4.65) with the Euler buckling load F_E, and Eqn. 4.64 becomes

$$\dot{a} = \frac{h^2\dot{\kappa}_o}{4(n_s + 2)}\left(\frac{Fa}{M_o}\right)^{n_s}\bigg/\left(1 - \frac{F}{F_E}\right).$$ (4.66)

Integrating this between the limits of $a = a_o$ at $t = 0$ and $a \to \infty$ at the time at which buckling becomes catastrophic, t_b, gives

$$t_b = 4\frac{(n_s + 2)}{(n_s - 1)}\left(\frac{h}{a_0}\right)^{n_s-1}\frac{(1 - F/F_E)}{h\dot{\kappa}_0(Fh/M_0)^{n_s}}$$ (4.67)

with $\dot{\kappa}_0$ and M_0 given by Eqns. (4.57b) and (4.57c). Rearranging, and writing $F = 2\sigma_2lb\cos\theta$ and $F_E = 2(\sigma_{el}^*)_2lb\cos\theta$ and using Eqn. (4.20)

$$\frac{(\sigma_{el}^*)_2}{E_s} = \frac{n^2\pi^2}{24}\frac{t^3}{lh^2}\frac{1}{\cos\theta}$$

gives

$$\frac{\dfrac{\sigma_2}{(\sigma_{el}^*)_2}}{\left(1 - \dfrac{\sigma_2}{(\sigma_{el}^*)_2}\right)^{1/n_s}} = \left[2\left(\frac{n_s + 2}{n_s - 1}\right)\left(\frac{h}{a_o}\right)^{n_s-1}\left(\frac{1}{t_b\dot{\epsilon}_{0s}}\right)\right]^{1/n_s}$$

$$\cdot \frac{2n_s}{2n_s + 1}\frac{\sigma_{0s}}{n^2\pi^2E_s}\left(\frac{t}{h}\right)^{\frac{1-n_s}{n_s}}.$$ (4.68)

The equation says that, for a given density, t/h, temperature (contained in $\dot{\epsilon}_{0s}$) and time of loading, t_b, there is a critical stress below which buckling will not occur and above which it will. The equation is complicated, but its behaviour can be understood by examining various limits. Note, first, that $\sigma_2/(\sigma_{el}^*)_2$ cannot have a value greater than 1 – if it did, the cell walls would buckle instantly. When the time-to-buckling is very short, the right-hand side of the equation is large (i.e. $\gg 1$) and this is only possible if

$$1 - \frac{\sigma_2}{(\sigma_{el}^*)_2} \ll 1$$

meaning that $\sigma_2 \approx (\sigma_{el}^*)_2$, that is, the critical stress is very close to that for elastic buckling.

At the other extreme, when the stress is much below the elastic-buckling stress (so that $F \ll F_E$ in Eqn. (4.67)), the stress at which creep buckling occurs, $(\sigma_{cb}^*)_2$, simplifies to:

$$(\sigma_{cb}^*)_2 = \frac{\sigma_{0s}}{24\cos\theta} \left(\frac{2n_s}{2n_s+1}\right) \left(\frac{n_s+2}{n_s-1}\right)^{1/n_s} \left(\frac{h}{a_0}\right)^{\frac{n_s-1}{n_s}} \left(\frac{2}{\dot\epsilon_{0s}t_b}\right)^{1/n_s} \left(\frac{t}{h}\right)^{\frac{2n_s+1}{n_s}} \left(\frac{l}{h}\right).$$

$$(4.69)$$

If, further, the exponent is large ($n > 5$, as it is for many metals) we find:

$$(\sigma_{cb}^*)_2 = \frac{\sigma_{0s}}{24\cos\theta} \left(\frac{h}{a_0}\right) \left(\frac{2}{\dot\epsilon_{0s}t_b}\right)^{1/n_s} \left(\frac{t}{h}\right)^2 \left(\frac{l}{h}\right). \tag{4.70}$$

This defines a 'threshold stress' for creep buckling. Inverting gives an expression for the time-to-buckling, t_b:

$$t_b = \frac{2}{\dot\epsilon_{0s}} \left[\frac{1}{24\cos\theta} \frac{h}{a_0} \left(\frac{t}{h}\right)^2 \left(\frac{l}{h}\right)\right]^{n_s} \frac{1}{(\sigma_2/\sigma_{0s})^{n_s}}. \tag{4.71}$$

This last equation illustrates that the time-to-buckling is very sensitive both to the size of the initial imperfection, a_0, and to the stress, σ_2.

When the time is long, but the creep exponent is small ($n_s = 1$, as it is for some polymers and occasional ceramics) it is necessary to re-integrate Eqn. (4.66) with a finite upper limit for a_u. This leads to the result, replacing equation (4.67):

$$t_b = \frac{12(1 - F/F_E)}{h\dot\kappa_0(Fh/M_0)} \ln\left(\frac{a_u}{a_0}\right) \tag{4.72}$$

or, replacing $\dot\kappa_0$, M_0, F, and σ_{el}^* in exactly the same way as we did for Eqn. (4.67) gives:

$$t_b = \frac{12}{n^2\pi^2} \frac{\sigma_{0s}}{\dot\epsilon_{0s}E_s} \frac{(1 - \sigma_2/(\sigma_{el}^*)_2)}{\sigma_2/(\sigma_{el}^*)_2} \ln\left(\frac{a_u}{a_0}\right) \tag{4.73}$$

where n arises from substitution of the elastic-buckling stress and is given by Table 4.1. Inverting the equation, with a given value of t_b:

$$\frac{(\sigma_{cb}^*)_2}{(\sigma_{el}^*)_2} = \frac{A}{1 + A} \tag{4.74}$$

with

$$A = \frac{12}{n^2\pi^2} \frac{\sigma_{0s}}{\dot\epsilon_{0s}E_s t_b} \ln\left(\frac{a_u}{a_0}\right).$$

In this limit, the creep-buckling stress is simply a fraction of the elastic-buckling stress σ_{el}^*. The fraction decreases as time increases, ultimately falling as $1/t_b$.

Note the danger implied by this: even the smallest load, if applied for a sufficiently long time, will cause buckling. There is no threshold below which buckling does not occur.

(g) Densification

At large compressive strains the opposing walls of the cells crush together and the cell-wall material itself is compressed. When this happens the stress–strain curve rises steeply in the way shown in the left-hand curves of Fig. 4.2. One might anticipate that the limiting strain ϵ_D would simply be equal to the porosity

$$p^* = 1 - \frac{(2 + h/l)t/l}{2\cos\theta(h/l + \sin\theta)}$$

(or, for regular hexagons $p^* = 1 - (2/\sqrt{3})(t/l)$ because this is the strain at which all pore space has been squeezed out. In reality, cell walls lock together at a smaller strain than this. Data are not available for honeycombs, but for foams (Chapter 5) it is found that the stress–strain curve is almost vertical when the instantaneous relative density $(1 - p^*)$ is about 0.7, giving:

$$\epsilon_D = 1 - 1.4 \frac{(2 + h/l)t/l}{2\cos\theta(h/l + \sin\theta)}. \tag{4.75}$$

The boundary between 'plateau behaviour' and 'densification' regimes shown in Fig. 4.5 defines the onset of densification; it occurs when the instantaneous relative density is about 0.5.

(h) Variability in cell structure

Our analysis of the mechanical behaviour of honeycombs has focussed on an array of identical prismatic hexagonal cells. Many of the properties can be described by calculating the response of a single unit cell to mechanical loads. One limitation of this approach is that it does not account for natural variations in microstructure, such as in the arrangement of cell walls, cell-wall thickness and cell-wall material properties. Several methods of estimating the effects of microstructural variability on mechanical properties have been described. Kraynik *et al.* (1991) analysed the elastic response of a two-dimensional liquid froth, modelling it as an array of hexagonal cells with varying sizes and shapes but which maintain an angle of 120° between the three cell edges at each node. They found that the elastic response was insensitive to such variations in microstructure. Cowin (1985) has related the elastic constants of a porous material to its relative density and fabric using nine unspecified scalar-valued functions of relative density which must be determined empirically. A third method, developed for multiple-phase composite materials, uses statistical characterization of the microstructure to account for microstructural variability (Torquato, 1991).

Eischen and Torquato (1993) have shown that this method can yield improved bounds on the elastic properties of composites compared to the Hashin–Strikman bounds. The applicability of Torquato's method for the analysis of porous materials of low volume fraction has not yet been demonstrated. Finally, homogenization theory can be used (Suquet, 1985; Hollister *et al.*, 1991, 1994). This method accounts for regional variations in microstructure by making detailed finite element models of subregions and then determining the properties of a structure comprised of many subregions by using a continuum finite element analysis.

Here, we use Voronoi honeycombs to represent a two-dimensional cellular solid with a non-periodic arrangement of the cell walls (Section 2.3c, Fig. 2.14) (Silva *et al.*, 1995). The Voronoi honeycombs are generated from a random set of points (separated by a given minimum distance) by constructing the perpendicular bisectors between the points. The cell walls of the honeycomb are of identical thickness. Their response to uniaxial compressive stress is calculated using finite element analysis, modelling each cell wall as a three-noded beam element which includes bending, axial and shear displacements. Finite element analyses of 20 different Voronoi honeycombs of the same relative density ($\rho^*/\rho_s = 0.15$) indicates that the mean value of the Young's moduli, the Poisson's ratio and the shear moduli are, respectively, 6%, 1% and 11% higher than those calculated using the analysis for the periodic (hexagonal) honeycomb of the same relative density (Eqns. (4.7), (4.11), (4.13), (4.14) and (4.17)), modified to account for shear and axial deformations); these differences are not statistically significant. The Young's moduli and the Poisson's ratios for loading in the X_1 and the X_2 directions were not statistically different, as expected for Voronoi honeycombs generated from initially random points. Similar analyses of a single Voronoi honeycomb structure with varying relative densities ($0.05 < \rho^*/\rho_s < 0.30$) indicates that there is no significant difference from the results for the periodic hexagonal honeycomb (Fig. 4.24). Variations in cell shape do not affect the elastic moduli of two-dimensional cellular solids, consistent with the results of Kraynik *et al.* (1991) for a two-dimensional liquid froth.

Anisotropic Voronoi honeycombs were generated by scaling ten isotropic Voronoi honeycomb structures ($0.05 < \rho^*/\rho_s < 0.30$) in one direction by a factor, S, ranging from 1–2 (Silva *et al.*, 1995). The elastic moduli of the anisotropic Voronoi honeycombs were then found from finite element analysis. The elastic moduli of periodic, hexagonal honeycombs of equivalent relative density and anisotropy scaling factor were calculated using the analytical results (Eqns. (4.7), (4.11), (4.13), (4.14) and (4.17), modified to account for shear and axial deformations). The elastic moduli of the anisotropic, non-periodic Voronoi honeycombs were within 15% of the analytical results for periodic, hexagonal honeycombs of equivalent relative density and anisotropy. Microstructural variability in cell shape appears to have little effect on the elastic moduli of honeycombs.

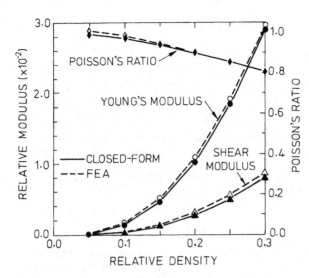

Figure 4.24 Relative moduli and Poisson's ratio versus relative density. (FEA results are shown as open symbols connected by dashed lines; analytical results are shown as filled symbols connected by solid lines.) FEA results were generated from models which originated from the same Voronoi diagram but which had different values of cell-wall thickness. Analytical results included the effect of shear and axial deformations (Appendix 4B).

In contrast, the strength of honeycombs is affected by variations in cell shape. Numerical analyses of the compressive stress–strain behaviour of Voronoi honeycombs indicates that both the elastic buckling and plastic yield strength are roughly 25% less than those of idealized honeycombs with hexagonal cells (Silva and Gibson, 1996). A sampling of the internal elastic strain within the cell walls in the Voronoi honeycomb indicates that the strains are normally distributed, with a wider distribution than in the idealized honeycombs: longer than average struts induce higher bending moments, increasing their internal stress, leading to a reduction in strength.

(i) Defects in honeycombs

Individual cell walls may sometimes be broken or missing in honeycombs: here we consider the effects of missing cell walls on their compressive behaviour. Numerical simulation of the stress–strain behaviour of both idealized honeycombs with hexagonal cells and Voronoi honeycombs indicates that both the Young's modulus and the plastic compressive strength decrease sharply with an increasing number of missing cell walls (Fig. 4.25): for example, the loss of only 5% of the cell walls results in reduction in modulus or strength of over 30% (Guo *et al.*, 1994, Silva and Gibson, 1996). Percolation theory suggests that when 35% of the cell walls are removed from a hexagonal network, there is a continuous path along the missing cell walls from one edge to the opposite edge of the material: the honeycomb has broken into two pieces (Stauffer and Aharony, 1992). Our numerical simulation results are consistent with this: both the modulus and the strength of the honeycomb degrade completely when 35% of the cell walls are removed.

Missing cell walls can also affect the pattern of failure. Observations on aluminium honeycombs indicate that in a honeycomb with regular, repeating cells, the removal of a single cell wall causes failure to initiate in the weakened cell and induces strain hardening as the intact cells collapse (Prakash et $al.$, 1996). Measurement of the strain distribution throughout the aluminium honeycombs indicates that the strain is concentrated in a localized deformation band, decaying to the far-field value in about four cell diameters. Multiple defects will interact if closer together than some critical influence distance. For loading in the X_1 direction, co-operative collapse can occur if the defects are aligned within the deformation band while for loading in the X_2 direction, the deformation band nucleates at one of the defects and proceeds to the other. Numerical simulations

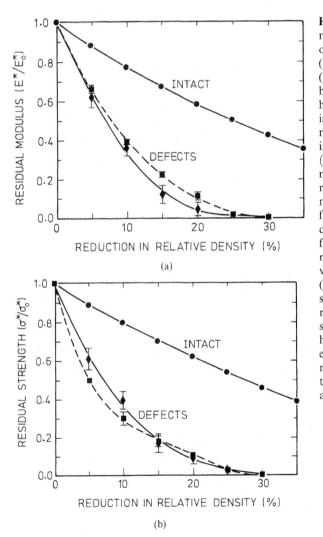

(a)

(b)

Figure 4.25 Effect of randomly located defects on (a) modulus and (b) strength of Voronoi (●, ◆) and periodic, hexagonal (■) honeycombs. Defect level is expressed as change in relative density (from an initial value $(\rho^*/\rho_s)_0$ of 0.15); modulus and strenth are residual values, normalized by their values for the intact case (no defects). Each data point for the defect curves represents the average value for three models (error bars denote one standard deviation). The residual modulus and strength of intact Voronoi honeycombs subjected to equivalent density reductions (due to uniform thinning of cell walls) are also shown.

of Voronoi honeycombs suggests that the removal of two adjacent cell walls is sufficient to cause the band of localization to shift to pass through the defects (Silva and Gibson, 1996).

The effect of filling the void within the cells has also been studied by filling some cells of an aluminium honeycomb with paraffin wax (Prakash *et al.*, 1996). Filled cells are stronger than unfilled cells: collapse initiates away from the filled cells so that the initiation of the compressive stress plateau remains unchanged but some strain hardening occurs at large strains. The interaction of filled cells which are closer together than some critical interaction distance produces shielding of the region between the cells, such that even at large strains, the region between the two filled cells is relatively undeformed.

4.4   The in-plane properties of honeycombs: biaxial loading

When a honeycomb is loaded biaxially – that is, in two perpendicular directions both contained in its plane (Fig. 4.26) – its *linear-elastic* response is straightforward: it can be calculated, by standard methods, from the moduli given in the

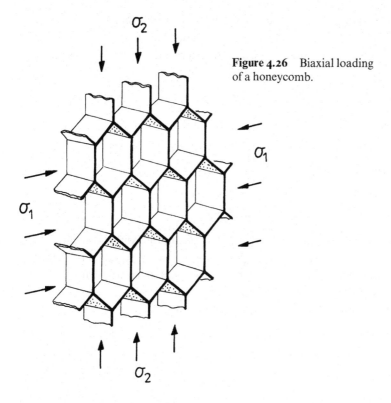

Figure 4.26   Biaxial loading of a honeycomb.

previous section. But the *elastic-buckling* mode changes, and the stress required to cause buckling in one direction is influenced by the stress applied in the other, so a new, biaxial relationship is needed to describe it. The *plastic* response changes more dramatically: in equal biaxial tension, for instance, the plastic collapse stress can be larger by a factor of 100 or more than in simple tension, giving the honeycomb an elongated *yield surface* which is quite different from that of the solid of which it is made. And *brittle fracture*, too, depends strongly on stress state, giving a *failure surface* which can also be extreme in shape.

We now examine each of these in turn, developing equations which describe the biaxial response and presenting the results schematically as yield and fracture surfaces.

(a) Linear-elastic response to biaxial stress

Consider the biaxial loading of the honeycomb shown in Fig. 4.26. The principal stresses are σ_1 and σ_2, and are positive when tensile, negative when compressive. The low in-plane stiffness of honeycombs, as we have seen, is due to cell-wall bending. In uniaxial loading the bending displacements are so much larger than the axial stretching or compression of the walls that these can be neglected. But in biaxial loading a combination of σ_1 and σ_2 can always be found for which the bending moments cancel. Then the axial extension or compression of the cell walls can no longer be neglected – it is the only remaining contribution.

The elastic response is found using the standard results:

$$\epsilon_1 = \frac{1}{E_1^*}(\sigma_1 - \nu_{12}^*\sigma_2)$$

$$\epsilon_2 = \frac{1}{E_2^*}(\sigma_2 - \nu_{21}^*\sigma_1)$$

(4.76)

and

$$\gamma_{12} = \frac{\tau_{12}}{G_{12}^*}$$

along with the expressions for the elastic moduli including axial and shear deformations in the cell wall, given in Appendix 4B. Under equal biaxial stress $(\sigma_1 = \sigma_2 = \sigma)$, the bending moments on the cell walls of an isotropic honeycomb with regular hexagonal cells cancel (Eqns. (4.3) and (4.8)). The elastic strains $(\epsilon_1 = \epsilon_2 = \epsilon)$ are given by:

$$\epsilon = \frac{\sigma}{E^*}(1 - \nu^*)$$

(4.77)

$$\epsilon = \sqrt{3}\,\frac{\sigma}{E_s(t/l)}.$$

(4.78)

The stiffness of the isotropic honeycomb under equal biaxial stress depends on t/l, rather than on $(t/l)^3$, reflecting the axial, rather than the bending, stiffness of the cell walls.

(b) Elastic buckling under biaxial loading

Under biaxial loading the mode of elastic buckling can change, depending on the stress state. Regular hexagonal honeycombs buckle in at least two modes, shown in Fig. 4.27; the first was analysed previously, for uniaxial loading, in Section 4.3(b), by relating the end constraint factor, n, in the Euler buckling equation (4.19):

$$P_{\text{crit}} = \frac{n^2 \pi^2 E_s I}{l^2}$$

to the rotational stiffness of the nodes. This factor depends on stress state (because a lateral compression aids bending of the adjacent members, reducing the rotational stiffness while a lateral tension does the opposite); and it depends on buckling mode (because different modes have different rotations associated

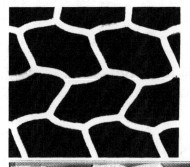

Figure 4.27 The larger photograph shows the buckling mode of an elastic honeycomb in biaxial compression. The small photograph shows, for comparison, the buckling mode in uniaxial compression.

with them). Here we find the combinations of biaxial stress which give each mode.

We begin by isolating the representative unit for each mode which, if repeated (with reflection), builds up the entire pattern (Fig. 4.28). We first calculate the axial loads and moments acting on each member when remote stresses σ_1 and σ_2 are applied to the material and identify the angles in terms of the rotation α at the node and the angles $\alpha - \beta$ through which each beam is bent. We then make use of the slope-deflection theorem (see Timoshenko and Young, 1965, for instance) to calculate the combination of stresses which maintain the buckled mode (Gibson *et al.*, 1989). For the beam AB,

$$\theta_A = \frac{M_{AB}l}{3EI}\psi(u) + \frac{M_{BA}l}{6EI}\phi(u)$$

$$\theta_B = \frac{M_{BA}l}{3EI}\psi(u) + \frac{M_{AB}l}{6EI}\phi(u)$$

(4.79)

where

$$\psi(u) = \frac{3}{2u}\left(\frac{1}{2u} - \frac{1}{\tan 2u}\right)$$

$$\phi(u) = \frac{3}{u}\left(\frac{1}{\sin 2u} - \frac{1}{2u}\right)$$

and

$$u = \frac{1}{2}\sqrt{\frac{P_{AB}}{EI}}.$$

Here P_{AB} is the axial load acting on the beam AB and M_{AB} and M_{BA} are the moments at its ends, causing the beam to bend through the angles θ_A and θ_B.

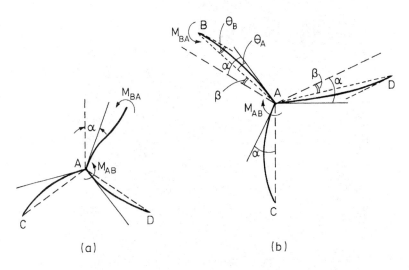

Figure 4.28 Modes of elastic buckling. (a) mode 1; (b) mode 2.

Applied to each of the three members in turn, this method gives a set of equations for the angles θ. The requirement of equilibrium at the central node and equilibrium of each member gives a further set of equations with the form (for Fig. 4.28(b), as an example):

$$2M_{AB} + M_{AC} = 0$$
$$M_{AB} - M_{BA} + P_{AB}\alpha l = 0.$$

This gives a set of equations from which the angles α and β and the moments M_{AB} etc. can be eliminated to give a relationship between the forces – and the remote stresses σ_1 and σ_2 – required to maintain each configuration. The final results for configurations (a) and (b) of Fig. 4.28 for regular hexagonal cells are, for mode 1:

$$\tan\left(\frac{1}{2t}\sqrt{\frac{-12\sqrt{3}l\sigma_2}{E_s t}}\right) \tan\left(\frac{1}{2t}\sqrt{\frac{-3\sqrt{3}l(3\sigma_1 + \sigma_2)}{E_s t}}\right) - \sqrt{\frac{3\sigma_1 + \sigma_2}{\sigma_2}} = 0$$

(4.80)

and, for mode 2:

$$\sqrt{\frac{3\sigma_1 + \sigma_2}{\sigma_2}} + \frac{\tan\left(\frac{1}{2t}\sqrt{\frac{-12(3\sigma_1 + \sigma_2)l/\sqrt{3}}{E_s t}}\right)}{\tan\left(\frac{1}{2t}\sqrt{\frac{-12\sigma_2 l/\sqrt{3}}{E_s t}}\right)} = 0.$$

(4.81)

Calculated values of σ_1 and σ_2 for regular hexagonal honeycombs are plotted in Fig. 4.29, where they are compared with experiment (Triantafillou et al., 1989). On the figure, the full symbols indicate that the first buckling mode was observed; open symbols, the second. The full lines represent Eqns (4.80) and (4.81). Simple compression in the X_1 direction does not cause buckling; the honeycomb folds up in a stable way. As a result, there is no intersection of the elastic-buckling envelope with the σ_1 axis; the true curve is asymptotic to the σ_1 axis. That is why the very end of the mode 2 curve is shown as a broken line, and why the data points (open symbols) deviate to the right. The figure documents the way in which the buckling mode switches, at the point $\sigma_1 = \sigma_2$, from mode 1 to mode 2, and shows the very large difference in the elastic-buckling stress that this produces.

Low relative density honeycombs can undergo large deformations before elastic buckling occurs, altering the equations of equilibrium at the initiation of buckling. This case has been analysed using a stiffness method by Zhang and Ashby (1992b) who find that mode 2 buckling is suppressed in honeycombs with double thickness vertical walls with $h/l = 1$ and $\theta > 7°$. Tests on paper–phenolic resin honeycombs (Nomex) confirm this result.

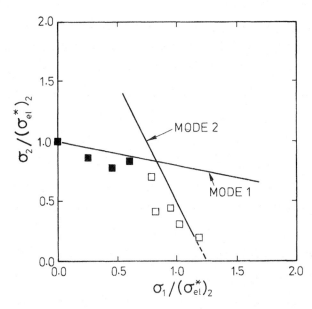

Figure 4.29 Data for the elastic buckling of a rubber honeycomb with regular hexagonal cells under biaxial loading. The lines indicate the calculated buckling surfaces corresponding to modes 1 and 2. Full symbols: mode 1; open symbols: mode 2.

Klintworth and Stronge (1988) suggest a third elastic-buckling mode in which the vertical members remain vertical and rigid while the inclined members of the honeycomb deform asymetrically with a sway component. By calculating the rotational stiffness of the nodes, they derive an equation for the elastic-buckling surface. Although the third mode of buckling may be geometrically possible, there have been no observations of it in tests on honeycomb materials.

(c) Plastic collapse and the yield surface

Biaxial loading has a major effect on the plastic collapse of a honeycomb (Fig. 4.30). If the two principal stresses σ_1 and σ_2 are of opposite sign the bending moment on the cell walls is increased, and collapse happens more easily. If they are of the same sign the moment caused by one is partly or wholly cancelled by the other and collapse by the formation of plastic hinges is made more difficult, or suppressed entirely. Then plastic collapse requires the axial yielding of the cell walls, at a stress level that may be 10 to 100 times larger than before. The combination of σ_1 and σ_2 which cause plastic collapse, plotted on axes of σ_1 and σ_2, is a closed surface called the *yield surface* for the honeycomb. Its shape is calculated as follows.

Consider the deformation of the cell shown in Fig. 4.30(a). It is subjected to principal stresses σ_1 and σ_2, positive when tensile, negative when compressive. The loads acting on an inclined cell wall are shown at (b). The net moment on the wall is

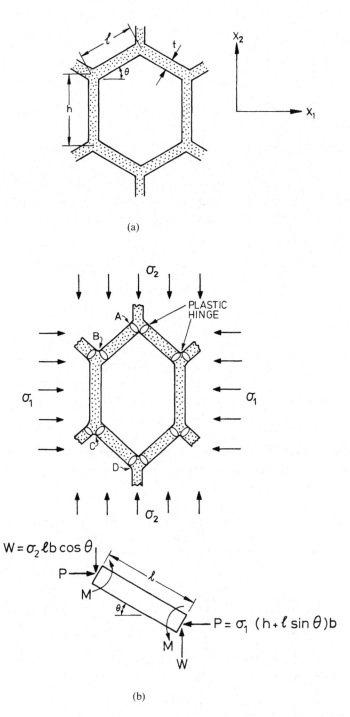

(a)

(b)

Figure 4.30 The plastic collapse of a honeycomb under biaxial loading. The forces and moments on a cell wall are shown at (b).

$$M = \pm \left[\frac{\sigma_1(h + l\sin\theta)bl\sin\theta - \sigma_2 l^2 b\cos^2\theta}{2} \right] \qquad (4.82)$$

and the axial stress is

$$\sigma_a = \frac{\sigma_1(h + l\sin\theta)\cos\theta + \sigma_2 l\sin\theta\cos\theta}{t}. \qquad (4.83)$$

The moment which will just cause plastic collapse of a beam is modified by an axial stress. When the beam is fully plastic the stress within it must equal σ_{ys} through the section (Fig. 4.31). The axial stress σ_a causes the neutral axis of the beam to move from the centre to a new location at a distance c from the top of the beam. This distance is calculated by equating the axial force in the beam to that caused by σ_a (see Fig. 4.31):

$$\sigma_{ys}cb - \sigma_{ys}(t - c)b = \sigma_a tb$$

giving

$$\frac{c}{t} = \frac{\sigma_a + \sigma_{ys}}{2\sigma_{ys}}.$$

The moment that causes full plasticity can then be calculated as

$$M_p = \frac{\sigma_{ys}bt^2}{4} \left(1 - \left(\frac{\sigma_a}{\sigma_{ys}} \right)^2 \right) \qquad (4.84)$$

(note that this reduces to Eqn. (4.23) when σ_a is zero, as it should). Plastic collapse of the honeycomb occurs when the moment given by Eqn. (4.82) exceeds M_p, giving the equation for the yield surface

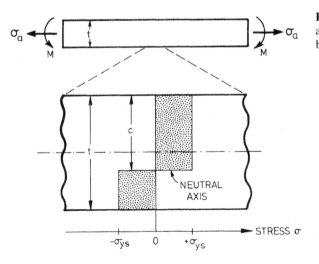

Figure 4.31   The forces in a cell wall subjected to bending and axial loads.

$$\pm \frac{\sigma_1(h/l + \sin\theta)\sin\theta - \sigma_2\cos^2\theta}{\sigma_{ys}(t/l)^2}$$

$$= \frac{1}{2}\left\{1 - \left(\frac{\sigma_1(h/l + \sin\theta)\cos\theta + \sigma_2\sin\theta\cos\theta}{\sigma_{ys}(t/l)}\right)^2\right\} \tag{4.85}$$

which correctly reduces to Eqns. (4.26a) and (4.28) for the appropriate uniaxial limits, when axial loads are neglected. For regular hexagons it reduces to

$$\pm\left(\frac{\sigma_1}{\sigma_{ys}} - \frac{\sigma_2}{\sigma_{ys}}\right) = \frac{2}{3}\left(\frac{t}{l}\right)^2\left\{1 - \left(\frac{3\sqrt{3}(\sigma_1/\sigma_{ys} + \frac{1}{3}\sigma_2/\sigma_{ys})}{4t/l}\right)^2\right\}. \tag{4.86}$$

It is helpful to rewrite this using the uniaxial compressive strength $\sigma_{pl}^* = \frac{2}{3}\sigma_{ys}(t/l)^2$ (Eqn. (4.26b)) as the normalizing factor:

$$\pm\left(\frac{\sigma_1}{\sigma_{pl}^*} - \frac{\sigma_2}{\sigma_{pl}^*}\right) = 1 - \left(\frac{\sqrt{3}}{2}\left(\frac{\sigma_1}{\sigma_{pl}^*} + \frac{1}{3}\frac{\sigma_2}{\sigma_{pl}^*}\right)\frac{t}{l}\right)^2. \tag{4.87}$$

This yield surface is plotted in Fig. 4.32, for $t/l = 0.1$. The surface has been truncated on the compressive side by an elastic-buckling line calculated from Eqns. (4.80) and (4.81), using $\sigma_{ys}/E_s = 1/50$, a value typical of many polymers. (For metals σ_{ys}/E_s is nearer $1/500$ and the buckling line lies further from the origin.) Note the elongated shape of the surface, which becomes more extreme as t/l is reduced: regular honeycombs are much stronger in equal biaxial tension than in uniaxial tension.

An anisotropic honeycomb, made up of irregular hexagonal cells, also has an elongated elliptical yield surface. But in this case the uniaxial yield strengths in the X_1 and X_2 directions are no longer equal. The extreme point on the yield surface, corresponding to pure axial load in the cell walls, is found by setting the moment produced from σ_1 and σ_2 to zero (Eqn. (4.82)). The ratio of σ_1 and σ_2 producing pure axial load is then:

$$\frac{\sigma_1}{\sigma_2} = \frac{\cos^2\theta}{(h/l + \sin\theta)\sin\theta} = \nu_{12}^*.$$

This is just equal to the Poisson's ratio for the honeycomb (Eqn. (4.13)). The net effect of anisotropy in a honeycomb is to rotate the yield surface; the angle of rotation, ψ, is given by:

$$\tan\psi = \frac{1}{\nu_{12}^*}.$$

The yield surface for an anisotropic honeycomb is shown in Fig. 4.33.

The failure surfaces in Figs. 4.32 and 4.33 are drawn with sharp corners at the transition between the elastic-buckling surface and the plastic-yield surface. In practice, interaction between the two failure modes leads to a more rounded intersection. Elastoplastic interactions can be calculated using the Rankine formula (Merchant, 1954; Horne, 1963):

$$\frac{1}{\lambda_c} = \frac{1}{\lambda_e} + \frac{1}{\lambda_p}$$

where λ_c, λ_e and λ_p are the combined elastic–plastic, the elastic and the plastic failure loads, respectively. The Rankine formula gives good estimates of the failure load when the elastic-buckling and plastic-collapse modes are geometrically similar; it gives conservative estimates when they are not. Elastoplastic interaction equations are given for several possible combinations of elastic and plastic failure modes by Gibson *et al.* (1989).

The case of commercially produced honeycombs with double thickness vertical walls has been analysed by modelling the vertical walls as rigid plates (Klintworth and Stronge, 1988). Although the rigid plate assumption leads to

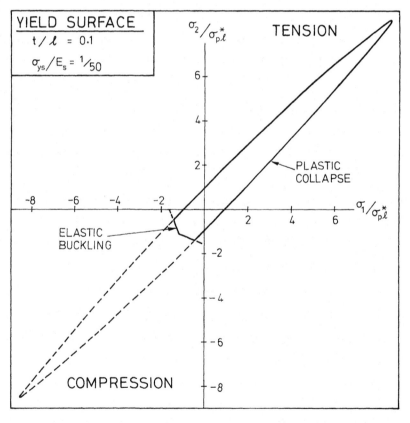

Figure 4.32   The yield surface for a regular hexagonal honeycomb with $t/l = 0.1$. The stresses are plotted in units of the uniaxial plastic collapse stress, σ_{pl}^*. The surface is truncated on the compressive side by elastic buckling.

somewhat different buckling and yielding modes, the analysis is similar to that outlined above. By considering the changes of cell geometry after initial yield, they then find the post-yield evolution of the plastic-collapse surfaces for compressive loading. As the initial collapse mode becomes more established, the other modes are suppressed. Strain softening, due to geometry changes, occurs, which generally produces localization of deformation.

Klintworth and Stronge have used their failure envelope and the associated flow law to describe the in-plane indentation of a ductile honeycomb by a plane punch (Klintworth and Stronge, 1989). The honeycomb initially yields at the edge of the indenter, with further loading resulting in a localized band of deformation that propagates into the honeycomb (Fig. 4.34). The orientation of the crushing band is found by requiring compatibility between the yielded and unyielded cells, or equivalently, requiring the plastic strain increment in the direction of the plane of localization to be zero and relating it to the normal and

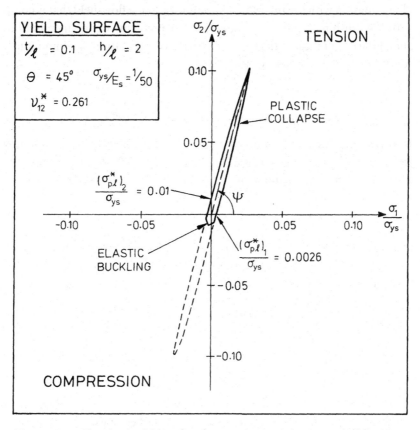

Figure 4.33 The plastic yield surface for an anisotropic honeycomb ($h/l = 2$; $\theta = 45°$: $t/l = 0.1$). The axes are normalized with respect to σ_{ys} rather than σ_{pl}^* as in Fig. 4.32. The effect of the anisotropy is to rotate the yield surface ellipse. The angle ψ is given by $\tan \psi = 1/\nu_{12}^*$.

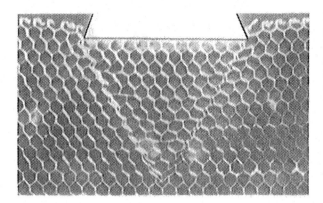

Figure 4.34 An aluminium honeycomb loaded by an indenter. (Klintworth and Stronge (1989) Fig. 11, courtesy of Elsevier.)

shear plastic strain increments. Lower and upper bounds for the indentation force are calculated by comparing the stress distribution beneath the indenter with the yield criteria of the honeycomb and by equating the external work-rate of indentation with the internal dissipation-rate of a flow-field that is compatible with the crushing bands, respectively. Measurements of the indentation stress and of the orientation of the band of localized deformation on aluminium honeycombs are well described by their analysis.

(d) Brittle failure and the fracture surface

Brittle failure occurs when the maximum surface stress in a cell wall, now subjected to both bending and axial loads, exceeds the modulus of rupture of the cell wall material σ_{fs}. A moment M produces a maximum surface stress

$$\sigma_{max} = \frac{6M}{bt^2}$$

in a wall of thickness t and depth b. Failure occurs when this, added to the axial stress σ_a, exceeds σ_{fs}. Using Eqns. (4.82) for M and (4.83) for σ_a gives the equation for the brittle-fracture surface:

$$\pm \left(\frac{\sigma_1(h/l + \sin\theta)\sin\theta - \sigma_2\cos^2\theta}{\sigma_{fs}(t/l)^2} \right) = \frac{1}{3}\left\{ 1 - \frac{\sigma_1(h/l + \sin\theta)\cos\theta + \sigma_2\sin\theta\cos\theta}{\sigma_{fs}(t/l)} \right\} \qquad (4.88)$$

which correctly reduces to Eqns. (4.32) and (4.33) for uniaxial brittle crushing in the appropriate limits. For regular hexagons this becomes

$$\pm\left(\frac{\sigma_1}{\sigma_{fs}} - \frac{\sigma_2}{\sigma_{fs}}\right) = \frac{4}{9}\left(\frac{t}{l}\right)^2\left\{ 1 - \frac{3\sqrt{3}}{4(t/l)}\left(\frac{\sigma_1}{\sigma_{fs}} + \frac{1}{3}\frac{\sigma_2}{\sigma_{fs}}\right) \right\}. \qquad (4.89)$$

As with plastic collapse it is helpful to normalize the stresses by the uniaxial crushing strength, $\sigma_{cr}^* = \frac{4}{9}\sigma_{fs}(t/l)^2$ (Eqn. (4.34)), giving

$$\pm\left(\frac{\sigma_1}{\sigma_{cr}^*} - \frac{\sigma_2}{\sigma_{cr}^*}\right) = 1 - \frac{1}{\sqrt{3}}\left(\frac{\sigma_1}{\sigma_{cr}^*} + \frac{1}{3}\frac{\sigma_2}{\sigma_{cr}^*}\right)\left(\frac{t}{l}\right). \tag{4.90}$$

This brittle-fracture surface is plotted in Fig. 4.35. In the quadrant of biaxial compression the surface may be truncated by elastic buckling or by the compressive failure of the walls themselves. In the tensile quadrant the surface is truncated, if the honeycomb contains cracks or flaws, by fast fracture (Section 4.3). Again, the surface is an elongated one, becoming more extreme in shape as t/l is reduced.

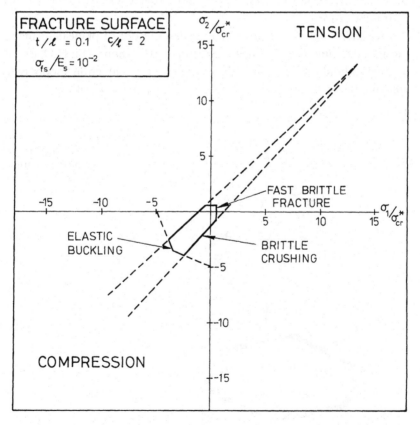

Figure 4.35 The brittle failure surface for a regular hexagonal honeycomb with $t/l = 0.1$. The stresses are plotted in units of the uniaxial crushing strength σ_{cr}^*. The surface may be truncated on the compressive side by elastic buckling or axial crushing, and on the tensile side by fast brittle fracture.

4.5   The out-of-plane properties of honeycombs

(a) Linear-elastic deformation

Honeycombs are often used as cores in sandwich panels. Skis, aircraft, and space vehicles, all today exploit this type of light, stiff structure. Sandwich panels are discussed in detail in Chapter 9. Here we note that the function of the honeycomb core is to carry normal and shear loads in planes containing the axis of the hexagonal prisms – the X_3 direction in Fig. 4.36. When loaded in this direction the cell walls are extended or compressed (rather than bent) and the moduli, for hexagonal honeycombs, are much larger than those calculated in Section 4.3 for in-plane loading. The plastic collapse strengths, too, are larger, and for the same reason: that axial, as well as bending, deformations are involved. In this section we outline the current level of understanding of the out-of-plane properties of honeycombs. Throughout, we assume a low density, so that $t \ll l$, and that all the walls have the same thickness, t; the extension to walls of differing thicknesses is straightforward.

Five additional moduli are needed to describe out-of-plane deformation, giving a total of nine for a complete description of the honeycomb. One is obvious: Young's modulus E_3^* for normal loading in the X_3 direction simply reflects the solid modulus, E_s, scaled by the area of the load-bearing section:

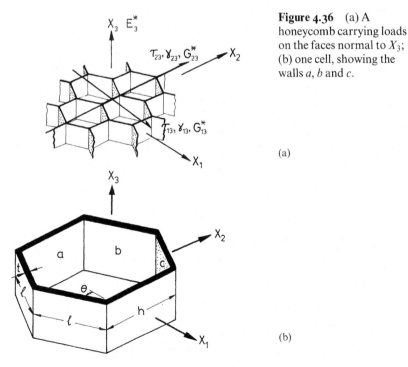

Figure 4.36 (a) A honeycomb carrying loads on the faces normal to X_3; (b) one cell, showing the walls a, b and c.

(a)

(b)

$$\frac{E_3^*}{E_s} = \left\{ \frac{h/l+2}{2(h/l+\sin\theta)\cos\theta} \right\} \frac{t}{l} = \frac{\rho^*}{\rho_s} \approx \frac{t}{l} \tag{4.91}$$

(the term in curly brackets merely gives a more precise description of the projected area of the cell walls normal to X_3; its value for regular hexagons is 1.15). Two more are straightforward: the two Poisson's ratios ν_{31}^* and ν_{32}^* are simply equal to those for the solid itself

$$\nu_{31}^* = \nu_{32}^* = \nu_s. \tag{4.92}$$

The Poisson's ratios ν_{13}^* and ν_{23}^* are then found from the reciprocal relations (Appendix, page 498, Eqn. A6):

$$\nu_{13}^* = \frac{E_1^*}{E_3^*}\nu_s \approx 0, \qquad \nu_{23}^* = \frac{E_2^*}{E_3^*}\nu_s \approx 0.$$

The shear moduli, however, are more complicated (Kelsey et al., 1958; Chang and Ebcioglu, 1961; Penzien and Didriksson, 1964). The stress distribution in a sheared honeycomb is not simple; each cell face suffers a non-uniform deformation because of the constraints imposed on it by its neighbours, and the initially plane honeycomb may not remain plane. Exact calculations are possible only by using numerical methods. But upper and lower bounds for the two shear moduli can be formulated relatively easily by using a simplification of the method used by Kelsey et al. (1958). This is done by calculating the strain energy associated, first, with a strain distribution which allows compatible deformation and, second, with a stress distribution which satisfies equilibrium. If the two coincide, then (it can be shown) the solution is exact. If not, the true solution lies between them.

To do this we make use of the theorems of minimum potential energy and of minimum complementary energy (Sokolnikoff, 1956; McClintock and Argon, 1966). The first theorem gives an upper bound. It states that the strain energy calculated from any postulated set of displacements which are compatible with the external boundary conditions and with themselves will be a minimum for the exact displacement distribution. Consider, then, a uniform shear, γ_{13}, caused by a shear stress τ_{13} acting on the face normal to X_3 in the X_1 direction (Fig. 4.36). Isolate a unit-element of the structure, as shown, which repeats exactly to build up the entire honeycomb. Almost all the elastic strain energy is stored in the shear displacements in the cell walls; the bending stiffnesses and the energies associated with bending are much smaller. The shear strains in walls a, b and c are

$$\gamma_a = 0$$
$$\gamma_b = \gamma_{13}\cos\theta \tag{4.93}$$
$$\gamma_c = \gamma_{13}\cos\theta.$$

The theorem can then be expressed as an inequality which, for shear in the X_1 direction, has the form:

$$\tfrac{1}{2}G^*_{13}\gamma^2_{13}V \le \frac{1}{2}\sum_i (G_s\gamma^2_i V_i) \qquad (4.94)$$

where G_s is the shear modulus of the cell wall material, γ_i are the shear strains in the three cell walls (Eqns. 4.93) and the summation is carried out over the three cell walls a, b and c, of volume V_a, V_b and V_c. Evaluating the sum gives:

$$\boxed{\frac{G^*_{13}}{G_s} \le \frac{\cos\theta}{h/l + \sin\theta}\left(\frac{t}{l}\right).} \qquad (4.95)$$

Repeating the calculation for a shear γ_{23} in the X_2 direction, we find the strains

$$\left.\begin{array}{l} \gamma_a = \gamma_{23} \\ \gamma_b = \gamma_{23}\sin\theta \\ \gamma_c = \gamma_{23}\sin\theta \end{array}\right\} \qquad (4.96)$$

and

$$\boxed{\frac{G^*_{23}}{G_s} \le \frac{1}{2}\frac{h/l + 2\sin^2\theta}{(h/l + \sin\theta)\cos\theta}\left(\frac{t}{l}\right).} \qquad (4.97)$$

The second theorem gives a lower bound for the moduli. It states, that, among the stress distributions that satisfy equilibrium at each point and are in equilibrium with the external loads, the strain energy is a minimum for the exact stress distribution. Expressed as an inequality, for shear in the X_1 direction

$$\frac{1}{2}\frac{\tau^2_{13}}{G^*_{13}}V \le \frac{1}{2}\sum_i \left(\frac{\tau^2_i}{G_s}V_i\right). \qquad (4.98)$$

Consider, first, loading in the X_1 direction. We postulate that an external stress τ_{13} induces a set of shear stresses τ_a, τ_b and τ_c in the three cell walls. By symmetry,

$$\tau_b = \tau_c$$

and as the wall a is loaded in simple bending, it carries no significant load ($\tau_a = 0$). Equilibrium requires that

$$2\tau_{13}l(h + l\sin\theta)\cos\theta = 2\tau_b tl\cos\theta. \qquad (4.99)$$

The inequality equation (4.98), together with this equilibrium equation, give a lower bound:

$$\boxed{\frac{G^*_{13}}{G_s} \ge \frac{\cos\theta}{(h/l + \sin\theta)}\left(\frac{t}{l}\right)} \qquad (4.100)$$

which is identical with the upper bound: the result is exact. For regular hexagons it reduces to

$$\frac{G_{13}^*}{G_s} = 0.577\left(\frac{t}{l}\right). \qquad (4.101)$$

For loading in the X_2 direction we again postulate that the external shear stress τ_{23} induces a set of shear stresses τ_a, τ_b and τ_c in the three cell walls. Symmetry again requires that

$$\tau_b = \tau_c.$$

Equilibrium in the X_3 direction at the nodes means that

$$\tau_a = \tau_b + \tau_c = 2\tau_b$$

and equilibrium with the external stress gives

$$2\tau_{23}l(h + l\sin\theta)\cos\theta = 2\tau_b tl\sin\theta + \tau_a th \qquad (4.102)$$

so that

$$\tau_b = \tau_{23}\cos\theta\frac{l}{t}. \qquad (4.103)$$

Combining these with the inequality (Eqn. (4.70)) gives

$$\boxed{\frac{G_{23}^*}{G_s} \geq \frac{h/l + \sin\theta}{(1 + 2h/l)\cos\theta}\left(\frac{t}{l}\right).} \qquad (4.104)$$

In this case the two bounds do not coincide for a general, anisotropic honeycomb, though they are close. But for regular hexagons the bounds *do* coincide; both Eqns. (4.97) and (4.104) reduce to:

$$\frac{G_{23}^*}{G_s} = 0.577\left(\frac{t}{l}\right). \qquad (4.105)$$

This is identical with the result for G_{13}^*, as it should be: regular hexagons are isotropic in the X_1–X_2 plane.

In calculating the upper and lower bounds we have assumed that the shear stresses are uniform within the cell walls. Using a finite element analysis of a unit cell ($t/l = 0.08$; $h/l = 1.0$; $0 < \theta < 30°$), Grediac (1993) has shown that the shear modulus in the X_2–X_3 plane depends on the width, b, of the honeycomb: G_{23}^* decreases from a value close to the upper bound at $b/l = 1$, to a value close to the lower bound at $b/l = 10$. The finite element results are well described by:

$$G_{23}^* = G_{23\,\text{lower}}^* + \frac{0.787}{(b/l)}\left(G_{23\,\text{upper}}^* - G_{23\,\text{lower}}^*\right). \qquad (4.106)$$

The out-of-plane moduli all scale as (t/l); that is, they depend linearly on the relative density. In this they differ from the in-plane moduli which scale as $(t/l)^3$. Roughly speaking, the out-of-plane moduli are larger than those in-plane by the factor $(l/t)^2$, or, for typical honeycombs, a factor of between 10 and

1000. The two out-of-plane shear moduli are equal for regular hexagons, as expected.

Measuring Young's modulus is easy, and generally confirms the simple proportionality with (t/l). Measuring shear moduli is more difficult. Data from extensive tests by Kelsey et al. (1958) on aluminium and on steel honeycombs (using three-point bend tests on samples of various densities and cell-wall angles) are replotted in Fig. 4.37, in a way which allows a comparison with Eqns. (4.101) and (4.105). They show a good agreement with theory.

(b) Non-linear elasticity: elastic buckling

The remaining out-of-plane properties are poorly investigated. Tests with rubber models show that an elastomeric honeycomb, compressed in the X_3 direction, will eventually buckle, the cell walls bulging in the periodic way shown in Fig. 4.38. The buckling of a panel (the cell wall) which is constrained along the two edges which lie parallel to the loading direction is a standard problem (see, for example, Timoshenko and Gere, 1961). The buckling load is determined by the second moment of inertia of the wall and by the width (l or h), not the height (b) of the panel. For the wall of length l the buckling load is:

$$P_{\text{crit}} = \frac{KE_s}{(1 - \nu_s^2)} \frac{t^3}{l}. \tag{4.107}$$

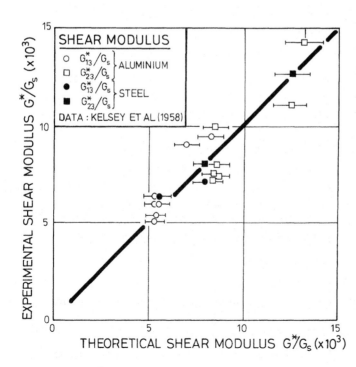

Figure 4.37 The out-of-plane shear moduli of metal honeycombs G_{13}^* and G_{23}^* compared with the theory developed in the text. The generally good agreement supports the use of Eqns. (4.101) and (4.105).

The constant, K, is another end constraint factor, analogous to the factor n used previously for a column. If the vertical edges in Fig. 4.38 are simply supported (that is, they are free to rotate), and the depth b is large compared with $l (b > 3l)$, then $K = 2.0$. If, instead, they are clamped, $K = 6.2$. In the honeycomb the cell wall is neither completely free nor rigidly clamped; as an approximation, we take the value $K = 4$. Then the elastic-buckling load is the sum of the loads carried by the individual cell walls, giving the elastic collapse stress

$$\frac{(\sigma_{el}^*)_3}{E_s} \approx \frac{2}{(1 - \nu_s^2)} \frac{(l/h + 2)}{(h/l + \sin\theta)\cos\theta} \left(\frac{t}{l}\right)^3 \tag{4.108}$$

which, for regular hexagons, and $\nu_s = 0.3$, becomes

$$\frac{(\sigma_{el}^*)_3}{E_s} = 5.2 \left(\frac{t}{l}\right)^3. \tag{4.109}$$

Note that this has the same form as the in-plane buckling stress (both depend on $(t/l)^3$), but that this is roughly 20 times larger. This analysis has been repeated for commercial honeycombs with double thickness walls of length h by Zhang and Ashby (1992a).

The elastic buckling of a honeycomb in out-of-plane shear can be analysed by combining the previous linear elastic out-of-plane shear stress analysis (Section 4.5a) with the shear buckling stress of a single cell wall, τ_{crit} (Roark and Young, 1975):

$$\tau_{crit} = \frac{CE_s}{(1 - \nu_s^2)} \left(\frac{t}{l}\right)^2 \tag{4.110}$$

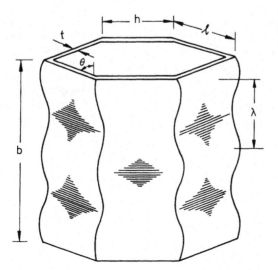

Figure 4.38 The axial elastic buckling of a hexagonal cell.

For loading in the X_1–X_3 plane, equilibrium requires (Eqn. 4.99):

$$2\tau_{13}l(h + l\sin\theta)\cos\theta = 2\tau_\mathrm{b}tl\cos\theta.$$

Elastic buckling occurs simultaneously in walls b and c; setting $\tau_\mathrm{b} = \tau_\mathrm{crit}$ we find:

$$\tau_{13}^* = \frac{CE_\mathrm{s}(t/l)^3}{(1 - \nu_\mathrm{s}^2)\left(\dfrac{h}{l} + \sin\theta\right)}. \tag{4.111}$$

For loading in the X_2–X_3 plane, equilibrium requires (Eqn. 4.102):

$$2\tau_{23}l(h + l\sin\theta)\cos\theta = 2\tau_\mathrm{b}tl\sin\theta + \tau_\mathrm{a}th. \tag{4.112}$$

The shear stress in wall a is twice that in walls b and c; we expect shear buckling to initiate in wall a, giving:

$$\tau_{23}^* = \frac{CE_\mathrm{s}(t/l)^3}{2(1 - \nu_\mathrm{s}^2)\cos\theta}. \tag{4.113}$$

In practice, the out-of-plane shear strength of honeycombs can only be measured by bonding the honeycomb to plates; for this reason, we assume that all four edges of each cell wall are essentially fixed, and take $C = 8.44$, a value intermediate to that for $b/l = 2$ and $b/l = \infty$. For commercial honeycombs, with double thickness cell walls of length h, the results are similar (Zhang and Ashby, 1992a). Data for the compressive (σ_3^*) and out-of-plane shear (τ_{13}^* and τ_{23}^*) strengths of paper–resin honeycombs (Nomex) are shown in Fig. 4.39. Agreement with the analysis is excellent.

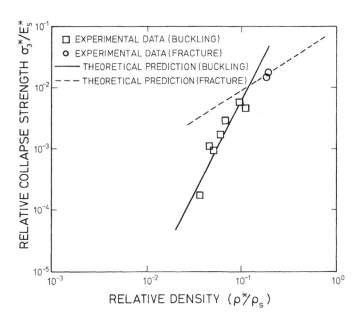

□ EXPERIMENTAL DATA (BUCKLING)
o EXPERIMENTAL DATA (FRACTURE)
—— THEORETICAL PREDICTION (BUCKLING)
--- THEORETICAL PREDICTION (FRACTURE)

Figure 4.39 (a) The failure stress in the X_s direction, σ_3^* normalized by E_s plotted against relative density. The solid line represents the equation for failure by buckling (4.109) while the dashed line represents the equation for failure by fracture (4.117).

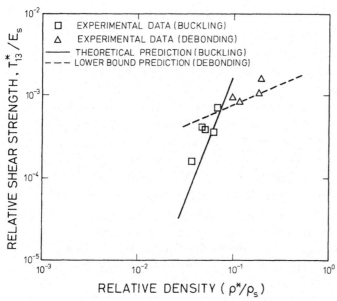

Figure 4.39 (b) The shear collapse stress τ_{13}^* normalized by E_s plotted against relative density. The solid line represents the equation for failure by buckling (4.111) while the dashed line represents failure by debonding.

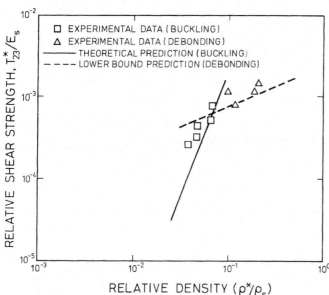

Figure 4.39 (c) The shear collapse stress, τ_{23}^* normalized by E_s plotted against relative density. The solid line represents the equation for failure by buckling (4.113) while the dashed line represents failure by debonding.

(c) Plastic collapse

If the net section stress in the plane normal to X_3 (roughly $\sigma_3(l/t)$), exceeds the yield strength σ_{ys} of the cell-wall material, then the cell walls will yield axially. This defines an upper limit for the plastic collapse strength of the honeycombs:

$$\frac{(\sigma_{pl}^*)_3}{\sigma_{ys}} = \frac{(h/l + 2)}{2\cos\theta(h/l + \sin\theta)}\frac{t}{l} = \frac{\rho^*}{\rho_s}. \tag{4.114}$$

This equation properly describes the axial strength in tension. But in compression this limit is seldom reached because plastic buckling occurs first.

The axial collapse of hexagonal honeycombs by plastic buckling has been treated by McFarland (1963) and, more recently and with more precision, by Wierzbicki (1983). The cells fold progressively with a wavelength λ which is often roughly equal to the cell side length l. Figure 4.40(a) shows a two-dimensional schematic of the buckling, but this conceals the problem of describing the displacement where the cell faces meet. Wierzbicki (1983) shows that a geometry of collapse is possible which involves very little extension or compression of the cell wall – it is achieved largely by bending. An isolated wall can collapse, as shown in Fig. 4.40(b), and in doing so plastic work πM_{p} per unit depth of cell wall is done against the plastic moment M_{p}. The unit cell has side length $(2l + h)$ associated with it, so the total work per cell is $\pi M_{\mathrm{p}}(2l + h)$. Equating this to the work done by the force P per cell (that is, per side length $(2l + h)$) in a displacement of $(\lambda/2 - 2t)$, which we approximate as $\lambda/2$, gives

$$\frac{P\lambda}{2} = \pi M_{\mathrm{p}}(2l + h).$$

Replacing P by $2\sigma_3 l(h + l \sin \theta) \cos \theta$ and λ by l, and with $M_{\mathrm{p}} = \sigma_{\mathrm{ys}} t^2/4$ per unit length, gives an estimate of the plastic buckling stress:

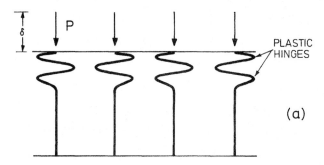

(a)

PLASTIC HINGES

Figure 4.40
(a) Schematic of the plastic buckling of a honeycomb, loaded in the X_3 direction: the walls fold in a periodic way; (b) the work done by the load P on one wall in a displacement $((\lambda/2) - 2t)$ is dissipated in the plastic hinge.

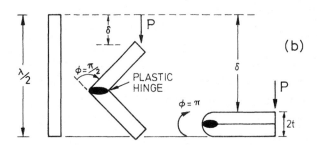

(b)

PLASTIC HINGE

$\phi = \frac{\pi}{2}$

$\phi = \pi$

$\frac{\lambda}{2}$

$2t$

$$\frac{(\sigma^*_{\text{pl}})_3}{\sigma_{\text{ys}}} \approx \frac{\pi}{4} \frac{(h/l+2)}{(h/l+\sin\theta)\cos\theta} \left(\frac{t}{l}\right)^2$$

which, for regular hexagons, reduces to

$$\frac{(\sigma^*_{\text{pl}})_3}{\sigma_{\text{ys}}} \approx 2\left(\frac{t}{l}\right)^2.$$

This simplified calculation, which ignores the necessity of compatibility at the cell edges, illustrates how a collapse stress which is less than that of Eqn. (4.114) may be possible (since it varies as $(t/l)^2$ instead of t/l). Wierzbicki identifies a compatible collapse mode which requires additional plastic hinges at the cell corners, and a limited amount of cell wall extension, also at the corners. By minimizing the collapse load with respect to the wavelength λ, Wierzbicki's method gives the stress for collapse for plastic buckling as

$$\boxed{\frac{(\sigma^*_{\text{pl}})_3}{\sigma_{\text{ys}}} = 5.6\left(\frac{t}{l}\right)^{5/3}} \tag{4.115}$$

for regular hexagons with a uniform wall thickness t, and

$$\frac{(\sigma^*_{\text{pl}})_3}{\sigma_{\text{ys}}} = 6.6\left(\frac{t}{l}\right)^{5/3} \tag{4.116}$$

for honeycombs in which two of the six cell walls are double. The stress for collapse by plastic buckling given by this equation is less than that for simple plastic compression (Eqn. 4.114) when $t/l < 0.1$, so for all but rather dense honeycombs the compressive collapse mode is by plastic buckling.

(d) Brittle failure

The strength of brittle honeycombs is important both when they bear loads and when they are exposed to temperature gradients since it is the strength which determines the maximum tolerable gradient. If the net section stress in the plane normal to X_3 exceeds the tensile fracture strength (roughly σ_{fs}) of the cell-wall material, a brittle honeycomb will fail in tension. This defines an upper limiting tensile strength

$$\frac{(\sigma^*_{\text{f}})_3}{\sigma_{\text{fs}}} = \frac{(h/l+2)}{2\cos\theta(h/l+\sin\theta)} \left(\frac{t}{l}\right) = \frac{\rho^*}{\rho_{\text{s}}}. \tag{4.117}$$

It is an upper limit because it describes a defect-free sample; if the honeycomb contains a defect (due to faulty manufacture or damage in use) it may fail at a lower stress. When a flaw which is large compared to the cell size propagates in the plane normal to the X_3 direction, extending its apparent area by A, it creates a fractured surface with an area which is ρ^*/ρ_{s} times smaller than this. We anticipate, then, that the toughness G^*_{c} of the flawed honeycomb should be less, by the factor ρ^*/ρ_s, than that of the solid of which it is made. The fracture toughness, $K_{IC}^* = (E_3^* G_{IC}^*)^{1/2}$, is given by

$$K_{IC} = K_{ICs}\left(\frac{\rho^*}{\rho_s}\right) \qquad\qquad (4.118)$$

and the fracture stress for a flaw of size $2a$ is

$$(\sigma_f^*)_3 = \frac{K_{ICs}}{\sqrt{\pi a}}\frac{\rho^*}{\rho_s}. \qquad\qquad (4.119)$$

Brittle solids are stronger in compression than in tension because compressive stresses close the small cracks or flaws which, ultimately, determine the strength. But even though they close, the flaws can shear and the shearing concentrates stress in a way which can still lead to fracture (Ashby and Hallam, 1986). The result, for axial loading, is that the crushing strength of a cell wall is about 12 times greater than its failure strength, σ_{fs}, in tension. The best estimate we can then make for the compressive strength of a brittle honeycomb, loaded in the X_3 direction, is

$$\frac{(\sigma_{cr}^*)_3}{\sigma_{fs}} \approx 12\left(\frac{t}{l}\right). \qquad\qquad (4.120)$$

4.6   Conclusions

Honeycombs deform by a number of different mechanisms. When loaded in-plane the cell walls at first bend, giving linear elasticity which can persist to strains as large as 10%. This is possible because the honeycomb is like a spring: the geometry allows a large distortion of the structure with only small strains in its members (the strain in the cell wall is less, by a factor of about t/l, than that of the honeycomb as a whole). When loaded out-of-plane the cell walls are extended or compressed or (sometimes) sheared, and the structure is much stiffer. Simple mechanics gives the full set of nine elastic constants required to describe a general orthotropic honeycomb. For regular honeycombs the number is reduced to five.

The complications start when linear elasticity breaks down. If the honeycomb is made of an elastomer its cell walls buckle in compression, giving a long, almost flat, plateau in the stress–strain curve; but in tension they rotate and stretch and the stress–strain curve rises steeply. If it is made of a plastic material the cell walls yield – by plastic bending for compressive and tensile in-plane loading and by plastic buckling or axial yield for compressive or tensile loading out-of-plane. This, too, gives a plateau in the stress–strain curve, though (unlike elastic buckling) the deformation associated with it is permanent. If the cell walls are made of a creeping material the honeycomb creeps; and if it is made of a brittle material it crushes and fractures in a brittle manner. The in-plane strength, in every instance, is lower than that for out-of-plane loading because bending predominates in the first case, axial deformation in the second. Each mechanism can be analysed, giving expressions for the strength for each sort of honeycomb. Comparisons with experiments give confidence that the mechanisms are correctly identified and analysed.

The response of honeycombs to biaxial loads is unusual. This is because the cell wall bending which completely dominates the in-plane properties of hexagonal cells for uniaxial loading can be suppressed completely by an appropriate choice of biaxial loads. Then the cell walls fail by axial yielding, or creep, or fracture, and the strength is much higher. This gives yield and brittle failure surfaces for honeycombs which are extremely elongated ellipses; this curious shape must be remembered when designing with honeycombs.

The results presented in this chapter provide a basis for mechanical design with honeycombs. Equally important, they establish an understanding of mechanisms – of their characteristics, and of techniques for analysing them – which underpins much of the later discussion of the properties of foams, to which we now proceed.

Appendix 4A: Elastic moduli of square and triangular honeycombs

Cells in honeycombs are not always hexagonal. For instance, ceramic and metal honeycombs used for heat exchangers and catalyst supports sometimes have square or triangular cells, either to increase the surface area or to give a stiffer structure. The in-plane elastic constants of these structures (Gulati, 1975) differ in a fundamental way from those with hexagonal cells.

The difference arises because the bending moments on the cell walls are equal to zero for certain directions of loading. When this is so, elastic deformation is possible only by the axial extension or compression of the cell walls, and the resulting moduli are larger, by a factor, typically, of $(l/t)^2$ than for directions in which bending is possible.

Square cells are shown in Fig. 4A.1. The Young's moduli E_1^* and E_2^* parallel to the cell walls are simply

$$\frac{E_1^*}{E_s} = \frac{E_2^*}{E_s} = \frac{t}{l}. \tag{4A.1}$$

Those in the diagonal direction, which we shall call E_{45}^*, are given by the methods used earlier as

$$\frac{E_{45}^*}{E_s} = 2\left(\frac{t}{l}\right)^3. \tag{4A.2}$$

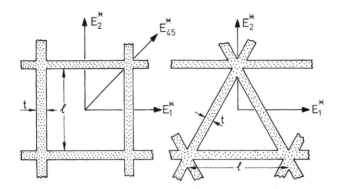

Figure 4A.1
Honeycombs made up
of square and triangular
cells, showing the axes
for which Young's
moduli are calculated.

This second result can be obtained from the limit of the equations for hexagonal
cells (Eqns. (4.7) and (4.11)) when $h = 0$ and $\theta = 45°$.

Triangular cells are even stiffer (Gulati, 1975). A glance at Fig. 4A.1 shows
that the cell walls are extended or compressed for any direction of loading. It follows that Young's modulus, for any in-plane direction of loading, will be proportional to t/l, not $(t/l)^3$. Simple analysis of this extension gives (Hunt, 1993):

$$\frac{E_1^*}{E_s} = \frac{E_2^*}{E_s} = 1.15 \frac{t}{l}. \tag{4A.3}$$

The value of the Young's modulus of a triangular alumina honeycomb, measured in a resonance test, is well described by this equation (Hunt, 1993). Any
imperfection in the triangular structure increases the bending contribution to
the deformation, reducing the modulus.

The moduli associated with the three cell shapes are most easily shown on a
polar diagram, Fig. 4A.2. The radial distance from the origin, in any direction,
shows the magnitude of E^* in that direction, for honeycombs of the same relative
density. Triangular cells are isotropic, and are, overall, the stiffest. Regular hexagonal cells with walls of constant thickness are also isotropic, but are much less
stiff. And square cells are highly anisotropic, having two stiff and two compliant
directions.

Appendix 4B: Small strain calculation of the moduli, including axial and shear deformations

In this appendix, we recalculate the moduli E_1^*, E_2^*, ν_{12}^*, ν_{21}^* and G_{12}^* including the
axial and shear deformation of the cell walls, which become significant at
$t/l > 0.2$ under uniaxial loading. In biaxial loading the bending moments resulting from loading in the X_1 and X_2 directions may counteract one another; then
the contribution of the axial deformations becomes significant at all relative densities.

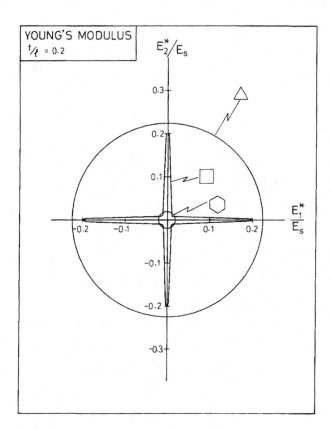

Figure 4A.2 A polar diagram showing Young's modulus as a function of direction for triangular, square and hexagonal cells. The triangular and hexagonal cells are isotropic; the square cells are extremely anisotropic. The triangular cells are much stiffer than the hexagonal cells; the square cells have moduli roughly between those two extremes.

Loading in the X_1 direction

From Fig. 4.8, we have

$$\delta = \frac{Pl^3 \sin \theta}{12 E_s I}.$$ (4B.1)

From Timoshenko and Goodier (1970), we find that we can write the shear deflection of the member as:

$$\delta_s = \frac{Pl^3 \sin \theta}{12 E_s I} (2.4 + 1.5\nu_s)(t/l)^2.$$ (4B.2)

An axial load of $P \cos \theta$ acts on the member and hence the axial deflection is:

$$\delta_a = \frac{Pl \cos \theta}{E_s t b}.$$ (4B.3)

The total deflection in the X_1 direction is then:

$$\delta_1 = \delta \sin \theta + \delta_s \sin \theta + \delta_a \cos \theta$$
$$= \frac{Pl^3 \sin^2 \theta}{12 E_s I} \left(1 + (2.4 + 1.5\nu_s + \cot^2 \theta)(t/l)^2 \right)$$ (4B.4)

and

$$\epsilon_{11} = \frac{\delta_1}{l\cos\theta} \tag{4B.5}$$

giving:

$$E_1^* = \frac{\sigma_{11}}{\epsilon_{11}}$$
$$= E_s\left(\frac{t}{l}\right)^3 \frac{\cos\theta}{\left(\dfrac{h}{l} + \sin\theta\right)\sin^2\theta} \frac{1}{1 + (2.4 + 1.5\nu_s + \cot^2\theta)(t/l)^2} \cdot \tag{4B.6}$$

The strain in the X_2 direction is:

$$\epsilon_{22} = -\frac{\delta_2}{h + l\sin\theta} = -\frac{(\delta\cos\theta + \delta_s\cos\theta - \delta_a\sin\theta)}{h + l\sin\theta}$$
$$= \frac{-Pl^3\sin\theta\cos\theta}{12E_sI(h + l\sin\theta)}\left(1 + (2.4 + 1.5\nu_s - 1)t^2/l^2\right) \tag{4B.7}$$

giving:

$$\nu_{12}^* = -\frac{\epsilon_{22}}{\epsilon_{11}} = \frac{\cos^2\theta}{\left(\dfrac{h}{l} + \sin\theta\right)\sin\theta} \frac{1 + (1.4 + 1.5\nu_s)(t/l)^2}{1 + (2.4 + 1.5\nu_s + \cot^2\theta)(t/l)^2} \cdot \tag{4B.8}$$

Loading in the X_2 direction

Reconsidering Fig. 4.8, we find the bending deflection of the inclined member is, as before:

$$\delta = \frac{Wl^3\cos\theta}{12E_sI}. \tag{4B.9}$$

The shear deflection of this member is then (Timoshenko and Goodier (1970)):

$$\delta_s = \frac{Wl^3\cos\theta}{12E_sI}(2.4 + 1.5\nu_s)t^2/l^2. \tag{4B.10}$$

The axial deflection of the inclined member is:

$$\delta_{a\text{-inclined}} = \frac{Wl\sin\theta}{E_s tb}. \tag{4B.11}$$

The axial deflection of the upright member is:

$$\delta_{a\text{-upright}} = \frac{2Wh}{btE_s}. \tag{4B.12}$$

The total deflection in the X_2 direction is then:

$$\delta_2 = \delta \cos\theta + \delta_s \cos\theta + \delta_{\text{a-inclined}} \sin\theta + \delta_{\text{a-upright}}$$

$$= \frac{Wl^3 \cos^2\theta}{12E_sI} \left[1 + (2.4 + 1.5\nu_s)\frac{t^2}{l^2} + \tan^2\theta\frac{t^2}{l^2} + \frac{2(h/l)}{\cos^2\theta}\frac{t^2}{l^2} \right]. \qquad (4\text{B}.13)$$

The strain in the X_2 direction is

$$\epsilon_{22} = \delta_2/(h + l\sin\theta) \qquad (4\text{B}.14)$$

giving:

$$E_2^* = \frac{\sigma_{22}}{\epsilon_{22}}$$

$$= \frac{\sigma_{22}(h + l\sin\theta)12E_sI}{Wl^3 \cos^2\theta\left[1 + \left(2.4 + 1.5\nu_s + \tan^2\theta + \frac{2(h/l)}{\cos^2\theta}\right)\frac{t^2}{l^2}\right]}$$

$$\boxed{E_2^* = E_s\left(\frac{t}{l}\right)^3 \frac{\left(\frac{h}{l} + \sin\theta\right)}{\cos^3\theta} \frac{1}{\left[1 + \left(2.4 + 1.5\nu_s + \tan^2\theta + \frac{2(h/l)}{\cos^2\theta}\right)\left(\frac{t}{l}\right)^2\right]}.}$$

$$(4\text{B}.15)$$

The strain in the X_1 direction is:

$$\epsilon_{11} = -\frac{\delta_1}{l\cos\theta} = -\frac{(\delta\sin\theta + \delta_s\sin\theta - \delta_{\text{a inclined}}\cos\theta)}{l\cos\theta}$$

$$= -\frac{Wl^2\sin\theta}{12E_sI}\left[1 + (2.4 + 1.5\nu_s - 1)\frac{t^2}{l^2}\right] \qquad (4\text{B}.16)$$

and:

$$\boxed{\nu_{21}^* = -\frac{\epsilon_{11}}{\epsilon_{22}}}$$

$$\boxed{= \frac{\sin\theta\left(\frac{h}{l} + \sin\theta\right)}{\cos^2\theta} \frac{1 + (1.4 + 1.5\nu_s)(t/l)^2}{1 + \left(2.4 + 1.5\nu_s + \tan^2\theta + \frac{2(h/l)}{\cos^2\theta}\right)(t/l)^2}.}$$

$$(4\text{B}.17)$$

Note that the reciprocal relation $E_1^*\nu_{21}^* = E_2^*\nu_{12}^*$ still applies and that the expressions reduce to Eqns. (4.7), (4.11), (4.13), and (4.14) for small t/l.

For the case of regular hexagonal cells ($h/l = 1$; $\theta = 30°$), the moduli reduce to:

$$E_1^* = E_2^* = \frac{4}{\sqrt{3}} E_s \left(\frac{t}{l}\right)^3 \frac{1}{1 + (5.4 + 1.5\nu_s)(t/l)^2} \qquad (4\text{B}.18)$$

and

$$\nu_{12}^* = \nu_{21}^* = \frac{1 + (1.4 + 1.5\nu_s)(t/l)^2}{1 + (5.4 + 1.5\nu_s)(t/l)^2}. \qquad (4\text{B}.19)$$

Shear loading in the X_1–X_2 plane

The shear modulus can be calculated in a similar manner. The previous calculation is modified by including shear deformations of the vertical strut of length h and axial and shear deformations of the oblique struts of length l. Note that, as before, bending of the oblique struts is accounted for by the rotation ϕ of the vertical strut. As before, the moment at the midpoint of each member is zero.

The horizontal deflection of the vertical strut h, including shear deformations, is:

$$u_s = \frac{Fh^2}{48E_sI} \left[l + 2h + 2h(2.4 + 1.5\nu_s)\left(t/h\right)^2\right]. \qquad (4\text{B}.20)$$

The shear strain is then

$$\gamma_h = \frac{Fh^2}{24E_sI} \left[1 + 2(h/l) + \frac{2}{(h/l)}(2.4 + 1.5\nu_s)(t/l)^2\right] \frac{1}{(h/l) + \sin\theta}. \qquad (4\text{B}.21)$$

The axial and shear deformation of the oblique struts is found as follows. Referring to Fig. (4B.1(a)), we first note that equilibrium requires that:

$$\tau = \frac{F}{2bl\cos\theta} \qquad (4\text{B}.22)$$

and

$$\tau = \frac{S}{(h + l\sin\theta)b} \qquad (4\text{B}.23)$$

or

$$S = \frac{F}{2}\left(\frac{h + l\sin\theta}{l\cos\theta}\right) \qquad (4\text{B}.24)$$

and

$$F = 2\tau bl\cos\theta. \qquad (4\text{B}.25)$$

The shear strain in the oblique members is then given by Figs. (4B.1(b) and (c)):

$$\gamma_l = \frac{2(u_{ss}\text{axial} + u_{ss}\text{shear})}{l\cos\theta} \qquad (4\text{B}.26)$$

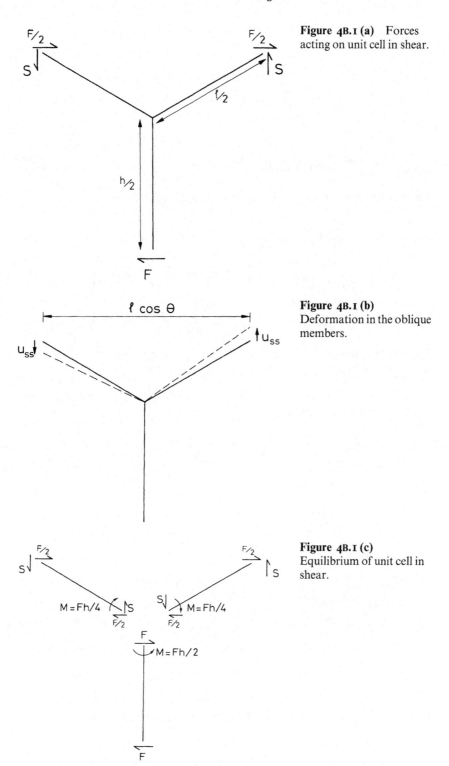

Figure 4B.1 (a)   Forces acting on unit cell in shear.

Figure 4B.1 (b) Deformation in the oblique members.

Figure 4B.1 (c) Equilibrium of unit cell in shear.

where

$$u_{ss}\text{axial} = \frac{Fh^2}{48E_sI}\left(\frac{t}{h}\right)^2[(h + l\sin\theta)\tan\theta + l\cos\theta]\sin\theta \tag{4B.27}$$

and

$$u_{ss}\text{shear} = \frac{Fh^2}{48E_sI}\frac{l}{(h/l)}(2.4 + 1.5\nu_s)(t/l)^2\cos\theta. \tag{4B.28}$$

Substituting Eqns. (4B.27) and (4B.28) into Eqn. (4B.26) gives:

$$\gamma_l = \frac{Fh^2}{24E_sI}\left(\frac{t}{l}\right)^2\frac{1}{h/l + \sin\theta}$$
$$\left\{\frac{1}{(h/l)^2}\left[\left(\frac{h}{l} + \sin\theta\right)\tan^2\theta + \sin\theta\right]\left(\frac{h}{l} + \sin\theta\right)\right.$$
$$\left.+ \frac{1}{(h/l)}(2.4 + 1.5\nu_s)\left(\frac{h}{l} + \sin\theta\right)\right\}. \tag{4B.29}$$

The total strain is then:

$$\gamma = \gamma_h + \gamma_l \tag{4B.30}$$

and the shear modulus is:

$$G = \frac{\tau}{\gamma} \tag{4B.31}$$

or

$$G_{12}^* = \frac{E_s(t/l)^3(h/l + \sin\theta)}{(h/l)^2\cos\theta}\frac{1}{F}$$

where

$$F = \left[1 + 2\left(\frac{h}{l}\right) + \left(\frac{t}{l}\right)^2\left\{\begin{array}{l}\frac{1}{h/l}(2.4 + 1.5\nu_s)(2 + h/l + \sin\theta) \\ + \frac{h/l + \sin\theta}{(h/l)^2}[(h/l + \sin\theta)\tan^2\theta + \sin\theta]\end{array}\right\}\right]$$

$$\tag{4B.32}$$

For regular hexagonal cells, the shear modulus becomes:

$$G_{12}^* = \frac{1}{\sqrt{3}}E_s(t/l)^3\frac{1}{1 + (3.30 + 1.75\nu_s)(t/l)^2}. \tag{4B.33}$$

Appendix 4C: The elastic buckling of a honeycomb

The plateau stress $(\sigma^*_{el})_2$ for an elastomeric honeycomb in compression is determined by the elastic buckling of its members. The buckling mode itself depends on stress state: Fig. 4.27 compared the modes observed for regular honeycombs in uniaxial and biaxial compression. In both cases buckling gives a regular, symmetric deformation of the cell walls, with a wavelength which is longer in uniaxial than in biaxial loading. The first is the easier to analyse (Gibson *et al.*, 1982), because in it all the joints rotate through an equal angle $\pm\phi$. Referring to Fig. 4.13, the midpoint D of the beam BE is a point of inflexion and thus carries zero bending moment. Equilibrium of the beam AB requires that

$$\frac{Pl\cos\theta}{2} = M_1 - M_2. \tag{4C.1}$$

The curvature of the beam is given by

$$\frac{d^2y}{dx^2} = -\frac{M}{E_s I}. \tag{4C.2}$$

Relating the end slope of the beam AB to the moments acting on it gives

$$M_1 + M_2 = \frac{4E_s I\phi}{l}. \tag{4C.3}$$

The member BE can then be considered as a column carrying an axial load P and constrained by a rotational spring of stiffness

$$K = \frac{M}{\phi} = \frac{4E_s I}{l}$$

at each end. Its behaviour is analysed by making use of the "elastic line" method (Fig. 4C.1). Consider a column of length λ pinned at both ends and subjected to its critical Euler buckling load

$$P_{cr} = \frac{\pi^2 E_s I}{\lambda^2}. \tag{4C.4}$$

The deflected shape of the column is described by

$$y = h_0 \sin\left(\frac{\pi x}{\lambda}\right)$$

where h_0 is the midspan deflection. The column BE of Fig. 4.13 can be considered to be part of this column. To determine the part we match the rotational stiffness at the ends of the column BE to that of the column of length λ at the point $x = L$, where L is to be determined. At $x = L$ the moment is

$$M = Ph_0 \sin\left(\frac{\pi L}{\lambda}\right) = \frac{\pi^2 E_s I}{\lambda^2} h_0 \sin\left(\frac{\pi L}{\lambda}\right) \tag{4C.5}$$

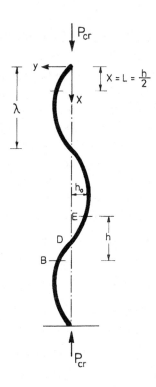

Figure 4C.1 The elastic line method for analysing the buckling of a honeycomb. The slope and curvature of the buckled wall is matched to the appropriate segment of the buckled column shown here.

and the rotation is

$$\phi = \frac{\mathrm{d}y}{\mathrm{d}x}\bigg|_{x=L} = \frac{h_0\pi}{\lambda}\cos\left(\frac{\pi L}{\lambda}\right). \tag{4C.6}$$

The rotational spring stiffness is

$$K = \frac{M}{\phi} = \frac{\pi E_s I}{\lambda}\tan\left(\frac{\pi L}{\lambda}\right). \tag{4C.7}$$

By relating λ to L and K we can find the critical load at which the column, of length L and constrained at its ends by springs of stiffness K, will buckle. Let

$$\beta = \frac{\pi L}{\lambda}$$

then

$$K = \frac{\beta E_s I}{L}\tan\beta.$$

For the hexagonal cells of Fig. 4.13, $L = h/2$ and $K = M/\phi = 4E_s I/l$, giving

$$\tan\beta = \frac{KL}{E_s I\beta} = \frac{2h}{l\beta}. \tag{4C.8}$$

The solution $\beta = \beta^*$ of this equation is found graphically, by plotting $\tan\beta$ and $2h/l\beta$ against β. This gives a relation between λ and L, which can be used to

Table 4C.1 Buckling constant for uniaxial compression

h/l	1.0	1.5	2.0
β^*	$0.343\,\pi$	$0.380\,\pi$	$0.403\,\pi$
n	0.686	0.760	0.806

determine the elastic-buckling load for the honeycomb. Noting that $\lambda = \pi h/2\beta^*$ we obtain the critical buckling load

$$P_{\text{crit}} = \frac{\pi^2 E_s I}{\lambda^2} = \frac{E_s b t^3 \beta^{*2}}{3h^2}. \tag{4C.9}$$

The critical buckling stress is given by

$$(\sigma_{\text{el}}^*)_2 = \frac{P_{\text{crit}}}{2lb\cos\theta} = \frac{E_s t^3 \beta^{*2}}{6lh^2\cos\theta}. \tag{4C.10}$$

Values for β^* for several h/l are given in Table 4C.1. In terms of the end constraint factor, n,

$$P_{\text{crit}} = \frac{n^2 \pi^2 E_s I}{h^2} \tag{4C.11}$$

and

$$(\sigma_{\text{el}}^*)_2 = \frac{n^2 \pi^2 E_s t^3}{24\cos\theta lh^2}. \tag{4C.12}$$

The analysis neglects the small change in cell-wall angle, θ, during the linear-elastic deformation that precedes buckling. It can be included, but results in changes of order 10%, which is less than the experimental error in measuring $(\sigma_{\text{el}}^*)_2$.

Appendix 4D: Mechanical properties of non-uniform commercial honeycombs

Honeycombs with double thickness vertical walls

Commercial honeycombs are commonly made by gluing strips of material periodically such that when pulled part, or *expanded*, hexagonal cells are formed (Fig. 4D.1). The resulting cells have four walls of length l of thickness t and two walls of length h of thickness $2t$. The relative density of the honeycomb is then given by:

$$\frac{\rho^*}{\rho_s} = \frac{t/l(h/l + 1)}{(h/l + \sin\theta)\cos\theta}. \tag{4D.1}$$

Because the in-plane uniaxial deformation is controlled by the walls of thickness t, the Young's moduli, E_1 and E_2, and the Poisson's ratios, ν_{12} and ν_{21} are

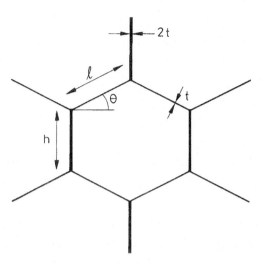

Figure 4D.1 Honeycomb with non-uniform thickness cell walls. The inclined walls have thickness t while the vertical walls have thickness $2t$.

the same as those for a honeycomb with uniform thickness walls (Eqns. (4.7) (4.11), (4.13) and (4.14)). Modifying the previous analysis of Sections 4.3 and 4.5 we find the results listed in Table 4D.1.

I-beam honeycombs

In 'I-beam' honeycombs (U.S. Patent Number 4632862, Eldim, Inc., Woburn MA, USA 12/30/86) a rectangular corrugation is introduced into the vertical walls to increase the mechanical stiffness and strength; the corrugated sheets are then spot welded together (Fig. 4D.2). The resulting honeycomb again has vertical walls of double thickness. The relative density of the I-beam honeycomb is:

$$\frac{\rho^*}{\rho_s} = \left(\frac{t}{l}\right) \frac{(1 + 2\cos\theta + h/l)}{\cos\theta(h/l + \sin\theta)}. \tag{4D.2}$$

The out-of-plane elastic moduli of the I-beam honeycomb are given in Table 4D.2.

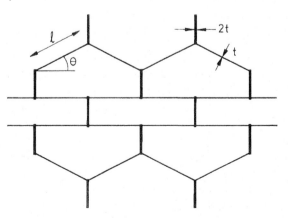

Figure 4D.2 I-beam honeycomb with all vertical walls double thickness.

Table 4D.1   Properties of honeycombs with double thickness vertical walls

In-plane properties

$$\frac{E_1^*}{E_s} = \left(\frac{t}{l}\right)^3 \frac{\cos\theta}{(h/l + \sin\theta)\sin^2\theta}$$

$$\frac{E_2^*}{E_s} = \left(\frac{t}{l}\right)^3 \frac{(h/l + \sin\theta)}{\cos^3\theta}$$

$$\nu_{12}^* = \frac{\cos^2\theta}{(h/l + \sin\theta)\sin\theta}$$

$$\nu_{21}^* = \frac{(h/l + \sin\theta)\sin\theta}{\cos^2\theta}$$

$$\frac{G_{12}^*}{E_s} = \left(\frac{t}{l}\right)^3 \frac{(h/l + \sin\theta)}{(h/l)^2\cos\theta(1 + 16h/l)}$$

$$\frac{(\sigma_{el}^*)_2}{E_s} = \frac{4}{3}\left(\frac{t}{l}\right)^3 \frac{\beta^{+2}}{(h/l)^2\cos\theta}$$

h/l	1.0	1.5	2.0
β^+	0.48	0.58	0.65

$$\frac{(\sigma_{pl}^*)_1}{\sigma_{ys}} = \left(\frac{t}{l}\right)^2 \frac{1}{2(h/l + \sin\theta)\sin\theta}$$

$$\frac{(\sigma_{pl}^*)_2}{\sigma_{ys}} = \left(\frac{t}{l}\right)^2 \frac{1}{2\cos^2\theta}$$

$$\frac{(\tau_{pl}^*)_{12}}{\sigma_{ys}} = \left(\frac{t}{l}\right)^2 \frac{1}{(h/l)\cos\theta}$$

Out-of-plane properties

$$\frac{E_3^*}{E_s} = \left(\frac{t}{l}\right) \frac{(1 + h/l)}{(h/l + \sin\theta)\cos\theta}$$

$$\nu_{31}^* = \nu_{32}^* = \nu_s$$

$$\nu_{13}^* = \nu_{23}^* \approx 0$$

$$\frac{G_{13}^*}{G_s} = \left(\frac{t}{l}\right) \frac{\cos\theta}{(h/l + \sin\theta)} \qquad \text{Kelsey et al. (1958)}$$

$$\left(\frac{t}{l}\right) \frac{(h/l + \sin\theta)}{(h/l + 1)\cos\theta} \le \frac{G_{23}^*}{G_s} \le \left(\frac{t}{l}\right) \frac{(h/l + \sin^2\theta)}{(h/l + \sin\theta)\cos\theta} \qquad \text{Kelsey et al. (1958)}$$

$$G_{23}^* \cong G_{23_{lower}}^* + \frac{0.787}{(b/l)}\left(G_{23_{upper}}^* - G_{23_{lower}}^*\right) \qquad \text{Grediac (1993)}$$

$$\frac{(\sigma_{el}^*)_3}{E_s} = \frac{K}{(1 - \nu_s^2)}\left(\frac{t}{l}\right)^3 \frac{\left(1 + 4\frac{l}{h}\right)}{\cos\theta(h/l + \sin\theta)} \qquad K = 5.73 \qquad \text{Zhang + Ashby(1992)}$$

Table 4D.1   *(cont.)*

Out-of-plane properties

$$\frac{(\sigma_{pl}^*)_3}{\sigma_{ys}} = \left(\frac{t}{l}\right) \frac{(1 + h/l)}{(h/l + \sin\theta)\cos\theta}$$ uniaxial tension

$$\frac{(\sigma_{pl}^*)_3}{\sigma_{ys}} = 6.6 \left(\frac{t}{l}\right)^{5/3}$$ Wierzbicki(1983), plastic buckling in uniaxial compression $(h/l = 1, \theta = 30°)$

$$\frac{(\tau_{13}^*)\text{el}}{E_s} = \left(\frac{t}{l}\right)^3 \frac{C}{(1 - \nu_s^2)(h/l + \sin\theta)}$$ $C = 8.44$ Zhang + Ashby(1992)

$$\frac{(\tau_{23}^*)\text{el}}{E_s} = \left(\frac{t}{l}\right)^3 \frac{C}{(1 - \nu_s^2)\cos\theta} \frac{(1 + \sin\theta)}{(h/l + \sin\theta)}$$ $C = 8.44$ Zhang + Ashby(1992)

Table 4D.2   Out-of-plane elastic moduli of I-beam honeycombs

$$\frac{E_3^*}{E_s} = \left\{ \frac{\frac{h}{l} + 2\cos\theta + 1}{\cos\theta \left(\frac{h}{l} + \sin\theta\right)} \right\} \frac{t}{l} = \frac{\rho^*}{\rho_s}$$

$$\frac{G_{13}^*}{G_s} = \frac{\cos\theta + 2}{\left(\frac{h}{l} + \sin\theta\right)} \frac{t}{l}$$

$$\frac{\frac{h}{l} + \sin\theta}{\left(\frac{h}{l} + 1\right)\cos\theta} \left(\frac{t}{l}\right) \leq \frac{G_{23}^*}{G_s} \leq \frac{\frac{h}{l} + \sin^2\theta}{\left(\frac{h}{l} + \sin\theta\right)\cos\theta} \left(\frac{t}{l}\right)$$

$$\nu_{13}^* = \nu_{23}^* = 0$$

$$\nu_{31}^* = \nu_{32}^* = \nu_s$$

References

Abd El-Sayed, F. K. (1976) Ph.D. thesis, University of Sheffield.

Abd El-Sayed, F. K., Jones, R. and Burgens, I. W. (1979) *Composites*, **10**, 209.

Abel, C., Ashby, M. F. and Cocks, A. C. F. (1995) Cambridge University Engineering Department Report. CUED Report No. CUED/MATS/25/95.

Ashby, M. F. (1983). *Met. Trans*, **14A**, 1755.

Ashby, M. F. and Hallam, S. D. (1986) *Acta Met.*, **34**, 497.

Chang, C. C. and Ebcioglu, I. K. (1961) *Trans ASME D*, December, p. 513.

Cowin, S. C. (1985) The relationship between the elasticity tensor and the fabric tensor. *Mech. Mater.* **4**, 137–47.

Davidge, R. W. (1979) *Mechanical Behaviour of Ceramics*. Cambridge University Press, Cambridge.

Eischen, J. W. and Torquato, S. (1993) Determining elastic behaviour of composites by the boundary element method. *J. Appl. Phys.* **74**, 159–70.

Evans, K. E. (1989) Tensile network microstructures exhibiting negative Poisson's ratios. *J. Phys. D: Appl. Phys.* **22**, 1870–6.

Gibson, L. J. (1981) Ph.D. thesis, Engineering Department, Cambridge University, Cambridge.

Gibson, L. J., Ashby, M. F., Schajer, G. S. and Robertson, C. I. (1982) *Proc. R. Soc. Lond.*, **A382**, 25.

Gibson, L. J., Ashby, M. F., Zhang, J. and Triantafillou, T. C. (1989) Failure surfaces for cellular materials under multiaxial loads I: modelling. *Int. J. Mech. Sci.* **31**, 635–63.

Grediac, M. (1993) A finite element study of the transverse shear in honeycomb cores. *Int. J. Solids Structures* **30**, 1777–88.

Gulati, S. T. (1975) Paper 750171, Automotive Engineering Congress and Exhibition Society of Automotive Engineers, Detroit, Michigan, 24–28 February, p. 157.

Guo, X. E., McMahon, T. A., Keaveny, T. M., Hayes, W. C. and Gibson, L. J. (1994) Finite element modelling of damage accumulation in trabecular bone under cyclic loading. *J. Biomechanics* **27**, 145–55.

Hollister, S. J., Fyhrie, D. P., Jepsen K. J., and Goldstein, S. A. (1991) Application of homogenization theory to the study of trabecular bone mechanics. *J. Biomechanics* **24**, 825–39.

Hollister, S. J., Brennan, J. M. and Kikuchi, N. (1994) A homogenization sampling procedure for calculating trabecular bone stiffness and tissue level stress. *J. Biomechanics* **27**, 433–44.

Horne, M. R. (1963) Elastic–plastic failure loads of plane frames. *Proc. Roy. Soc. Lond.* **A274**, 343.

Huang, J. S. and Gibson, L. J. (1991) Fracture toughness of brittle honeycombs. *Acta Metall. Mater.* **39**, 1617–26.

Hunt, H. E. M. (1993) The mechanical strength of ceramic honeycomb monoliths as determined by simple experiments. *Trans. Inst. Chem. Eng.* **71A**, 257–66.

Kelsey, S., Gellatly, R. A. and Clark, B. W. (1958) *Aircraft Engineering*, October, p. 294.

Klintworth, J. W. and Stronge, W. J. (1988) Elasto-plastic yield limits and deformation laws for transversely crushed honeycombs. *Int. J. Mech. Sci.* **30**, 273–92.

Klintworth, J. W. and Stronge, W. J. (1989) Plane punch indentation of a ductile honeycomb. *Int. J. Mech. Sci.* **31**, 359–78.

Kraynik, A. M., Reinelt, D. A. and Princen, H. M. (1991) The nonlinear elastic behavior of polydisperse hexagonal foams and concentrated emulsions. *J. Rheol.* **35**, 1235–53.

Lakes, R. S. (1991) Deformation mechanisms in negative Poisson's ratio materials: structural aspects. *J. Mat. Sci.* **26**, 2287–92.

McClintock, F. A. and Argon, A. S. (1966) *Mechanical Behaviour of Materials*. Addison Wesley, Reading, Mass., Section 10.2, p. 360.

McFarland, R. K. (1963) *AIAA Journal*, **1**, 1380.

Merchant, W. (1954) The failure load of rigid jointed frameworks as influenced by stability. *Struct. Eng.* **32**, 185.

Milton, G. (1992) Composite materials with Poisson's ratios close to −1, *J. Mech. Phys. Solids*, **40**, 1105–37.

Papka, S. D. and Kyriakides, S. (1994) In-plane compressive response and crushing of honeycomb. *J. Mech. Phys. Solids* **42**, 1499–532.

Patel, M. R. and Finnie, I. (1970) *J. Mat.*, **5**, 909.

Penzien, J. and Didriksson, T. (1964) *AIAA J.*, **2**, 531.

Poirier, C., Ammi, M., Bideau, D. and Troadec, J. P. (1992) Experimental study of the geometrical effects in localization of deformation. *Phys. Rev. Lett.* **68**, 216–19.

Prakash, O., Bichebois, P., Brechet, Y., Louchet, F. and Embury, J. D. (1996) A note on the deformation behaviour of two-dimensional model cellular structures. Submitted to *Phil. Mag.*

Roark, R. J. and Young, W. C. (1975) *Formulas for Stress and Strain*. McGraw-Hill, London.

Roark, R. J. and Young, W. C. (1976) *Formulas for Stress and Strain*, 5th edn. McGraw-Hill, London, p. 96.

Rothenburg, L., Berlin, A. A. and Bathurst, R. J. (1991) Materials with negative Poisson's ratio, *Nature*. **354**, 470–72.

Silva, M. J., Gibson, L. J. and Hayes, W. C. (1995) The effects of non-periodic microstructure on the elastic properties of two-dimensional cellular solids. *Int. J. Mech. Sci.*, **37**, 1161–77.

Silva, M. J. and Gibson, L. J. (1996) The effects of non-periodic microstructure and defects on the compressive strength of two-dimensional cellular solids. *Int. J. Mech. Sci*, in press.

Sokolnikoff, I. S. (1956) *Mathematical Theory of Elasticity*, 2nd edn. McGraw-Hill, New York, p. 382.

Stauffer, D. and Aharony, A. (1992) *Introduction to Percolation Theory* Second Edition Taylor and Francis. London.

Stronge, W. J. and Shim, V. P.-W. (1988) Microdynamics of crushing in cellular solids. *J. Eng. Mater. Tech.* **110**, 185–90.

Suquet, P. M. (1985) Elements of homogenization for inelastic solid mechanics, in *Homogenization Techniques for Composite Media* 193–225, Eds. E. Sanches-Palencia and A. Zaoui, Springer-Verlag, Berlin.

Timoshenko, S. P. and Gere, J. M. (1961) *Theory of Elastic Stability*, 2nd edn. McGraw-Hill, Tokyo, p. 46.

Timoshenko, S. P. and Goodier, J. N. (1970) *Theory of Elasticity*. 3rd edn. McGraw-Hill, New York.

Timoshenko, S. P. and Young, D. H. (1965) *Theory of structures*. McGraw-Hill, New York.

Torquato, S. (1991) Random heterogeneous media: microstructure and improved bounds on effective properties. *App. Mech. Rev.* **44**, 37–76.

Triantafillou, T. C., Zhang, J., Shercliff, T. L., Gibson, L. J., and Ashby, M. F. (1989) Failure surfaces for cellular materials under multiaxial loads II: Comparison of models with experiment. *Int. J. Mech. Sci.* **31**, 665–78.

Warren, W. E. and Kraynik, A. M. (1987) *Mechanics of Materials*, **6**, 27.

Weibull, W. (1951) Statistical distribution function of wide application. *J. Appl. Mech.* **18**, 293.

Wierzbicki, T. (1983) *Int. J. Impact Engng.*, **1**, 157.

Wojciechowski, K. W. (1989) two-dimensional isotropic system with a negative Poisson ratio. *Phys. Lett. A*, **137**, 60–4.

Wojciechowski, K. W. and Branka, A. C. (1989) Negative Poisson ratio in a two-dimensional isotropic solid. *Phys. Rev. A*, **40**, 7222–5.

Zhang, J. and Ashby, M. F. (1992a) The out-of-plane properties of honeycombs. *Int. J. Mech. Sci.* **34**, 475–89.

Zhang, J. and Ashby, M. F. (1992b) Buckling of honeycombs under in-plane biaxial stress. *Int. J. Mech. Sci.* **34**, 491–509.

Chapter 5

The mechanics of foams: basic results

Introduction and synopsis

Almost any solid can be foamed. Techniques now exist for making three-dimensional cellular solids out of polymers, metals, ceramics and even glasses. Man-made foams, manufactured on a large scale, are used for absorbing the energy of impacts (in packaging and crash protection) and in lightweight structures (in the cores of sandwich panels, for instance); their efficient use requires a detailed understanding of their mechanical behaviour. Even when the primary use is not mechanical – when the foam is used for thermal insulation, or flotation, or as a filter, for example – the strength and fracture behaviour are still important. Nature, too, uses cellular materials on a large scale. Often the primary function is mechanical, as with wood (to support the tree) or cancellous bone (to give the animal a light, stiff, frame). In other cases it may not be: the shape of the cells in leaf and stalk, or in cork and sponge, may be dictated by the need to optimize fluid transport or thermal insulation or surface area, but even here the mechanical properties are important because the cells still support the structure. And there is the consuming subject of food. Bread – The Staff of Life – and many other starch-based foods are foams. Foaming with yeast or CO_2 gives the tough, leathery starch a crisp crunchiness which is attractive to bite on and chew (mechanical operations); but it also makes transporting the product more difficult because its crushing strength is reduced.

So there is every reason for wishing to understand the mechanics of foams. The current level of this understanding is set out in this chapter. In it, the properties of a foam are related to its structure and to the properties of the material of which the cell walls are made. The salient structural features of a foam (Chapter 2) are its *relative density*, ρ^*/ρ_s, the degree to which the cells are *open* or *closed*, and their *shape anisotropy ratios* R_{12} and R_{13}. The crucial cell-wall properties (Chapter 3) are the solid density, ρ_s, the Young's modulus, E_s, the yield strength, σ_{ys}, the fracture strength, σ_{fs}, and the creep parameters n_s, $\dot{\epsilon}_{os}$ and σ_{os}.[†] Foam properties are analysed in terms of these parameters and are compared with, and calibrated against, experimental data to give equations suitable for design. Factors such as the strain-rate, the temperature, anisotropy and multiaxial loading all influence the properties, too; they are discussed in Chapter 6.

There is a considerable, but sometimes rather confused, literature on the mechanics of foams; reviews of it can be found in the articles by Suh and Skochdopole (1980), Suh and Webb (1985) and Weaire and Fortes (1994) and in the books edited by Wendle (1976), Hilyard (1982) and Hilyard and Cunningham (1994). Part of the confusion relates to the understanding of mechanisms: some postulated mechanisms (which are then analysed to give equations for moduli or strengths) do not correspond to reality. Part relates to geometric complexity: a full mechanical analysis of any of the three-dimensional cells described in Chapter 2, allowing for all possible cell orientations, is a gory business. But recent work has done much to elucidate the mechanisms (though some mysteries remain); a procedure has been developed to analyse them which combines simple mechanics with scaling ideas; and most of the results of the analysis have been tested by careful experimentation. We now describe the mechanics of foams, using these developments.

5.2　Deformation mechanisms in foams

Figure 5.1 shows schematic *compressive* stress–strain curves for an elastomeric, an elastic–plastic and a brittle foam. Like those for honeycombs, they show *linear elasticity* at low stresses followed by a long *collapse plateau*, truncated by a regime of *densification* in which the stress rises steeply. Careful observations using optical and scanning microscopy, some of which are shown in Fig. 5.2, strongly suggest that the mechanism of deformation associated with each regime parallels that for honeycombs.

Linear elasticity is controlled by *cell wall bending* and, if the cells are closed, by *cell face stretching*; Young's modulus, E^*, is the initial slope of the stress–strain curve. When loading is compressive the plateau is associated with *collapse* of

---

[†]Throughout this book symbols with a subscript 's' refer to properties of the solid while those with a superscripted asterisk '*' refer to the foam properties.

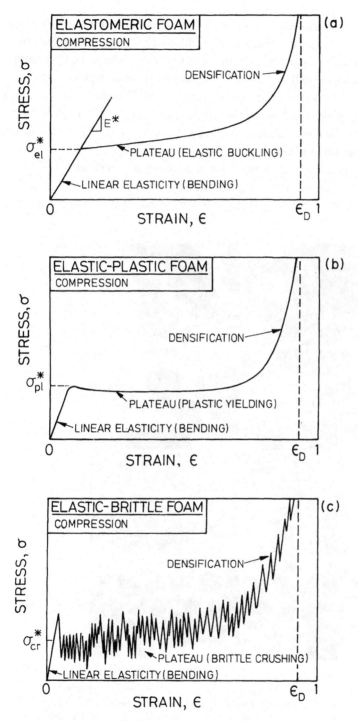

Figure 5.1. Schematic compressive stress–strain curves for foams, showing the three regimes of linear elasticity, collapse and densification: (a) an elastomeric foam; (b) an elastic-plastic foam; (c) an elastic-brittle foam.

the cells – by elastic buckling in elastomeric foams (rubbers, for example); by the formation of *plastic hinges* in a foam which yields (such as a metal); and by *brittle crushing* in a brittle foam (such as a ceramic). When the cells have almost completely collapsed opposing cell walls touch and further strain compresses the solid itself, giving the final region of rapidly increasing stress. Families of stress–strain curves for elastic, plastic and brittle foams are shown in Fig. 5.3. Increasing the relative density of the foam increases Young's modulus, raises the plateau stress and reduces the strain at which densification starts.

The *tensile* response of each kind of foam is shown schematically in Fig. 5.4; again, it resembles that for honeycombs. The initial linear elasticity is caused by cell wall bending, plus stretching if the cells are closed. In the elastomeric foam

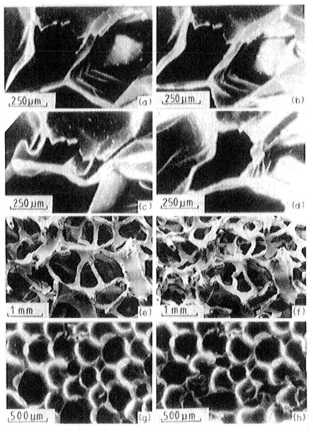

Figure 5.2 Micrographs showing the modes of deformation corresponding to each regime of behaviour in compression: (a-d) elastomeric polyethylene: (a) unloaded, (b) cell-wall bending during loading in the linear-elastic regime, (c) cell-wall buckling during loading in the non-linear elastic regime, (d) recovery of elastic deformation on unloading; (e,f) elastic-plastic copper foam: (e) unloaded, (f) cell-wall yield from loading in the plastic regime; (g, h) elastic-plastic polyurethane foam: (g) unloaded, (h) cell yielding and fracture.

larger strains rotate the cell edges towards the tensile axis, increasing the stiffness of the structure. The cell walls in the plastic foam, too, rotate towards the tensile axis (by plastic bending), giving a yield point followed by a rising stress–strain curve which is ultimately truncated by fracture. In the brittle foam a crack nucleates at a weak cell wall or pre-existing flaw and propagates catastrophically, giving fast brittle fracture. Some foams (rigid polymer foams are examples) are plastic in compression but brittle in tension. This is because of the stress-concentrating effect of a crack, which can cause cell wall failure and fast fracture when loaded in tension, but which is less damaging in compression.

As with honeycombs, the mechanical properties of foams depend on those of the solid cell-wall material of which the foam is made. Direct measurement of cell-wall moduli using tensile or bend tests is difficult due to the small size of the specimens and the variation in their cross-section along the length (Ryan and Williams, 1989; Choi et al., 1990). Both machined and unmachined specimens have been used: machined specimens have a uniform cross-section but may be altered microstructurally by the machining process. Friction and adhesives have been used to grip tensile specimens. Both tensile and bend specimens can give extraneous deformations (either by slip or by local deformation at the loading points) resulting in an underestimate of the true modulus. The most reliable method appears to be the use of ultrasound (Ashman and Rho, 1988; Rho et al., 1993).

The strut strength of open-cell ceramic foams has been estimated from both mechanical tests and fractography. Brezny et al. (1989a) attempted to measure the strut strength of open-cell ceramic foams by threading a thin wire around a single horizontal strut in a block of foam that was clamped to the crosshead of a testing machine, giving three-point bending, with ends assumed to be rigidly clamped. In practice, the degree of constraint at the ends depends on the stiffness of the adjacent struts: the failure stress in a rigidly constrained strut is half that in one with no constraint. Difficulties in estimating the size of the central void within the strut led to additional errors of roughly ±50%. In another study, Brezny and Green (1989b) calculated fracture strengths from fractographic analysis of broken cell walls, from which the incipient crack-size was inferred and combined with a known macroscopic fracture toughness (using $\sigma_{fs} = K_{IC}/\sqrt{\pi a}$). The method is, at best, approximate, but is better than nothing.

Superimposed on the deformation of the cell walls is the effect of the fluid contained within the cells. In man-made foams this is usually a gas, though sometimes foams are deliberately saturated with liquids. In nature, liquid-filled foams are common: the cells of leaves and stalks are filled with sap; those of cancellous bone contain marrow. When a closed-cell foam is compressed, the cell fluid is compressed, too, giving an additional restoring force which can be calculated from Boyle's law. If, instead, the cells are open and interconnected, deformation forces the fluid to flow from cell to cell, doing viscous work, and this

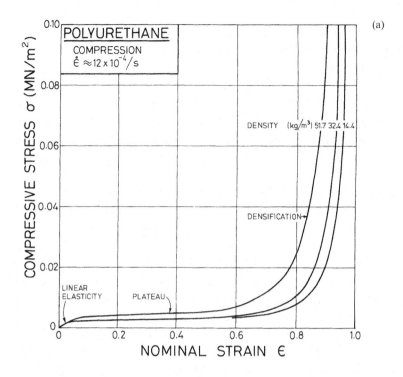

(a)

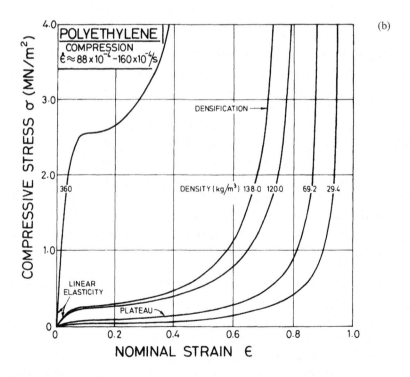

(b)

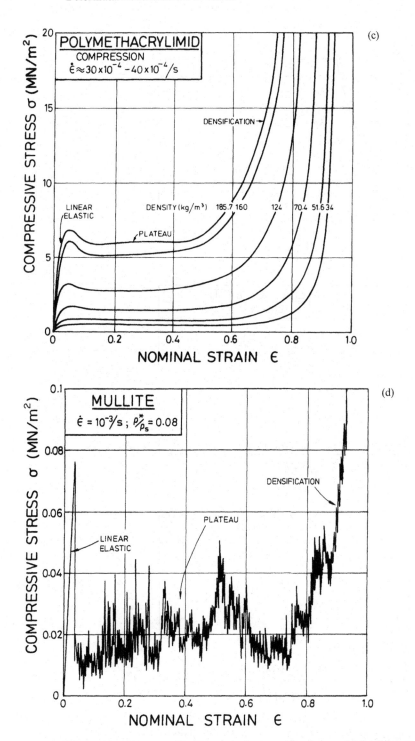

Figure 5.3 Compressive stress-strain curves for foams: (a) flexible polyurethane, (b) polyethylene, (c) polymethacrylimid and (d) mullite. Increasing the relative density of the foam increases the Young's modulus and plateau stress, and decreases the final strain at densification.

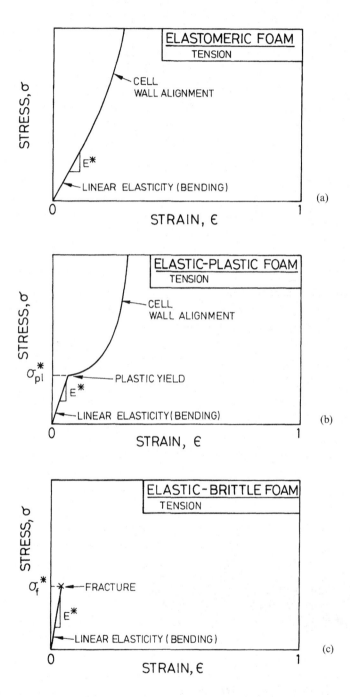

Figure 5.4 Schematic tensile stress-strain curves for foams, showing the regimes of behaviour: (a) elastomeric foam; (b) elastic-plastic foam; (c) elastic-brittle foam.

generates a force which must also be overcome. This, and time-dependent properties (including creep and viscoelasticity) we leave to Chapter 6.

Each of these mechanisms is analysed in the following sections, giving equations which describe the mechanical characteristics of each class of foam.

5.3 Mechanical properties of foams: compression

Most applications of foams cause them to be loaded in compression, the subject of this section. Their behaviour in tension is discussed in Section 5.4.

(a) Linear elasticity

The linear elastic behaviour of a foam is characterized by a set of moduli. Two moduli are needed to describe an isotropic foam; usually they are chosen from the Young's modulus, E^*, the shear modulus, G^*, the bulk modulus, K^*, and the Poisson's ratio, ν^*. More are required for one that is not isotropic: five if the structure is axisymmetric; nine if it is orthotropic. Here we give expressions for E^*, G^*, K^*, and ν^*, for an isotropic foam in terms of the cell-wall modulus, E_s, and the relative density, ρ^*/ρ_s of the foam itself (anisotropy is discussed in Chapter 6). The literature contains numerous attempts to do this (Gent and Thomas, 1959, 1963; Ko, 1965; Chan and Nakamura, 1969; Patel and Finnie, 1970; Lederman, 1971; Menges and Knipschild, 1975; Barma et al., 1978; Gibson and Ashby, 1982; Christensen, 1986; and Warren and Kraynik, 1988). Much of it makes difficult reading, partly because of a confusion about the mechanism of linear-elastic deformation, and partly because authors often select a particular cell shape (connected struts forming a rhombic dodecahedron, for example) and seek a full analysis of its response to stress.

The mechanism of linear elasticity depends on whether the cells are open or closed. At low relative densities, open-cell foams deform primarily by cell-wall bending (Ko, 1965; Patel and Finnie, 1970; Menges and Knipschild, 1975; Gibson and Ashby, 1982; Warren and Kraynik, 1988). As the relative density increases ($\rho^*/\rho_s > 0.1$) the contribution of simple extension or compression of the cell walls becomes more significant (Warren and Kraynik, 1988). Fluid flow through an open-cell foam usually only contributes to the elastic moduli if the fluid has a high viscosity or the strain-rate is exceptionally high; this effect is described in Chapter 6. In closed-cell foams the cell edges both bend and extend or contract, while the membranes which form the cell faces stretch, increasing the contribution of the axial cell-wall stiffness to the elastic moduli (Gent and Thomas, 1959, 1963; Patel and Finnie, 1970; Lederman, 1971; Green, 1985; and Christensen, 1986). If the membranes do not rupture, the compression of the cell fluid which is trapped within the cells also increases their stiffness. Each of the mechanisms which contribute to the linear-elastic response of foams is shown schematically in Fig. 5.5. We analyse the response of both open- and closed-cell foams to load using simple mechanics and avoiding difficult geometry by using scaling laws.

Open cells

At the simplest level an open-cell foam can be modelled as a cubic array of members of length l and square cross-section of side t (Fig. 5.6). Adjoining cells are staggered so that their members meet at their midpoints. The cell-shapes in real foams are, of course, more complex than that of Fig. 5.6. But if they deform and fail by the same mechanisms, their properties can be understood using dimensional arguments which omit all constants arising from the specific cell geometry. The relative density of the cell, ρ^*/ρ_s, and the second moment of area of a member, I, are related to the dimensions t and l by

$$\frac{\rho^*}{\rho_s} \propto \left(\frac{t}{l}\right)^2 \tag{5.1}$$

and

$$I \propto t^4. \tag{5.2}$$

Young's modulus for the foam is calculated from the linear-elastic deflection of a beam of length l loaded at its midpoint by a load F. Standard beam theory (e.g. Timoshenko and Goodier, 1970) gives this deflection, δ, as proportional to

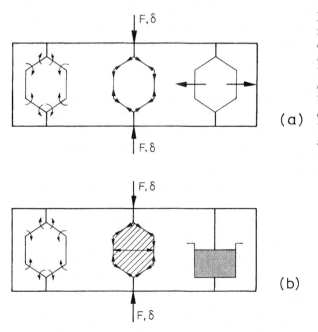

(a)

(b)

Figure 5.5 The mechanisms of deformation in foams. (a) open-cell foams – cell-wall bending + cell-wall axial deformation + fluid flow between cells. (b) closed-cell foams – cell-wall bending + edge contraction and membrane stretching + enclosed gas pressure.

Fl^3/E_sI, where E_s is the Young's modulus for the material of the beam. When a uniaxial stress is applied to the foam so that each cell edge transmits a force F (Fig. 5.7) the edges bend, and the linear-elastic deflection of the structure as a whole is proportional to Fl^3/E_sI. The force, F, is related to the remote compressive stress, σ, by $F \propto \sigma l^2$ and the strain ϵ is related to the displacement, δ, by $\epsilon \propto \delta/l$. It follows immediately that Young's modulus for the foam is given by

$$E^* = \frac{\sigma}{\epsilon} = \frac{C_1 E_s I}{l^4}$$

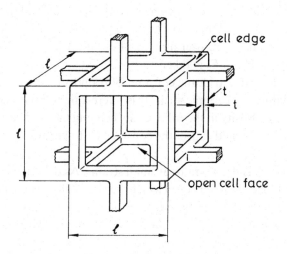

Figure 5.6 A cubic model for an open-cell foam showing the edge length, l, and the edge thickness, t.

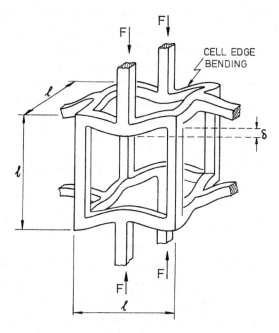

Figure 5.7 Cell edge bending during linear-elastic deformation.

from which (using Eqns. (5.1) and (5.2))

$$\frac{E^*}{E_s} = C_1 \left(\frac{\rho^*}{\rho_s}\right)^2 \quad \text{(open cells)} \tag{5.3}$$

where C_1 includes all of the geometric constants of proportionality. Any equiaxed cell shape leads to this result; the only difference lies in the constant C_1. Data, discussed later, show that $C_1 \approx 1$.

Some ceramic foams are made by infiltrating a slurry of ceramic powder into an open-cell polymer foam and then firing the ceramic. The firing process yields a porous ceramic strut and burns off the polymer, leaving a central void in the strut in place of the polymer. Measurement of the Young's moduli of such alumina foams indicate that the data are well fitted by Eqn. (5.3) with $C_1 = 0.3$ (Hagiwara and Green, 1987). Here, E_s is taken to be the Young's modulus of the fully dense solid making up the strut; the constant of proportionality, C_1, is reduced by the porosity within the ceramic strut. It is interesting to note that a hollow tubular strut with no porosity in the solid resists bending more efficiently than a fully filled strut; then we would expect the constant C_1 to be greater than one.

This, it must be emphasized, is the modulus at small strains. As the elastic distortion increases, the axial load on a cell edge (which we shall call P) increases, too. If P reaches the Euler load, P_{crit}, for the edge, it buckles; that is a problem we analyse in Section 5.3b. But even before it buckles the axial load exerts an additional moment (which, till now, we have neglected) on the bent edge. In compression this beam–column interaction lowers the modulus, E^*; in tension it increases it. So the part of the stress–strain curve that we have called 'linear-elastic' is not truly linear, but is concave downwards; and Young's modulus, if measured at finite strain, is smaller in compression than in tension.

The shear modulus is calculated in a similar way. If a shear stress, τ, is applied to a foam, the cell members again respond by bending. Since the bending deflection, δ, is proportional to $Fl^3/E_s I$, and the overall stress, τ, and strain, γ, are proportional to F/l^2 and δ/l, respectively, we find

$$G^* = \frac{\tau}{\gamma} = \frac{C_2 E_s I}{l^4}$$

from which

$$\frac{G^*}{E_s} = C_2 \left(\frac{\rho^*}{\rho_s}\right)^2 \quad \text{(open cells)} \tag{5.4}$$

for the equiaxed shape. Data shown below suggest that $C_2 \approx 3/8$.

Poisson's ratio, ν^*, is the negative ratio of the lateral to the axial strain. Since both are proportional to a bending deflection per cell length, their ratio is a constant, and for this reason Poisson's ratio is solely a function of cell geometry and is independent of density. For a material which is linear-elastic and isotropic

$$G = \frac{E}{2(1 + \nu)}$$

from which

$$\nu^* = \frac{C_1}{2C_2} - 1 = C_3. \tag{5.5}$$

As with honeycombs, foams with a negative Poisson's ratio can be made by inverting the cell walls to make a structure which is the equivalent, in three dimensions, of the inverted shape shown in Fig. 4.3(b). Here, the effect arises (as pointed out by Gibson *et al.*, 1982) from the change of cell geometry associated with the flexural deformation of the cell walls. The inverted cell shape (Fig. 5.8 and Fig. 5.9(a)) can be produced from conventional foams in a variety of ways (Lakes, 1987; Friis, Lakes and Park, 1988; Lakes, 1991; and Choi and Lakes, 1992). Thermoplastic foams can be triaxially compressed at higher temperatures and then cooled in the deformed state so that the ribs of each cell permanently

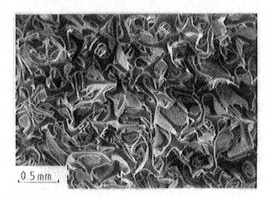

Figure 5.8 A foam with $\nu^* = -0.5$. Poisson's ratio is negative because the structure has re-entrant walls, the three-dimensional equivalent of Fig. 4.3b (Lakes, 1987).

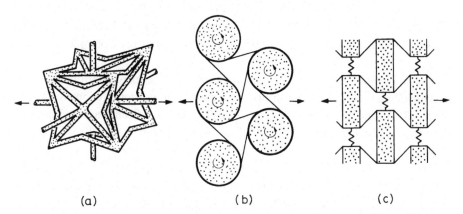

    (a)   (b)   (c)

Figure 5.9 Microstructures giving negative Poisson's ratios. (a) Inverted 're-entrant' cell shape. (b) Solid cylinders or spheres attached to each other by elastic strips. (c) Nodes, connected by tensile springs, constrained by hinged inextensible rods.

penetrate inwards. Thermosetting silicone rubber foams can be triaxially compressed during the curing process. And conventional metal foams can be plastically deformed sequentially in each of three orthogonal directions. The re-entrant cells that result give Poisson's ratios lying between 0 and −0.8.

Negative Poisson's ratios can arise in other ways (Almgren, 1985; Lakes, 1987, 1988, 1991, 1992a,b, 1993; Lakes *et al.*, 1988, Evans, 1989; 1990; Evans *et al.*, 1992; Evans and Caddock, 1989; Caddock and Evans, 1989; Warren, 1990; Wei, 1992). Two are shown in Figs. 5.9(b) and (c). The first (Lakes, 1991) shows solid cylinders or spheres attached to each other by thin elastic strips or wires. When the structure is stretched in the direction of the arrows, the ligaments unwrap from the cylinders or spheres, causing them to rotate; in doing so, the structure expands in the other direction(s). If the strips, when unloaded, have curvature (they conform to the cylinders, for example) the structure is elastic: on removing the load, it returns to its original shape. Microstructures like this have been produced in microporous polytetrafluoroethylene (PTFE), made by rapidly heating and drawing previously sintered PTFE particles (Evans, 1989; Caddock and Evans, 1989; and Evans and Caddock, 1989) and in microporous ultra-high molecular weight polyethylene (UHMPE), made by sintering of a powder followed by extrusion (Alderson and Evans, 1992). The second structure (Fig. 5.9(c)) is an example of a family of open tensile networks of nodes, linked by simple tensile (rather than flexural) springs, constrained by hinged inextensible rods or threads (Almgren, 1985, Evans, 1989). These constraints force a lateral expansion of the structure when the material is stretched axially.

Foams with negative Poisson's ratios present a number of fascinating possibilities, though it is not yet clear that they can be usefully realized. One is the idea that a negative Poisson's ratio matrix could give a continuous, unidirectional fibre composite a higher transverse Young's modulus without reducing its longitudinal modulus (Nkansah, Evans and Hutchinson, 1993). Other potential applications include plate bending, indentation and improved fracture resistance, because all depend on the factor $E/(1 − \nu^2)$ (recall that for isotropic materials $−1 < \nu < 0.5$).

The analysis for E^*, G^* and ν^* contains a number of approximations. The way in which density is calculated double-counts the cell corners, for instance; and the axial and shear displacements of the cell walls have been neglected because they are usually small compared with the bending displacements. These approximations have been examined by Gibson and Ashby (1982), who show that the errors they introduce at large values of ρ^*/ρ_s fortuitously cancel; more exact calculations still give expressions which are well approximated by the simple Eqns. (5.3), (5.4) and (5.5). Experimental results, discussed below, support this finding.

The contributions of both bending and axial cell-wall deformations to the linear-elastic response have also been analysed by Warren and Kraynik (1988) who considered a tetrahedral unit cell randomly oriented with respect to the principal directions of strain. Under uniaxial or shear loading they find that bending is the dominant mechanism of cell-wall deformation at low relative densities; their results for the Young's modulus and the shear modulus reduce to those given by Eqns. (5.3) and (5.4). Under pure hydrostatic loading, bending is suppressed in an ideal isotropic foam: the cell walls simply contract uniaxially, leading to a linear dependence of the bulk modulus on relative density (Christensen, 1986; Warren and Kraynik, 1988):

$$\frac{K^*}{E_s} = \frac{1}{9}\left(\frac{\rho^*}{\rho_s}\right)$$

In practice, any imperfections in the foam (such as non-uniformities in relative density or initially bent cell walls) induce bending of the cell walls, reducing the bulk modulus to a value much less than this.

We have also assumed that the cellular material can be treated as a continuum; that is, that the cell size is small relative to the size of a test specimen or a component. Lakes (1983) shows that classical elasticity, with its continuum assumption, fails to describe the elastic behaviour of foam specimens with a ratio of specimen diameter to cell size of less than 20. The experiments of Brezny and Green (1990) confirm this: they report that the flexural modulus and strength of reticulated vitreous carbon foams decrease when the ratio of specimen diameter to cell size is less than 20.

Comparison with data for open-cell foams

Data for Young's modulus, shear modulus and Poisson's ratio, for foams with a wide range of densities, are shown in Figs. 5.10, 5.11 and 5.12. The figures include data for rigid polymers, elastomers, metals and glasses. The normalizing properties ρ_s and E_s are given in Table 5.1. In these figures, open-cell foams are plotted as open symbols, closed-cell foams as filled symbols. The solid line on each figure shows the equation for open-cell foams. The open-cell data are well fitted by taking $C_1 = 1$, $C_2 = 3/8$ and $C_3 = 0.33$, so that, to an adequate approximation,

$$\frac{E^*}{E_s} \approx \left(\frac{\rho^*}{\rho_s}\right)^2 \tag{5.6a}$$

$$\frac{G^*}{E_s} \approx \frac{3}{8}\left(\frac{\rho^*}{\rho_s}\right)^2 \tag{5.6b}$$

$$\nu^* \approx \frac{1}{3}. \quad \text{(open cells)} \tag{5.6c}$$

The larger scatter in the data for Poisson's ratio (Fig. 5.12) reflects, in part, the variability in ν in foams of different cell geometry (cell shape, as discussed above, can change ν from 0.5 to -0.7) and, in part, the inherent difficulty in measuring this property.

Table 5.1 Cell-wall properties

Material	Symbol	Reference	$\rho_s(\mathrm{Mg/m^3})$	$l(\mathrm{mm})$	$E_s(\mathrm{GN/m^2})$	$\sigma_{ys}(\mathrm{MN/m^2})$	$\sigma_{fs}(\mathrm{MN/m^2})$
Polymers							
Polyethylene	PE	Gibson and Ashby (1982)	0.91 (h)	—	0.20 (h)	—	—
		Maiti et al. (1984a)	1.20 (s)	—	0.70 (s)	—	—
Polymethacrylimid	PMA	Maiti et al. (1984a, b)	1.20 (t)	0.30 (t)	3.60 (t)	120 (s)	120 (t)
Polypropylene copolymer	PPC	Moore et al. (1974)	0.90 (e)	—	1.13 (e)	—	—
Polystyrene, rigid	PS	Matonis (1964)	1.05 (i)	—	1.38 (b)	79 (b)	—
		Chan and Nakamura (1969)	1.05 (f)	—	1.40 (f)	—	—
		Baxter and Jones (1972)	1.02 (j)	—	2.65 (i)	—	—
Polystyrene acrylonitrile	PSA	Moore et al. (1974)	1.07 (l)	—	3.67 (l)	—	—
Polyurethane, flexible	PU(F)	Gibson and Ashby (1982)	1.20 (i)	—	0.0450 (d)	—	—
		Maiti et al. (1984a)	1.10 (s)	—	0.0450 (d)	—	—
Polyurethane, rigid	PU(R)	Traeger (1967)	1.20 (i)	—	1.60 (g)	127 (g)	130 (s)
		Patel and Finnie (1970)	1.20 (i)	—	1.60 (g)	127 (g)	130 (s)
		Phillips and Waterman (1974)	1.20 (i)	—	1.60 (g)	127 (g)	130 (s)
		Fowlkes (1974)	1.20 (i)	0.50 (k)	1.60 (g)	127 (g)	130 (s)
		Wilsea et al. (1975)	1.20 (i)	—	1.60 (g)	127 (g)	130 (s)
		McIntyre and Anderton (1979)	1.20 (i)	—	1.60 (g)	127 (g)	130 (s)
		Gibson and Ashby (1982)	1.20 (i)	—	1.60 (g)	127 (g)	130 (s)
Polyvinylchloride	PVC	Brighton and Meazey (1973)	1.40 (i, m)	—	3.00 (i, m)	49 (i, m)	—
Rubber latex	RL	Gent and Thomas (1959)	1.00 (s)	—	0.0026 (a)	—	—
		Lederman (1971)	1.00 (s)	—	0.0026 (a)	—	—

Metals

Aluminium	Al	2.70 (v)	—	69.00 (n)	52 (n)	—	Thornton and Magee (1975a)
Aluminium, 7% magnesium	Al, 7% Mg	2.70 (v)	—	69.00 (n)	229 (n)	—	Thornton and Magee (1975a)
Aluminium 7075 alloy	Al (7075)	2.70 (v)	—	69.00 (n)	342 (n)	—	Thornton and Magee (1975a)
Zinc (at −196°C)	Zn	7.13 (v)	—	—	—	207 (o)	Thornton and Magee (1975b)

Ceramics

Glass	G	2.51 (c)	—	75.00 (c)	—	65 (s)	Walsh et al. (1965 (s))
		2.50 (p)	1.00 (q)	69.00 (q)	—	65 (s)	Morgan et al. (1981)
		2.50 (p)	—	69.00 (q)	—	65 (s)	Pittsburg-Corning Inc. (1982)
		2.50 (p)	1.25 (r)	69.00 (q)	—	65 (s)	Zwissler and Adams (1983)
Mullite	M	3.20 (s)	—	145.00 (s)	—	4 (s)	Maiti et al. (1984a)
Zirconia	Z	5.60 (v)	—	200.00 (w)	—	75 (w)	Ashby et al. (1986)

Notes: (a) Gent and Thomas (1959); (b) Matonis (1964); (c) Walsh et al. (1965); (d) Lazan (1968); (e) Bonnin et al. (1969); (f) Chan and Nakamura (1969); (g) Patel and Finnie (1970); (h) Billmeyer (1971); (i) Roff and Scott (1971); (j) Baxter and Jones (1972); (k) Fowlkes (1974); (l) Moore et al. (1974); (m) Harper (1975); (n) Thornton and Magee (1975a); (o) Thornton and Magee (1975a); (p) Oliver (1980); (q) Morgan et al. (1981); (r) Zwissler and Adams (1983); (s) Maiti et al. (1984a); (t) Maiti et al. (1984b); (u) Green (1985); (v) Weast (1985); (w) Ashby et al. (1986).

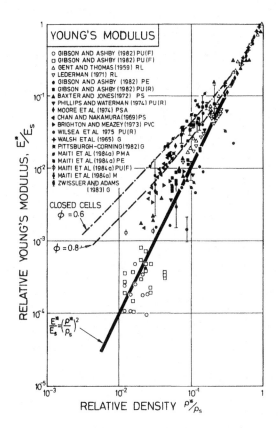

Figure 5.10 Data for the relative Young's modulus of foams, E^*/E_s, plotted against relative density, ρ^*/ρ_s. The solid line represents the theory for open-cell foams. The two dashed lines represent the theory for closed-cell foams (Eqn. (5.13)) with $\phi = 0.8$ and $\phi = 0.6$.

Closed cells

Foams with closed cells are more complicated. When foams are made from liquid components (as many are), surface tension can draw the material into the cell edges, leaving only a thin membrane across the faces of the cell, which ruptures easily. Then, although the foam has cells which are initially closed, its stiffness derives entirely from that of the cell edges, and its moduli are identical with those of an open-cell foam. But not all closed-cell foams are like this. Some polymers and glasses give foams in which a substantial fraction of the solid is contained in the cell faces, which now contribute to the stiffness; and many natural cellular materials (like leaves) have closed cells with thick cell faces. Then the moduli depend on density in a slightly different way.

Consider a closed-cell foam in which a fraction ϕ of the solid is contained in the cell edges, which have a thickness, t_e; the remaining fraction $(1 - \phi)$, is in the faces which have a thickness t_f (Fig. 5.13). The fraction ϕ is given by Eqn. (2.15) and ratios t_e/l and t_f/l are related to the density by Eqns. (2.18) and (2.19).

Young's modulus for the closed-cell foam is the sum of the three contributions shown in Fig. 5.5(b). The first is the contribution of cell-edge bending: almost

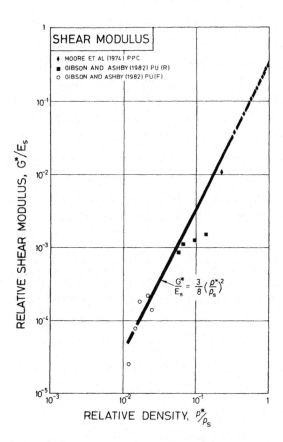

Figure 5.11 Data for the relative shear modulus of foams, G^*/E_s, plotted against relative density, ρ^*/ρ_s. The solid line represents the theory for open-cell foams.

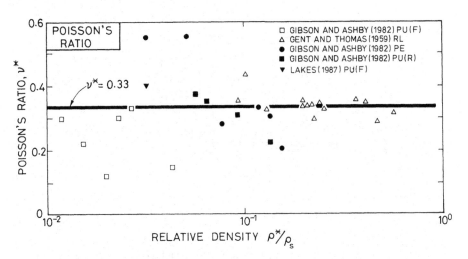

Figure 5.12 Data for Poisson's ratio of foams, ν^* (normal, not re-entrant) plotted against relative density, ρ^*/ρ_s. Poisson's ratio is independent of relative density. The average value of the data points is about $1/3$.

always[†], it is the same as the modulus calculated earlier (Eqn. (5.3)), multiplied by the factor $(0.86\phi)^2$ (Eqn. (2.19b)).

The second contribution is that caused by the compression of the cell fluid (Gent and Thomas, 1963; and Skochdopole and Rubens, 1965). We calculate it by considering a sample of foam of volume, V_0, and relative density, ρ^*/ρ_s, the cells of which contain a gas. If the sample is compressed axially by a strain ϵ, its volume decreases from V_0 to V, where

$$\frac{V}{V_0} = 1 - \epsilon(1 - 2\nu^*). \tag{5.7}$$

The gas occupies the cell space and is excluded from the volume occupied by the solid cell edge and faces, so its volume decreases from V_g^0 to V_g where

$$\frac{V_g}{V_g^0} = \frac{1 - \epsilon(1 - 2\nu^*) - \rho^*/\rho_s}{1 - \rho^*/\rho_s}. \tag{5.8}$$

The contribution to the modulus is calculated from Boyle's law. If the initial gas pressure is p_0 (usually atmospheric pressure) then the pressure, p, after a strain ϵ is given by

$$pV_g = p_0 V_g^0.$$

The pressure which must be overcome by the applied stress is

$$p' = p - p_0.$$

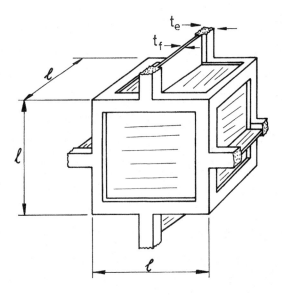

Figure 5.13 A cubic cell model for a closed-cell foam, showing the edge thickness, t_e, and the face thickness, t_f.

[†] The cell faces, too, have a bending stiffness, but because they are usually thin, it is seldom important. For the faces, $I \propto l t_f^3$ and $\rho^*/\rho_s \propto t_f/l$, giving an additional bending stiffness contribution of $E_s(1 - \phi)^3(\rho^*/\rho_s)^3$ (Gibson and Ashby, 1982). Because it is always small, we shall ignore it.

Using the previous equation we find

$$p' = \frac{p_0 \epsilon (1 - 2\nu^*)}{1 - \epsilon(1 - 2\nu^*) - \rho^*/\rho_s}.$$ (5.9)

The contribution to the modulus (taking the limit at small ϵ) is

$$E_g^* = \frac{\mathrm{d}p'}{\mathrm{d}\epsilon} = \frac{p_0(1 - 2\nu^*)}{(1 - \rho^*/\rho_s)}.$$ (5.10)

When p_0 is atmospheric pressure (0.1 MPa), this contribution is small: it contributes to the modulus of closed-cell elastomeric foams but does not change that of other foams much. On the other hand, if p_0 is much larger than atmospheric pressure, or the cell fluid is not air but is a liquid (when the almost-incompressible response of the fluid requires additional cell-wall stretching (Warner and Edwards, 1988) then the contribution of the cell fluid cannot be neglected. And it is important in another regard, discussed further on: it modifies the shape of the elastic collapse plateau. But first we examine the third contribution to the stiffness of closed-cell foams.

This derives from membrane stresses in the cell faces (Patel and Finnie, 1970; Green, 1985; Christensen, 1986). When a closed-cell foam is loaded the bending of the cell edges causes the cell faces to stretch (Fig. 5.14). The stretch direction when the loading is compressive is at 90° to that when the loading is tensile, but the magnitude is similar in both cases. It is calculated as follows. The force F causes the cell edge to deflect by δ (Fig. 5.14). The structure is linearly elastic, so work $\frac{1}{2}F\delta$, is done against the restoring force caused by cell edge bending and face stretching. The first of these is proportional to $\frac{1}{2}S\delta^2$ where S is the stiffness of the cell edge ($S \propto E_s I/l^3$). The second is proportional to $\frac{1}{2}E_s\epsilon^2 V_f$ where ϵ is the strain caused by stretching of a cell face, and V_f is the volume of solid in a cell face ($\epsilon \propto \delta/l$ and $V_f \propto l^2 t_f$, where, temporarily, we distinguish the thickness of the edges, t_e, from that of the faces, t_f). Thus

$$\frac{1}{2}F\delta = \frac{\alpha E_s I \delta^2}{l^3} + \beta E_s \left(\frac{\delta}{l}\right)^2 l^2 t_f.$$

Using $I \propto t_e^4$ and $E^* \propto (F/l^2)/(\delta/l)$ gives

$$\frac{E^*}{E_s} = \alpha' \frac{t_e^4}{l^4} + \beta' \frac{t_f}{l}.$$

Substituting for t_e and t_f from Eqns. (2.19), gives

$$\frac{E^*}{E_s} = C_1 \phi^2 \left(\frac{\rho^*}{\rho_s}\right)^2 + C_1'(1 - \phi)\frac{\rho^*}{\rho_s}$$ (5.11)

(closed cells, including membrane stresses)

where α, β, α', β', C_1 and C_1' are simply constants of proportionality. This equation describes the combined effect of cell-edge bending and cell-face stretching. To it should be added the contribution

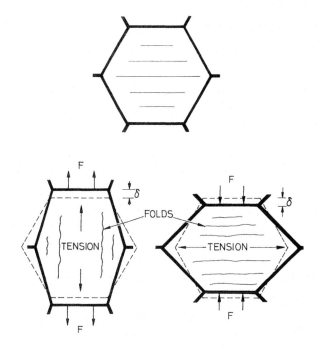

Figure 5.14   Stretching of the faces of a closed-cell foam in compression and tension.

$$\frac{p_0(1 - 2\nu^*)}{E_s(1 - \rho^*/\rho_s)}$$

when the compression of the gas is important. It remains to determine the constants C_1 and C_1'. The limits of $\phi = 1$ (when Eqn. (5.11) must reduce to the first of Eqns. (5.6)) and of $\phi = 0$ (when we expect $E^*/E_s = \rho^*/\rho_s$) suggest $C_1 \approx C_1' \approx 1$. Data, reviewed below, support this. Note that finite element analysis of periodic, closed cell tetrakaidecahedral cells with faces of uniform thickness gives (Simone and Gibson, 1998)

$$\frac{E^*}{E_s} = 0.32\left[\left(\frac{\rho^*}{\rho_s}\right)^2 + \left(\frac{\rho^*}{\rho_s}\right)\right]$$

for relative densities less than 0.2, which is identical to Eqn (5.11), with $C_1 = 0.69, C_1' = 1$ and $\phi = 0.68$.

The shear modulus G^*, of a closed-cell foam, like its Young's modulus, is influenced by cell-face stretching. The argument exactly parallels that just developed for E^*, giving

$$\frac{G^*}{E_s} = C_2\phi^2\left(\frac{\rho^*}{\rho_s}\right)^2 + C_2'(1 - \phi)\frac{\rho^*}{\rho_s}. \tag{5.12}$$

(closed cells, including membrane stresses)

A pure shear produces no volume change, so there is no contribution of gas pressure to the shear modulus.

Poisson's ratio for closed cells, as with open cells, is the ratio of two strains, and thus depends on the details of the cell shape but not on the relative density.

Comparison with data for closed cells

Data for E^* were shown in Fig. 5.10. The solid symbols, which describe closed-cell foams, lie above the solid line for open cells, as expected from Eqn. (5.11). The extent of the deviation (according to the equation) is determined by the fraction $(1 - \phi)$ of solid which is contained in the cell faces. Taking $C_1 = 1$ and either $\phi = 0.8$ or $\phi = 0.6^{\dagger}$ (20% and 40% of the solid is in the faces, respectively) gives the two broken lines on Fig. 5.10. They adequately describe the data.

Data for G^* are less extensive (Fig. 5.11, solid symbols): they are adequately described by the second of Eqns. (5.6). We expect an influence of face stretching when ϕ is less than 0.95, but it is not detectable here.

Data for ν^* for closed-cell foams (Fig. 5.11, solid symbols) like those for open cells, show no significant trend with density, as expected.

If groups of data for the modulus are taken separately, each agrees well with the appropriate equation over the entire density range, but there is some variation in the intercept, C_1, at $\rho^*/\rho_s = 1$. There are several possible explanations for this: the variable fraction, ϕ, of solid in the cell faces; the variable geometry of the foams (which determines the constants C_1, etc.) and the uncertainty in the value of the solid Young's modulus, E_s. The last is the most likely: the value of E_s is rarely known with precision for polymer foams because it depends on the degree of polymer–chain alignment, on chemical changes brought about by the foaming agent and on the gradual ageing and oxidation of the polymer. But despite the scatter, there is every indication that the moduli are adequately described by

$$\frac{E^*}{E_s} \approx \phi^2 \left(\frac{\rho^*}{\rho_s}\right)^2 + (1 - \phi)\frac{\rho^*}{\rho_s} + \frac{p_0(1 - 2\nu^*)}{E_s(1 - \rho^*/\rho_s)} \qquad (5.13a)$$

$$\frac{G^*}{E_s} \approx \frac{3}{8}\left\{\phi^2\left(\frac{\rho^*}{\rho_s}\right)^2 + (1 - \phi)\frac{\rho^*}{\rho_s}\right\} \qquad (5.13b)$$

$$\nu^* \approx \frac{1}{3}. \qquad (5.13c)$$

(closed cells, including membrane stresses and gas pressure)

---

†Measurements of ϕ for rigid polyurethane foams show that it is typically 0.8 (Reitz et al., 1984).

(b) Non-linear elasticity and densification

Linear elasticity, of course, is limited to small strains, typically 5% or less. Elastomeric foams can be stretched or compressed to much larger strains than this. The deformation is still recoverable (and thus elastic), but it is non-linear. In compression (the more common mode of loading) the stress–strain curve shows an extensive plateau at a stress, σ_{el}^*, the *elastic collapse stress* (Fig. 5.1(a)), and it is this stress which is important in the design of cushions, packaging and foam-based systems for damping vibration.

Elastic collapse in foams, like that in honeycombs (Chapter 4), is caused by the elastic buckling of cell walls (Gent and Thomas, 1959; Matonis, 1964; Chan and Nakamura, 1969; Patel and Finnie, 1970; Menges and Knipschild, 1975; Barma *et al.*, 1978; Gibson and Ashby, 1982; Christensen, 1986). The literature suffers a little from confusion between the elastic and the plastic collapse of a foam (plastic collapse is discussed further on), and from models which rely too heavily on specific and rather improbable features of the cell geometry. The development given here avoids these problems.

The elastic collapse stress and the post-collapse behaviour depend on whether the foam has open or closed cells. Open-cell foams, like the polyurethanes of Fig. 5.3(a), collapse at almost constant load, giving a long flat plateau. In closed-cell foams, like the polyethylenes of Fig. 5.3(b), the compression of the gas within the cells, together with the membrane stresses which appear in the cell faces, give a stress–strain curve which rises with strain. We examine the simpler case of open-cell foams first.

Open-cell foams

When an elastomeric open-cell foam is loaded in compression the cell walls first bend, and then buckle (Fig. 5.15). The critical load at which a cell edge of length l, Young's modulus E_s and second moment of area, I, buckles is given by Euler's formula

$$F_{crit} = \frac{n^2 \pi^2 E_s I}{l^2}.$$

(5.14)

The factor n^2 describes the degree of constraint at the ends of the column. If this load is reached in a layer of cells spanning the cross-section they will buckle, initiating the elastic collapse of the foam. The stress at which this happens, σ_{el}^*, is given by

$$\sigma_{el}^* \propto \frac{F_{crit}}{l^2} \propto \frac{E_s I}{l^4}.$$

(5.15)

Using the relationships $I \propto t^4$ and $\rho^*/\rho_s \propto (t/l)^2$ appropriate to open cells gives

$$\frac{\sigma_{el}^*}{E_s} = C_4 \left(\frac{\rho^*}{\rho_s}\right)^2 \qquad \text{(open cells)}$$

(5.16)

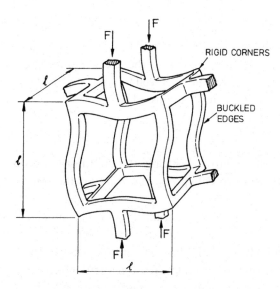

Figure 5.15 Elastic buckling in the cell walls of an open-cell foam.

RIGID CORNERS

BUCKLED EDGES

where C_4 contains all the constants of proportionality.

As with the moduli, refinements are possible. Gibson and Ashby (1982) consider the extra complications of geometry which arise when the density is not small (when cell corners account for a significant part of the volume) and find the slightly modified results:

$$\frac{\sigma_{el}^*}{E_s} = C_4' \left(\frac{\rho^*}{\rho_s}\right)^2 \left(1 + \left(\frac{\rho^*}{\rho_s}\right)^{1/2}\right)^2.$$ (5.17)

(open cells, including density correction)

The correction is insignificant when $\rho^*/\rho_s < 0.3$, but has the effect of making the foam slightly stronger when the density is larger than this.

Comparison with data for open-cell foams

Data for the elastic collapse (or plateau) stress, for elastomeric foams with a wide range of densities, are plotted in Fig. 5.16. Open symbols are for foams with open cells, filled symbols for closed cells. The data are fitted approximately by the simple Eqn. (5.16) with $C_4 = 0.05$ (full line); the fit is rather better with the more refined Eqn. (5.17), with $C_4' = 0.03$ (broken line). In summary

$$\frac{\sigma_{el}^*}{E_s} \approx 0.05 \left(\frac{\rho^*}{\rho_s}\right)^2 \qquad \text{(open cells)} \qquad (5.18a)$$

$$\frac{\sigma_{el}^*}{E_s} \approx 0.03 \left(\frac{\rho^*}{\rho_s}\right)^2 \left(1 + \left(\frac{\rho^*}{\rho_s}\right)^{1/2}\right)^2. \qquad (5.18b)$$

(open cells, including density correction)

The strain at which elastic buckling occurs is given by the elastic-buckling stress divided by the Young's modulus; taking $C_1 = 1$ and $C_4 = 0.05$, we obtain a critical elastic-buckling strain of 0.05, consistent with the stress–strain curves in Fig. 5.3(a) and (b). In the post-collapse regime open-cell foams show an almost horizontal plateau, truncated by densification, discussed below.

Closed-cell foams

When foams with closed cells collapse elastically the fluid in the cells is compressed. This can change both the collapse stress, σ_{el}^*, and the post-collapse behaviour.

Consider the initiation of collapse in a fluid-filled foam. The initial fluid pressure is p_0, and atmospheric pressure is p_{at}. If p_0 is greater than p_{at}, the pressure difference $p_0 - p_{at}$ puts the cell edges and faces in tension. They cannot buckle until the applied stress has overcome both this tension and the buckling load of the cell edges given by Eqn. (5.18) (that of the faces is negligible), so that

$$\frac{\sigma_{el}^*}{E_s} = 0.05 \left(\frac{\rho^*}{\rho_s} \right)^2 + \frac{p_0 - p_{at}}{E_s}. \tag{5.19}$$

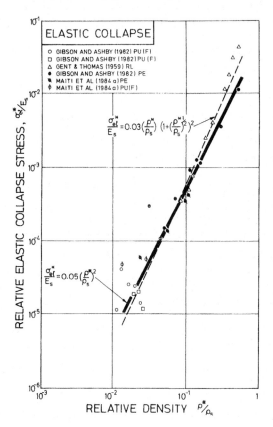

Figure 5.16 Data for the elastic collapse stress for foams, σ_{el}^*, normalized by the solid modulus, E_s. The solid line represents the theory for open-cell foams; the dashed line that for closed-cell foams.

In man-made foams p_0 is usually about equal to p_{at} and the gas has little effect on the initiation of collapse; but in natural cellular materials, which form the leaves and stalks of plants, this is not so. The turgor (or cell) pressure in a leaf is often as large as 6 atmospheres, so that the internal pressure raises the collapse stress considerably – indeed, it largely controls it.

The post-collapse behaviour is modified, too. In an open-cell foam the post-collapse plateau is almost horizontal. In a closed-cell foam it rises. Part of the reason for this is that the enclosed gas is compressed as the cells collapse, creating a restoring pressure which was calculated earlier as Eqn. (5.9)

$$p' = \frac{p_0\epsilon(1 - 2\nu^*)}{1 - \epsilon(1 - 2\nu^*) - \rho^*/\rho_s} \approx \frac{p_0\epsilon}{1 - \epsilon - \rho^*/\rho_s}.$$

The approximation is possible because, in the post-collapse regime, $\nu^* \approx 0$. Then, the post-collapse curve is described by

$$\frac{\sigma^*}{E_s} = 0.05\left(\frac{\rho^*}{\rho_s}\right)^2 + \frac{p_0\epsilon}{E_s(1 - \epsilon - \rho^*/\rho_s)}. \tag{5.20}$$

(closed cells, including gas pressure contribution)

The second term on the right-hand side of the equation causes the 'plateau' stress to rise with strain.

As already said, the post-yield curve of closed-cell foams does not have a flat plateau, but rises with increasing slope. It is not always clear whether the rising curve is caused by the compression of the pore gas or by membrane stresses. The most studied system is that based on polyethylene, which is available as closed-cell foams with a wide range of densities. Zhang and Ashby (1988) investigated the shape of the stress–strain curves of a number of polyethylene foams: their results are described in Chapter 6 in the section on high strain-rates. The raw stress–strain curves (Fig. 6.12) have the typical rising shape. But when the contribution of the gas pressure, $p_0\epsilon/(1 - \epsilon - \rho^*/\rho_s)$, is subtracted out (as it is in Fig. 6.12) the corrected curve has exactly the shape found for open-cell foams: a long, horizontal plateau rising only when densification starts. Their results suggest that the rising part of the curve is primarily caused by gas compression and that membrane stresses are much less important. But they cannot be dismissed altogether.

Comparison with data for closed-cell foams

If Eqn. (5.20) was the whole story, plots of the post-collapse stress, σ^* against $\epsilon/(1 - \epsilon - \rho^*/\rho_s)$ should give straight lines with a slope p_0 and an intercept of σ_{el}^*. Figure 5.17 shows such a plot for a different series of closed-cell polyethylene foams (Maiti et al., 1984a). The lines are straight, and their intercepts are equal to σ_{el}^*. But the slopes, which should all be equal to about p_0 (that is, roughly 0.1 MN/m^2), differ. The lower two are close to the expected value, but the slope

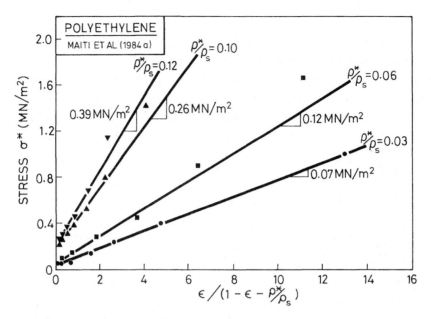

Figure 5.17 The plateau stress, σ^*, plotted against the remaining gas volume, $\epsilon/(1 - \epsilon - \rho^*\rho_s)$, for closed-cell polyethylene.

clearly increases with density. We conclude that compression of the cell gas is an important contribution to the slope of the post-collapse stress–strain curve, and for low-density foams it is the dominant contribution. But as the density increases the faces carry load in some additional way. Intuitively, this is hardly surprising: as the edges buckle, the faces must be folded and pulled in such a way that they carry some lateral tensile stresses, roughly as shown in Fig. 5.14 and because they are relatively stiff in tension this will cause the stress–strain curve to rise above the level needed merely to compress the gas. But calculating this is difficult, and it remains an incompletely understood aspect of foam deformation; a semi-empirical formulation for it is given in Section 5.5.

The stress σ^*_{el} for the initiation of collapse is better understood. Figure 5.16 showed data for both open and closed cells. The closed cells (full symbols) are adequately described by Eqn. (5.19) with $p_0 = p_{at}$ (as we expect for man-made foams). In summary,

$$\frac{\sigma^*_{el}}{E_s} \approx 0.05 \left(\frac{\rho^*}{\rho_s}\right)^2 + \frac{(p_0 - p_{at})}{E_s} \qquad \text{(closed cells)} \qquad (5.21\mathrm{a})$$

or

$$\frac{\sigma^*_{el}}{E_s} \approx 0.03 \left(\frac{\rho^*}{\rho_s}\right)^2 \left(1 + \left(\frac{\rho^*}{\rho_s}\right)^{1/2}\right)^2 + \frac{(p_0 - p_{at})}{E_s}. \qquad (5.21\mathrm{b})$$

(closed cells, including density correction and gas pressure contribution)
The post-collapse behaviour is truncated by densification, which we describe next.

Densification

At large compressive strains the opposing walls of the cells crush together and the cell wall material itself is compressed. When this happens the stress–strain curve rises steeply, tending to a slope of E_s (though this is so much larger than E^* that the stress–strain curve looks vertical) at a limiting strain of ϵ_D (Fig. 5.1(a)). One might expect that this limiting strain would simply be equal to the porosity $(1 - \rho^*/\rho_s)$, because this is the strain at which all the pore space has been squeezed out. In reality the cell walls jam together at a rather smaller strain than this. Experimental data for ϵ_D are plotted in Fig. 5.18. They are well described by

$$\epsilon_D = 1 - 1.4\left(\frac{\rho^*}{\rho_s}\right).$$

(5.22)

(c) Plastic collapse and densification

Foams made from materials which have a plastic yield point (rigid polymers, or metals, for instance) *collapse plastically* when loaded beyond the linear-elastic regime. Plastic collapse, like elastic buckling, gives a long horizontal plateau to

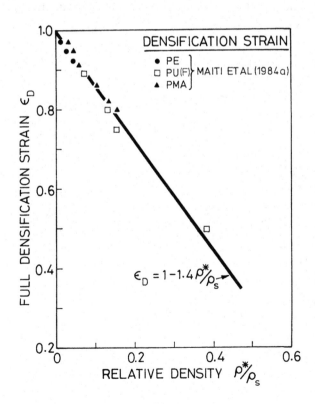

Figure 5.18 The densification strain, ϵ_D, plotted against the relative density, ρ^*/ρ_s.

the stress–strain curve, though the strain is no longer recoverable. Again, like elastic buckling, the failure is localized in a band transverse to the loading direction which propagates throughout the foam with increasing strain (Vaz and Fortes, 1993). By analogy with ductile honeycombs, we expect that this localization is associated with a local maximum in the stress–strain curve of a unit cell. The long stress plateau is exploited in foams for crash protection and energy-absorbing systems. The use of foams for energy absorption is described in Chapter 8; a case study in the selection of foams for a crash helmet is described in Chapter 13.

Plastic collapse in an open-cell foam occurs when the moment exerted on the cell walls exceeds the fully plastic moment (Thornton and Magee, 1975a,b; Gibson and Ashby, 1982) creating plastic hinges like those shown in Fig. 5.19. As always, closed-cell foams are more complicated; in them, the plastic-collapse load may be affected by the stretching as well as the bending of the cell walls, and by the presence of a fluid within the cells. We examine the simpler case of open cells first.

Open-cell foams

Plastic collapse occurs when the moment exerted by the force F (Fig. 5.19) exceeds the fully plastic moment of the cell edges. For a beam with a square section of side t, this moment is

$$M_\mathrm{p} = \frac{1}{4}\sigma_\mathrm{ys}t^3 \tag{5.23}$$

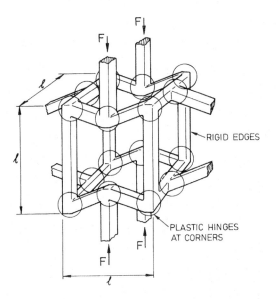

Figure 5.19   The formation of plastic hinges in an open-cell foam.

where σ_{ys} is the yield strength of the cell wall material. If the force F has a component normal to the cell edge (of length l), the maximum bending moment is proportional to Fl. The stress on the foam, as before, is proportional to F/l^2. Combining the results gives the plastic-collapse strength of the foam:

$$\sigma_{pl}^* \propto \frac{M_p}{l^3}.$$

Using the equation $\rho^*/\rho_s \propto (t/l)^2$ for open cells we obtain

$$\frac{\sigma_{pl}^*}{\sigma_{ys}} = C_5 \left(\frac{\rho^*}{\rho_s}\right)^{3/2} \qquad \text{(open cells)} \qquad (5.24)$$

where, as always, the constant C_5 contains all the constants of proportionality. Refinements are again possible but have very little effect on the result. At higher densities the dimensions of the cell corners must be subtracted from the length of the beam, and corrections must be made to the equation for the density. When this is done (Gibson and Ashby, 1982) the last equation becomes

$$\frac{\sigma_{pl}^*}{\sigma_{ys}} = C_5' \left(\frac{\rho^*}{\rho_s}\right)^{3/2} \left(1 + \left(\frac{\rho^*}{\rho_s}\right)^{1/2}\right). \qquad (5.25)$$

(open cells, including density correction)

But as Fig. 5.20 shows, the correction is insignificant.

Two limits on this behaviour are worth noting. At large relative densities ($\rho^*/\rho_s > 0.3$) the beam-bending concept breaks down. The cell walls are now so short and squat that they yield axially (in compression or tension) before they bend. Above this density the material is better thought of as a solid with holes in it, not as a foam. At the other extreme – that of very low relative density – it is possible for elastic collapse to precede plastic collapse, even though the subsequent large-strain behaviour (and thus the plateau stress) may be determined by plasticity. This will happen when

$$\sigma_{el}^* < \sigma_{pl}^*$$

that is (using Eqns. (5.18a) and (5.24), and, anticipating the next section, taking $C_5 = 0.3$), when

$$0.05 E_s \left(\frac{\rho^*}{\rho_s}\right)^2 < 0.3\sigma_{ys} \left(\frac{\rho^*}{\rho_s}\right)^{3/2}$$

or

$$\frac{\rho^*}{\rho_s} < 36 \left(\frac{\sigma_{ys}}{E_s}\right)^2. \qquad (5.26)$$

For elastomers this criterion is always satisfied. For 'rigid' polymers, $\sigma_{ys}/E_s \approx 1/30$, with the result that, for densities below about 0.04, elastic collapse precedes (and thus triggers) plastic collapse. For metals, $\sigma_{ys}/E_s \approx 1/300$, so that plastic collapse dominates at all densities.

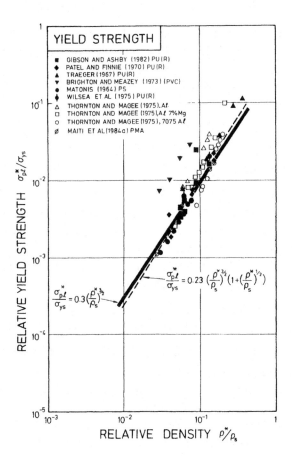

Figure 5.20 Data for the yield strength of foams, normalized by the yield strength of the solid, $\sigma^*_{pl}/\sigma_{ys}$, plotted against relative density, ρ^*/ρ_s. The solid line represents the theory for open-cell foams; the dashed line represents that for open-cell foams, including the density correction.

Comparison with data for open cells

Data for the plastic-collapse strength and for the plastic indentation of foams (discussed further on) are plotted in Fig. 5.20. Open symbols are for open cells; full symbols are for closed cells. They are adequately described by Eqn. (5.24) with $C_5 = 0.3$ (full line, Fig. 5.20); Eqn. (5.25) with $C'_5 = 0.23$ does marginally better (broken line). Each set of data follows the theory well, but with differing values for C_5, probably because of the errors in the value taken for yield strength of the solid, which is used as a normalizing parameter. In summary:

$$\frac{\sigma^*_{pl}}{\sigma_{ys}} \approx 0.3 \left(\frac{\rho^*}{\rho_s}\right)^{3/2} \qquad \text{(open cells)} \qquad (5.27a)$$

or

$$\frac{\sigma^*_{pl}}{\sigma_{ys}} \approx 0.23 \left(\frac{\rho^*}{\rho_s}\right)^{3/2} \left(1 + \left(\frac{\rho^*}{\rho_s}\right)^{1/2}\right). \qquad \text{(open cells,} \qquad (5.27b)$$

including density correction)

Closed-cell foams

Closed cells have membranes spanning their faces. Plastic collapse causes the membranes to crumple in the compression direction; being thin, the force required to crumple them is small. But at right angles to this direction the membranes are stretched, and the plastic work required to extend them contributes significantly to the yield strength of the foam. This is assessed by calculating the work done in an increment of deformation and equating it to the plastic dissipation in bending and stretching the cell walls.

Consider the collapse of a cell like that shown in Fig. 5.21. The cell edges and faces have thicknesses t_e and t_f, respectively; they are related to the fraction of solid in the edges, ϕ, by the equations derived earlier (Eqns. (2.19)). A compressive plastic displacement δ of one cell allows the applied force F to do work $F\delta$. The angle of rotation at the four plastic hinges is then proportional to δ/l, and the plastic work done at these hinges is proportional to $M_p\delta/l$. The cell face is stretched by a distance which is again proportional to δ, doing work which scales as $\sigma_{ys}\delta t_f l$. Equating these gives

$$F\delta = \alpha M_p \frac{\delta}{l} + \beta\sigma_{ys}\delta t_f l$$

where α and β are constants. Replacing F by σl^2, and M_p by $\sigma_{ys}t_e^3/4$ (Eqn. (5.23) with $t = t_e$) gives

$$\frac{\sigma_{pl}^*}{\sigma_{ys}} = \frac{\alpha}{4}\left(\frac{t_e}{l}\right)^3 + \beta\left(\frac{t_f}{l}\right).$$

Finally, using Eqns. (2.19) we obtain

$$\frac{\sigma_{pl}^*}{\sigma_{ys}} = C_5\left(\phi\frac{\rho^*}{\rho_s}\right)^{3/2} + C_5''(1-\phi)\left(\frac{\rho^*}{\rho_s}\right) \tag{5.28}$$

(closed cells, including membrane stresses)

where C_5 and C_5'' contain the earlier constants. In the limit of $\phi = 1$ (open cells), we already know that $C_5 = 0.3$; in the other limit of $\phi = 0$, finite element analysis

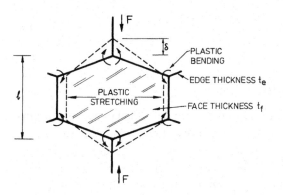

Figure 5.21 Plastic stretching of the cell faces of a closed-cell foam.

PLASTIC BENDING

EDGE THICKNESS t_e

PLASTIC STRETCHING

FACE THICKNESS t_f

of tetrakaidecahedral cells suggests $C_5'' = 0.44$ for $\rho^*/\rho_s < 0.2$. The density correction factor $(1 + (\rho^*/\rho_s)^{1/2})$ can be applied to the first term if desired (see 'open cells', above), but its influence is small.

This equation describes the combined effect of the plastic bending of cell edges and plastic stretching of their faces. To it must be added the contribution of the fluid contained in the cells. If the cell fluid is a gas at atmospheric pressure the contribution is usually small (because $\sigma_{ys} \gg p_{at}$). But if the cells contain fluid at a pressure p_0 much larger than atmospheric, this pressure must be added to the right-hand side of Eqn. (5.28).

Comparison with data for closed cells

The limited data for the plastic collapse of closed-cell foams are plotted in Fig. 5.20 as solid symbols. The scatter is considerable, due largely to uncertain values of σ_{ys}. The data are adequately described by Eqn. (5.27a); there is no evidence here for the influence of membrane stresses or of cell fluid. This may be because the cell faces rupture before full plastic collapse: then the closed-cell foam will behave like one with open faces. Or it may be because, in these foams, $\sigma \approx 1$, and the contribution of the faces, though present, is small. It should not, in our judgement, be ignored even though direct evidence for it is lacking. The best current description of the collapse strength of closed-cell, plastic foams is then

$$\frac{\sigma_{pl}^*}{\sigma_{ys}} \approx 0.3 \left(\phi \frac{\rho^*}{\rho_s} \right)^{3/2} + 0.4(1 - \phi) \frac{\rho^*}{\rho_s} + \frac{p_0 - p_{at}}{\sigma_{ys}}. \tag{5.29}$$

(closed cells, including membrane stresses and gas pressure contribution)

Densification

As with elastic collapse, large plastic strains in compression cause the cell walls to crush together, and make the stress–strain curves rise steeply to a limiting strain ϵ_D. Figure 5.18 contains data for both elastic and plastic foams. Both follow the same straight line, well described by

$$\epsilon_D = 1 - 1.4 \left(\frac{\rho^*}{\rho_s} \right). \tag{5.30}$$

(d) Plastic indentation

Unlike dense solids, which are incompressible when deformed plastically, foams change their volume when compressed. The cells of the foam collapse as the foam is squeezed, so that axial compression produces very little lateral spreading once collapse has begun (the ratio of the lateral spreading to the axial compression is, typically, 0.04; Shaw and Sata, 1966; Rinde, 1970; Gibson and Ashby,

1982). Because of this the indentation hardness of a foam is less than that of a dense solid with the same yield strength. More general multiaxial stress effects are discussed further in Chapter 6; but since the indentation hardness is an important design parameter (particularly in packaging applications) we include it here.

When a foam is compressed by a flat punch the cells beneath the punch collapse in the direction of punching, but expand sideways hardly at all. Because of this the material in the plastic zone is not constrained by its surroundings in the way that the zone in a dense solid would be. An analysis of this problem (Wilsea *et al.*, 1975) shows that, for relative densities less than about 0.3, the indentation pressure, or 'hardness', H, of the foam is simply

$$H = \sigma_{pl}^* \tag{5.31}$$

(instead of the value $H = 3\sigma_{ys}$ which characterizes a dense solid). The result is obvious: the lack of sideways expansion makes the indentation test exactly like a compression test. As the effective Poisson's ratio (the negative ratio of the lateral strain to the axial compressive strain) increases, the hardness increases from σ_{pl}^* to $3\sigma_{pl}^*$ as shown in Fig. 5.22 (Shaw and Sata, 1966).

Comparison with data for indentation hardness

Two experimental studies confirm Eqn. (5.31) (Shaw and Sata, 1966; Wilsea *et al.*, 1975). The data for the second study are included in Fig. 5.20, and conform well with the simple compression data.

(e) The brittle crushing strength and densification

Brittle foams (ceramics, glasses and some brittle polymers are examples) collapse by yet another mechanism: brittle crushing (Rusch, 1970; Morgan *et al.*, 1981;

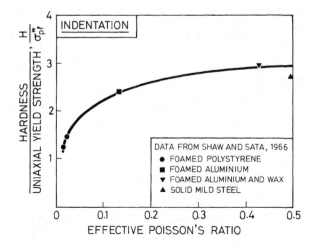

Figure 5.22 The ratio of the indentation hardness to the uniaxial yield strength as a function of the effective Poisson's ratio during plastic yielding. For foams with an effective Poisson's ratio close to zero, multiaxial stresses are not produced by indentation; as a result, H/σ_{pl}^* is about 1. In fully dense solids the effective Poisson's ratio during plastic yielding is 0.5; indentation produces multiaxial stresses which increase the hardness to $H = 3\sigma_{ys}$.

Ashby, 1983; Maiti *et al.*, 1984a; Kurauchi *et al.*, 1984). The low crushing strength of refractory brick limits the loads which can be applied to it, and the low crushing strength of glass and polymer foams can be a problem when they are used as insulation which must also support load.

Ceramic foams are often made by coating an open-cell polymer foam with a ceramic slurry and then firing the material to give an open-cell ceramic foam. When they are, the polymer pre-form burns off leaving a central void in the ceramic strut (Lange and Miller, 1987). Here we first model the brittle-crushing strength of open-cell ceramic foams with solid cell walls; we then modify the analysis to account for struts with a central tubular void.

Open cells

Let the modulus of rupture[†] of the cell-wall material be σ_{fs}. A cell wall will then fail (Fig. 5.23) when the moment acting on it exceeds

$$M_{\mathrm{f}} = \tfrac{1}{6}\sigma_{\mathrm{fs}}t^{3}. \qquad (5.32)$$

As before, a force F acting with a component normal to the wall of length l, exerts a moment which is proportional to Fl. The stress on the foam, as before, is proportional to F/l^{2}. Combining these results gives the brittle collapse stress σ_{cr}^{*} as

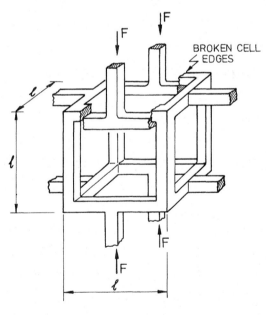

BROKEN CELL
EDGES

Figure 5.23 Cell-wall fracture during crushing of a brittle open-cell foam.

[†]The modulus of rupture of an elastic beam, loaded in bending, is the maximum surface tensile stress in the beam at the moment of failure. This maximum stress is related to the moment by Eqn. (5.32). The modulus of rupture for a brittle solid is close to, but usually a little larger than, the fracture stress in simple tension. It is, of course, related to the fracture toughness of the cell-wall material, K_{ICs}, and the size of defects within the cell wall, a_{s} by $\sigma_{\mathrm{fs}} = (YK_{\mathrm{ICs}})/(\sqrt{\pi a_{\mathrm{s}}})$.

$$\sigma_{cr}^* \propto \frac{M_f}{l^3}$$

from which (using $\rho^*/\rho_s = (t/l)^2$),

$$\frac{\sigma_{cr}^*}{\sigma_{fs}} = C_6 \left(\frac{\rho^*}{\rho_s} \right)^{3/2}. \qquad \text{(open cells)} \qquad (5.33)$$

The analysis can be modified for open-cell foams with a box-like central tubular void of size t_i within the struts of thickness, t (Fig. 5.24). Assuming that the cross-sectional shape of the void and the strut are similar, the relative density of the foam is:

$$\frac{\rho^*}{\rho_s} \propto \left(\frac{t}{l} \right)^2 \left[1 - \left(\frac{t_i}{t} \right)^2 \right]$$

and the moment of inertia is:

$$I \propto (t^4 - t_i^4).$$

Following the above analysis we find the brittle-crushing strength to be:

$$\sigma_{cr}^* = C_6 \sigma_{fs} (\rho^*/\rho_s)^{3/2} \frac{1 + (t_i/t)^2}{\sqrt{1 - (t_i/t)^2}} \qquad (5.34)$$

As t_i/t increases, the foam becomes stronger. In practice, the process which gives rise to hollow tubular struts also gives porous cell walls, reducing the modulus of rupture of the cell-wall material, σ_{fs}, below that of the fully dense material.

The brittle-crushing strength (Eqns. (5.33) and (5.34)) has been found assuming that the modulus of rupture of the strut is constant. In practice, it varies for brittle materials, following a Weibull distribution (as described in Chapter 4).

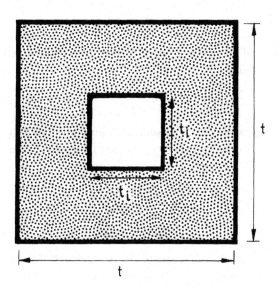

Figure 5.24 Cross-section of a strut with a box-like central tubular void.

Measurement of the modulus of rupture of the cell walls of brittle foams, by direct mechanical testing and by fractography, reveals that the Weibull modulus is low: between 1 and 3 for alumina foams (Brezny, *et al.*, 1989a; Brezny and Green, 1990) and between 4.5 and 6.2 for reticulated vitreous carbon foams (Huang and Gibson, 1991).

Brittle materials that follow the Weibull distribution exhibit a size effect: big specimens fail at lower stresses than small ones simply because it is more probable that they contain a larger pre-existing crack. Two specimens of volume V_1 and V_2, loaded in uniaxial tension, will fail at stresses of σ_1 and σ_2, respectively, according to:

$$\frac{\sigma_1}{\sigma_2} = \left(\frac{V_2}{V_1}\right)^{1/m_w}$$

where m_w is the Weibull modulus. The ratio of the moduli of rupture of the cell wall for two open-cell foams of cell sizes l_1 and l_2 with strut thicknesses t_1 and t_2 are then:

$$\frac{\sigma_{fs,1}}{\sigma_{fs,2}} \propto \left(\frac{l_2 t_2^2}{l_1 t_1^2}\right)^{1/m_w} \propto \left(\frac{l_2}{l_1}\right)^{3/m_w} \left(\frac{t_2}{l_2}\frac{l_1}{t_1}\right)^{2/m_w}.$$

The corresponding ratio of the crushing strengths is:

$$\frac{\sigma_{cr,1}^*}{\sigma_{cr,2}^*} = \left(\frac{l_2}{l_1}\right)^{\frac{3}{m_w}} \left(\frac{t_2}{l_2}\frac{l_1}{t_1}\right)^{\frac{2}{m_w}-3}$$

or

$$\frac{\sigma_{cr,1}^*}{\sigma_{cr,2}^*} = \left(\frac{l_2}{l_1}\right)^{\frac{3}{m_w}} \left(\frac{\rho_2^*}{\rho_1^*}\right)^{\frac{1}{m_w}-\frac{3}{2}}. \tag{5.35}$$

For brittle open-cell foams of the same relative density, the crushing strength decreases with increasing cell size.

Closed cells

The derivation of the crushing strength for closed-cell foams parallels that for the plastic-collapse strength so closely that there is no need to repeat it. We expect that

$$\frac{\sigma_{cr}^*}{\sigma_{fs}} = C_6\left(\phi\frac{\rho^*}{\rho_s}\right)^{3/2} + C_6''(1-\phi)\frac{\rho^*}{\rho_s}. \tag{5.36}$$

(closed cells, including membrane stresses)

As the fraction of solid in the cell faces increases, the dependence of the strength on density should move from a power of $3/2$ to a linear one.

Comparison with data for brittle foams

The data for the crushing of brittle foams are shown in Fig. 5.25. The data are limited but generally support the dependence on relative density raised to the power 3/2. The constant of proportionality is difficult to estimate from the data; instead, we note that the derivation for the brittle-crushing strength exactly parallels that for the plastic-collapse strength with the fracture moment $M_f = \sigma_{fs} t^3/6$ replacing the plastic moment $M_p = \sigma_{ys} t^3/4$. We expect the constant of proportionality for the brittle-crushing strength, then, to be 2/3 of that for the plastic-collapse strength, or $C_6 = 0.2$. There is wide variation in the intercepts of the data sets, suggesting that the values used for the cell-wall modulus of rupture, σ_{fs}, may be inappropriate, probably as a result of the difficulty in estimating the maximum flaw size within the cell wall. Observations of tensile fracture (described later) give additional support for the model which led to this result, which can be regarded as the best current description. In summary:

$$\frac{\sigma_{cr}^*}{\sigma_{fs}} \approx 0.2 \left(\frac{\rho^*}{\rho_s}\right)^{3/2} \qquad \text{(open cells)} \qquad (5.37a)$$

$$\frac{\sigma_{cr}^*}{\sigma_{fs}} = 0.2 \left(\phi\frac{\rho^*}{\rho_s}\right)^{3/2} + (1-\phi)\left(\frac{\rho^*}{\rho_s}\right) \qquad (5.37b)$$

(closed cells, including membrane stresses)

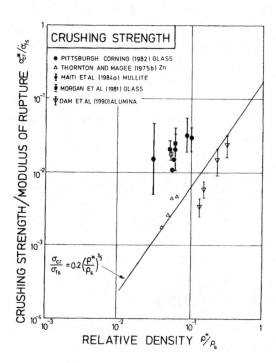

Figure 5.25 Data for the crushing strength of brittle foams. The data for crushing strength, σ_{cr}^*, are normalized with respect to the modulus of rupture of the solid, σ_{fs}, and plotted against relative density. The solid line represents the theory for open-cell foams.

Tests on reticulated vitreous carbon foams (Brezny and Green, 1990) confirm the cell size dependence resulting from the Weibull distribution of cell-wall moduli of rupture (Eqn. 5.35): for foams with constant relative density and with a strut Weibull modulus $m_w = 3$, the compressive crushing strength decreases linearly with increasing cell size.

Both visual observation and acoustic emission data suggest that these foams, which have solid struts, fail catastrophically by strut failure, as the model suggests. Similar observations and measurements on alumina–mullite foams, which have hollow struts with a large number of processing defects suggest that they fail in a more progressive way, perhaps by linking of pre-existing microcracks (Brezny and Green, 1993).

(f) The indentation of brittle foams

Indentation tests on brittle foams with open cells show an unusual and surprising effect. The indentation pressure, instead of being equal to the 'compressive strength' (which we will define more precisely in a moment), is found to depend strongly on the size of the indenter (Ashby et al., 1986; Dam et al., 1990). Figure 5.26 shows a typical force-displacement curve (this one for the indentation of a zirconia foam). The force rises to a peak and then falls back. It fluctuates wildly, sometimes falling almost to zero and sometimes rising to maxima significantly lower than the initial peak. Figure 5.27 summarizes the results of a series of tests with different indenter sizes; it shows the maximum indenter force F_m divided by the indenter area A against the indenter area. (F_m is defined as the mean of the highest 10 maxima, excluding the first peak, which is always the largest and which is plotted separately.) It can be seen that there is a strong 'size effect', and that F_m/A is inversely proportional to $A^{1/2}$. The size effect persists to values of A which are much larger than the area of a single cell. It can be understood as follows.

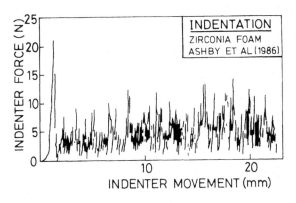

Figure 5.26 Indentation of a brittle zirconia foam.

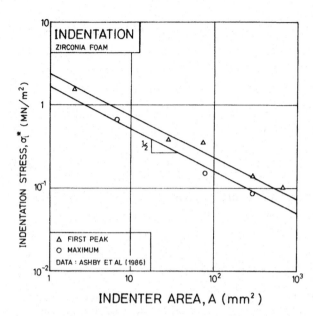

Figure 5.27 Data for the dependence of the indentation stress on the indenter area for a brittle foam.

Imagine an open-cell foam to be idealized as a network of rods of length l and thickness t made from a linear-elastic material of Young's modulus E_s which fractures when the tensile stress locally reaches σ_{fs}. The deformation process involves a sequence of fracture events (Fig. 5.28). In each event one rod in contact with the indenter bends and then breaks. The broken fragment falls off, and drops through the interstices of the foam away from the contact area. This clearing mechanism has a significant effect: if the fragments did not drop clear they would accumulate in the contact region and redistribute the force (and that is what does happen in dense open-cell foams and foams with closed cells). Consider the life history of one contact point (Fig. 5.28). The indenter comes into contact with the tip of one rod-like cell edge and deflects it elastically until it breaks. The force to break it is of the order of $\sigma_{fs}t^3/6l$ (Eqn. 5.32 with $M = Fl$). The deflection δ_c at which the rod breaks is $\sigma_{fs}l^2/E_st$, and if the fracture strain σ_{fs}/E_s is small (as it is in a brittle material), then δ_c is small by comparison with l. When the cell edge breaks off it does so at a distance of order l below the surface, so that the foam surface develops an irregularity of amplitude $l/2$. The important

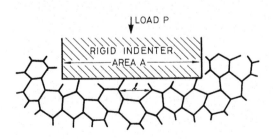

Figure 5.28 Contact of the cell edges with the indenter.

consequence of this is that, at any instant, the probability of a cell edge touching the indenter face is

$$p = \frac{\delta_c}{l}. \tag{5.38}$$

Each cell edge exerts a force of order $\sigma_{fs} t^3/6l$ on the punch. The indenter covers an area A, much larger than l^2, and the number of cells within this area is

$$n = \frac{A}{l^2}. \tag{5.39}$$

Then, over the area A, the probability of a given number of contacts is:

$$P \text{ (0 contacts)} = (1 - p)^n$$
$$P \text{ (1 contact)} \ = np(1 - p)^{n-1}$$
$$P \text{ (2 contacts)} = \tfrac{1}{2}n(n - 1)p^2(1 - p)^{n-2}$$

and so on. This is a binomial distribution. The mean number of active contacts is np, and the standard deviation is $[np(1 - p)]^{1/2}$ or $(np)^{1/2}$ if p is small. An estimate of the *maximum* number of active contacts is

$$\text{(mean)} + m \text{ (standard deviation)} = np + m(np)^{1/2}$$

where m is a multiplier of the order of 3. The corresponding maximum force is the maximum number of active contacts multiplied by the force at each, giving an indentation pressure of

$$\sigma_i^* = \frac{F_m}{A} = C_7 \sigma_{fs} \left(\frac{t}{l}\right)^3 \frac{\delta_c}{l} \left\{ 1 + m \left(\frac{l^3}{A\delta_c}\right)^{1/2} \right\}. \tag{5.40}$$

There is then a size effect, and when $m^2 l^3/A\delta_c$ is larger than 1, the maximum indentation force per unit area is inversely proportional to $A^{1/2}$, as observed. For the zirconia foam of Fig. 5.27 l is about 1 mm, δ_c/l is about 0.001 (corresponding to σ_{fs}/E_s of about 10^{-4}) and t/l about 0.1, so if m is taken as 3, the area A which makes $m^2 l^3/A\delta_c$ equal to 1 is 9000 mm².

If all the cell edges contact the indenter, we measure the true, full crushing strength of the foam: and this can be arranged by putting a compliant facing on to the punch. Then the probability of a contact, δ_c/l, is unity and, in the limit of large A we recover the result for the compressive strength (Eqn. (5.37a))

$$\frac{\sigma_{cr}^*}{\sigma_{fs}} = C_7 \left(\frac{t}{l}\right)^3 \approx C_7 \left(\frac{\rho^*}{\rho_s}\right)^{3/2}$$

from which we expect that $C_7 \approx 0.2$. The final result has a different value of the constant C_7 (arising from the difficulty in estimating σ_{fs}); it agrees well with the data of Fig. 5.27:

$$\frac{\sigma_i^*}{\sigma_{\text{fs}}} = 0.65 \left(\frac{\rho^*}{\rho_s}\right)^{3/2} \frac{\delta_c}{l} \left\{1 + m\left(\frac{l^3}{A\delta_c}\right)^{1/2}\right\}. \tag{5.41}$$

This effect is important when foams are clamped. A rigid clamp or bolt will damage a brittle foam easily (because δ_c/l is small), and the maximum clamping pressure is limited. If a compliant facing is inserted between the clamp and the foam, clamping loads which are as much as 10 times larger can be supported.

(g) Fatigue

Repeated compression damages some flexible foams – a drawback when they are used for cushions or seat padding. Flexible polyurethane foam, particularly, suffers from 'flex-fatigue': a loss of strength and a permanent decrease in volume, caused, it is thought, by changes in the chemical structure (Beals *et al.*, 1965; Kane, 1965; *Plastics Technology*, 1964). Tensile fatigue is a different problem, leading to fracture rather than a slow change of properties; it is discussed in Section 5.4(f).

5.4   Mechanical properties of foams: tension

Structural foams, and foam cores in sandwich structures, are subjected to tension as well as compression. In most cases the tensile behaviour is described by a simple adaptation of results developed in the last section (see Fig. 5.4). Fracture, however, is quite different: tensile stress can cause a fast, brittle fracture whereas compression causes progressive crushing. This section summarizes the basic results for tension.

(a) Linear elasticity

The small strain linear-elastic modulus of a foam in tension is the same as that in compression. The moduli of open-cell foams are determined by cell-edge bending; those of closed-cell foams by edge bending, face stretching and the enclosed gas pressure. They are given, to an adequate approximation, by Eqns. (5.6) and (5.13).

(b) Non-linear elasticity

At strains of more than a few per cent the stiffness of elastomeric foams increases. The buckling which gives non-linear elasticity in compression is not possible in tension. Instead, cell edges which lie, initially, at an angle to the tensile axis, rotate toward this axis, and the bending moment acting on them decreases (Fig. 5.29).

Initially, the elastic response is dominated by cell-edge bending, giving the initial value of E^* described earlier. But as the cell walls rotate the stiffness rises. Both the simple model sketched in Fig. 5.29 and experiments on real foams suggest that as the edges become aligned, stretching, rather than bending, dominates the deformation. The non-linearity introduced into the stress–strain curve by these effects has been analysed using a tetrahedral element to model the struts intersecting at a node in an open-cell foam (Warren and Kraynik, 1991). They find, for uniaxial tension:

$$\frac{\sigma^*}{E_s} = 1.1\left(\frac{\rho^*}{\rho_s}\right)^2 \epsilon + 3.74\left(\frac{\rho^*}{\rho_s}\right)^2 \epsilon^2 + 0.0343\left(\frac{\rho^*}{\rho_s}\right)\epsilon^3. \qquad (5.42)$$

(c) Plastic collapse

A plastic foam yields in tension by the same mechanism, and at essentially the same stress, as in compression (Eqns. (5.27) and (5.29)). The post-yield behaviour, however, is different. In compression the rotation of the cell walls causes the bending moment to remain constant or even increase a little, giving a long, horizontal plateau. In tension the rotation (sketched in Fig. 5.29) reduces the bending moment so that after a strain of about $1/3$ the cell walls are substantially aligned with the tensile axis; then further strain requires the plastic extension of the cell walls themselves. The stress σ^* rises over this strain interval from the value σ^*_{pl} calculated earlier (Eqns. (5.27) and (5.29)) to the value

$$\frac{\sigma^*_A}{\sigma_{ys}} \approx \frac{\rho^*}{\rho_s}. \qquad (5.43)$$

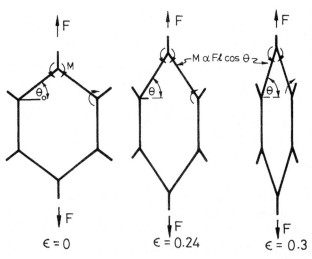

$\epsilon = 0$ $\epsilon = 0.24$ $\epsilon = 0.3$

Figure 5.29 Alignment of the cell edges during tensile loading. The limiting strain is about 0.3.

(d) Brittle fracture

Brittle fracture in tension is quite different from that in compression, and requires the development of new results. In compression the foam crushes progressively; in tension it fails by the propagation of a single crack. Brittle foams (and this includes most 'rigid' polymer foams) are linear-elastic in tension right up to fracture, so tensile failure can be treated by the methods of linear-elastic fracture mechanics. We therefore seek an expression for the fracture toughness of the foam, K_{IC}^*, in terms of the fracture strength of the cell walls, σ_{fs}, and the relative density, ρ^*/ρ_s (Fowlkes, 1974; McIntyre and Anderton, 1979; Morgan *et al.*, 1981; Zwissler and Adams, 1983; Ashby, 1983; Maiti *et al.*, 1984b; and Green, 1985).

Open cells

On the scale of the dimensions of the cells, a crack in a foam extends in a discrete way. Each time the row of cell walls along a crack front fractures, the crack advances by one cell width (Fig. 5.30).

When the foam is loaded the cell walls deform elastically. The load is transmitted through the foam as a set of discrete forces and moments acting on the cell edges. These can be calculated from the stress fields of a crack in the equivalent lin-

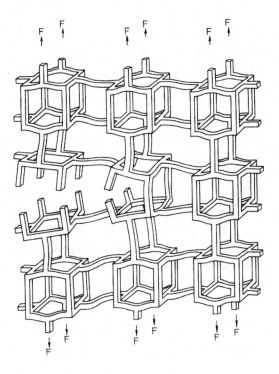

Figure 5.30 Propagation of a crack through a brittle open-cell foam.

ear-elastic continuum, just as we do on a smaller scale in replacing the discrete bonds between atoms by a continuum.

A crack of length $2a$ in an elastic solid, lying normal to a remote tensile stress, σ^∞, creates a singular stress field:

$$\sigma = \frac{CK_I^*}{\sqrt{2\pi r}} = \frac{C\sigma^\infty \sqrt{\pi a}}{\sqrt{2\pi r}} \tag{5.44}$$

at a distance r from its tip; C is a constant, with a value near unity, which copes with specimen shape. Consider the first unbroken cell edge, which we take to be $l/2$ beyond the tip of the crack. It is subjected to a force:

$$F = \int_0^l \frac{CK_I^*}{\sqrt{2\pi r}} l \, \mathrm{d}r = \sqrt{2} C\sigma^\infty l^2 \sqrt{\frac{a}{l}}. \tag{5.45}$$

The edges fail when the bending moment, proportional to Fl, exceeds the fracture moment $M_f = \sigma_{fs} t^3 / 6$ (Eqn. (5.32)), or when the total skin stress exceeds the tensile fracture strength σ_{fs}. The bending moment on the cell wall is proportional to Fl, or $\sigma^\infty l^3 \sqrt{a/l}$, giving the failure condition (Ashby, 1983; Maiti *et al.*, 1984b)

$$\sigma^\infty \propto \left(\frac{t}{l}\right)^3 \frac{\sigma_{fs}}{\sqrt{a/l}}. \tag{5.46}$$

For open-cell foams the relative density ρ^*/ρ_s is related to the cell dimension by $\rho^*/\rho_s \propto (t/l)^2$. Then, using Eqn. (5.44):

$$K_{IC}^* = C_8 \sigma_{fs} \sqrt{\pi l} \left(\frac{\rho^*}{\rho_s}\right)^{3/2} \quad \text{(open cells)} \tag{5.47}$$

where C_8 contains all the constants of proportionality.

In using Eqn. (5.44) we have assumed that the foam can be treated as a continuum. Fracture toughness tests on reticulated vitreous carbon foams with varying ratios of semi-crack length to cell size indicate that this assumption is valid for $a/l > 10$ (Huang and Gibson, 1991).

The analysis can be modified for open-cell foams with a central void of size t_i in the struts of thickness, t (Fig. 5.24). Analogous to the compressive crushing strength, we find:

$$K_{IC}^* = C\sigma_{fs} \sqrt{\pi l} (\rho^*/\rho_s)^{3/2} \frac{1 + (t_i/t)^2}{\sqrt{1 - (t_i/t)^2}}. \tag{5.48}$$

The fracture toughness depends on the modulus of rupture of the solid cell-wall material which, for brittle materials, exhibits a size effect. Using the same argument as in Section 5.3(e) we find (Huang and Gibson, 1991):

$$\frac{K_{IC,1}^*}{K_{IC,2}^*} = \left(\frac{l_1}{l_2}\right)^{\frac{1}{2} - \frac{3}{m_w}} \left(\frac{t_1}{l_1}\frac{l_2}{t_2}\right)^{3 - \frac{2}{m_w}}$$

or

$$\frac{K^*_{IC,1}}{K^*_{IC,2}} = \left(\frac{l_1}{l_2}\right)^{\frac{1}{2}-\frac{3}{m_w}} \left(\frac{\rho^*_1}{\rho^*_2}\right)^{\frac{3}{2}-\frac{1}{m_w}}. \tag{5.49}$$

For materials of constant relative density, the fracture toughness increases with cell size if $m_w > 6$, it remains constant if $m_w = 6$ and it decreases if $m_w < 6$.

Comparison with data for fracture toughness

The fracture toughnesses of brittle polymer foams have been measured, using conventional methods, by Fowlkes (1974), McIntyre and Anderson (1979), Morgan et al. (1981), Zwissler and Adams (1983), Maiti et al. (1984b) and Brezny and Green (1989b). These data are plotted in the way suggested by Eqn. (5.47) in Fig. 5.31. The full line has a slope of $3/2$. The data are consistent with Eqn. (5.47) with the constant of proportionality $C_8 = 0.65$. Brezny and Green's data lie on a slope of $3/2$, but fall somewhat lower than the rest of the data; they suggest a value of C_8 for their material of between 0.13 and 0.23.

Both the fracture toughness and the brittle-crushing strength of brittle foams depend on the modulus of rupture of the cell wall in the same way. Reticulated vitreous carbon foams oxidized in air at 500°C for up to 120 minutes to form oxidation pits on the strut surfaces, reducing σ_{fs}, show a drop in both fracture toughness and crushing strength of about 65% (Brezny and Green, 1991).

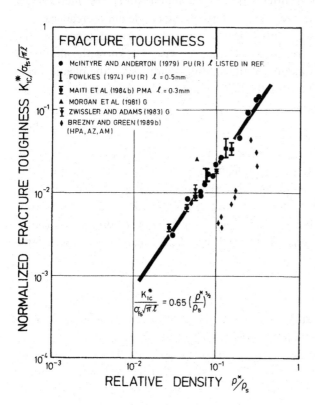

Figure 5.31 Data for the fracture toughness of foams. The fracture toughness of the foam, K^*_{IC}, is normalized with respect to the cell wall modulus of rupture and the cell size, $\sigma_{fs}\sqrt{\pi l}$, and plotted against relative density, ρ^*/ρ_s. The solid line indicates the theory for open-cell foams.

The Weibull distribution of cell-wall moduli of rupture leads to a cell size effect (Eqn. (5.49)). Data for the log of fracture toughness of reticulated vitreous carbon (RVC) foams of constant relative density ($m = 5$) decreases with the log of cell size raised to the power -0.1, consistent with Eqn. (5.49). The fracture toughness of RVC foams with $m = 6$ is independent of cell size, again consistent with Eqn. (5.49) (Huang and Gibson, 1991). Brezny and Green (1990b) also report that the fracture toughness of the same material is independent of cell size, but do not report the Weibull modulus for their material.

The results show that the fracture toughness falls rapidly with decreasing density in foams. Many polymer and ceramic foams have values of K_{IC}^* which are less than $0.1\,\mathrm{MN/m^{3/2}}$, making them exceptionally brittle materials.

(e) Fatigue

The tensile fatigue of foams becomes important when they are used structurally (as cores for sandwich panels, for instance) or when they are subjected to repeated thermal shock (as in the insulation of liquid-gas containers, for example).

When a true solid is subjected to cyclic loads so that, for at least part of the cycle, the solid is in tension, it is found that cracks nucleate and grow inward from its surface. One crack eventually becomes dominant and grows increasingly rapidly, extending by successively larger steps on each cycle until it reaches a critical length and propagates catastrophically. Some engineering components are, initially, perfect and crack-free; then the engineer wishes to know the *number of cycles to failure*, N_f, as a function of the loading conditions – the plastic strain range, $\Delta\epsilon_{pl}$, for instance; it is described empirically by Basquin's law

$$N_f = \left(\frac{\Delta\epsilon}{\Delta\epsilon_f}\right)^{-u} \tag{5.50}$$

where $\Delta\epsilon_f$ is the strain which will cause failure in one half-cycle, and u is a constant – typically, about 2. Other components are flawed from the start: manufacturing defects, or processing damage, has created small cracks of length a, and a cyclic stress induces a stress intensity ΔK_I at each crack, causing it to grow. Then the engineer wishes to know the *crack growth per cycle*, da/dN, as a function of the loading conditions – measured by ΔK, for instance; it is well fitted by the Paris law

$$\frac{da}{dN} = C\Delta K_I^m \tag{5.51}$$

where C and m are constants. Data for both equations are available for solid metals (Hertzberg, 1983) and polymers (Hertzberg and Manson, 1980).

Fatigue in foams is modelled by an extension of the method used earlier to calculate the fracture toughness (Fleck and Parker, 1988). The cell wall just ahead of a crack of length a is flexed repeatedly until it breaks; when it does, the crack advances by one cell diameter. The force on this cell wall was calculated earlier (Eqn. (5.45)): it leads to a stress of order F/t^2 and a corresponding strain, for cyclic loading, of

$$\Delta\epsilon = \frac{\Delta F}{t^2 E^*} = C\frac{\Delta\sigma^\infty}{E^*}\left(\frac{l}{t}\right)^2\sqrt{\frac{a}{l}}.$$

The number of cycles required to break the wall is given by Eqn. (5.50)

$$N_f = \left(\frac{\Delta\epsilon}{\Delta\epsilon_f}\right)^{-u} = \left[\frac{C}{\Delta\epsilon_f}\frac{\Delta K_I}{\sqrt{\pi l}}\frac{1}{E^*}\left(\frac{l}{t}\right)^2\right]^{-u}$$

and the crack advance per cycle is

$$\frac{da}{dN} = \frac{l}{N_f} = l\left(\frac{C}{\Delta\epsilon_f}\frac{\Delta K_I}{\sqrt{\pi l}}\frac{1}{E^*}\left(\frac{l}{t}\right)^2\right)^u$$

or, substituting for E^*,

$$\frac{da}{dN} = C'(\Delta K_I)^u \tag{5.52}$$

where

$$C' = l\left(\frac{C}{\Delta\epsilon_f E_s\sqrt{\pi l}}\left(\frac{\rho_s}{\rho^*}\right)^3\right)^u. \tag{5.53}$$

The model leads directly to the Paris law, and predicts a strong dependence on relative density.

Comparison with fatigue data

Studies of fatigue-crack growth in foams (Fleck and Parker, 1988) only partly support this view. Beam-shaped samples of a number of foams, with a single edge-crack, and with the standard dimensions of three-point bend fracture toughness samples, were subjected to cyclic bending. The crack advance was measured and its rate, da/dN, plotted against ΔK, as shown in Fig. 5.32. The straight lines correspond to the Paris law; that far, the model is substantiated. But Fleck *et al.* find that n has a value of about 10 for rigid polyurethane, 13 for phenolic, and 15 for polyisocyanurate foams, regardless of density (though all had relative densities below 0.2). These exponents are higher than those found in solid polymers, suggesting that the mechanism of fatigue-crack growth in foams may differ from that in true solids. The constant C' depends on density (Fig. 5.33), varying roughly as $(\rho^*/\rho_s)^{3/2}$. An increase in the mean tensile stress σ_{mean} for a given stress amplitude $\Delta\sigma$ causes an increase in crack growth rate, suggesting that it is the maximum K_I, as well as the amplitude ΔK_I, which is important in crack growth. If correct, these conclusions are important for safe design with foams; but there is an obvious need for more detailed studies of the fatigue of foams.

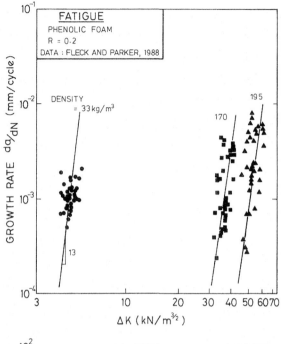

Figure 5.32 Fatigue-crack growth rates for three densities of rigid phenolic foam (data from Fleck and Parker, 1988).

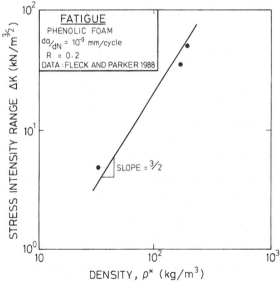

Figure 5.33 The stress intensity amplitude required to give a crack growth rate of 10^{-3} mm/cycle, plotted against density for the three foams of Fig. 5.32.

5.5　Summary of mechanical behaviour of foams: stress–strain maps

We have seen that when an elastomeric foam is compressed, it first deforms in a linear-elastic way; then its cells buckle to give non-linear elasticity; and, finally, the cells collapse completely and the stress rises rapidly as the faces and edges are forced together. A plastic foam behaves in a similar way, except that linear elasticity is now followed by plastic collapse instead of elastic buckling. With brittle foams, progressive crushing can again lead to a plateau which ends when the material is completely crushed. We have seen, too, that each of these processes can be modelled to give equations which relate foam properties to the relative density (ρ^*/ρ_s) and to the properties of the material of which the foam is made.

The relations, summarized in the boxes in the text, suggest a normalization which brings the properties of foams with the same relative densities into coincidence. Then the behaviour of an entire family of foams can be shown as a *stress–strain map* (Ashby, 1983), of which Figs 5.34, 5.35, 5.37 and 5.38 are examples. Each map has axes of normalized compressive stress σ/E_s, and compressive strain, ϵ. It shows the *fields* in which each mode of deformation (linear elasticity by cell-wall bending, elastic collapse by elastic buckling, plastic collapse by the formation of plastic hinges and so forth) is dominant. Superimposed on the fields are *stress–strain contours* for constant (initial) relative density. A map provides a convenient summary of the properties of a family of foams which is particularly helpful in the selection of foams for load-bearing applications. We will illustrate the method by constructing maps for the family of elastomeric foams, and for a family of rigid (plastic) foams.

(a) Elastomeric foams

Figures 5.34 and 5.35 show maps for elastomeric foams. The first shows the experimental stress–strain curves for polyethylene foams; the second shows

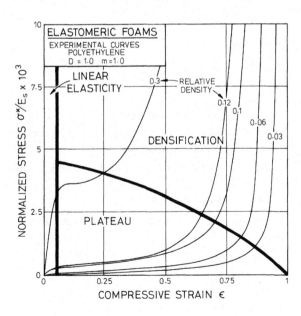

Figure 5.34 A stress–strain map for elastomeric foams. It shows the data for polyethylene (Fig. 5.3b), normalized with respect to the solid cell wall modulus, E_s. The construction of the field boundaries is described in the text.

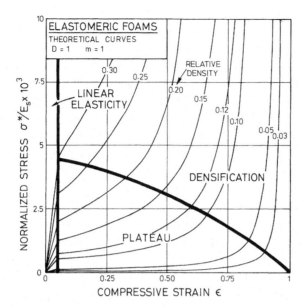

Figure 5.35 A stress–strain map for elastomeric foams, constructed entirely from equations developed in the text.

curves constructed from the equations described below. Both are divided into fields within which a given mechanism of deformation is dominant. The field boundaries (heavy lines) are derived from the equations developed earlier, in the following way.

The first boundary separates the linear-elastic regime from that of elastic buckling (which, of course, is non-linear). The equation of the field boundary is given by setting the linear-elastic stress ($\sigma = E^* \epsilon$) equal to the buckling stress σ_{el}^*. For open cells (Eqns. (5.6a) and (5.18a)) the result is

$$\epsilon = 0.05 \tag{5.54a}$$

or, using the more exact equation (Eqn. (5.18b))

$$\frac{\sigma}{E_s} = \epsilon \left\{ \left(\frac{\epsilon}{0.03}\right)^{1/2} - 1 \right\}^4. \tag{5.54b}$$

The equivalent result for closed cells is found by using Eqns. (5.11) and (5.21) instead; it is very similar in form but contains the fraction of solid in the cell edges, ϕ, and the cell pressure, p_0, as additional parameters.

The post-buckling behaviour has been studied by Gent and Thomas (1959), Rusch (1970), Ashby (1983) and Maiti *et al.* (1984a), all of whom propose equations describing the shape of the stress–strain curve. None are very satisfactory. The best approach is a semi-empirical one. Densification is complete and the stress–strain curve becomes almost vertical when the strain reaches the densification strain, ϵ_D (Eqn. (5.30), Fig. 5.18) where

$$\epsilon_D = 1 - 1.4 \left(\frac{\rho^*}{\rho_s}\right).$$

When data for the post-buckling stress σ^* are plotted against various trial functions of ϵ it is found that the non-linear behaviour is best described in two segments, which can be written as:

$$\frac{\sigma^*}{\sigma^*_{el}} = 1 \quad \text{when} \quad \epsilon \le \epsilon_D\left(1 - \frac{1}{D}\right) + \epsilon^*_{el} \tag{5.55a}$$

$$\frac{\sigma^*}{\sigma^*_{el}} = \frac{1}{D}\left(\frac{\epsilon_D}{\epsilon_D - \epsilon}\right)^m \quad \text{when} \quad \epsilon > \epsilon_D\left(1 - \frac{1}{D}\right) + \epsilon^*_{el} \tag{5.55b}$$

where D and m are constants for a given class of foams. This is best understood by looking at Fig. 5.36 which shows data for two elastomeric foams with σ^*/σ^*_{el} plotted, on a log scale, against $\epsilon_D/(\epsilon_D - \epsilon)$. The first branch describes the plateau; the second, densification. The two join at the point $\epsilon = \epsilon_D(1 - (1/D))$. The constant m is the slope of the line; the constant D is found from its intercept with the strain axis, at which point

$$D = \frac{\epsilon_D}{\epsilon_D - \epsilon}.$$

For the polyethylene foam, $m = 1$ and $D = 1$; for the polyurethane, $m = 1$ and $D = 1.55$.

Once buckling starts, the stress is related to the strain by this equation. Foams with a relative density greater than 0.3 show no real plateau (Fig. 5.3); this suggests, as a criterion for the onset of densification, that the strain be sufficient to compress the foam to an instantaneous relative density of about 0.3. This strain is

$$\epsilon_c = 1 - \frac{1}{0.3}\frac{\rho^*}{\rho_s}. \tag{5.56}$$

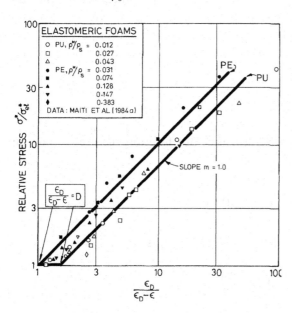

Figure 5.36 The post-buckling stress of elastomeric foams, σ^*, normalized by the elastic collapse stress, σ^*_{el}, plotted against $\epsilon_D/(\epsilon_D - \epsilon)$. The equations of the two lines are described by Eqns. (5.55a) and (5.55b).

Using this criterion with Eqns. (5.30), (5.55b) and (5.18a) and taking $m = 1$ gives the equation of the boundary between buckling and densification:

$$\frac{\sigma^*}{E_s} = \frac{0.0045}{D}(1 - \epsilon_c)(1 + 0.72\epsilon_c). \tag{5.57}$$

Figure 5.34 shows data of Fig. 5.3 for polyethylene foams, replotted on normalized axes, and with field boundaries constructed from Eqns. (5.54a) and (5.57). Figure 5.35 is a mechanism map constructed entirely from the model-based equations. The contours are stress–strain curves for relative densities between 0.01 and 0.3. They show a linear-elastic regime (Eqn. (5.6)) and a plateau corresponding to elastic buckling, truncated by a regime of densification (Eqn. (5.55b)). Within the field of elastic buckling the material exists in two states at almost the same stress: the linear-elastic state and the densified state; it is like the p–V response of an ideal gas in which gas and liquid states can co-exist. It is often found that the foam deforms, in compression, by the formation of densified bands embedded in a matrix which is still in the linear-elastic regime. As the strain is increased the bands thicken at almost constant stress until the entire material has reached the dense state.

(b) Plastic foams

Plastic foams, like the elastic ones, show three regions: linear elasticity, plastic collapse and densification – though now the strain beyond the linear-elastic regime is not recoverable. Figures 5.37 and 5.38 are a pair of maps, one showing experimental stress–strain curves for plastic foams, the other based on the theory alone. Mechanism field boundaries are superimposed on the stress–strain curves.

The boundary between the linear-elastic field and that for plastic collapse is given by equating the linear-elastic stress ($\sigma = E^*\epsilon$) to that for plastic collapse (Eqn. (5.27)) giving

$$\frac{\sigma^*}{E_s} = \left(\frac{0.3\sigma_{ys}}{E_s}\right)^4 \frac{1}{\epsilon^3}. \tag{5.58}$$

The post-yield behaviour is modelled, as before, by fitting the stress–strain curve to the equation

$$\frac{\sigma^*}{\sigma^*_{pl}} = 1 \quad \text{when} \quad \epsilon \leq \epsilon_D\left(1 - \frac{1}{D}\right) \tag{5.59a}$$

$$\frac{\sigma^*}{\sigma^*_{pl}} = \frac{1}{D}\left(\frac{\epsilon_D}{\epsilon_D - \epsilon}\right)^m \quad \text{when} \quad \epsilon > \epsilon_D\left(1 - \frac{1}{D}\right) \tag{5.59b}$$

with ϵ_D given, as before, by Eqn. (5.30). Figure 5.39 shows data for polymethacrylimid foams, plotted (as before) to allow m and D to be determined. The

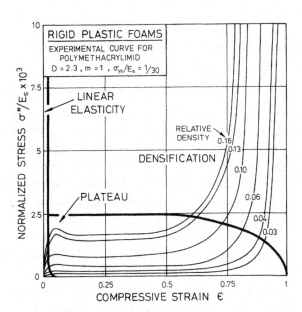

Figure 5.37 A stress–strain map for plastic foams. It shows the data for polymethacrylimid (Fig. 5.3c) normalized with respect to the solid cell wall modulus, E_s. The construction of the field boundaries is described in the text.

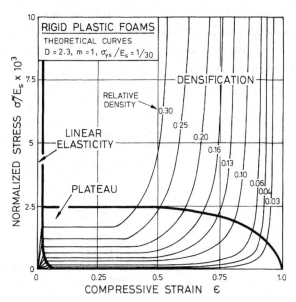

Figure 5.38 A stress-strain map for plastic foams, constructed entirely from the equations developed in the text.

data, which cover a wide range of densities, are approximately described by $m = 1 \pm 0.4$ and $D = 2.3$.

The onset of densification in plastic foams occurs later than in elastomeric ones: the stress–strain curve does not really start to rise until the foam has been compressed to a relative density of about 0.5. We therefore replace Eqn. (5.56) by

$$\epsilon_c = 1 - \frac{1}{0.5}\frac{\rho^*}{\rho_s}.$$

Then, following the same procedure as before with $m = 1$ (but using Eqn. (5.27a) for σ^*_{pl}) the boundary between plastic collapse and densification is given by

$$\frac{\sigma^*}{E_s} = \frac{0.106}{D} \frac{\sigma_{ys}}{E_s}(1 - \epsilon_c)^{1/2}(1 + 2.33\epsilon_c). \tag{5.60}$$

Figure 5.37 shows the data of Fig. 5.3 for rigid plastic foams, replotted on normalized axes, and with field boundaries constructed from Eqns. (5.58) and (5.60), with $\sigma_{ys}/E_s = 1/30$. Figure 5.38 is a stress–strain map constructed entirely from the model-based equations. The contours as before are stress–strain curves for relative densities between 0.01 and 0.3. They show a linear-elastic regime (Eqn. (5.6)) and plateau regime corresponding to plastic collapse, truncated by a regime of densification (Eqn. (5.59)). As before, two states of strain co-exist at almost the same stress, so that complete collapse of part of the structure can occur while the rest is still elastic; the bands of dense material broaden with increasing strain.

The figure shows the overall response of isotropic, plastic foams in compression. It is less general than the map for elastomeric foams because it must be constructed for a particular value of σ_{ys}/E_s (in this instance 0.033, a value typical of many polymers), but it still provides a compact summary of the mechanical properties of a class of foams. The comparison between Figs 5.37 and 5.38 shows the level of precision with which foam properties can be modelled.

(c) Brittle foams

Brittle foams show linear-elastic behaviour to fracture. In compression the foam crushes at constant stress (Eqn. (5.37)) and since the crushing equation has the

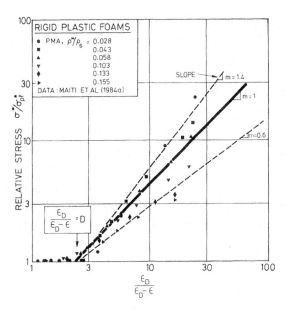

Figure 5.39 The post-yield stress of plastic foams, σ^*, plotted against $\epsilon_D/(\epsilon_D - \epsilon)$. The equations of the two lines are described by Eqns. (5.59a) and (5.59b).

same form as that for plastic collapse, the behaviour will resemble that of Figs. 5.37 and 5.38. If the foam is contained, it will densify at the strain given approximately by Eqn. (5.59), with σ_{ys}/E_s replaced by σ_{fs}/E_s.

5.6 Conclusions

The response of foams to stress is best analysed by combining the beam theory already used to analyse honeycombs (Chapter 4), with scaling laws. Expressions for the moduli, the elastic, plastic and brittle collapse stresses, the fracture toughness and the strain at which densification occurs, derived by examining the way in which the cell edges respond to stress, give a good description of the behaviour of real open-cell foams. The response of closed-cell foams is a little more complicated: the deformation of the cell faces, as well as the edges, and the response of the fluid (usually a gas) within the cells, must be included to give a complete description. Some unusual results are found for the indentation strengths of foams: those which deform plastically are found to have an indentation strength equal to the yield strength, rather than $3\sigma_{pl}^*$; and those which are brittle and have open cells are found to have an indentation strength which depends on the size of the indenter. The results of the analysis can be summarized in stress–strain diagrams which indicate the regimes of stress and strain for which each deformation mechanism is dominant. Refinements to include the influence of strain-rate, of temperature, of multiaxial loading and of anisotropy are given in the next chapter.

References

Alderson, K. L. and Evans, K. E. (1992) *Polymer*, **33**, 4435–8.

Almgren, R. (1985) An isotropic three-dimensional structure with Poisson's ratio $= -1$, *J. Elast.*, **15**, 427–30.

Ashby, M. F. (1983) *Met. Trans.*, **14A**, 1755.

Ashby, M. F., Palmer, A. C., Thouless, M., Goodman, D. J., Howard, M., Hallam, S. D., Murrell, S. A. F., Jones, H., Sanderson, T. J. O. and Ponter, A. R. S. (1986) *18th Annual Offshore Technology Conference*, Houston, Texas, OTC, p. 399.

Ashman, R. B. and Rho, J. Y. (1988) Elastic modulus of trabecular bone material. *J. Biomech.*, **21**, 177–81.

Barma, P., Rhodes, M. B. and Salover, R. (1978) *J. Appl. Phys.*, **49**, 4985.

Baxter, S. and Jones, T. T. (1972) *Plast. Polym.*, **40**, 69.

Beals, B., Dwyer, F. J. and Kaplan, M. A. (1965) *J. Cell. Plastics*, **1**, 32.

Billmeyer, F. W. (1971) *Textbook of Polymer Science*, 2nd edn. Wiley Interscience, New York.

Bonnin, M. J., Dunn, C. M. R. and Turner, S. (1969) *Plast. Polym.*, **37**, 517.

Brezny, R. Green, D. J. and Dam, C. Q. (1989a) Evaluation of strut strength in open-cell ceramics. *J. Amer. Ceramic Soc.* **72**, 885–9.

Brezny, R. and Green, D. J. (1989b) Fracture behaviour of open-cell ceramics. *J. Amer. Ceramic Soc.*, **72**, 1145–52.

Brezny, R. and Green, D. J. (1990) The effect of cell size on the mechanical behaviour of cellular materials. *Acta. Metall. Mater.*, **38**, 2517–26.

Brezny, R. and Green, D. J. (1991) Factors controlling the fracture resistance of brittle cellular materials. *J. Amer. Ceramic Soc.*, **74**, 1061–5.

Brezny, R. and Green, D. J. (1993) Uniaxial strength behaviour of brittle cellular materials. *J. Amer. Ceramic Soc.*, **76**, 2185–92.

Brighton, C. A. and Meazey, A. E. (1973) Expanded PVC, in *Expanded Plastics*. QMC Industrial Research Unit, London.

Caddock, B. D. and Evans, K. E. (1989) Microporous materials with negative Poisson's ratios: I. microstructure and mechanical properties, *J. Phys. D: Appl. Phys.*, **22**, 1877–82.

Chan, R. and Nakamura, M. (1969) *J. Cell. Plast.*, **5**, 112.

Choi, J. B. and Lakes, R. S. (1992) Non-Linear properties of metallic cellular materials with a negative Poisson's ratio, *J. Mat. Sci.*, **27**, 5375–81.

Choi, K., Kuhn, J. L., Ciarelli, M. J. and Goldstein, S. A. (1990) The elastic moduli of human subchondral, trabecular and cortical bone tissue and the size-dependency of cortical bone modulus. *J. Biomech.*, **23**, 1103–13.

Christensen, R. M. (1986) Mechanics of low density materials. *J. Mech. Phys. Solids* **34**, 563–78.

Dam, C. Q., Brezny, R. and Green, D. J. (1990) Compressive behaviour and deformation-mode map of an open-cell alumina. *J. Mat. Res.*, **5**, 163–71.

Evans, K. E. (1989) Tensile network microstructures exhibiting negative Poisson's ratios, *J. Phys. D: Appl. Phys.*, **22**, 1870–6.

Evans, K. E. and Caddock, B. D. (1989) Microporous materials with negative Poisson's ratios: II. mechanisms and interpretation *J. Phys. D: Appl. Phys.*, **22**, 1883–7.

Evans, K. E. (1990) Tailoring the negative Poisson ratio, *Chem. and Indus.*, October, 654–7.

Evans, K. E., Nkansah, M. A. and Hutchinson, I. J. (1992) Modelling negative Poisson ratio effects in network-embedded composites, *Acta Metall. Mater.*, **40**, 2463–9.

Fleck, N. A. and Parker, P. (1988) Part II project report. Cambridge University Engineering Department.

Fowlkes, C. W. (1974) *Int. J. Fracture*, **10**, 99.

Friis, E. A., Lakes, R. S. and Park, J. B. (1988) Negative Poisson's ratio polymeric and metallic foams, *J. Mat. Sci.*, **23**, 4406–14.

Gent, A. N. and Thomas, A. G. (1959) *J. Appl. Polymer Sci.*, **1**, 107.

Gent, A. N. and Thomas, A. G. (1963) *Rubber Chem. Technol.*, **36**, 597.

Gibson, L. J. and Ashby, M. F. (1982) *Proc. Roy. Soc.*, **A382**, 43.

Gibson, L. J., Ashby, M. F., Schajer, G. S., and Robertson, C. I. (1982) *Proc. Roy. Soc. Lond.*, **A382**, 25.

Goretta, K. C., Brezny, R., Dam, C. Q. and Green, D. J. (1990) High temperature mechanical behaviour of porous open-cell alumina. *Mat. Sci. Eng.*, **A124**, 151–8.

Green, D. J. (1985) *J. Amer. Ceramic Soc.*, **68**, 403.

Hagiwara, H. and Green, D. J. (1987) Elastic behaviour of open-cell alumina. *J. Amer. Ceramic Soc.*, **70**, 811–15.

Harper, C. A. (1975) *Handbook of Plastics and Elastomers*. McGraw-Hill, New York.

Hertzberg, R. W. (1983) *Deformation and Fracture of Engineering Materials*, 2nd edn. Wiley, New York.

Hertzberg, R. W. and Manson, J. A. (1980) *Fatigue of Engineering Plastics*. Academic Press, New York.

Hilyard, N. C. (ed.) (1982) *Mechanics of Cellular Plastics*. Applied Science, Barking.

Hilyard, N. C. and Cunningham, A. (1994) *Low Density Cellular Plastics* Chapman and Hall, London.

Huang, J. S. and Gibson, L. J. (1991) *Acta Metallurgica et materialia*, **39**, 1627–36.

Kane, R. P. (1965) *J. Cell. Plast.*, **1**, 217.

Ko, W. L. (1965) *J. Cell. Plast.*, **1**, 45.

Kurauchi, T., Sato, N., Kamigaito, O. and Komatsu, N. (1984) *J. Mat. Sci.*, **19**, 871.

Lakes, R. S. (1983) Size effects and micromechanics of a porous solid. *J. Mat. Sci.*, **18**, 2572–80.

Lakes, R. (1987) Foam structures with a negative Poisson's ratio, *Science*, **235**, 1038–40.

Lakes, R. S., Park, J. B. and Friis, E. A. (1988) Materials with negative Poisson's ratios: dependence of properties on structure, *Proc.*

Amer. Soc. for Composites, 3rd Technical Conference, Seattle, September, pp. 527–33.

Lakes, R. S. (1988) Cosserat micromechanics of structured media: Experimental methods. *Proc. Amer. Soc. for Composites. 3rd Technical Conference, Seattle, Sept. 1988.*

Lakes, R. S. (1991) Deformation mechanisms in negative Poisson's ratio materials: structural aspects, *J. Mat. Sci.,* **26,** 2287–92.

Lakes, R. S. (1992a) Saint-Venant end effects for materials with negative Poisson's ratios, *J. of Appl. Mech.,* **59,** 744–6.

Lakes, R. S. (1992b) No contractile obligations, *Nature,* **358,** 713–14.

Lakes, R. S. (1993) Strongly Cosserat elastic lattice and foam materials for enhanced toughness. *Cellular Polymers,* **12,** 17–30.

Lange, F. F. and Miller, K. T. (1987) Open-cell, low density ceramics fabricated from reticulated polymer substrates. *Adv. Ceramic Mater.,* **2,** 827–31.

Lazan, B. J. (1968) *Damping of Materials and Members in Structural Mechanics.* Pergamon Press, Oxford.

Lederman, J. M. (1971) *J. Appl. Polymer. Sci.,* **15,** 693.

Maiti, S. K., Gibson, L. J. and Ashby, M. F. (1984a) *Acta Metal.,* **32,** 1963.

Maiti, S. K., Ashby, M. F. and Gibson, L. J. (1984b) *Scripta Metal.,* **18,** 213.

Matonis, V. A. (1964) *Soc. Plast. Eng. J.,* September, p. 1024.

McIntyre, A. and Anderton, G. E. (1979) *Polymer,* **20,** 247.

Menges, G. and Knipschild, F. (1975) *Polymer Eng. Sci.,* **15,** 623.

Moore, D. R., Couzens, D. H. and Iremonger, M. J. (1974) *J. Cell. Plast.,* **10,** 135.

Morgan, J. S., Wood, J. L. and Bradt, R. C. (1981) *Mat. Sci. Eng.,* **47,** 37.

Nkansah, M. A., Evans, K. E. and Hutchinson, I. J. (1993) Modelling the effects of negative Poisson's ratios in continuous-fibre composites, *J. Mat. Sci.,* **28,** 2687–92.

Oliver, D. S. (1980) *The Use of Glass in Engineering.* Design Council Guide 05, Oxford University Press, Oxford.

Patel, M. R. and Finnie, I. (1970) *J. Mater.,* **5,** 909.

Phillips, P. J. and Waterman, N. R. (1974) *Polymer Eng. Sci.,* **4,** 67.

Pittsburgh-Corning Inc. (1982) Foamglass Data Sheets.

Plastics Technology, (1964), **8,** 26.

Reitz, D. W. Schnetz, M. A. and Glicksman, L. R. (1984) *J. Cell. Plast.,* **20,** 104.

Rho, J. Y., Ashman, R. B. and Turner, C. H. (1993) Young's modulus of trabecular and cortical bone material: ultrasonic and microtensile measurements. *J. Biomech.,* **26,** 111–19.

Rinde, J. A. (1970) *J. Appl. Polymer Sci.,* **14,** 1913.

Roff, W. F. and Scott, J. R. (1971) *Fibres, Films, Plastics and Rubbers – A Handbook of Common Polymers.* Butterworths, London.

Rusch, K. C. (1970) *J. Appl. Polymer Sci.,* **14,** 1263.

Ryan, S. D. and Williams, J. L. (1989) Tensile testing of rodlike trabeculae excised from bovine femoral bone. *J. Biomech.,* **22,** 351–5.

Shaw, M. C. and Sata, T. (1966) *Int. J. Mech. Sci.,* **8,** 469.

Simone, A.E. and Gibson, L. J. (1998) *Acta Mat.,* **46,** 2139.

Skochdopole, R. E. and Rubens, L. C. (1965) *J. Cell. Plastics,* **1,** 91.

Suh, K. W. and Skochdopole, R. E. (1980) Foamed plastics, in *Encyclopedia of Chemical Technology,* vol. 2, 3rd edn. Wiley, New York, p. 82.

Suh, K. W. and Webb, D. D. (1985) Cellular materials, in *Encyclopedia of Polymer Science,* vol. 3, 2nd edn. Wiley, New York, p. 1.

Thornton, P. H. and Magee, C. L. (1975a) *Met. Trans.,* **6A,** 1253.

Thornton, P. H. and Magee, C. L. (1975b) *Met. Trans.,* **6A,** 1801.

Timoshenko, S. P. and Goodier, J. N. (1970) *Theory of Elasticity,* 3rd edn. McGraw-Hill, New York.

Traeger, R. K. (1967) *J. Cell. Plast.,* **3,** 405.

Vaz, M. F. and Fortes, M. A. (1993) Characterization of deformation bands in the compression of cellular materials. *J. Mater. Sci. Lett.,* **12,** 1408–10.

Walsh, J. B., Brace, W. F. and England, A. W. (1965) *J. Amer. Ceram. Soc.,* **48,** 605.

Warner, M. and Edwards, S. F. (1988) *Europhys. Lett.,* **5,** 623–8.

Warren, T. L. (1990) Negative Poisson's ratio in a transversely isotropic foam structure, *J. Appl. Phys.,* **67,** 7591–4.

Warren, W. E. and Kraynik, A. M. (1988) The linear elastic properties of open-cell foams. *J. Appl. Mech.*, **55**, 341–6.

Warren, W. E. and Kraynik, A. M. (1991) The nonlinear elastic behaviour of open-cell foams. *J. Appl. Mech.*, **58**, 376–81.

Weaire, D. and Fortes, M. A. (1994) Stress and strain in liquid and solid foams. *Adv. Phys.*, **43**, 685–738.

Weast, R. C. (ed.) (1985) *Handbook of Chemistry and Physics*. Chemical Rubber Co., Boca Raton, Fla.

Wei, G. (1992) Negative and conventional Poisson's ratios of polymeric networks with special microstructures, *J. Chem. Phys.*, **96**, 3226–33.

Wendle, B. C. (ed.) (1976) *Engineering Guide to Structural Foams*. Technomic Publishing Co., Westport, Conn.

Wilsea, M., Johnson, K. L. and Ashby, M. F. (1975) *Int. J. Mech. Sci.*, **17**, 457.

Zhang, J. and Ashby, M. F. (1988) CPGS Thesis, Engineering Department, Cambridge, UK.

Zwissler, J. G. and Adams, M. A. (1983) in *Fracture Mechanics of Ceramics*, vol. 6, (ed. Bradt, R. C.). Plenum Press, New York, p. 211.

Chapter 6

The mechanics of foams: refinements

6.1   Introduction and synopsis

In Chapter 5 we described the uniaxial behaviour of isotropic foams. Expressions for their elastic moduli and compressive strengths, and their resistance to tensile fracture and fatigue, were derived assuming that the temperature and rate of loading corresponded to those for which the cell wall properties are tabulated (typically, $T = 20°C$ and $\dot{\epsilon} = 10^{-3}/s$). In practice, foams are often anisotropic. And in many engineering applications, they are loaded in more than one direction and at high rates or different temperatures.

Military, aerospace, automative and packaging applications often require a knowledge of foam properties at high rates of deformation and at temperatures other than room temperature. Here the modulus and failure-mechanism diagrams of Chapter 3 become useful. The moduli and strength of rigid foams (metals as well as polymers) decrease linearly as the temperature rises. Increasing the strain-rate does not affect their moduli but increases their strength. Semi-rigid foams (those used at a temperature close to the glass temperature T_g of the base polymer) are more complicated: both their moduli and strength increase with strain-rate, sometimes dramatically. Elastomeric foams are different again: their moduli *increase* slightly with increasing temperature, but are almost independent of strain-rate.

These effects are directly related to the temperature and strain-rate dependence of the cell-wall properties (Chapter 3): they are inherent properties of the material of which the foam is made. But almost always the cells of a foam contain a fluid: a gas like air or nitrogen, or a liquid such as water or (in biological cells) natural fluids. When the foam is deformed the cell fluid is compressed (if the cells are closed) or made to flow from one cell to another (if they are open), and this introduces a new dependence on temperature and strain-rate. Both inherent and pore-fluid effects are reviewed in Section 6.2.

Most man-made foams are anisotropic because of the way in which they are made: foaming agents, for instance, cause the foam to rise, producing cells which are elongated in the rise direction. Natural foams, too, are almost always anisotropic: the cell walls in cancellous bone, for example, grow in response to load, thickening and aligning themselves along the directions of principal stress. Foams with structural anisotropy have properties which are anisotropic, too. Man-made foams are commonly twice as stiff in the rise direction as in the plane normal to it because the shorter walls normal to the rise direction bend less easily. The strength and toughness are both anisotropic, too, for similar reasons. They are analysed in Section 6.3; experiments, though limited, support the analysis.

Finally, there is the problem of multiaxial loading. In complex packaging, and in the use of foams in thermal insulation, the material is subjected to biaxial or triaxial loading. The elastic response is straightforward and is handled in conventional ways. But the strength is more complicated. The mechanism of failure depends on the state of stress, as well as on the cell-wall properties. In uniaxial tension it is the bending stresses which cause yield or fracture, while in hydrostatic tension it is the axial stress. This difference leads to hydrostatic strengths which are of the order of 10 times the uniaxial, giving a highly elongated failure surface. And in compression an additional failure mode, elastic buckling, must be considered. We deal with this in Section 6.4 by developing yield and fracture criteria, and using them to construct surfaces in stress space which describe the combination of principal stresses which will cause collapse or fast fracture.

6.2   The effect of temperature and strain-rate

The stiffness and strength of a foam depends, as one might expect, on the temperature T and the rate at which it is strained $\dot{\epsilon}$. Most foams are used near ambient temperature, so the temperature range between $-20°C$ and $+40°C$ is the most interesting one; but there are applications – as in thermal insulation, catalytic converters, or heat exchangers – when the range of temperature is much wider. The range of strain-rates is wide, too. In most conventional uses, foams are not subjected to strain-rates much above $1/s$, but when they are used for crash-protection or high-performance packaging it may be necessary to design for rates as high as $10^5/s$.

It helps to distinguish two separate contributions to the temperature- and rate-dependence of foam properties. The first derives from the solid of which the foam is made: the foam inherits the temperature- and strain-rate dependence of the solid of the cell walls (Chapter 3). We call this the *inherent* temperature- and rate-dependence. The second derives from the fluid which fills the cells. When the foam is deformed, the pore fluid (which may be a gas or a liquid) either deforms or is forced to flow from cell to cell. The compressibility of a gas in a closed-cell foam depends on temperature, so a new temperature-dependence is introduced by its presence. And in an open-cell foam, pore fluid is expelled or drawn into the foam as it deforms; the work done against the viscosity of the fluid introduces a new strain-rate dependence.

(a) The inherent temperature- and strain-rate dependence of foam properties

Elastic and viscoelastic response

We saw in Chapter 5 that the *moduli* of a foam are proportional to the Young's modulus of the cell walls, E_s. For polymers well below their glass temperatures, T_g, and for metals and ceramics well below their melting points, T_m, the modulus varies more or less linearly with temperature, and its rate-dependence is so small that it can usually be forgotten (even in polymers, an increase of strain-rate by a factor of 10^6 increases the modulus by a factor of only 1.1 (Hoge and Wasley, 1969; Rinde and Hoge, 1971). A convenient approximation (Eqns. (3.2) and (3.14)) is:

$$E_s = E_s^0 \left(1 - \alpha_m \frac{T}{T_g} \right) \tag{6.1a}$$

for polymers, and

$$E_s = E_s^0 \left(1 - \alpha_m \frac{T}{T_m} \right) \tag{6.1b}$$

for metals and ceramics. Here E_s^0 is the modulus at $0°K$ and α_m is a constant for which data are available; typically $\alpha_m = 0.5 \pm 0.2$.

Typical data for the temperature-dependence of the compressive Young's modulus for a set of rigid polyurethane foams, all made from the same solid, are shown in Fig. 6.1. The data lie on straight lines converging to a common point and are well described by Eqn. (6.1a) with $\alpha_m = 0.65$.

Polymers near their glass temperature (as many are, at room temperature) are viscoelastic: the modulus then depends more strongly on both temperature and strain-rate or time of loading. A foam made from a viscoelastic polymer is itself viscoelastic. Its response to a given history of loading is found by an extension of the method developed for honeycombs in Chapter 4, Section 4.3(e). The cell edges are treated as viscoelastic beams loaded in bending, and their deflection and deflection-rate are calculated. The load per beam, P, is related to the stress by $P \propto \sigma l^2$, and the strain, ϵ, and strain-rate, $\dot{\epsilon}$, of the foam are related to the deflection and deflection-rate of the beam by $\epsilon \propto \delta/l$ and $\dot{\epsilon} \propto \dot{\delta}/l$, leading to the *differential equation for viscoelastic deformation* of an open-cell foam:

$$\dot{\epsilon} + \frac{E_{s2}}{\eta_s}\epsilon = \frac{1}{C_1(\rho^*/\rho_s)^2}\left(\frac{\dot{\sigma}}{E_{s1}} + \frac{\sigma}{\eta_s}\left(\frac{E_{s1}+E_{s2}}{E_{s1}}\right)\right) \quad (6.2)$$

with $C_1 \approx 1$. (That for a closed-cell foam requires the substitution of Eqn. (5.13a) in place of $C_1(\rho^*/\rho_s)^2$.) In this equation, E_{s1} is the unrelaxed modulus of the solid, $E_{s1}E_{s2}/(E_{s1}+E_{s2})$ is its relaxed modulus and η_s is its viscosity. Note that foam properties enter only as a multiplying factor on the right-hand side; except for this the equation is identical with that for the solid polymer itself.

The response of the foam to any given history of loading is found by integrating this equation, applying appropriate boundary conditions. But the simplicity of the equation, referred to above, means that we can use the well-known solutions for solid viscoelastic bodies (see, for example, Powell, 1983); all that is necessary is to replace the stress σ in the solution by $\sigma/C_1(\rho^*/\rho_s)^2$ (or the equivalent factor for closed cells). Thus the response to a steady uniaxial stress, σ, applied for a time, τ (so that $\dot{\sigma} = 0$) is:

$$\epsilon = \frac{\sigma}{C_1(\rho^*/\rho_s)^2}\frac{E_{s1}+E_{s2}}{E_{s1}E_{s2}}\left(1 - \exp - \frac{E_{s2}\tau}{\eta_s}\right). \quad (6.3)$$

This has the same relaxation time as the solid, but has strains which are larger, by the factor $1/C_1(\rho^*/\rho_s)^2$, at any given time, τ. The creep compliance of a linear

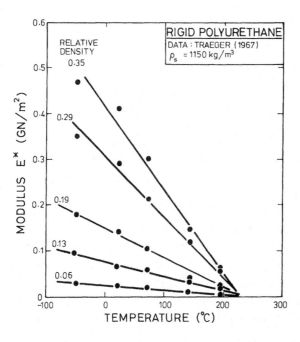

Figure 6.1 The temperature-dependence of Young's modulus, E^*, for rigid polyurethane foams with relative densities between 0.06 and 0.35.

viscoelastic foam, $J^*(t) = \epsilon(t)/\sigma$, is simply that of the solid multiplied by the factor E_s/E^*. Data for the creep of rigid polyurethane foams are well described in this way (Huang and Gibson, 1991).

The response to a steady displacement (so that $\dot{\epsilon} = 0$) is

$$\sigma = C_1(\rho^*/\rho_s)^2 E_{s1}\epsilon\left\{1 - \frac{E_{s1}}{E_{s1} + E_{s2}}\left(1 - \exp-\frac{(E_{s1} + E_{s2})}{\eta_s}\tau\right)\right\} \qquad (6.4)$$

which again has the same relaxation time as the solid, but has stresses which are less, by the factor $C_1(\rho^*/\rho_s)^2$, at any given time, τ. Other solutions (such as that for a sinusoidally varying stress) are adapted in the same way.

Plastic yielding

The plastic collapse strength of a foam, σ_{pl}^*, is directly proportional to the yield strength, σ_{ys}, of the solid of which it is made. As the solid is heated the yield strength falls; increasing strain-rate causes a slight increase in yield strength. Foams made of the solid should show a similar response; giving, by analogy with Eqns. (3.9) and (3.16):

$$\sigma_{pl}^* = (\sigma_{pl}^*)^0\left(1 - \frac{AT}{T_g}\ln\frac{\dot{\epsilon}_0}{\dot{\epsilon}}\right) \qquad (6.5a)$$

for rigid polymers and

$$\sigma_{pl}^* = (\sigma_{pl}^*)^0\left(1 - \frac{AT}{T_m}\ln\frac{\dot{\epsilon}_0}{\dot{\epsilon}}\right) \qquad (6.5b)$$

for metals and ceramics, where $(\sigma_{pl}^*)^0$ is the yield strength at $0°K$ and A and $\dot{\epsilon}_0$ are material properties. The equations predict a linear decrease in yield strength and increasing temperature, and a linear increase with increase in log $\dot{\epsilon}$. Data in the literature (Lacey, 1965; Traeger, 1967; Hoge and Wasley, 1969; Rinde and Hoge, 1971) confirm this behaviour. Figure 6.2 shows the decrease in σ_{pl}^* of polyurethane foams with increasing temperature: up to T_g (roughly 100°C) it is linear. Figure 6.3 is typical of the way in which the compressive stress–strain curve changes with strain-rate; the plateau stress σ_{pl}^* rises, and, almost always, a yield point appears or, if present at low strain-rates, becomes more pronounced. Figure 6.4 shows the variation of yield strength of the same foams with the log of strain-rate: it shows the typical linear increase. The data for rigid polyurethane foams are adequately described by Eqn. (6.5a) with $A \approx 0.1$, $\dot{\epsilon} \approx 10^4/s$, and $T_g = 100°C$.

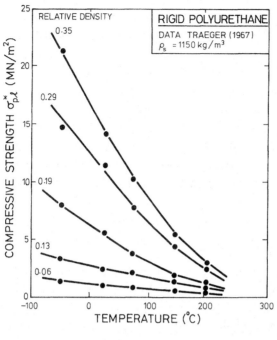

Figure 6.2 The temperature-dependence of the strength σ_{pl}^* for rigid polyurethane foams with relative densities between 0.06 and 0.35.

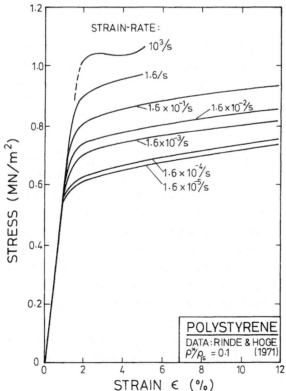

Figure 6.3 Compressive stress-strain curves for a rigid polystyrene foam at a number of strain-rates between 10^{-5}/s and 10^3/s.

The dependence of σ_{pl}^* on density is unchanged by temperature or strain-rate. Figure 6.5 shows Traeger's data for the strength σ_{pl}^* of rigid polyurethane, plotted against density. It follows the usual equation

$$\sigma_{pl}^* = C_5 \sigma_{ys} \left(\frac{\rho^*}{\rho_s} \right)^{3/2}$$

(Eqn. (5.27)) at all temperatures.

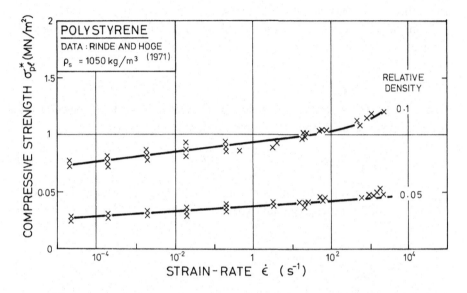

Figure 6.4 The strain-rate dependence of the strength σ_{pl}^* for rigid polystyrene foams at room temperature, for two relative densities: 0.05 and 0.1.

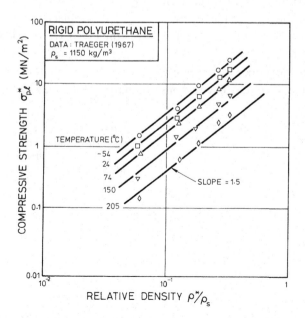

Figure 6.5 The variation of the strength σ_{pl}^* with density at various temperatures for rigid polyurethane. The underlying dependence on density is not changed by change of temperature.

Creep and creep buckling

At and above their glass temperature T_g, polymers show slow, permanent, time-dependent deformations, or *creep*. Metals and ceramics creep, too, though the rate of creep is significant only when the temperature is greater than about 0.3 of the melting temperature, T_m. Creep is a problem when foams carry loads for long periods of time – refractory brick in furnace walls and polymer foams used in structural applications are examples. The creep-rate is calculated following the method described in Section 4.3(f) with the modification that in an open-cell foam the cell wall has area t^2 and length l.

Consider the creep of cell walls loaded as in Fig. 6.6. Let the creep-rate of the cell wall material be described (as in Chapters 3 and 4) by Eqn. (3.17):

$$\dot{\epsilon} = \dot{\epsilon}_{0s} \left(\frac{\sigma}{\sigma_{0s}} \right)^{n_s} \tag{6.6}$$

where $\dot{\epsilon}_{0s}$, σ_{0s} and n_s are creep constants. The deflection-rate of the cell wall is found, following the method of Chapter 4 (Eqn. (4.58)) to be:

$$\dot{\delta} = \frac{1}{4(n_s + 2)} \dot{\epsilon}_{0s} \frac{l^2}{t} \left(\left(\frac{2n_s + 1}{4n_s} \right) \cdot \frac{Fl}{\sigma_{0s} t^3} \right)^{n_s} \tag{6.7}$$

which correctly reduces to the result for plastic failure in the limit of $n = \infty$. As before, the load F is proportional to σl^2 and the strain-rate $\dot{\epsilon}$ of the foam is proportional to $\dot{\delta}/l$. For open-cell foams, this gives

$$\frac{\dot{\epsilon}^*}{\dot{\epsilon}_{0s}} = \frac{C_9}{(n_s + 2)} \left(\frac{C_{10}(2n_s + 1)}{n_s} \frac{\sigma}{\sigma_{0s}} \right)^{n_s} \left(\frac{\rho_s}{\rho^*} \right)^{\frac{3n_s + 1}{2}} \tag{6.8}$$

where C_9 and C_{10} are constants. They are found by considering two limits. At the limit of $n_s = \infty$ (with σ_{0s} replaced by σ_{ys}) the equation reduces identically to that for plastic-collapse (Eqn. (5.27a)) from which we find that $C_{10} = 1.7$. At the other limit of $n_s = 1$ (with $\sigma/\dot{\epsilon}^*$ replaced by E^* and $\sigma_{0s}/\dot{\epsilon}_{0s}$ replaced by E_s) the equation reduces exactly to that for linear elasticity (Eqn. (5.3)), from which we learn that $C_9 C_{10} = 1$ or $C_9 = 0.6$.

Closed cells, as always, introduce extra complications in the calculation. As the cell edges bend, the faces stretch. The response is analysed by equating the

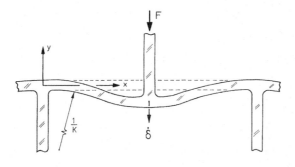

Figure 6.6 A cell wall, loaded in bending. The wall creeps, giving a deflection-rate $\dot{\delta}$.

work-rate $F\dot{\delta}$ to the energy dissipation-rate in bending and stretching the cell walls; in its details it parallels the calculation given earlier for the plastic-collapse stress for closed cells. The result is

$$\frac{\dot{\epsilon}}{\dot{\epsilon}_{0s}} = \left\{ \frac{\sigma/\sigma_{0s}}{\frac{1}{C_{10}} \left(\frac{n_s + 2}{C_9} \right)^{1/n_s} \left(\frac{n_s}{2n_s + 1} \right) \left(\phi \frac{\rho^*}{\rho_s} \right)^{\frac{3n_s+1}{2n_s}} + \frac{2}{3}(1 - \phi) \frac{\rho^*}{\rho_s}} \right\}^{n_s}.$$

(6.9)

When all the solid is in the cell edges ($\phi = 1$), the equation reduces to Eqn. (6.8). But when the faces are of uniform thickness, with no thickening at the edges ($\phi = 0$), it predicts a creep-rate which scales as $(\rho^*/\rho_s)^{-n_s}$.

Consider now the buckling of the vertical cell wall shown in Fig. 6.7(a). It has length l, and section t^2. As with honeycombs, we assume that the strut has some imperfections which induce bending; such imperfections can be represented by an initial mid-point deflection, a_0. Repeating the analysis described in Section 4.3(f), again using a constitutive law for flexural response which includes terms describing both stiffness and creep (Eqn. (4.57a), we obtain the rate at which the mid-point deflects sideways (Eqn. (4.66):

$$\dot{a} = \frac{\frac{l^2 \dot{\kappa}_0}{4(n_s + 2)} \left(\frac{Fa}{M_0} \right)^{n_s}}{\left(1 - \frac{F}{F_E} \right)}.$$

(6.10)

Initially, at time $t = 0$, $a = a_0$. Catastrophic buckling occurs at time $t = t_b$ at which point $a \rightarrow \infty$. Integrating between these two limits gives the time at which buckling becomes catastrophic:

$$t_b = 4 \frac{(n_s + 2)}{(n_s - 1)} \left(\frac{l}{a_0} \right)^{n_s - 1} \frac{(1 - F/F_E)}{l \dot{\kappa}_0 (Fl/M_0)^{n_s}}$$

(6.11)

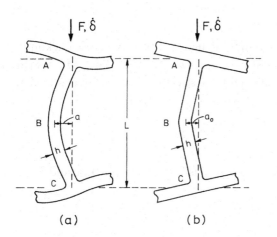

(a)　　　　(b)

Figure 6.7 The creep-buckling of a cell wall. Because creep is non-linear, the sinusoidal shape of (a) is quickly replaced by one in which most of the curvature is at the mid-point, as shown at (b).

with $\dot{\kappa}_0$ and M_0 given, as before, by:

$$\dot{\kappa}_0 = \frac{2\dot{\epsilon}_{0s}}{t}$$

and

$$M_0 = \left(\frac{2n_s}{2n_s + 1}\right)\frac{\sigma_{0s}t^3}{4}.$$

Rearranging, and writing $F \propto \sigma l^2$ and $F_E \propto \sigma_{el}^* l^2$ and using $\sigma_{el}^* = 0.05E_s(\rho^*/\rho_s)^2$ we obtain:

$$\frac{\dfrac{\sigma}{\sigma_{el}^*}}{\left(1 - \dfrac{\sigma}{\sigma_{el}^*}\right)^{1/n_s}} = \left[2\left(\frac{n_s + 2}{n_s - 1}\right)\left(\frac{l}{a_0}\right)^{n_s - 1}\left(\frac{1}{t_b\dot{\epsilon}_{0s}}\right)\right]^{1/n_s}\left(\frac{2n_s}{2n_s + 1}\right)$$

$$\cdot \left(\frac{\sigma_{0s}}{0.2C_{12}E_s}\right)\left(\frac{\rho^*}{\rho_s}\right)^{\frac{1-n_s}{2n_s}}. \tag{6.12}$$

The equation says that, for a given density (ρ^*/ρ_s), temperature (contained in $\dot{\epsilon}_{0s}$) and time of loading, t_b, there is a critical stress below which buckling will not occur and above which it will.

Like the analogous result for honeycombs, the behaviour of this equation is best understood by examining various limits. Note, first, that σ/σ_{el}^* cannot have a value greater than 1 – if it did, the cell walls would buckle instantly. When the time-to-buckling is very short, the right-hand side of the equation is large (that is, much greater than 1) and this is only possible if

$$1 - \frac{\sigma}{\sigma_{el}^*} \ll 1$$

meaning that

$$\sigma = \sigma_{el}^* = 0.05E_s\left(\frac{\rho^*}{\rho_s}\right)^2$$

that is, the critical stress is equal to that calculated earlier for elastic buckling.

At the other extreme, when the stress is much below the buckling stress (so that $F \ll F_E$ in Eqn. (6.11)), the equation for the stress σ_{cb}^* at which creep buckling occurs simplifies to:

$$\sigma_{cb}^* \propto \left(\frac{2n_s}{2n_s + 1}\right)(\sigma_{0s})\left(\frac{l}{a_0}\right)^{\frac{n_s - 1}{n_s}}\left[\frac{2(n_s + 2)}{n_s - 1}\frac{1}{t_b\dot{\epsilon}_{0s}}\right]^{\frac{1}{n_s}}\left(\frac{\rho^*}{\rho_s}\right)^{\frac{3n_s + 1}{2n_s}}. \tag{6.13}$$

If, further, the exponent is large ($n_s > 5$, as it is for many metals) we find:

$$\sigma_{cb}^* \propto \sigma_{0s}\left(\frac{l}{a_0}\right)\left[\frac{2}{t_b\dot{\epsilon}_{0s}}\right]^{\frac{1}{n_s}}\left(\frac{\rho^*}{\rho_s}\right)^{\frac{3}{2}}. \tag{6.14}$$

This defines a 'threshold stress' for creep buckling. Inverting gives an expression for the time-to-buckling, t_b, at a given applied stress, σ:

$$t_b \propto \left[\frac{2}{\dot{\epsilon}_{0s}}\right]\left(\frac{l}{a_0}\right)^{n_s}\left(\frac{\sigma_{0s}}{\sigma}\right)^{n_s}\left(\frac{\rho^*}{\rho_s}\right)^{\frac{3n_s}{2}}. \tag{6.15}$$

This last equation illustrates that – for large n_s – the time-to-buckling is very sensitive both to the size of the initial imperfection, a_0, and to the stress, σ.

When the time is long, but the creep exponent is small ($n_s = 1$, as it is for many polymers and occasional ceramics) it is necessary to re-integrate Eqn. (6.10) with $n_s = 1$ and a finite upper limit for $a(a_u)$. This leads to the result:

$$t_b \propto \frac{1}{l\dot{\kappa}_0}\frac{(1 - F/F_E)}{(Fl/M_0)}\ln\left(\frac{a_u}{a_0}\right) \tag{6.16}$$

or, replacing $\dot{\kappa}_0$, M_0, F, and σ^*_{el} in exactly the same way as we did in deriving Eqn. (6.12) from (6.11):

$$t_b \propto \frac{\sigma_{0s}}{E_s\dot{\epsilon}_{0s}}\left(\frac{1 - \sigma/\sigma^*_{el}}{\sigma/\sigma^*_{el}}\right)\ln\left(\frac{a_u}{a_0}\right). \tag{6.17}$$

Inverting the equation with a given value of t_b gives:

$$\frac{\sigma^*_{cb}}{\sigma^*_{el}} \propto \frac{B}{1 + B} \quad \text{with} \quad B = \frac{\sigma_{0s}}{E_s\dot{\epsilon}_{0s}t_b}\ln\left(\frac{a_u}{a_0}\right) \tag{6.18}$$

where σ_{cb} is the stress at which creep buckling occurs. In this limit, the creep-buckling stress is simply a fraction of the elastic-buckling stress σ^*_{el}. The fraction decreases as time increases, ultimately falling as $1/t_b$. Note the danger implied by this: even the smallest load, if applied for sufficiently long, will cause buckling. There is no threshold below which buckling does not occur.

Comparison with data for creep of foams

Although many polymer, metal and ceramic foams have been tested in the creep range (Brown, 1960; Hart et al., 1973: Thornton and Magee, 1975; Campbell, 1979, Goretta et al., 1990) few of the studies are sufficiently detailed to allow a test of the equations developed above. Flexible foams suffer a permanent 'compression set' when loaded for long periods of time, which must be caused by creep (Frisch and Saunders, 1972; Benning, 1969; Meinecke and Clark, 1972; Terry, 1971, 1976). Brown (1960), whose data for the creep of a polystyrene foam are shown in Fig. 6.8, distinguishes two sorts of creep in this material: short-term and long-term. Short-term creep gives strain in times of the order of a few hours, and is found at all stress levels. Long-term creep requires a time scale of a few days, and is found only when the stress exceeds a threshold level. Brown's data suggested that the threshold stress is roughly proportional to the relative density.

The measurements of Goretta *et al.* (1990), shown in Figs. 6.9 and 6.10 are more discriminating. They characterise the compressive creep of an open-cell Al_2O_3 in the temperature range 1300–1500°C. At low stresses, creep occurred by diffusional flow (see Section 3.3(d)), with a creep exponent n_s of 1 ± 0.1, and an activation energy within the range commonly reported for creep of bulk alumina. This is consistent with Eqn. (6.8), which – for a creep exponent of 1 ± 0.1 – predicts a dependence of creep-rate on density such that

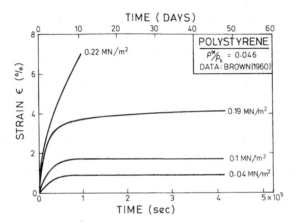

Figure 6.8   Creep curves for polystyrene foam (data from Brown, 1960).

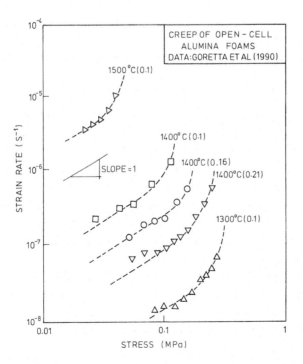

Figure 6.9   Creep-rates of open-cell alumina foams, plotted against stress. At low stresses, $\dot{\varepsilon} \propto \sigma$; at higher stresses the strain-rate accelerates, and creep buckling is possible (data from Goretta *et al.*, 1990).

$$\frac{\dot{\epsilon}}{\dot{\epsilon}_{0s}} \alpha \left(\frac{\rho_s}{\rho^*}\right)^{2\pm0.15} .$$

Goretta *et al.* observe an exponent of 1.8. At higher stresses the creep-rate accelerated rapidly (Fig. 6.9), suggesting the possibility of creep buckling. Further experimental studies are needed to explore this and other aspects of the creep of foams more fully.

Dynamic crushing of cellular solids

At very high strain-rates, dynamic (inertial) effects drive the compression strength upwards. Recent studies of the dynamic crushing of cellular structures and foams are particularly associated with the names of Reid, Stronge, Fortes and Wierzbicki (Reid and Reddy (1983), Reid and Bell (1984), Reid *et al.* (1993a, b), Klintworth and Stronge (1988), Klintworth (1989), Stronge (1991), Shim *et al.* (1992), Stronge and Shim, (1987, 1988), Fatima Vaz, *et al.* (1995), Abramowicz and Wierzbicki (1988)). Detailed analyses can be found in their work. Here we summarise the physical origins of the effect.

Three features of dynamic crushing influence the overall force–displacement response. The first is *localisation*: the concentration of deformation, at a given instant, into a thin layer or band, often adjacent to the loading face. It is caused by geometric softening – the change in cell shape with deformation – and it leads to local strain-rates which are much larger than the apparent, nominal, strain-rate. The second is *micro-inertia*: inertia associated with rotation and lat-

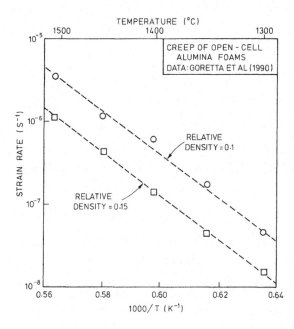

Figure 6.10 Creep-rates of open-cell alumina foams, plotted against temperature. Both densities show the same activation energy, in agreement with the model.

eral motion of cell walls when they buckle. Micro-inertia tends to suppress the more compliant buckling modes so it increases the crushing stress (particularly the initial peak) and diffuses the crushing wavefront. The last is *densification* which – as always – causes the stress to rise steeply when cell walls come into contact. This leads to 'shock' enhancement at very high strain-rates due to cell collapse that is overdriven during densification.

Impact forces transmitted through cellular materials exhibit several stages, depending on the value of the *impact intensity factor*, Q, defined by

$$Q = \frac{\text{Kinetic energy of impacting body}}{\text{Energy to crush one layer of cells}}.$$

If Q is not much larger than unity, cells near the impact surface experience small deformation in a symmetric mode that is relatively stiff. At larger values of Q the impacting body transmits sufficient energy to layers of cells at the contact surface that these cells are driven into the range of large deformations; then strain-softening in more compliant asymmetric (buckling) modes of deformation becomes prevalent. At $Q > 6$, impact results in crushing that initiates at the impact surface and proceeds sequentially through one layer of cells after another.

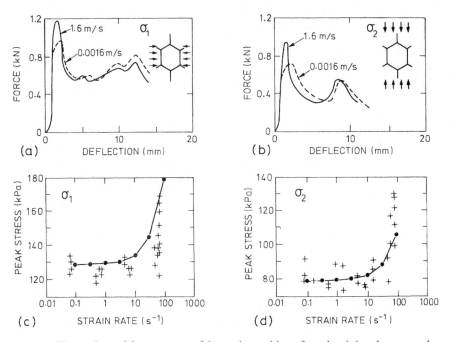

Figure 6.11 Measurement of dynamic crushing of an aluminium honeycomb with a cell size of 5mm. The upper graphs (a), (b) show the force–displacement curves for two crushing speeds and two directions of loading. The lower graphs (c), (d) show the peak stress, plotted against local strain-rate in the deformation band, and compare it (full lines) with micro-inertial hardening theory (reproduced from Stronge, 1991, Fig. 4, courtesy of Elsevier).

Data for the high-rate crushing of an aluminium honeycomb are shown in Fig. 6.11. After initiation of crushing in a single band of cells, static crushing proceeds by one layer after another being compressed into the band; it is a highly non-homogeneous, locally unstable mode of deformation. At high rates, the stress in the initial phase of crushing for an elasto-plastic cellular solid is enhanced by two dynamic effects: (a) as impact velocity increases the cells near the impact surface are more tightly compressed during densification causing the crushing stress to increase; and (b) the highly localized nature of crushing results in very large strain-rates in the crushing band and consequent enhancement of the symmetric (bending) modes of deformation and suppression of the asymmetric (buckling) ones by micro-inertia. The first of these dynamic sources of stress enhancement has been described as a 'shock' wave associated with the final strain-hardening or densifying phase of cell deformation (Reid and Bell, 1984). For impact speeds v large enough to cause substantial densification, this results in a nominal crushing stress that increases with v (Shim et al., 1992). The second dynamic source of stress enhancement is micro-inertia from buckling of cell walls. By suppressing the more compliant asymmetric modes of cell deformation, the crushing bands are diffused and the crushing stress plateau level is increased (Klintworth, 1989). Micro-inertia thus plays a significant role in controlling distribution of crushing in lightweight open-celled foams and honeycombs.

In each cell, crushing initiates when the stress state equals the collapse stress for the easiest mode of deformation (Klintworth and Stronge, 1988). As a cell begins to deform, large changes in geometry decrease the stress required to continue deformation; at an intermediate phase of deformation of a honeycomb there can be a reduction in the crushing stress by a factor as large as $1/2$ (Stronge and Shim, 1987). The extent of strain-softening during the intermediate phase is greatest for asymmetric modes and least for symmetric modes of cell deformation. Hence if micro-inertia decreases the compliance to the extent that deformation occurs in a symmetric rather than an asymmetric mode, the transmitted forces are substantially increased. In either case, if deformation continues in a layer of cells, interference between cell surfaces and increasing constraint from neighbouring deformed cells causes strong strain-hardening during the last phase of densification.

Because of strain-softening, dynamic crushing is localized in a thin deforming band where strain-rates are very large. In this band, buckling of cell walls is the usual mechanism for plastic cell collapse; buckling motion is resisted however, by inertia of the cell walls. Effects of micro-inertia increase the crushing stress and diffuse the localized deforming region. The upper part of Fig. 6.11, taken from Stronge (1991), shows the force–displacement curve for an aluminium honeycomb loaded in the 1 and 2 directions, for two impact velocities. The lower part of the figure shows the peak stress plotted against the *local* strain-rate. There is a small increase in the initial peak stress at the higher compression velocity of 1.6 m/s, when localisation resulted in a local strain-rate in a layer of cells of order $10^2 \, s^{-1}$. A large deflection analysis of cell collapse was used to calculate the enhancement of peak stress due to micro-inertia (Klintworth, 1989). For this honeycomb, micro-inertial enhancement was increasingly important at strain-rates above $10^2 \, s^{-1}$. In Fig. 6.11 the data points represent measurements of crushing stress for 'plane strain' compression in transverse principal directions 1 and 2. Although this data contains considerable scatter it does show the same trend as the theory.

In summary, special features, unique to cellular solids, drive up the crushing strength at high compression velocities. They derive from localisation, which leads to high local strain-rates, and micro-inertia, which derives from the inertial energy contained in the rotations and asymmetric deflections associated with buckling modes of deformation.

(b) Temperature- and strain-rate dependence of foam properties caused by pore fluid

The cells of almost all foams contain a fluid. In the case of man-made foams it is usually a gas; in natural cellular solids it is often a liquid.

Closed-cell foams containing a gas

When a closed-cell foam is deformed the cell fluid is compressed or expanded. The effect of this on the moduli and the collapse strength was described in Section 5.3. A gas at atmospheric pressure adds, at most, $0.1 \, MN/m^2$ to the Young's modulus of the foam, a contribution so small that it can be neglected for all but elastomeric foams of low density $(\rho^*/\rho_s < 0.1)$. A gas at higher-than-atmospheric pressure, of course, adds more; and changing the temperature changes the gas pressure and thus the contribution to the modulus, at least temporarily, until gas diffusion in or out of the sample equalizes the pressures again. If the gas pressure at the initial temperature T_0 is p_0, then the argument of Section 5.3 (treating the gas as ideal) gives the contribution of the gas to the modulus E^* at a temperature T (from Eqn. (5.10) and Boyle's law) of:

$$E_g^* = (E_g^*)_0 \left(\frac{T}{T_0} \right) \tag{6.19}$$

where $(E_g^*)_0$ is the gas contribution to the modulus at T_0. This, too, is usually too small to be of importance except in low-density elastomeric foams.

The effect of the gas on the rest of the stress–strain curve of a closed-cell elastomeric foam is much more pronounced.[†] The analysis of Section 5.3 led to an

---

[†]The cell walls of closed-cell foams made of materials which yield or fracture may rupture before the gas pressure becomes significant; then the compressibility of the cell fluid does not affect their plateau stress.

equation for the post-collapse curve of a foam containing gas at an initial pressure p_0 as (Eqn. 5.20):

$$\sigma^* = \sigma_{el}^* + \frac{p_0 \epsilon}{(1 - \epsilon - \rho^*/\rho_s)} .$$ (6.20)

Recent work (Zhang and Ashby, 1988) confirms the accuracy of this result, and gives further insight into the various contributions to the stress–strain curve. A range of low-density polyethylene foams with closed cells were compressed at strain-rates varying from 2×10^{-3}/s to 50/s. One set of stress–strain curves is shown in Fig. 6.12a. The second term in Eqn. (6.20) is plotted on the figure (using $p_0 = 1$ atmosphere). When this is subtracted out, the residual stress–strain curves (Fig. 6.12b) have the classical shape associated with cellular materi-

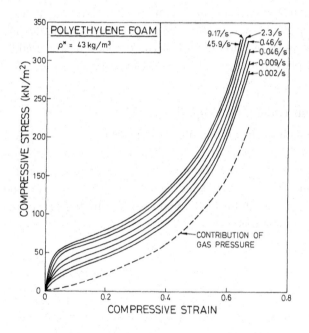

Figure 6.12 (a) Stress–strain curves for a closed-cell polyethylene foam at strain-rates from 2×10^{-3} to 50/s. The broken line shows the contribution of the gas pressure in the cells which, because the gas remains isothermal, is independent of strain-rate. (b) The stress–strain curves of (a) after subtraction of the contribution of the gas.

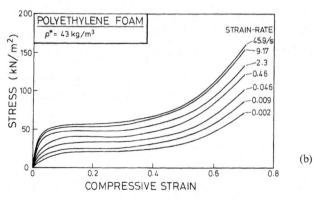

als. The plateau stress, normalized by that at 10^{-2}/s, is plotted against strain-rate in Fig. 6.13, where it is compared with data for solid low-density polyethylene, similarly normalized. The figure shows that the strain-rate dependence of the foam properties derives directly from that of the cell-wall material. At first sight, one might have expected an additional strain-rate dependence caused by the transition from isothermal to adiabatic compression of the gas itself as the strain-rate is increased. But, as Zhang and Ashby point out, the gas is always isothermal because of its intimate contact with the cell-wall solid which has a thermal capacity far larger than that of the gas.

The role of the pore gas can also be determined experimentally by loading the specimen in a sealed chamber of known volume and measuring the resulting pressure changes within the chamber (Rehkopf et al., 1996). Subtraction of the gas contribution from the total stress–strain curve again produces the classic horizontal stress plateau. Repeated loading damages the cell membranes allowing the gas to escape: after 10 cycles of load the gas contribution becomes negligible, giving a horizontal stress plateau in the total stress–strain curve.

Closed-cell foams containing a liquid

A closed-cell foam containing an incompressible liquid within the cells behaves differently. The volume of the cell is conserved by cell-wall stretching which compensates for reductions in cell volume resulting from cell-wall bending; the stretching is of the same order as the bending deformation (Warner and Edwards, 1988). The Young's modulus of the foam is given by the sum of the stretching and bending contributions:

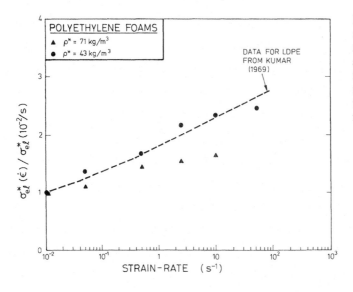

Figure 6.13 The plateau stress, σ_{el}^*, plotted against strain-rate. The figure shows that the inherent foam strength has the same rate dependence as that of the solid of which it is made.

$$\frac{E^*}{E_s} = \left(\frac{\rho^*}{\rho_s}\right)\left[1 + \left(\frac{\rho^*}{\rho_s}\right)\right]. \tag{6.21}$$

An initial pressure within the liquid pre-stresses the cell wall in tension. Additional uniaxial stress requires the same compensation of volume changes from bending deformations by cell-wall stretching as in the unpressurized case. Young's modulus is independent of initial cell pressure.

The cell-wall stretching affects failure, too: we expect that the compressive strength of the foam will scale with relative density in the same way as the modulus (i.e. roughly linearly). The initial pressure, p, of the liquid does affect the strength: it is reduced by a value equal to the initial pre-stress in the cell wall, roughly $p(\rho_s/\rho^*)$.

Open-cell foams containing a fluid

Cell fluids contribute to the strength of open-cell foams in a completely different way. When the foam is compressed, the fluid it contains is squeezed out; when extended, fluid is drawn in. The fluid has a viscosity, so work is done forcing it through the interconnected porosity of the foam. The faster the foam is deformed the more work is done; the pore fluid introduces a strongly strain-rate dependent contribution to the strength.[†]

There are a number of ways of analysing this (Kosten and Zwikker, 1939; Gent and Rusch, 1966a, b; Hilyard and Kanakkanatt, 1970; Warner and Edwards, 1988). The simplest is to treat the foam as a porous medium, characterized by an *absolute permeability*, K; then fluid flow through it is described by Darcy's law (Darcy, 1856):

$$u = \frac{K}{\mu}\frac{\mathrm{d}p}{\mathrm{d}x} \tag{6.22}$$

where u is the velocity of the fluid, μ is its dynamic viscosity and $\mathrm{d}p/\mathrm{d}x$ is the pressure gradient. The units of K are m^2, and those for dynamic viscosity are Ns/m^2. Typical values are listed in Table 6.1. Note that at sufficiently high Reynolds numbers inertial effects, neglected in Darcy's law, become important (Philipse and Schram, 1991).

[†] If the flow is turbulent, further work must be done to compensate for the loss of kinetic energy. But turbulence is possible only when a critical Reynolds number $\mathbb{Re} = \rho_l u d/\mu$ is exceeded (where d is the diameter of the flow channel – in this case, the cell size; u is the fluid velocity, ρ_l the density of the liquid and μ its dynamic viscosity). Experiments show that the critical Reynolds number for the transition between laminar and turbulent flow depends on the tortuosity of the channel. For flow through a straight pipe the transition occurs at $\mathbb{Re} = 2100$; for flow through soils (with large tortuosity and low porosity) it is at $\mathbb{Re} = 10$. We assume that the transition for foams lies between these two values, at a Reynolds number of about 100. All published data for deformation of fluid-filled foams are in the range $\mathbb{Re} < 100$, so we neglect the contribution of turbulence.

Table 6.1 Dynamic viscosities at 20°C

Fluid	Dynamic viscosity (Ns/m^2)
CO_2	1.48×10^{-5}
Air	1.85×10^{-5}
Water	1.0×10^{-3}
Ethyl alcohol	1.2×10^{-3}
Light machine oil	0.1
Glycerine	1.4
Water/glycerine mixtures	0.001 to 1.0

For a permeable material with pores of average diameter d, the permeability (Brace, 1977) is

$$K = Ad^2(1 - \rho^*/\rho_s)^{3/2} \tag{6.23}$$

where the constant A generally takes the empirical value 0.4, and the term $(1 - \rho^*/\rho_s)^{3/2}$ describing the porosity can usually be taken as equal to 1 because the relative densities of foams are always low. The equation describes the observed permeabilities of foams well (Gent and Rusch, 1966a). Foams typically have permeabilities in the range 10^{-10} to 10^{-8} m^2.

The contribution of viscous flow to foam strength is calculated in the way illustrated by Fig. 6.14. It shows a block of foam of base L, height H and unit depth, compressed at a velocity V. The flux of fluid, q, through each of the two vertical faces is

$$q = \frac{VL}{2H} = \frac{\dot{\epsilon}L}{2} \tag{6.24}$$

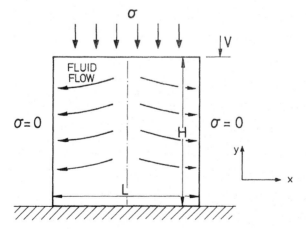

Figure 6.14 The compression of a fluid-filled, open-cell foam. The pressure gradient, of order σ/L, expels the fluid through the two vertical faces.

where H is the current height and $\dot{\epsilon} = V/H$ is the strain-rate. The *average* flux across any vertical internal surface is one-half of this, because the liquid flows outwards from the centre of the block. Inserting this factor of $\frac{1}{2}$, and substituting the result into Darcy's law, gives

$$\frac{\dot{\epsilon}L}{4} = -\frac{K}{\mu}\frac{\mathrm{d}p}{\mathrm{d}x}. \tag{6.25}$$

The pressure gradient across the foam of Fig. 6.14 is proportional to

$$\frac{\mathrm{d}p}{\mathrm{d}x} \approx -\frac{\sigma}{L}.$$

The 'size' of the pores, d, is obviously proportional to the cell edge-length l at the start of deformation, but as the foam is compressed the pores become narrower. Gent and Rusch (1966a) suggest that

$$d \propto l(1 - \epsilon)^{1/2}. \tag{6.26}$$

Substituting these into Eqn. (6.25), with K defined by Eqn. (6.23), gives the contribution of pore fluid to the strength of open-cell foams:

$$\sigma_g^* = \frac{C\mu\dot{\epsilon}}{1-\epsilon}\left(\frac{L}{l}\right)^2 \tag{6.27}$$

where the various constants of proportionality have been combined in the constant C, of order unity. The equation tells us that the contribution of a viscous pore fluid to the strength σ^* of an open-cell foam is proportional to the strain-rate $\dot{\epsilon}$, to the viscosity of the fluid, μ, and to the reciprocal of the cell size, squared. This contribution to the strength has a temperature dependence which is the same as that of the fluid viscosity; for many fluids this is exponential:

$$\mu = \mu_0 \exp Q/RT$$

where Q is the activation energy for viscous flow and R is the gas constant.

When a polyurethane sponge is saturated with water and squeezed quickly between the hands, one can feel, without difficulty, the added resistance to compression caused by the expulsion of the water. But the strain-rate in such an experiment is quite high – greater than $1/s$ – and there have been very few properly instrumented studies, extending to high strain-rates, to characterize the effect quantitatively. Results of experiments on flexible, open-cell polyurethane foams are shown in Figs. 6.15 and 6.16 (Tyler and Ashby, 1986). When filled with air and compressed at strain-rates between 10^{-3} and $20/s$ the foams showed no detectable rate-dependence of the plateau stress, σ^*, either at the beginning of the plateau ($\epsilon = 0.05$) or at two points further along the stress–strain curve (Fig. 6.15). But when filled with water–glycerine mixtures (to give viscosities between 10^{-3} and $1\,\mathrm{Ns/m}^2$), a transition to a linear dependence on strain-rate was found (Fig. 6.16). The broken lines on Fig. 6.16 are plotted using Eqn.

(6.27) with a cell size of 0.4 mm and a sample size, $2L$, of 80 mm. The simple model gives a good description of the observations.

Viscous effects of this sort depend strongly on the cell size – or, more precisely, on the size of the apertures which connect one cell wall to the next.

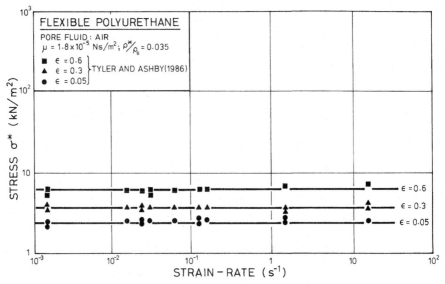

Figure 6.15   The strain-rate dependence of the plateau strength of an elastomeric foam. The stress corresponding to three points on the stress–strain curve are plotted, corresponding to $\epsilon = 0.05$ (where $\sigma^* = \sigma_{el}^*$), $\epsilon = 0.3$ and $\epsilon = 0.6$.

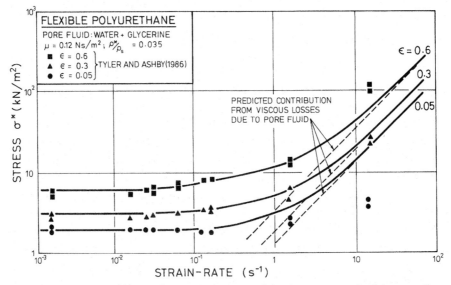

Figure 6.16   The strain-rate dependence of the plateau strength of the same foam as Fig. 6.15, when saturated with a water-glycerine mixture, at the same three strains. The broken lines are a plot of eqn. (6.27).

Common polymer foams (like that described in the last paragraph) have cell sizes of around 0.5 mm; then gases flow so freely through the foam that the viscous contribution to the strength is negligible even at impact strain-rates. But techniques are evolving for producing microporous foams with connected porosity, such as Polyhipe (Gregory, 1987); they have a cell size of order 20 μm and an aperture diameter of about 10 μm (Fig. 6.17). Equation (6.27) then shows that the contribution to the strength, even from a gas, is important at modest strain-rates; and the recovery time of the foam when compressed and released (which can be calculated from Eqn. (6.27)) is several seconds.

6.3   Anisotropy of foam properties

Most foams are anisotropic. Polymer foams made by pouring the polymer plus a hardener and a foaming agent into an open mould (so that it rises like a loaf of bread) usually have cells which are elongated in the rise direction. Those made by spraying (which is like laying down succesive layers of pressurized shaving cream) also rise and in these, too, the cells are usually elongated in the spray direction. And natural cellular materials like bone, wood and coral, which have evolved so as to meet natural needs in an efficient way, are always anisotropic, sometimes strongly so. A description of foam properties in terms of the (isotropic) properties of the solid of which it is made and the relative density does not allow for this. To progress further we must examine the origins of the anisotropy and ways of characterizing and modelling it.

Anisotropy can arise in two quite different ways, which can superpose. The more obvious is *structural anisotropy*: direction-dependent foam properties directly attributable to the shape of the cells, as in the open-mould foaming mentioned above. The other arises from *material anisotropy* in the properties of the cell wall itself – the walls of the cells of many natural materials, notably woods, are examples. The two are, in principle, independent, but, at least in nature, they are almost always combined to maximize some aspect of structural efficiency.

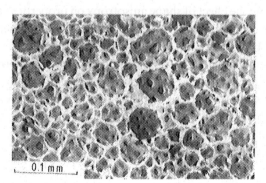

0.1 mm

Figure 6.17 A microporous elastomeric foam, Polyhipe. The cell diameter is 20 μm, and the cell-wall apertures are less than 10 μm in diameter. Ordinary polymer foams have a cell size at least 10 times larger than this.

A rigorous treatment of anisotropy poses considerable theoretical difficulties. A framework for describing structural anisotropy in terms of the *fabric* – a second-rank tensor characterizing the geometric arrangement of the porous or cellular microstructure – is developed by Cowin (1985). The tensor – if its components can be determined – is helpful in the analysis of the linear-elastic response of anisotropic foams, but the extension of this approach to non-linear response (buckling, plasticity, fracture) has not been demonstrated; nor does it include cell-wall material anisotropy. Attempts to deal with this (e.g. Asterly, 1996) generally assume an ideal, simplified shape for the cells (long, parallel hexagons, for instance) and seek, using numerical methods, to compute the mechanical response of the cellular array as a function of anisotropic cell-wall properties. The approach allows simulation of specific cell geometries, but lacks the generality of Cowin's approach. In this section we attempt a less ambitious approach, but one which gives some insight into the origins of the anisotropy and its sensitivity to changes in cell shape.

The anisotropy in cell shape is conveniently measured by the ratio of the largest cell dimension to the smallest; we have called this the shape–anisotropy ratio, R (see Chapter 2, Section 2.5). The value of R for polymer foams is typically about 1.3 (Fig. 6.18a); it varies from 1 for an isotropic foam to 10 or more for one which is very anisotropic, like the pumice shown in Fig. 6.18b. The properties depend strongly on R: the anisotropy in Young's modulus varies at least as fast as R^2, and that of strength, as R. So the anisotropy of foam properties cannot be ignored in engineering design.

In this section we analyse anisotropy in axisymmetric foam structures; then the shape anisotropy is characterized by a single value of R. Many foams are approximately axisymmetric, but some (depending on how they are made) are orthotropic: then all three principal dimensions of the cell differ, and two values of R, or more elaborate schemes such as that of Harrigan and Mann (1984), are needed to characterize it (Section 2.5). The method used here is easily generalized to the orthotropic case: the required equations for modulus ratios, etc., are given by Huber and Gibson (1988). Generalization to yet lower symmetry is more difficult, and we shall not attempt it.

The methods of analysing the effect of shape anisotropy on properties are simple extensions of those developed in Chapter 5. They build on the earlier work of Patel (1969), Patel and Finnie (1969, 1970) and of Gibson and Ashby (1982), who recognized the critical importance of bending and buckling in determining foam properties: the non-linear relations between foam properties and density demonstrated in Chapter 5 all derive from this. The method differs fundamentally from those adapted by Kanakkanatt (1973), Mehta and Colombo (1976), Cunningham (1981) and Hilyard (1982), who implicitly assume that all foam properties depend linearly on relative density.

(a) Linear elasticity: the moduli

Figure 6.19 shows an idealized open cell, typifying that in an anisotropic foam
with axisymmetry: the rise direction is parallel to X_3. Let the Young's modulus
of the foam in the rise direction be E_3^*, and that in the two directions normal to
this be E_1^*. We seek to calculate the ratio E_3^*/E_1^* in terms of the anisotropy ratio,
R. The figure shows how this is done by a slight modification of the method used
in Chapter 5, Section 5.3a. A load in the X_3 direction is carried by the four
beams of length l, which respond by bending. The force F on each beam is proportional to $\sigma_3 l^2$, and the deflection δ_3 of each is proportional to $Fl^3/E_s I$
(where, as always, E_s is the modulus of the solid and I the second moment of
area of the cell edge – see Section 5.3). The strain ϵ_3 is related to the displacement
δ_3 by $\epsilon_3 = \delta_3/h$, where h is the height of the cell. Assembling these, we find,

$$E_3^* = \frac{\sigma_3}{\epsilon_3} = \frac{CE_s Ih}{l^5} = CE_s \left(\frac{t}{l}\right)^4 \frac{h}{l} \tag{6.28}$$

where C is a constant of proportionality. The equation reduces to the result given
in Section 5.3 when $h = l$, as, of course, it should.

A load in the X_1 direction is carried by two beams of length l and two of length
h. The deflection, δ_1, of both sets must be equal, so the load carried by the longer

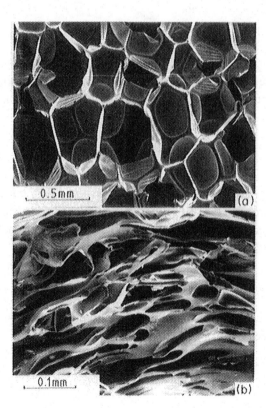

Figure 6.18 (a) An
axisymmetric rigid
polyurethane foam
$(\rho^*/\rho_s = 0.027)$ sectioned
in a plane containing the
rise direction. The shape–
anisotropy ratio, R, for this
foam is 1.2. (b) A section
through volcanic pumice,
a natural glass foam.
Deformation of the foam
while still viscous has
distorted the cells, giving a
shape–anisotropy ratio R
of 6.

0.5mm

(a)

0.1mm

(b)

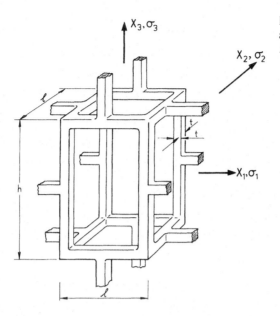

Figure 6.19 An axisymmetric unit cell with $R = 1.5$.

beams is less than that carried by the shorter ones; the first is proportional to $E_s I / h^3$, the other to $E_s I / l^3$. The total force is $\sigma_1 h l$, and the strain $\epsilon_1 = \delta_1 / l$. Assembling these, we obtain

$$E_1^* = \frac{\sigma_1}{\epsilon_1} = \frac{C E_s I}{2h} \left(\frac{1}{l^3} + \frac{1}{h^3} \right) = \frac{C E_s}{2} \left(\frac{t}{l} \right)^4 \frac{l}{h} \left(1 + \left(\frac{l}{h} \right)^3 \right) \tag{6.29}$$

which, again, reduces to the isotropic result of Section 5.3 when $h = l$. Taking the ratio of these two equations, and writing $R = h/l$ gives the *Young's modulus anisotropy ratio*:

$$\frac{E_3^*}{E_1^*} = \frac{2R^2}{(1 + (1/R)^3)}. \tag{6.30}$$

The modulus ratio depends strongly on anisotropy: cells with a shape anisotropy of 2 have a modulus anisotropy of nearly 8. When cells are closed and membrane stresses are important, they contribute to anisotropy such that an additional term

$$(1 - \phi) \frac{2R}{1 + (1/R)}$$

appears in Eqn. (6.30).

A similar analysis for the shear modulus shows that it is much less sensitive to anisotropy in cell shape. We find

$$\frac{G_{31}^*}{G_{12}^*} = \frac{2}{1 + R}. \tag{6.31}$$

Note that these results for the elastic moduli are special cases of Cowin's more general relationship between the elastic moduli and the fabric of porous materials (Cowin, 1991).

Poisson's ratio is the negative ratio of a lateral to an applied strain. As a result, it is independent of relative density; it depends only on the cell geometry. Dimensional arguments of the type used to calculate the Young's modulus and the shear modulus offer no insight into its dependence on cell geometry. We do not attempt a calculation of it here.

(b) Non-linear elasticity: elastic buckling

Some understanding of the ratio of the elastic-buckling stresses, $(\sigma_{el}^*)_3/(\sigma_{el}^*)_1$, is given by extending the methods of Sections 4.3b and 5.3b. For loading in the X_3 direction (Fig. 6.19) cell walls buckle when the load on them exceeds the Euler load $F_{cr} = n_3^2\pi^2 E_s I/h^2$ (Eqn. (5.14)). The load $F \propto \sigma_3 l^2$ giving

$$(\sigma_{el}^*)_3 = \frac{Cn_3^2\pi^2 E_s I}{h^2 l^2}$$

where n_3 describes the rotational stiffness for this direction of loading, and C is again a constant of proportionality. For loading in the X_1 direction, $F_{cr} = n_1^2\pi^2 E_s I/l^2$ and $F \propto \sigma_1 hl$ giving:

$$(\sigma_{el}^*)_1 = \frac{Cn_1^2\pi^2 E_s I}{hl^3}.$$

Taking the ratio of these two and inserting $R = h/l$ gives the *elastic collapse ratio*

$$\frac{(\sigma_{el}^*)_3}{(\sigma_{el}^*)_1} = \frac{n_3^2}{n_1^2}\frac{1}{R}. \tag{6.32}$$

Two effects compete in determining the ratio. The longer columns in the X_3 direction, if unconstrained at their ends, would buckle more easily than the shorter ones in the X_1 direction. But the rotational constraint at the ends of the long columns is greater than that at the shorter ones, tending to stabilize them. Further, elongation of the cells increases the angle θ at which the edges meet, and the factor $\cos\theta$ in Eqn. (4.20) (or its three-dimensional equivalent) changes in a way which delays buckling in the X_3 direction. Experiments, described below, show that the competing effects almost cancel, giving a weak dependence on R.

(c) Plastic collapse

Plastic collapse is analysed in a similar way, using the methods of Section 5.3c. Collapse occurs when two plastic hinges form in each cell edge (Fig. 5.19), and this requires that the fully-plastic moment M_p of the cell wall is exceeded. For loading in the X_3 direction (Fig. 6.19) four edges of length l must collapse. The moment M on each is proportional to Fl and the force F is proportional to $\sigma_3 l^2$, giving

$$(\sigma_{pl}^*)_3 = \frac{CM_p}{l^3}$$

where C, once again, is a constant of proportionality. For loading in the X_1 direction, on the other hand, two edges of length l and two of length h must collapse. The first pair support a force proportional to M_p/l, while the other pair support a force proportional to M_p/h. The total force is $\sigma_1 lh$, giving

$$(\sigma_{pl}^*)_1 = \frac{CM_p}{2lh}\left(\frac{1}{l} + \frac{1}{h}\right).$$

Taking the ratio and substituting $R = h/l$ leads to the *plastic collapse ratio*

$$\frac{(\sigma_{pl}^*)_3}{(\sigma_{pl}^*)_1} = \frac{2R}{1 + \left(\dfrac{1}{R}\right)}. \tag{6.33}$$

The cells are thus stronger in the rise direction, although the anisotropy in strength is not as large as that in stiffness (Eqn. (6.30)): cells with a shape anisotropy of 2 should have a strength anisotropy of around 2.6.

(d) Brittle crushing

The argument for the crushing strength (Section 5.3e) parallels that for plastic collapse. The result is Eqn. (6.33) with σ_{pl}^* replaced by σ_{cr}^*.

(e) The fracture toughness

The fracture toughness of an anisotropic foam depends on the direction in which the crack propagates. This is best defined with two subscripts, the first indicating the normal to the crack plane, the second the direction of crack propagation: thus $(K_{IC}^*)_{13}$ is the fracture toughness associated with a crack on the plane normal to X_1, advancing in the X_3 direction; the crack front lies parallel to X_2. Repeating the argument of Section 5.4d, we find

$$(K_{IC}^*)_{31} = (K_{IC}^*)_{32} = C\sigma_{fs}\left(\frac{t}{l}\right)^3\sqrt{l}$$

$$(K_{IC}^*)_{21} = (K_{IC}^*)_{12} = C\sigma_{fs}\left(\frac{t}{l}\right)^3\left(\frac{l}{h}\right)\sqrt{l}$$

$$\tag{6.34}$$

and

$$(K_{IC}^*)_{23} = (K_{IC}^*)_{13} = C\sigma_{fs}\left(\frac{t}{l}\right)^3\left(\frac{l}{h}\right)^2\sqrt{h}$$

where C is again a constant of proportionality. The ratios of the fracture toughnesses for a given crack length a, are:

$$\frac{(K_{IC}^*)_{31}}{(K_{IC}^*)_{21}} = R; \quad \frac{(K_{IC}^*)_{31}}{(K_{IC}^*)_{23}} = R^{3/2}; \quad \frac{(K_{IC}^*)_{21}}{(K_{IC}^*)_{23}} = R^{1/2}. \tag{6.35}$$

(f) Comparison with data for anisotropic foams

Anisotropy in the moduli, strength and toughness of foams is a common observation. In polymer foams, there is general agreement that the Young's modulus is greatest in the rise direction – that is, the direction of greatest elongation of the cells (Mal'tsev, 1968; Patel, 1969; Kanakkanatt, 1973; Gupta et al., 1986), and that the plastic collapse strength of a rigid plastic foam, too, is greatest in this direction (Patel, 1969; Gupta et al., 1986). But despite the frequent reports of anisotropy, there are very few which contain enough information on cell shape and on properties in the three principal directions to allow a critical comparison with the theory. In two studies, Gupta et al. (1986) and Huber and Gibson (1988) measured both the mean intercept lengths and the stress–strain curves of foams in the three principal directions. The foams were roughly axisymmetric with a shape anisotropy ratio, R, of between 1.2 and 1.6; micrographs and the characterization chart for one of the rigid polyurethane foams were shown in Fig. 2.19 and Table 2.4. Typical load-deflection curves for flexible and rigid polyurethane foams loaded in the three principal directions are shown in Fig. 6.20. The data for the Young's modulus, the elastic and plastic collapse stresses and the fracture toughness of the foams are compared with the model in Figs. 6.21 to 6.24. The agreement for Young's modulus (Eqn. (6.30)), the property most sensitive to cell shape, is excellent. The plastic collapse strength ratio, too, is well described by Eqn. (6.33) although the model underestimates the data slightly (Fig. 6.22). The elastic collapse stress is least dependent on R (Fig. 6.23), perhaps for the reason outlined earlier: shape anisotropy elongates some of the cell walls, but also constrains their ends more rigidly, so that their buckling load scarcely changes. Measurements of the fracture toughness, shown in Fig. 6.24, were made only in the least sensitive direction of loading (for which the model suggests $(K_{IC}^*)_{12}/(K_{IC}^*)_{13}$ varies as $R^{1/2}$); additional data for foams with a larger shape anisotropy ratio are needed to confirm this prediction.

It is clear, then, that foams with elongated cells are strongly anisotropic in their mechanical properties. It is not uncommon for rigid polymer foams to have a shape–anisotropy ratio of 1.5 which, both theory and experiment show, leads to a modulus ratio of 3.5 and a plastic strength ratio of 1.8 Flexible foams, too, can be anisotropic, although their elastic collapse strengths are less sensitive to cell elongation. In engineering design with foams, anisotropy in properties should not be ignored.

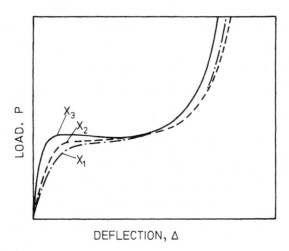

Figure 6.20 Load–deflection curves measured parallel to the three principal axes of the structure for: (a) an elastomeric foam, (b) a rigid (plastic) foam.

(a)

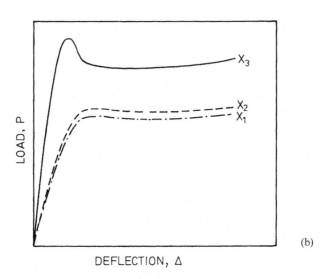

(b)

6.4   Multiaxial loading

When foams are used in critical load-bearing applications, their response to multiaxial stresses becomes important. In their use as insulation, too, the large temperature changes subject the foam to biaxial tension and compression. And in designing lightweight sandwich structures, or in critical packaging applications (like that involved in transporting nuclear or chemical waste, for instance), the engineer again requires precise information about the way in which the foam will behave under biaxial or triaxial loading.

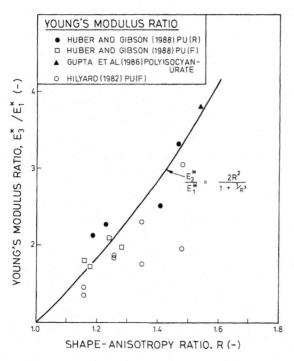

Figure 6.21 The ratio of the moduli E_3^*/E_1^*, plotted against the shape-anisotropy ratio, R.

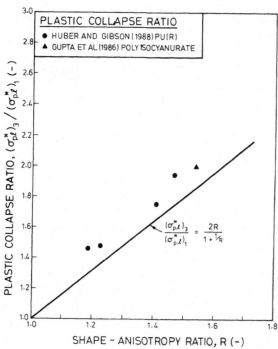

Figure 6.22 The ratio of the plastic collapse strengths $(\sigma_{pl}^*)_3/(\sigma_{pl}^*)_1$ plotted against the shape-anisotropy ratio, R.

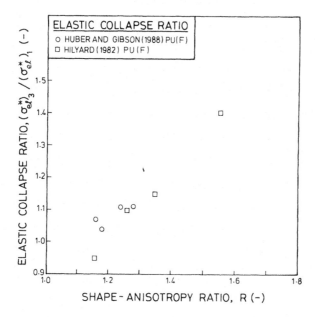

Figure 6.23 The ratio of the elastic collapse strengths $(\sigma_{el}^*)_3/(\sigma_{el}^*)_1$ plotted against the shape–anisotropy ratio, R. The dependence is much weaker than for E^* or σ_{pl}^*.

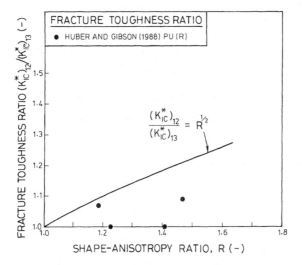

Figure 6.24 The ratio of the fracture toughnesses $(K_{IC}^*)_{12}/(K_{IC}^*)_{13}$ plotted against the shape–anisotropy ratio, R.

One way of presenting this is as a *failure criterion*: it is an equation relating the principal stresses at failure and can be thought of as the equivalent, for foams, of the Von Mises or Tresca yield criteria for metals, or the Coulomb criterion for soils. Sometimes a simple equation describing failure can be derived (see below); but more often the *mechanism* of failure itself depends on the stress state, and two or more equations are needed. Then it is helpful to plot a *failure surface* in stress space which shows the combination of stresses which cause failure; it is the inner envelope of the (intersecting) surfaces for the individual mechanisms. In this section we derive failure criteria and failure surfaces for foams, and compare the results with experiments. We start with the elastic response to multiaxial loading, and then consider the surfaces for yield, brittle failure, and elastic collapse. In each case we consider the triaxial loading of the cell shown in Fig. 6.25. The normal stresses are σ_1, σ_2 and σ_3, and are positive when tensile, negative when compressive.

(a) Elastic deformation

While the foam is linear-elastic it obeys Hooke's law. Then the complete constitutive equation for elastic deformation is the standard one (Appendix, Eqn. (A8)). If the principal axes of stress coincide with those of the cell shape:

$$\epsilon_1 = \frac{1}{E_1^*}(\sigma_1 - \nu_{12}^*\sigma_2 - \nu_{13}^*\sigma_3)$$

$$\epsilon_2 = \frac{1}{E_2^*}(\sigma_2 - \nu_{23}^*\sigma_3 - \nu_{21}^*\sigma_1). \tag{6.36}$$

$$\epsilon_3 = \frac{1}{E_3^*}(\sigma_3 - \nu_{31}^*\sigma_1 - \nu_{32}^*\sigma_2)$$

But, for a complete description, a correction is needed. It arises because, in calculating Young's modulus, the axial extension or compression and shear deformations of the cell edges were neglected. The low Young's modulus of foams in uniaxial compression (Eqn. (5.6)),

$$E^* = E_s \left(\frac{\rho^*}{\rho_s}\right)^2$$

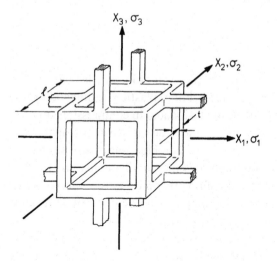

Figure 6.25 A cell subjected to multiaxial loading.

is caused, as we have seen, by cell-edge bending (Section 5.3). When the contributions of axial and shear deformations, proportional to $Fl/E_s t^2$, are added, it becomes

$$E^* = E_s \left(\frac{\rho^*}{\rho_s}\right)^2 \left\{\frac{1}{1 + \rho^*/\rho_s}\right\}. \qquad (6.37)$$

Under triaxial loading, a set of stresses σ_1, σ_2, and σ_3 can be found such that the bending moment on the cell edges caused by one of them is cancelled by the other two. For an isotropic foam, bending is eliminated when the deviatoric stress is zero and $\sigma_1 = \sigma_2 = \sigma_3$. Then only the axial deformation remains, leading to a bulk modulus, K^*, of (Christensen, 1986, Warren and Kraynik, 1988):

$$K^* = \frac{1}{9} E_s \left(\frac{\rho^*}{\rho_s}\right). \qquad (6.38)$$

In practice, any imperfections in the foam (such as non-uniformities in relative density or initially bent cell walls) induce bending of the cell walls, reducing the bulk modulus to a value much less than this. The Poisson's ratio for the isotropic foam is then given by

$$\nu^* = \frac{1}{2} - \frac{E^*}{6K^*} \qquad (6.39)$$

and the linear-elastic behaviour of the isotropic foam is obtained from:

$$\epsilon_1 = \frac{1}{E^*}(\sigma_1 - \nu^*(\sigma_2 + \sigma_3)) \qquad (6.40a)$$

$$\epsilon_2 = \frac{1}{E^*}(\sigma_2 - \nu^*(\sigma_1 + \sigma_3)) \qquad (6.40b)$$

$$\epsilon_3 = \frac{1}{E^*}(\sigma_3 - \nu^*(\sigma_1 + \sigma_2)). \qquad (6.40c)$$

(b) Failure surfaces

'Failure', in the most general sense, is the breakdown of linear elasticity. Three distinct mechanisms can cause this: elastic buckling, plastic yielding and brittle crushing or fracture. Yielding is possible in tension and in compression, and for any ratios of principal stresses; so the yield surface (the surface in principal stress space, with σ_1, σ_2 and σ_3 as axes, showing the combination of stresses which cause yield) is a closed shape. Brittle fracture is the same: any loading path in stress space will, ultimately, cause fracture if no other mechanism intervenes, so the fracture surface, too, is closed. Elastic collapse is different: it can occur only in compressive stress states, not in tension, so that the surface describing the combination of stresses which cause elastic collapse can truncate the yield or the fracture surface in the compressive quadrant but it cannot, by itself, give a complete failure criterion. For this reason we treat it last. The failure surface for each mechanism is developed in turn, following the approach of Gibson *et al.* (1989).

(c) The yield surface

We have seen in Chapter 5 that, under uniaxial loading, plastic collapse occurs by bending, when (Eqn. (5.27a)) the collapse stress is:

$$\sigma_{pl}^* = 0.3\sigma_{ys}\left(\frac{\rho^*}{\rho_s}\right)^{3/2}. \tag{6.41}$$

If the three principal stresses are all equal, the cell walls of an isotropic foam (like that sketched in Fig. 6.25) are subjected to simple tension or simple compression only; there is no bending. Then (ignoring the possibility of buckling for the moment) plastic collapse occurs at a much higher stress than before: it is when the stress in the cell wall exceeds the yield stress, σ_{ys}, that is, when

$$\sigma_1 = \sigma_2 = \sigma_3 = \frac{\sigma_{ys}}{3}\left(\frac{\rho^*}{\rho_s}\right). \tag{6.42}$$

The factor of 3 arises because the stress in any one direction loads only one third of the cell walls of the unit cell. The hydrostatic loading of a random distribution of cell walls in a sphere produces the same result; it is not sensitive to cell geometry.

Under any other combination of loads, the cell walls suffer both bending and axial loading. The fully plastic moment of a beam of square section, l, in simple bending was used in the last chapter (Eqn. (5.23)) to calculate the yield strength of an open-cell foam; it is simply

$$M_p = \tfrac{1}{4}\sigma_{ys}t^3.$$

When an axial stress σ_a, either tensile or compressive, is added, the moment which will cause plastic collapse is reduced, as described in Chapter 4 (Eqn. 4.84). Neglecting geometry changes, the result is

$$M_p = \tfrac{1}{4}\sigma_{ys}t^3\left(1 - \left(\frac{\sigma_a}{\sigma_{ys}}\right)^2\right). \tag{6.43}$$

We obtain the equation for the yield surface by substituting for M_p and σ_a in this equation. The average axial stress σ_a in a cell wall is proportional to the mean principal stress σ_m:

$$\sigma_a = \frac{C\sigma_m}{(t/l)^2} \tag{6.44}$$

where

$$\sigma_m = \frac{\sigma_1 + \sigma_2 + \sigma_3}{3}$$

and C is a geometric constant of proportionality. The average bending moment, is proportional to the deviatoric stress σ_d:

$$M \propto l^3 \sigma_d = l^3 \{\tfrac{1}{2}[(\sigma_1 - \sigma_2)^2 + (\sigma_2 - \sigma_3)^2 + (\sigma_3 - \sigma_1)^2]\}^{1/2}. \tag{6.45}$$

Inserting these in Eqn. (6.43), noting that $\rho^*/\rho_s \propto (t/l)^2$ and assembling the constants of proportionality into two constants α and β gives

$$\frac{\sigma_d}{\sigma_{ys}} = \pm \alpha \left(\frac{\rho^*}{\rho_s}\right)^{3/2} \left[1 - \left(\frac{\sigma_m}{\beta(\rho^*/\rho_s)\sigma_{ys}}\right)^2\right]. \tag{6.46}$$

The constants are obtained by examining the limits of yield in pure hydrostatic tension and simple axial compression. In pure hydrostatic tension ($\sigma_d = 0$) collapse occurs when (Eqn. (6.42)):

$$\sigma_m = \frac{\sigma_{ys}}{3} \left(\frac{\rho^*}{\rho_s}\right)$$

from which $\beta = 1/3$. In uniaxial compression $\sigma_m = \sigma_1/3$ and $\sigma_d = \sigma_1$. Substituting these values into Eqn. (6.46) and making use of Eqn. (6.41) gives:

$$\alpha[1 - 0.09(\rho^*/\rho_s)] = 0.3.$$

For all interesting values of relative density (less than about 0.3) we find $\alpha \sim 0.3$. The yield criterion for foams then becomes:

$$\frac{\sigma_d}{0.3\left(\dfrac{\rho^*}{\rho_s}\right)^{3/2}\sigma_{ys}} = \pm\left[1 - \left(\frac{3\sigma_m}{(\rho^*/\rho_s)\sigma_{ys}}\right)^2\right]. \tag{6.47}$$

It is helpful to rewrite this, using the uniaxial compressive strength, Eqn. (6.41), as the normalizing factor, giving:

$$\pm\frac{\sigma_d}{\sigma_{pl}^*} + 0.81\left(\frac{\rho^*}{\rho_s}\right)\left(\frac{\sigma_m}{\sigma_{pl}^*}\right)^2 = 1. \tag{6.48}$$

The yield surface is an ellipsoid, extended along the direction of pure hydrostatic tension and compression. It is shown, in two sections, in Figs. 6.26 and 6.27. The first is for axisymmetric loading ($\sigma_2 = \sigma_3$); the surface becomes increasingly elongated as the density ρ^*/ρ_s decreases, because the yield strength in simple compression is less than that in hydrostatic tension by the factor $0.9(\rho^*/\rho_s)^{1/2}$ (Eqn. (6.48)). The second is the section for plane stress (or biaxial) loading ($\sigma_3 = 0$). It is shown on the same scales as Fig. 6.26. In the tension–tension and compression–compression quadrants it is almost a surface of constant maximum principal stress. But in the tension–compression quadrants it is more nearly a surface of constant shear stress.

Constitutive equations, describing the post-yield behaviour of an isotropic elastic–perfectly-plastic open-cell foam, have been developed by Triantafillou and Gibson (1990b). Description of the constitutive equations is beyond the

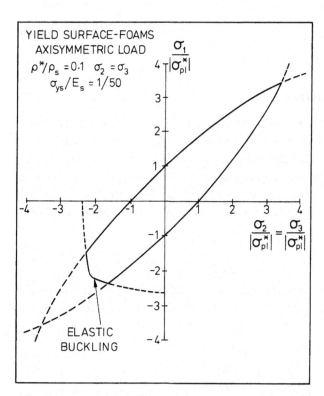

Figure 6.26 The yield surface for a plastic foam, for axisymmetric loading. The surface, which is constructed for a relative density of 0.1, is truncated on the compression side by elastic buckling (taking $\sigma_{ys}/E_s = 0.02$).

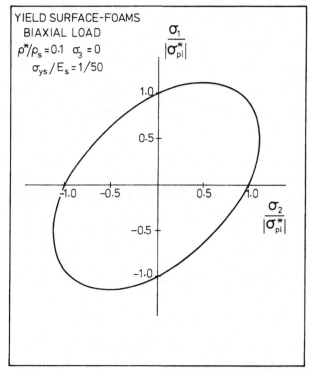

Figure 6.27 The yield surface for a plastic foam, for plane stress. It is an oblique section through the elongated ellipsoid of which Fig. 6.26 is a different section.

scope of this book; the reader is referred to their paper for more detailed information.

(d) Brittle fracture

If the cell of Fig. 6.25 is made of a brittle material the failure surface is that of brittle crushing or fracture. We first assume that the foam does not contain macroscopic flaws, and relax this condition later. Then cell walls fail when the maximum surface stress in the wall exceeds the modulus of rupture, σ_{fs}. The maximum surface stress in a beam of section t^2 subject to a bending moment M (Eqn. 5.32) is $\sigma = 6M/t^3$. When an axial stress σ_a is superimposed, the failure condition becomes:

$$\frac{6M}{t^3} + \sigma_a = \sigma_{fs}. \tag{6.49}$$

Noting that σ_a and M are related to σ_m and σ_d as before (Eqns. (6.44) and (6.45)), replacing $(t/l)^2$ by ρ^*/ρ_s and aggregating the constants of proportionality gives the failure criterion:

$$\frac{\sigma_d}{\sigma_{fs}} = \pm\alpha'\left(\frac{\rho^*}{\rho_s}\right)^{3/2}\left[1 - \frac{\sigma_m}{\beta'(\rho^*/\rho_s)\sigma_{fs}}\right]. \tag{6.50}$$

The constants α' and β' are again determined by examining the limits in pure hydrostatic tension and simple compression. In hydrostatic tension ($\sigma_d = 0$) failure occurs when:

$$\sigma_m = \frac{\sigma_{fs}}{3}\left(\frac{\rho^*}{\rho_s}\right)$$

from which $\beta' = 1/3$. In uniaxial compression $\sigma_d = \sigma_1$ and $\sigma_m = \sigma_1/3$ and the uniaxial compressive strength is $0.2(\rho^*/\rho_s)^{3/2}\sigma_{fs}$. Substituting in Eqn. (6.50) gives

$$\alpha' = \frac{0.2}{1 + 0.2(\rho^*/\rho_s)^{1/2}}.$$

For typical values of $(\rho^*/\rho_s) = 0.1$, we find $\alpha' = 0.19$. The failure surface for brittle tensile failure of the cell wall is then:

$$\frac{\sigma_d}{\sigma_{fs}} = \pm 0.19\left(\frac{\rho^*}{\rho_s}\right)^{3/2}\left[1 - \frac{3\sigma_m}{\sigma_{fs}(\rho^*/\rho_s)}\right]. \tag{6.51}$$

Again, it is helpful to normalize with respect to the uniaxial crushing strength σ_{cr}^*:

$$\pm\frac{\sigma_d}{\sigma_{cr}^*} + 0.6\left(\frac{\rho^*}{\rho_s}\right)^{1/2}\frac{\sigma_m}{\sigma_{cr}^*} = 1. \tag{6.52}$$

Figures 6.28 and 6.29 show the same two sections through the failure surface as before: one for axisymmetric loading, the other for biaxial (plane stress) loading.

As with the yield surface, the brittle-fracture surface is elongated in the direction of pure hydrostatic tension. The section describing biaxial loading gives a maximum principal stress criterion in two quadrants and a maximum shear stress criterion in the other two.

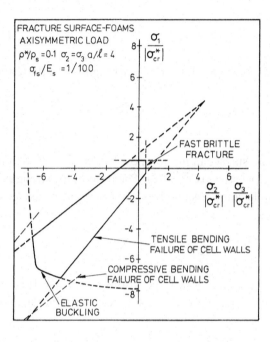

Figure 6.28 The fracture surface for a brittle foam, for axisymmetric loading. The surface, which is constructed for a relative density of 0.1, is truncated on the tension side by fast, flaw-nucleated fracture; and it closes on the compression side because of the elastic buckling of the cell walls.

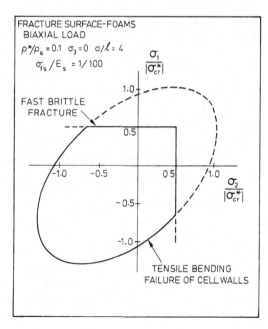

Figure 6.29 The fracture surface for a brittle foam, for plane stress. It is an oblique section through the elongated surface of which Fig. 6.28 is a different section.

When the foam contains flaws or defects (as most foams do) the fracture strength in tension is less than the crushing strength in compression, and the failure surface becomes more asymmetric. Fast fracture by flaw propagation was analysed in Section 5.4d: failure occurs when the maximum principal tensile stress becomes equal to:

$$\sigma_{fr}^* = \frac{C K_{IC}^*}{\sqrt{\pi a}}$$

where the initial flaw size is $2a$ and K_{IC}^* is the fracture toughness of the foam (Eqn. 5.47):

$$K_{IC}^* = C_8 \sigma_{fs} \sqrt{\pi l} \left(\frac{\rho^*}{\rho_s} \right)^{3/2} \tag{6.53}$$

with $C_8 \approx 0.65$. The failure surface in tension now becomes a box bounded by surfaces of constant principal stress (Fig. 6.28). The ratio of the strength in simple tension to that in simple compression gives a measure of the asymmetry of the surface. It is given by:

$$\frac{\sigma_{fr}^*}{\sigma_{cr}^*} = \sqrt{\frac{l}{a}}. \tag{6.54}$$

Elastomeric foams fail in tension at very large strains by tearing. It is rare that this mode of failure is of interest, and we shall not discuss the failure surface for it.

(e) Elastic buckling

In compressive stress states, foams may fail by elastic buckling. In an open-cell foam, four cell edges meet symmetrically at each vertex in a tetrahedral arrangement. When loaded in uniaxial compression, one of each set of four – the one most nearly aligned with the axis of compression – buckles. It is restrained at its end by the other three which (as explained earlier) tend to prevent rotation. Their effectiveness is measured by the *rotational stiffness* – the restoring moment per unit rotation – caused by the three restraining edges. In a foam under a uniaxial compressive stress, buckling is averaged over all cell orientations, and occurs when the external stress exceeds (Eqn. (5.18a))

$$\sigma_{el}^* = 0.05 E_s \left(\frac{\rho^*}{\rho_s} \right)^2.$$

The effect of multiaxial loads is to change the rotational stiffness of the vertices. The edge which buckles first is still the one carrying the largest axial compression. But compression in the other two directions helps bend the three restraining edges and this reduces their rotational stiffness; tension in the other two directions increases it.

The problem is complicated by the fact that – at least in honeycombs – the buckling mode itself changes with stress state (see Fig. 4.27). It is not known whether a similar change takes place in foams.

Triantafillou *et al.* (1989) calculate the elastic-buckling surface for one assumed mode; their results are summarized in Table 6.2. The elastic-buckling surface is shown truncating the yield and fracture surfaces in Figs. 6.26 and 6.28. It is convex and lies close to the surface of maximum principal stress. Its position relative to the other surfaces depends on the ratio of σ_{ys}/E_s or σ_{fs}/E_s. In practice, the transition between the failure surfaces for elastic buckling and plastic yielding will be smooth, like that for the transition between Euler buckling and yield with changing slenderness ratio for a simple column. This smooth transition is the result of interaction between elastic buckling and plastic collapse modes; it can roughly be described by the Rankine formula, Section 4.4(c). Elastoplastic interactions are extremely difficult to quantify for foams, due to their complex structure; we do not attempt this here.

(f) Failure in anisotropic foams

Consider first the yield surface of an anisotropic foam. The uniaxial plastic-collapse stress, σ_{pl}^*, now depends on direction. Let the values in the three principal directions be $(\sigma_{pl}^*)_1$, $(\sigma_{pl}^*)_2$ and $(\sigma_{pl}^*)_3$; the yield surface intercepts the three axes at these points. We start from the result for the isotropic case (Eqn. (6.48) which we write in the form:

$$\frac{\sigma_d}{\sigma_{pl}^*} = \pm \left\{ 1 - \left(\frac{\sigma_m}{\sigma_h^*}\right)^2 \right\} \tag{6.55}$$

where $\sigma_{pl}^* = 0.3\sigma_{ys}(\rho^*/\rho_s)^{3/2}$ and $\sigma_h^* = \sigma_{ys}(\rho^*/\rho_s)/3$.

We postulate that we can replace the strength–normalized deviatoric stress σ_d/σ_{pl}^* by the quantity

Table 6.2 End restraint, n, for elastic buckling of foams

Load condition	n^2	σ^*/σ_{el}^*†
Uniaxial compression $\sigma_1 = \sigma, \sigma_2 = \sigma_3 = 0$	0.41	1.00
Biaxial compression $\sigma_1 = \sigma_2 = \sigma, \sigma_3 = 0$	0.36	0.88
Hydrostatic compression $\sigma_1 = \sigma_2 = \sigma_3 = \sigma$	0.34	0.83
$\sigma_1 = \sigma, \sigma_2 = \sigma_3 = -\sigma/8$	0.42	1.02
$\sigma_1 = -\sigma/2, \sigma_2 = \sigma_3 = \sigma$	0.37	0.90

†*Note:* σ^* is the value of σ at failure.

$$\tilde{\sigma}_{\mathrm{D}} = \left\{ \frac{1}{2} \left[\left(\frac{\sigma_{11}}{(\sigma_{\mathrm{pl}}^*)_1} - \frac{\sigma_2}{(\sigma_{\mathrm{pl}}^*)_2} \right)^2 + \left(\frac{\sigma_{22}}{(\sigma_{\mathrm{pl}}^*)_2} - \frac{\sigma_{33}}{(\sigma_{\mathrm{pl}}^*)_3} \right)^2 + \left(\frac{\sigma_{33}}{(\sigma_{\mathrm{pl}}^*)_3} - \frac{\sigma_{11}}{(\sigma_{\mathrm{pl}}^*)_1} \right)^2 \right] \right.$$

$$\left. + \left[\left(\frac{\tau_{12}}{(\tau_{\mathrm{pl}}^*)_{12}} \right)^2 + \left(\frac{\tau_{23}}{(\tau_{\mathrm{pl}}^*)_{23}} \right)^2 + \left(\frac{\tau_{31}}{(\tau_{\mathrm{pl}}^*)_{31}} \right)^2 \right] \right\}^{1/2} \tag{6.56}$$

where $(\sigma_{\mathrm{pl}}^*)_i$ is the plastic-yield strength of the foam for uniaxial loading in the i direction and $(\tau_{\mathrm{pl}}^*)_{12}$, $(\tau_{\mathrm{pl}}^*)_{23}$ and $(\tau_{\mathrm{pl}}^*)_{31}$ are the plastic shear strengths of the orthoptropic material (Triantafillou, 1989); for isotropic materials $\sigma_{\mathrm{pl}}^* = \sqrt{3}\tau_{\mathrm{pl}}^*$. Equation (6.56) is a generalization of the yield criterion for orthotropic materials, due to Hill (1950). When expression (6.56) is equal to unity, the foam deforms plastically by a cell-wall bending mechanism. When it is zero there is no tendency to yield by this mechanism; and when this is true, the normal stresses at yield are related by the equations

$$\frac{\sigma_{11}}{(\sigma_{\mathrm{pl}}^*)_1} = \frac{\sigma_{22}}{(\sigma_{\mathrm{pl}}^*)_2} = \frac{\sigma_{33}}{(\sigma_{\mathrm{pl}}^*)_3} . \tag{6.57}$$

Under these conditions, plastic failure is only possible by the axial stretching of the cell walls. This suggests that the quantity $\sigma_{\mathrm{m}}/\sigma_{\mathrm{h}}^*$ be replaced by the quantity

$$\tilde{\sigma}_{\mathrm{m}} = \frac{1}{3} \left[\frac{\sigma_{11}}{(\sigma_{\mathrm{pl}}^*)_1} + \frac{\sigma_{22}}{(\sigma_{\mathrm{pl}}^*)_2} + \frac{\sigma_{33}}{(\sigma_{\mathrm{pl}}^*)_3} \right] \frac{\sigma_{\mathrm{pl}}^*}{\sigma_{\mathrm{h}}^*} \tag{6.58}$$

giving a simple transformation of the axes from σ_1 to $\sigma_{11}\left[\sigma_{\mathrm{pl}}^*/(\sigma_{\mathrm{pl}}^*)_1\right]$ etc. Assembling these results gives the anisotropic yield criterion:

$$\tilde{\sigma}_{\mathrm{D}} = \pm(1 - \tilde{\sigma}_{\mathrm{M}}^2). \tag{6.59}$$

Note that this equation reduces identically to our starting equation when all three uniaxial strengths are the same, and that it corresponds to a distortion of the original yield envelope in the ratios of the uniaxial yield strengths.

The argument for brittle fracture parallels that for yield. The elastic-buckling surface is roughly approximated by normalizing each stress axis by the uniaxial elastic-buckling stress in that direction.

Constitutive equations for an elastic–perfectly-plastic open-cell foam have been developed by Triantafillou and Gibson (1990b). They are beyond the scope of this book; the interested reader is referred to their paper.

(g) Indentation

The special case of indentation (Shaw and Sata, 1966; Wilsea et al., 1975) has been described for both plastic and brittle foams in Chapter 5, Sections 5.3c and 5.3d. We note here that indentation of a foam, unlike that of a dense solid, is not a true multiaxial test. The near-zero 'effective Poisson's ratio' during plastic collapse or brittle crushing means that the material beneath the indenter suffers simple compression, not multiaxial compression. Because of this, little can be learnt about multiaxial response from indentation tests.

(h) Comparison with data for multiaxial loading

Both biaxial and axisymmetric triaxial tests have been performed on a variety of foams which fail by some combination of the elastic-buckling, plastic-yielding, brittle-crushing and brittle-fracture mechanisms described above (Shaw and Sata, 1966; Patel, 1969; Zaslawsky, 1973; Triantafillou *et al.*, 1989; and Triantafillou and Gibson, 1990a). The first three studies carried out biaxial tests, and all three find surfaces which are only partially consistent with the picture developed above. Figure 6.30 shows Patel's data for an almost isotropic polyurethane, obtained by applying axial tension or compression together with torsion and internal pressure to tubular samples, and by the biaxial tension of cruciform samples. The specimens fractured when the stress state was tensile (open circles) but yielded when it was compressive (full circles), so the failure surface is the inner envelope of two intersecting surfaces: that in the compression quadrants is a yield surface, while that in the tension quadrants is, presumably, a fast-fracture surface. Rather similar results were obtained by Shaw and Sata (1966) for polystyrene, and by Zaslawsky (1973) for a rigid polyurethane. All these tests imposed plane stress (biaxial loading), so the expected fracture surface is that for plane stress (Figs. 6.27 and 6.29) not that for axisymmetric loading (Figs. 6.26 and 6.28). The experimental surface is best described as one of constant maximum principal stress, and this is consistent with that expected of fast fracture in the tensile–tensile quadrant and with the yield surface in the compres-

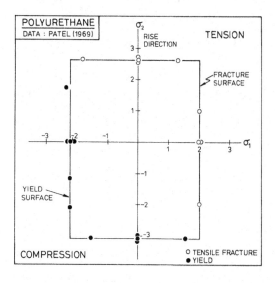

Figure 6.30 An experimental plane-stress failure surface for a rigid polyurethane foam, replotted from the data of Patel (1969). The foam yields in compression but fractures in tension, so the surface is really the inner envelope of two intersecting surfaces.

sion–compression quadrant. But in the tension–compression quadrants the expected tendency towards a surface of constant shear stress is not observed.

Further biaxial tests on rigid polyurethane, giving more data points in the tensile–compressive quadrants, follow the constant shear stress surface (Eqn. (6.59)) more closely (Fig. 6.31) (Triantafillou *et al.*, 1989).

Triaxial tests (σ_1, $\sigma_2 = \sigma_3$) have been performed on cylindrical specimens in standard soil-mechanics triaxial cells. Typical results for triaxial loading are shown in Figs. 6.32 and 6.34. The rigid polyurethane (Fig. 6.32) fails by elastic

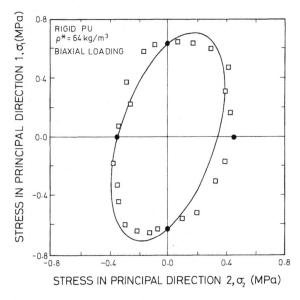

Figure 6.31 Data for the failure of rigid polyurethane foam loaded biaxially. The solid line represents the yield criteria (Eqn. (6.59)).

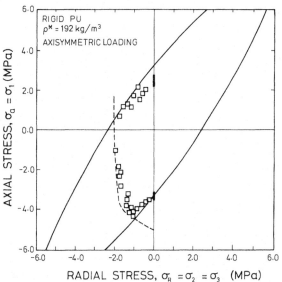

Figure 6.32 Data for the failure of rigid polyurethane foam under axisymmetric loading. The solid line represents the yield failure criteria (Eqn. (6.59)) while the dashed line represents the equation for failure by elastic buckling.

buckling in most of the compression–compression quadrant with a transition to plastic yielding at high axial compressive stresses. Plastic yielding also occurs under combined axial tension and radial compression. The aluminium foam fails by plastic yielding in radial tension (Fig. 6.33). The data suggest that the point at which the two yield surfaces intersect in the tensile quadrant is significantly lower than that given by the model (Eqn. (6.49)); this may be due to irregularities in the cell structure which give rise to bending, as well as axial deformations, in the cell walls. The reticulated vitreous carbon foam fails by fast brittle fracture in the tensile quadrant, by brittle crushing in the tensile/compressive quadrants and by both brittle crushing and elastic buckling in the compressive quadrant (Fig. 6.34). The data are well described by the models (Eqns. (6.51, 6.52)).

6.5   Conclusions

The temperature- and strain-rate dependence of the mechanical properties of foams arise from two sources. The first, the inherent dependence, is just that of the cell-wall material itself. The second is caused by the presence of a gas or liquid inside the cell. In closed-cell foams, the compressibility of the cell fluid increases the modulus and plateau stress; this effect introduces a temperature-dependence (through Boyle's law) but no additional rate-dependence. In open-cell foams, the cell fluid is expelled when the foam is compressed; the additional stress required to do this, derived from Darcy's law, increases with strain-rate and depends strongly on cell size. Experiments on fluid-filled foams are well described by the equations developed in Section 6.2.

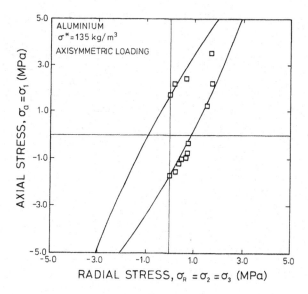

Figure 6.33 Data for the failure of aluminium foam under axisymmetric loading. The solid line represents the yield failure criterion (Eqn. (6.59)).

Foams are almost always anisotropic: the principal dimensions of the cells dif
fer. Many are axisymmetric: then, their properties in one plane are isotropic,
but differ in the direction normal to this plane. A simple extension of the ideas
developed in Chapter 5 for estimating the moduli, strengths and fracture toughness of isotropic foams describes how the shape–anisotropy of the cells affects
material properties. Again, experimental data are adequately described by the
equations developed in Section 6.3.

In some applications foams are not loaded in uniaxial compression or tension, but are subject to triaxial loading. Then it is helpful to plot a failure surface in stress space which identifies the combination of stresses which cause
failure. The yield and fracture surfaces developed in Section 6.4 extend the
ideas of Chapter 4 for honeycombs. In pure shear the cell walls are subjected
to bending, but carry no axial loads; while in pure hydrostatic tension they
carry axial loads but no bending moments; and for other combinations of
load, bending and axial loads coexist. The response of a cell wall to axial
loads differs from that in bending, and the shapes of the failure envelopes for
yielding and fracture reflect this. That predicted for plastic yield is an elongated
ellipse; the degree of elongation increases as the relative density decreases,
because the plastic-collapse stress in pure shear is less than that in pure tension
by a factor of $(\rho^*/\rho_s)^{1/2}$. That expected for brittle fracture is an elongated hexagon. Both envelopes may be truncated in the hydrostatic compression sector
by the elastic-buckling surface.

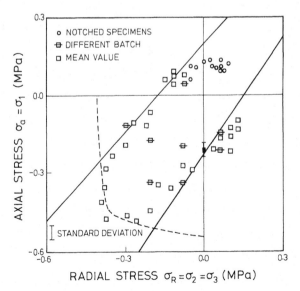

Figure 6.34 Data for the
brittle failure of reticulated
vitreous carbon foams. The
solid line represents failure
by brittle crushing (Eqn.
(6.51)) while the dashed
line represents failure by
elastic buckling.

References

Abramowicz, W. and Wierzbicki, T. (1988), *Int. J. Mech. Sci.*, **30**, 263–71.

Asterly, J. (1996), Private communication, to be published.

Benning, C. J. (1969) in *Plastic Foams*, Vol. 1. Marcel Dekker, New York.

Brace, W. F. (1977) *J. Geophys. Res.*, **82**, 3343.

Brown, W. B. (1960) *Plastics Prog.*, p. 149.

Campbell, G. A. (1979) *J. Appl. Polymer Sci.*, **24**, 709.

Christensen, R. M. (1986) Mechanics of low density materials. *J. Mech. Phys. Solids*, **34**, 563–78.

Cowin, S. C. (1985), The relationship between the elasticity tensor and the fabric tensor, *Mech. Mater.*, **4**, 137–147.

Cowin, S. C. (1991) A note on the anisotropy and fabric of highly porous materials. *J. Mater. Sci.* **26**, 5155–7.

Cunningham, A. (1981) *Polymer*, **22**, 882.

Darcy, H. (1856) *Les Fontains Publique de Ville de Dijon*, p. 150.

Fatima, Vaz, M., Faria, L. and Fortes, M. A. (1995) *Int. J. Engng.*, **16**, 253–27.

Frisch, K. C. and Saunders, J. H. (1972) in *Plastic Foams*, Vol. 1. Marcel Dekker, New York.

Gent, A. N. and Rusch, K. C. (1966a) *Rubber Chem. Tech.*, **39**, 38.

Gent, A. N. and Rusch, K. C. (1966b) *J. Cell. Plast.*, **2**, 46.

Gibson, L. J. and Ashby, M. F. (1982) *Proc. Roy. Soc.*, **A382**, 43.

Gibson, L. J., Ashby, M. F., Zhang, J. and Triantafillou, T. C. (1989) Failure surfaces for cellular materials under multiaxial loads – I. Modelling. *Int. J. Mech. Sci.*, **31**, 635–63.

Goretta, K. C., Brezny, R., Dam, C. Q., Green, D. J., De Arellano-Lopez, A. R. and Dominguez-Rodriguez, A. (1990) High temperature mechanical behaviour of porous open-cell Al_2O_3. *Mater. Sci. Eng.*, **A124**, 151–8.

Gregory, D. (1987) Polyhipe Patent, UK Patent Office.

Gupta, S., Watson, B. Beaumont, P. W. R. and Ashby, M. F. (1986) Final Year Project, Cambridge University Engineering Department.

Harrigan, T. P. and Mann, R. W. (1984) *J. Mat. Sci.*, **19**, 761.

Hart, G. M., Balazs, C. F. and Clipper, R. B. (1973) *J. Cell. Plast.*, **9**, 139.

Hill, R. (1950) *The Mathematical Theory of Plasticity*. Clarendon Press, Oxford.

Hilyard, N. C. (1982) editor, *Mechanics of Cellular Plastics*. Applied Science Publishers, London.

Hilyard, N. C. and Kanakkanatt, S. V. (1970) *J. Cell. Plast.*, **6**, 87.

Hoge, K. G. and Wasley, R. J. (1969) *J. Appl. Polymer Sci.*, **12**, 97.

Huang, J. S. and Gibson, L. J. (1991) Creep of polymer foams. *J. Mat. Sci.*, **26**. 637–47.

Huber, A. T. and Gibson, L. J. (1988) *J. Mater. Sci.*, **23**, 3031–40.

Kanakkanatt, S. V. (1973) *J. Cell. Plast.*, **9**, 50.

Klintworth, J. W. (1989) Dynamic Crushing of Cellular Solids, PhD Thesis, Engineering Department, Cambridge University.

Klintworth, J. W. and Stronge, W. J. (1988) *Int. J. Mech. Sci.*, **30**, 273.

Kosten, C. W. and Zwikker, C. (1939) *Rubber Chem. Technol.*, **12**, 105.

Lacey, R. M. (1965) *High Speed Testing*, **5**, 99.

Mal'tsev, K. I. (1968) *J. Soviet Phys – Acoustics*, **13**, 391.

Mehta, B. S. and Colombo, E. A. (1976) *J. Cell. Plast.*, **12**, 59.

Meinecke, E. A. and Clark, R. E. (1972) *Mechanical Properties of Polymeric Foams*. Technomic Publishing, Stamford, CT, USA.

Patel, M. R. (1969) Ph.D. thesis, University of California at Berkeley, Calif, USA.

Patel, M. R. and Finnie, I. (1969) Lawrence Livermore Laboratory Report UCRL-13420.

Patel, M. R. and Finnie, I. (1970) *J. Mat.*, **5**, 909.

Philipse, A. P. and Schram, H. L. (1991) Non-Darcian airflow through ceramic foams. *J. Amer. Ceramic Soc.*, **74**, 728–32.

Powell, P. C. (1983) *Engineering with Polymers*. Chapman and Hall, London.

Rehkopf, J. D., McNeice, G. M. and Brodland, G. W. (1996) Fluid and matrix components of polyurethane foam behaviour under cyclic compression. ASME *J. Eng. Mater. and Tech.*, **118**, 58–62.

Reid, S. R. and Reddy, T. Y. (1983), *Int. J. Impact Engng*, **1**, 85–106.

Reid, S. R. and Bell, W. W. (1984) in *Mechanical Properties of Materials at High Rates of Strain*, editor J. Harding, Institute of Physics, London, p. 471–8.

Reid, S. R., Bell, W. W. and Barr, R. (1993a), *Int. J. Impact Engng*, **1**, 175–91.

Reid, S. R., Reddy, T. Y. and Peng, C. (1993b) in *Structural Crashworthiness and Failure*, editors J. Jones and T. Wierzbicki, Elsevier Applied Science.

Rinde, J. A. and Hoge, K. G. (1971) *J. Appl. Polymer Sci.*, **15**, 1377.

Shaw, M. C. and Sata, T. (1966) *Int. J. Mech. Sci.*, **8**, 469.

Shim, V. P. W., Yap, K. Y. and Stronge, W. J. (1992), *Int. J. Impact Engng*, **12**, 585–602.

Stronge, W. J. (1991) in *Proc. 6th Int. Conf., Mechanical Behaviour of Materials*, Kyoto editors J. Masahiro and T. Inoue, pp. 377–82.

Stronge, W. J. and Shim, V. P. W. (1987), *Int. J. Mech. Sci.*, **29**, 381–406.

Stronge, W. J. and Shim, V. P. W. (1988), *J. Engng. Mat. and Tech.*, **110**, 185–90.

Terry, S. M. (1971) *J. Cell. Plast.*, **7**, 229.

Terry, S. M. (1976) *J. Cell. Plast.*, **12**, 156.

Thornton, P. H. and Magee, C. L. (1975) *Met. Trans.*, **6A**, 1253.

Traeger, R. K. (1967) *J. Cell. Plast.*, **3**, 405.

Triantafillou, T. C. (1989) Ph.D. Thesis. Department of Civil Engineering, MIT, Cambridge, MA.

Triantafillou, T. C., Zhang, J., Shercliff, T. L., Gibson, L. J. and Ashby, M. F. (1989) Failure surfaces for cellular materials under multiaxial loads – II. Comparison of models with experiment. *Int. J. Mech. Sci.*, **31**, 665–78.

Triantafillou, T. C. and Gibson, L. J. (1990a) Multiaxial failure criteria for brittle foams. *Int. J. Mech. Sci.*, **32**, 479–96.

Triantafillou, T. C. and Gibson, L. J. (1990b) Constitutive modelling of elastic–plastic open-cell foams. *J. Eng. Mech.*, **116**, 2772–8.

Tyler, C. J. and Ashby, M. F. (1986) Project Report, Cambridge University Engineering Department.

Warner, M. and Edwards, S. F. (1988) A scaling approach to elasticity and flow in solid foams. *Europhys. Lett.*, **5**, 623–8.

Warren, W. E. and Kraynik, A. M. (1988) The linear elastic properties of open-cell foams. ASME *J. App. Mech.*, **55**, 341–7.

Wilsea, M., Johnson, K. L. and Ashby, M. F. (1975) *Int. J. Mech. Sci.*, **17**, 457.

Zaslawsky, M. (1973) *Exp. Mech.*, No. 2, Feb., p. 70.

Zhang, J. and Ashby, M. F. (1988) CPGS Thesis, Engineering Department, Cambridge, UK.

Chapter 7

Thermal, electrical and acoustic properties of foams

7.1   Introduction and synopsis

Foams have unique thermal, electrical and acoustic properties. Among these are: exceptionally low thermal conductivity, making them a prime choice for thermal insulation; very low dielectric loss, allowing transmission of microwaves without attenuation or scattering; and the ability to absorb sound, suiting them as materials for noise abatement.

In this chapter we survey the thermal, electrical and acoustic properties of foams. Where possible, the underlying physical understanding of the behaviour is emphasized, since it is this which allows a degree of predictive modelling of foam properties. Case studies are used to illustrate some of the results.

7.2   Thermal properties

More foam is used for thermal insulation than for any other purpose. Closed-cell foams have the lowest thermal conductivity of any conventional non-vacuum insulation. Several factors combine to limit heat flow in foams: the low volume fraction of the solid phase; the small cell size which virtually suppresses convection and reduces radiation through repeated absorption and reflection at the cell walls; and the poor conductivity of the enclosed gas. This low thermal conductivity is exploited, at one extreme of sophistication, in the insulation for liquid oxygen rocket tanks and, at the other, in disposable cups for hot drinks. The frozen-

283

food industry relies on it: the double skins of refrigerated truck and railway cars are filled with foam. And vast quantities of liquefied natural gas are transported around the world in tankers lined with foam.

The specific heat per unit volume, too, is low for foams, making them especially attractive for structures with a low thermal mass; this is important, for instance, in ultra-low-temperature research to minimize the consumption of refrigerant.

The coefficient of thermal expansion of most foams is roughly the same as that of the solid from which they are made. But their moduli are much smaller, and because of this, the thermal stresses generated by a temperature gradient are much smaller, too, giving them good thermal-shock resistance. This is exploited in ceramic foams used as heat shields and ablative coatings: a ceramic foam is much less likely to suffer thermal spalling than the solid from which it is made.

Firebrick, which typically has a relative density of 0.3, is a good example of a ceramic foam which exploits all of these properties. The melting point is high; the low thermal conductivity reduces heat loss by conduction; the low heat capacity minimizes the energy required to get it up to temperature; and the thermal-shock resistance prevents spalling if there are sudden changes in temperature. But this is just one example. Whenever good thermal insulation is required, it is worth considering the use of foams. We now examine their thermal properties in more detail, relating them to relative density and to structure.

(a) Melting or softening point

The melting or softening point of a foam is the same as that of the solid from which it is made; data are given in Chapter 3 (Tables 3.1, 3.4 and 3.5). Pure metals and ceramics have sharp melting points, T_m. Alloys and two-phase mixtures melt over a range of temperatures but the onset of melting (the solidus temperature) still corresponds to a sharp discontinuity in mechanical properties. Glasses and polymers are different: well below the glass temperature, T_g, they progressively soften and become viscoelastic.

The way in which mechanical properties vary with temperature is summarized in the deformation maps shown in Chapter 3. The reader should refer to these for a broad description of the influence of temperature on modulus and strength.

(b) Thermal conductivity

Foams are remarkable for their good thermal insulation. There is considerable literature on the subject, from which a fairly complete picture emerges (Doherty et al., 1962; Griffin and Skochdopole, 1964; Gorring and Churchill, 1961; Guenther, 1962; Patten and Skochdopole, 1962; Skochdopole, 1961; Toohy, 1961; Schuetz and Glicksman, 1983; Reitz et al., 1983; and Glicksman, 1994).

Two material properties characterize heat conduction in solids: the thermal conductivity, λ, and the thermal diffusivity, a. The thermal conductivity is defined by Fourier's law: the heat flux, q (the amount of heat flowing across a unit area per unit time), induced by a temperature gradient, ∇T, is:

$$q = -\lambda \nabla T. \tag{7.1}$$

The units of λ are J/m s K or W/m K.[†] Typical values are listed in Table 7.1. Problems of steady-state conduction (those for which the temperature profile does not change with time) are generally solved by the use of Eqn. (7.1). Non-steady heat conduction is analysed by considering the difference in heat entering and leaving a small element; the difference remains within the element causing its temperature to change with time, τ:

$$\rho C_{\rm p} \frac{\partial T}{\partial \tau} = \frac{\partial}{\partial x}\left(\lambda \frac{\partial T}{\partial x}\right). \tag{7.2}$$

The product of the density, ρ, and the specific heat, $C_{\rm p}$, gives the thermal capacity per unit volume of the material (the heat required to increase the temperature of a unit volume by one degree K). If the properties ρ, $C_{\rm p}$ and λ are constant, the non-steady-state heat conduction equation can be rewritten:

$$\frac{\partial T}{\partial \tau} = a\frac{\partial^2 T}{\partial x^2} \tag{7.3}$$

where a, the thermal diffusivity of the material, is:

$$a = \frac{\lambda}{\rho C_{\rm p}} \tag{7.4}$$

Its units are the same as those for any diffusion coefficient: m^2/s. Typical values of thermal diffusivity are also given in Table 7.1; it can be calculated if λ is known.

The thermal conductivity of a foam can be thought of as having four contributions: conduction through the solid, $\lambda_{\rm s}^*$; conduction through the gas, $\lambda_{\rm g}^*$; convection within the cells, $\lambda_{\rm c}^*$; and radiation through the cell walls and across the cell voids, $\lambda_{\rm r}^*$ (McIntire and Kennedy, 1948; Schuetz and Glicksman, 1983). Then

$$\lambda^* = \lambda_{\rm s}^* + \lambda_{\rm g}^* + \lambda_{\rm c}^* + \lambda_{\rm r}^*. \tag{7.5}$$

An example helps to establish the relative sizes of the contributions. A closed-cell polystyrene foam with a relative density of 0.025, and containing air at 1 atmosphere, has a thermal conductivity of 0.04 W/m K. Not much of this comes from the solid. The contribution $\lambda_{\rm s}^*$ is the product of the conductivity of the fully dense solid, $\lambda_{\rm s}$, and its volume fraction $(\rho^*/\rho_{\rm s})$ multiplied by an efficiency factor which allows for the tortuous shape of the cell walls (Schuetz and Glicksman, 1983; Glicksman, 1994). Using data from Table 7.1 with an

---

[†] Some sources give data in Btu/ft hr F. The conversion factor is 1 Btu/ft hr F = 1.732 W/m K.

Table 7.1   Thermal conductivities and diffusivities

Material	Thermal conductivity λ(W/m K)	Thermal diffusivity a (m^2/s)
Copper (solid)	384[a]	8.8×10^{-5} [a]
Aluminium (solid)	230[a]	8.9×10^{-5} [a]
Alumina (solid)	25.6[a]	8.2×10^{-6} [a]
Glass (solid)	1.1[a]	4.5×10^{-7} [a]
Polyethylene (solid)	0.35[a]	1.7×10^{-7} [a]
Polyurethane (solid)	0.25[c]	
Polystyrene (solid)	0.15[a]	1.0×10^{-7} [a]
Air	0.025[a]	–
Carbon dioxide	0.016[a]	–
Trichlorofluoromethane (CCl_3F)	0.008[a]	–
Oak ($\rho^*/\rho_s = 0.40$)	0.150[a]	–
White pine ($\rho^*/\rho_s = 0.34$)	0.112[a]	–
Balsa ($\rho^*/\rho_s = 0.09$)	0.055[a]	–
Cork ($\rho^*/\rho_s = 0.14$)	0.045[a]	–
Polystyrene foam ($\rho^*/\rho_s = 0.025$)	0.040[b]	1.1×10^{-6} [b]
Polyurethane foam ($\rho^*/\rho_s = 0.02$)	0.025[b]	9.0×10^{-7} [b]
Polystyrene foam ($\rho^*/\rho_s = 0.029$–0.057)	0.029–0.035[d]	
Polyisocyanurate foam, (CFC-11) ($\rho^* = 32$ kg/m^3)	0.020[d]	
Phenolic foam, (CFC-11, CFC-113) ($\rho^* = 48$ kg/m^3)	0.017[d]	
Glass foam ($\rho^*/\rho_s = 0.05$)	0.050[d]	
Glass wool ($\rho^*/\rho_s = 0.01$)	0.042[d]	
Mineral fibre ($\rho^*/\rho_s = 4.8$–32 kg/m^3)	0.046[d]	

All values for room temperature.

References

[a] Handbook of Chemistry and Physics, 66th edn (1985–6) Chemical Rubber Co. ed. R. C. Weast.

[b] Patten, G. A. and Skochdopole, R. E. (1962) *Mod. Plast.*, **39**, 149.

[c] Schuetz, M. A. and Glicksman, L. R. (1983) *Proc. SPI 6th International Technical/ Marketing Conference*, pp. 332–40.

[d] Glicksman, L. R. (1994) Heat transfer in foams, in *Low Density Cellular Plastics* ed. Hilyard, N. C. and Cunningham, A. Chapman and Hall.

efficiency factor of 2/3 gives $\lambda_s^* = 0.003$ W/m K. The biggest contribution is that of conduction through the gas: the product of the conductivity of air, λ_{air}, and its volume fraction in the foam $(1 - \rho^*/\rho_s)$ gives $\lambda_g^* = 0.024$ W/m K. The sum, 0.027 W/m K, already accounts for most of the observed conductivity.

The rest comes from convection and radiation. Convection is important only when the Grashof number (which describes the ratio of the buoyant force driving convection to the viscous force opposing it) is greater than about 1000 (Holman, 1981). The Grashof number is defined by:

$$Gr = \frac{g\beta\Delta T_c l^3 \rho^2}{\mu^2} \tag{7.6}$$

where g is the acceleration due to gravity (9.81 m/s^2), β is the volume coefficient of expansion for the gas (for an ideal gas, $\beta = 1/T$), ΔT_c is the temperature difference across one cell, l is the cell size, and ρ and μ are the density and dynamic viscosity of the gas, respectively. Setting $Gr = 1000$ defines the minimum cell size for convection. Using data appropriate to air at 1 atmosphere (i.e. $\rho = 1$ kg/m^3; $\mu = 2 \times 10^{-5}$ Ns/m^2; $\Delta T_c = 10°$C; and $T = 300°$K) gives $l = 10$ mm. The result is not sensitive to the precise values of the variables. The cell sizes in real foams are a factor of 10 or more smaller than this, so convection should be suppressed completely. The result is confirmed by experiment (King, 1932; Skochdopole, 1961; Traeger, 1967; and Baxter and Jones, 1972).

Radiation, on the other hand, does contribute to heat transfer through foams. The heat flux passing by radiation from a surface of temperature T_1 to one at a lower temperature T_0, with vacuum in between them, is described by Stefan's law:

$$q_r^0 = \beta_1 \sigma (T_1^4 - T_0^4) \tag{7.7}$$

where σ is Stefan's constant (5.67×10^{-8} W/m^{2}K^4) and β_1 is a constant, less than unity, which describes the emissivity of the surfaces. If a foam is inserted between two surfaces, the flux of heat is reduced because radiation is absorbed by the solid and reflected by the cell walls. The attenuation is described approximately by Beer's law:

$$q_r = q_r^0 \exp -(K^* t^*) \tag{7.8}$$

where K^* is the extinction coefficient for the foam (which has units of m^{-1}) and t^* is the thickness of the foam. In the simplest case, for optically thin walls and struts ($t < 10\mu$), the extinction coefficient of the foam is simply that of the solid, K_s, times the relative density of the foam, ρ^*/ρ_s. Dividing this by the temperature gradient (and using the approximations $dT/dx \approx (T_1 - T_0)/t^*$ and $T_1^4 - T_0^4 \approx 4\Delta T \bar{T}^3$ where $\bar{T}$ is the mean temperature, $(T_1 + T_0)/2$), gives the radiative contribution to the foam conductivity:

$$\lambda_r^* = 4\beta_1 \sigma \bar{T}^3 t \exp -\left(K_s \frac{\rho^*}{\rho_s} t^*\right). \tag{7.9}$$

The radiative contribution increases as ρ^*/ρ_s decreases, rising steeply as the foam density approaches zero.

In general, the cell struts and walls may not be optically thin: then the extinction coefficient of the foam is not simply $K_s(\rho^*/\rho_s)$. Glicksman (1994) has found the extinction coefficient for the cell edges (assuming that they act as black bodies) to be:

$$K_{edge} = 4.10\sqrt{\frac{\phi(\rho^*/\rho_s)}{l}}. \tag{7.10}$$

For the cell faces he finds:

$$K_{faces} = \frac{3.46}{l} \frac{(1 - \exp -(2K_s t_f))}{(1 + \exp -(2K_s t_f))} \tag{7.11}$$

where, as before, ϕ is the volume fraction of solid in the cell edges, l is the cell size and t_f is the face thickness. For a closed-cell foam with uniformly thick faces, t_f is proportional to $(\rho^*/\rho_s)l$; for pentagonal dodecahedral cells:

$$t_f = 0.29(1 - \phi)(\rho^*/\rho_s)l.$$

Glicksman (1994) also suggests the use of the Rossland equation to calculate radiative heat transfer:

$$\lambda_r = \frac{16}{3K^*} \sigma T^3 \tag{7.12}$$

with K^* given by the sum of K_{edges} and K_{faces}. The radiative contribution to the thermal conductivity of a foam increases with increasing cell size and decreases with increasing relative density.

The relative contributions of λ_s^*, λ_g^*, and λ_r^* for four polyurethane foams, calculated from these equations, are shown in Fig. 7.1. The competition between the contributions is illustrated by Fig. 7.2, which shows how the total thermal conductivity (the sum of the four contributions) changes with density. It is a minimum at a relative density (for closed-cell foams) of between 0.03 and 0.07, at which point the conductivity is only a little larger than that of the air contained in the cells. The only way to reduce it further is to replace the air with a gas of a lower conductivity such as trichlorofluoromethane, CCl_3F; in closed-cell polymer foams with CCl_3F in the cells about half of the heat transfer is by conduction through the gas and half is by conduction through the solid and radiation. At lower relative densities, the conductivity increases, partly because of the increasing transparency of the cell walls to radiation and partly because, at very low densities, the cell walls may rupture (Guenther, 1962).

Heat transfer increases with cell size (Fig. 7.3). This is partly because radiation is reflected less often in a foam with large cells and partly because, for cells of more than 10 mm or so in diameter, cell convection starts to contribute.

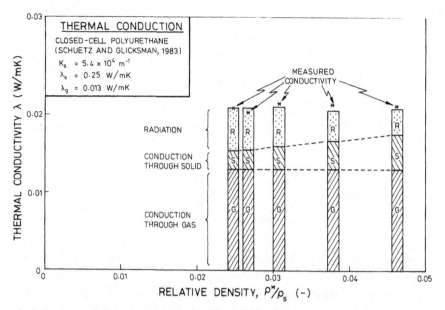

Figure 7.1   The contributions of conduction through the gas, conduction through the solid and radiation to the thermal conductivity of closed-cell rigid polyurethane foams. As the relative density of the foams increases, the contribution from conduction increases while that from radiation decreases.

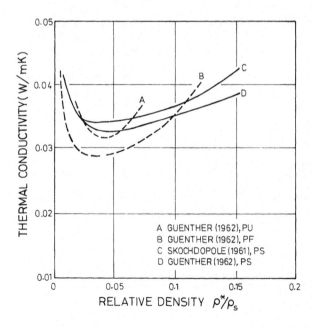

Figure 7.2   Thermal conductivity of foams as a function of their relative density (polystyrene, PS; phenolformaldehyde, PF; polyurethane, PU; after Griffin and Skochdopole, 1964).

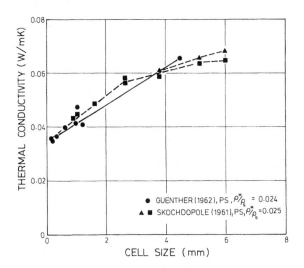

Figure 7.3 Thermal conductivity of polystyrene (PS) foams as a function of their cell size. The relative densities of the two foams are about 0.025, near the value which minimizes thermal conduction; after Griffin and Skochdopole, 1964.

Other factors, such as the fraction of open cells, affect heat transfer, too, but to a lesser extent. Temperature itself changes the conductivity, and in a complicated way. Thermal conduction through solid polymers and glasses, and through gases, decreases with decreasing temperature. Radiation, too, becomes less effective at low temperatures. The result is that heat transfer through most foams falls steeply with decreasing temperature. Figure 7.4 shows typical data for polystyrene and polyurethane foams.

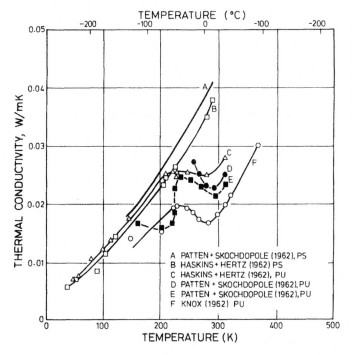

Figure 7.4 Thermal conductivity of foams as a function of temperature (polystyrene, PS; polyurethane, PU; after Griffin and Skochdopole, 1964).

The age of a foam can affect its thermal conductivity. Because the thermal conductivity of foams depends largely on the conductivity of the gas within the cells, many foams are blown using low conductivity gases. Over a period of time, the enclosed gas diffuses out of the cells while air diffuses in, increasing the thermal conductivity of the foam by up to 50%, as shown in Fig. 7.5.

Chlorofluorocarbons (CFCs), with their extremely low thermal conductivity have been widely used as blowing agents for foams (for instance, trichlorofluoromethane CCl_3F or CFC-11, $\lambda = 0.008$ W/mK). After they diffuse out of a closed-cell insulation foam, they remain stable in the lower part of the atmosphere until they reach the stratosphere (10–50 km above the Earth's surface). At that point, solar UV radiation decomposes CFCs, freeing chlorine ions which act as catalysts in the destruction of ozone. Depletion of the ozone layer increases the transmission of ultraviolet radiation, with adverse effects on plants and animals (for instance, increased incidence of skin cancer in humans). For this reason, the use of CFCs is now being phased out: production is to be eliminated in developed countries by the year 2000 and in developing countries by the year 2010 (McFarland, 1992). Research into replacement gases focuses on

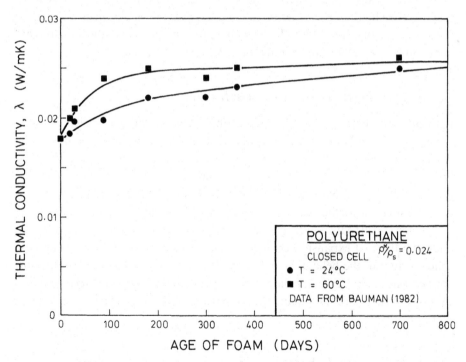

Figure 7.5 Thermal conductivity of foams as a function of age and temperature. The low-conductivity blowing agent gas diffuses out of the cells over time while air (of higher conductivity) diffuses in. Increasing temperature increases the rate at which the low-conductivity blowing agent gas diffuses out; after Schuetz and Glicksman, 1983.

other compounds in the fluorocarbon family which either do not contain chlorine (such as hydrofluorocarbons, HFCs) or which do contain chlorine but decompose in the lower atmosphere, rather than the stratosphere (such as hydrochlorofluorocarbons or HCFCs). Because some small fraction of the chlorine in HCFCs may still reach the stratosphere, they, too, can contribute to ozone depletion (although much less than CFCs); they are seen as transitional compounds to be used until better replacement gases are developed (McFarland, 1992). The thermal conductivity of rigid polyurethane foams blown with two HCFCs currently under study, HCFC-141b and HCFC-123, is roughly 3% higher than those blown with CFC-11 (Cunningham *et al.*, 1989).

(c) Specific heat

The specific heat, C_p, is the energy required to raise a unit mass of a material by a unit temperature; it is measured in J/kg K. For a two-phase system, it is simply the sum of that of each phase multiplied by its weight fraction. Even in low-density foams the weight fraction of gas is small, so that the specific heat of the foam is essentially equal to that of the solid of which it is made. Specific heats of polymers are high: of the order of 2000 J/kg K. Those for ceramics and glasses are lower: usually about 750 J/kg K; while those for metals are the lowest of all: about 30 J/kg K. Data can be found in Tables 3.1, 3.4 and 3.5.

(d) Thermal expansion coefficient

If a foam is thought of as a framework, then it is clear that the coefficient of linear expansion of the framework is the same as that of the solid of which it is made. All but the lightest foams do, in fact, behave in this way. For this reason, polymeric foams have the highest expansion coefficients: typically about 10^{-4}/K. Those for metallic foams are less: usually about 10^{-5}/K. All ceramic foams have the smallest coefficients: often as little as 3×10^{-6}/K.

If a closed-cell foam is either cooled to very low temperatures or heated to high ones, the thermal expansion coefficient may no longer equal that of the base solid because of the behaviour of the gas it contains. Cooling the foam to low temperatures may freeze it, collapsing the cells. Heating expands the gas so that the resulting pressure can expand the foam itself. Both effects lead to exaggerated thermal strains.

(e) Thermal-shock resistance

When a solid is exposed to a sudden change of surface temperature, thermal stresses appear within it which, if sufficiently large, can cause cracking and spalling; that is, loss of material from the surface as flakes or chips. If a block of the solid, initially at a uniform temperature T_1, is suddenly cooled (by dropping it into water, for instance) its surface temperature will drop to T_2, contracting the surface layers of the block and producing a thermal strain of $\alpha \Delta T$, where α is the thermal expansion coefficient and ΔT is the difference between T_1 and T_2. But the surface is firmly bonded to the underlying mass of the block, constraining it to its original dimensions and producing a thermal stress of

$$\sigma = \frac{E\alpha\Delta T}{1 - \nu} \tag{7.13}$$

in the surface. When this stress exceeds the fracture strength of the material, σ_f, it cracks and spalls. The thermal-shock resistance is measured by the temperature drop which will just cause cracking:

$$\Delta T_c = \frac{\sigma_f(1 - \nu)}{E\alpha}. \tag{7.14}$$

The modulus and fracture strength of foams are discussed in Chapter 5. The modulus (Eqn. (5.6a)) is approximately $E^* = E_s(\rho^*/\rho_s)^2$, the fracture strength (which for a defect-free foam is equal to the crushing strength, Eqn. (5.37a)) is $\sigma_f^* = 0.2(\rho^*/\rho_s)^{3/2}\sigma_{fs}$, and the thermal expansion coefficient, as we have seen, is the same as that of the solid. Thus:

$$\Delta T_c^x = \frac{0.2\Delta T_{cs}}{(\rho^*/\rho_s)^{1/2}} \tag{7.15}$$

where $\Delta T_{cs} = \sigma_{fs}/E_s\alpha_s$ is the thermal-shock resistance of the fully dense solid. The point to note is that the thermal-shock resistance increases as the density of the foam decreases: this is because the network of struts which make up a low-density foam can accommodate the thermal strain by bending, as described in Chapters 4 and 5. This property of foams is exploited in firebrick and other low-density refractories. Data for the themal-shock resistance of various solids and foams are listed in Table 7.2.

(f) Case study: the optimization of foam density for thermal insulation

There is an optimum foam density for any given insulation problem. It depends on the temperature difference across the insulation, and on its thickness: the optimum density for a disposable coffee cup is different from that for cavity-wall insulation. That there is an optimum can be seen by noting that a relatively dense foam conducts well because of the large volume fraction of solid it contains, while one of extremely low density allows radiation to pass through it easily. The optimum lies in between, and for polymer foams is usually in the range $\rho^*/\rho_s = 0.01$–0.1.

Table 7.2 Thermal-shock resistances

Material	Thermal-shock resistance[a,b] (°K)
Metals	High; typically 1000
Soda glass	84
Borosilicate glass	280
Pottery, brick, porcelain	220
Alumina	150
Silicon carbide	300
Stabilized zirconia, sialons	500
Polymethylmethacrylate (dense)	280
Polystyrene (dense)	350
Firebrick (90% alumina; 10% chromia) ($\rho^*/\rho_s = 0.82$)	100
Firebrick (73% chromia; 26% magnesia) ($\rho^*/\rho_s = 0.83$)	400

[a]Calculated from data for E, σ_{fs} and α given in Chapter 3.
[b]Bandypadhyay, G., Chen, J., Kennedy, C. R. and Diercks, D. R. (1983) *J. Mat. Energy Systems*, **4**, 234.

The problem is illustrated by Fig. 7.6. We wish to minimize heat transfer through the foam, subject to the constraint that the foam thickness, t, is fixed (it is the wall thickness of the cup or the cavity depth of the wall). The effective thermal conductivity of the foam is the sum of the contributions described in Section 7.2(b): conduction through the solid, conduction through the gas, the radiative transfer. (We assume the cells are sufficiently small to suppress convection.) Using an efficiency factor of 2/3, the effective thermal conductivity is:

$$\lambda^* = \frac{2}{3}\left(\frac{\rho^*}{\rho_s}\right)\lambda_s + \left(1 - \frac{\rho^*}{\rho_s}\right)\lambda_g + 4\beta_1\sigma\bar{T}^3 t \exp-\left(K_s\frac{\rho^*}{\rho_s}t\right) \qquad (7.16a)$$

where β_1, σ and K_s are defined in Section 7.2(b). We wish to minimize λ^*, subject to the constraint:

$$t = \text{constant}. \qquad (7.16b)$$

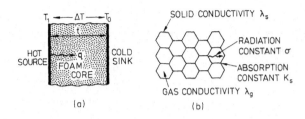

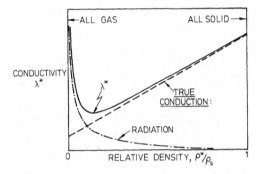

Figure 7.6 Heat passes through a plastic foam by conduction (through the solid and the gas) and by radiation; convection is suppressed. There is an optimum density for the foam, giving the lowest overall rate of heat transfer.

This is done by setting

$$\frac{\partial \lambda^*}{\partial(\rho^*/\rho_s)} = 0$$

giving the *optimum foam density* for minimum conduction:

$$\left(\frac{\rho^*}{\rho_s}\right)_{\text{opt}} = \frac{1}{K_s t} \ln\left\{\frac{4 K_s \beta_1 \sigma t^2 \bar{T}^3}{\frac{2}{3}\lambda_s - \lambda_g}\right\}. \qquad (7.17)$$

The optimum density decreases as the thickness, t, increases: and it increases as the mean temperature, $\bar{T}$, increases. Thus the best foam for insulating a coffee cup is denser than that for cavity-wall insulation; and the wisest choice for cryogenic insulation is a foam with a very low density indeed. More specifically, a coffee cup with a 3 mm thick wall, and made from a polymer with the thermal constants listed in Table 7.3, is best made of a foam of relative density 0.08, while insulation for a cavity wall, of thickness 50 mm, should have a relative density of about 0.02.

7.3   Electrical properties

Foams are not, usually, chosen for their electrical properties, but these often play a secondary role. A high electrical resistance can be important for structural foam enclosures, for insulating coatings, and for mounting panels for electrical components. High resistivity may not always be desirable: a degree of conductiv-

Table 7.3   Data for optimization case study

Extinction coefficient of solid polymer, K_s	$5.67 \times 10^4 \, \text{m}^{-1}$
Emissivity factor, β_1	0.5
Conductivity of solid polymer, λ_s	$0.22 \, \text{W/m K}$
Conductivity of gas, λ_g	$0.02 \, \text{W/m K}$
Mean temperature, $\bar{T}$	$300°\text{K}$
Stefan's constant, σ	$5.67 \times 10^{-8} \, \text{W/m}^2\text{K}^4$

ity is sought to dissipate static electricity, or for electrical screening in, for example, computer cases. Dielectric properties become of critical importance in structures that must transmit microwaves. Foams are particularly useful in this regard; the protective dome of a radar guidance system is frequently a sandwich structure with a foam core chosen for its low dielectric loss. For these and other reasons, we have included a discussion of the electrical properties of foams, although, for the most part, they have been poorly characterized.

(a) Electrical resistivity

When a potential gradient is applied to all but perfectly insulating materials, a current flows. The *bulk resistivity R* is the potential gradient divided by the current flowing per unit area, that is, it is the resistance of a unit cube of material with unit potential-difference between a pair of faces. The resistivity varies over an immense range; from near 10^{-8} Ωm for good conductors, to more than 10^{16} Ωm for good insulators.

Most polymers have resistivities which lie at the upper end of this range. A few – doped polyacetylene, for instance – are intrinsic conductors, and all can be given a degree of conductivity by filling them with graphite or other conducting powders. Oxide ceramics, too, are usually insulators; a few, when doped, are semiconductors. Carbide ceramics (NbC, ZrC, WC etc.), like metals, are electronic conductors. A cellular material inherits many of the electrical characteristics of the solid of which it is made, modified, as described below, by a function of its relative density.

When a conducting material is foamed, its resistivity increases. We expect that the electrical conductivity increases linearly with relative density, analogous to the thermal conductivity. The constant of proportionality reflects the reduction in conductivity due to the tortuosity of the path in the foam. Limited data, illustrated by Fig. 7.7, suggest that the electrical conductivity can be described using the same constant of proportionality as that for thermal conductivity (2/3) (solid line). The resistivity R is similarly described by:

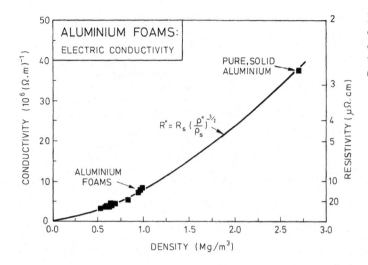

Figure 7.7 The electrical conductivity and resistivity of aluminium foams, plus that for solid aluminium (data from Mepura, 1995).

$$R^* \propto \frac{R_s}{(\rho^*/\rho_s)} \qquad (7.18)$$

where R_s is the resistivity of the solid and ρ^*/ρ_s its relative density. The behaviour can be qualitatively understood by noting that, as the density falls, the average cross-section available for conduction decreases and the tortuosity of the current path increases, both raising the resistivity. A quantitative theory consistent with Eqn. (7.18) is lacking.

Polymeric foams are, generally, good insulators. Polymers with polar groups as part of their molecular structure may absorb water – up to 10% in some cases – with a pronounced effect on resistivity. Woods are a particular example of this. Oven-dry wood, like most polymers, is an excellent insulator, but its conductivity increases rapidly with moisture content (Fig. 7.8). Data are insufficient to establish a density-dependence for the resistivity of woods, but it might be expected that, parallel to the grain, the resistivity would be inversely proportional to the amount of solid in the section (Eqn. (7.18)).

(b) Dielectric constant

When a material is placed in an electric field, it becomes polarised. Charges appear at the surfaces which tend to screen the interior from the external field. The tendency to polarise is measured by the *dielectric constant*, ϵ. It is a measure of the electric energy per unit volume stored in the material when placed in an electric field. The exceptionally low dielectric constant of polymer foams makes them attractive for structural panels and domes which must transmit microwaves (radomes).

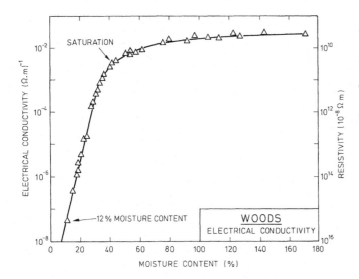

Figure 7.8 The electrical conductivity and resistivity of woods as a function of moisture content (12% under normal room conditions). Data from Kollman and Côté (1968) and Stamm (1929).

Polymers with non-polar structures, such as polyethylene and polypropylene, have low dielectric constants (values: about 3); those which contain polar groups, such as the nylons, have higher dielectric constants (values: about 10), augmented by their tendency to absorb water which raises ϵ further. Foaming reduces the dielectric constant simply because ϵ scales as the fraction of space filled by solid or:

$$\epsilon^* = 1 + (\epsilon_s - 1)\left(\frac{\rho^*}{\rho_s}\right) \qquad (7.19)$$

where ϵ_s is the dielectric constant of the solid.

Woods, like polymer foams, have dielectric constants which decrease linearly with density. Figure 7.9 illustrates both this and the effect of absorbed water. The water molecule carries a dipole moment, so the presence of absorbed water in wood (or any other polymer) increases both the dielectric constant and the loss factor (see below).

(c) Dielectric strength

The *dielectric strength*, V_c (usual units: MV/m) is the electric potential gradient at which the resistance of an insulator breaks down, and a damaging surge of current flows through the specimen. It is measured by increasing, at a uniform rate, an alternating potential (60 Hz) applied across a plate of the material until breakdown occurs. The dielectric strengths of foams are poorly characterized. For woods V_c depends on temperature, moisture content and (to a small extent) the thickness of the sample itself (Kollman and Côté, 1968).

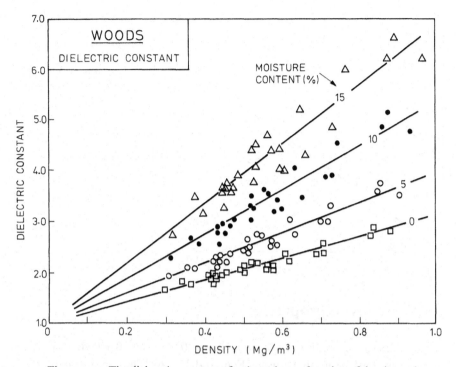

Figure 7.9 The dielectric constant of redwoods as a function of density and moisture content (under normal room conditions the moisture content is about 12%). Data from Kollman and Côté (1968) and Skaar (1948).

(d) Loss factor

When an alternating voltage is applied to a 'perfect' dielectric, current flows that is 90° out of phase with the voltage. No insulating material is perfect, with the result that the current lags behind the voltage by something more than 90°. The dissipation factor (identical with the 'power factor' when its value is less than 0.1) is the tangent of the angle by which the current deviates from the ideal of 90°. It is equivalent to the ratio of the current dissipated in heat to the current transmitted, so good dielectric materials have small dissipation factors. The dissipation factor depends on temperature, frequency, and on moisture content.

The *loss factor* D_ϵ (important for transparency to microwaves) is the dissipation factor multiplied by the dielectric constant. It increases linearly with density (Fig. 7.10), simply because the dielectric constant does so, and is well described by

$$D_\epsilon^* = D_{\epsilon s}\left(\frac{\rho^*}{\rho_s}\right) \tag{7.20}$$

where $D_{\epsilon s}$ is the loss factor of the solid of which the cellular solid is made.

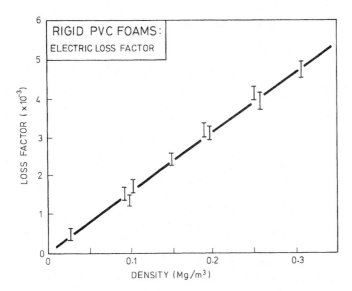

Figure 7.10 The loss factor, D_ε^*, of rigid PVC foams measured at 12 GHz, as function of density (data from DIAB 1995). The value of $D_{\varepsilon s}$ for solid PVC is 0.02.

7.4   Acoustic properties

The human ear responds to frequencies from 20 to about 20 000 Hz, corresponding to wavelengths in air between 17 m and 17 mm. The vibrations that cause sound produce a change of air pressure in the range 10^{-4} Pa (low amplitude sound) to 10 Pa (the threshold of pain). It is usual to measure this on a relative, logarithmic scale, with units of *decibels* (dB). The decibel scale compares two sound intensities using the *threshold of hearing* as the reference level (0 dB).

The acoustic properties of cellular solids are important in their uses for sound management in buildings, and as sound-boards and reflectors in musical instruments (Lauriks, 1994). Their suitability for these applications is described below; but first a little background on sound velocity and wavelength.

(a) Sound velocity and wavelength

The sound velocity in a long rod of a solid material (such that the thickness is small compared with the wavelength is

$$v_l = \sqrt{\frac{E}{\rho}} \tag{7.21}$$

where E is Young's modulus and ρ is the density. If the thickness of the rod is large compared with the wavelength, the velocity, instead is

$$v_B = \sqrt{\frac{E(1-\nu)}{(1-\nu-2\nu^2)\rho}} \tag{7.22}$$

where ν is Poisson's ratio. Shear waves propagate with a velocity

$$v_s = \sqrt{\frac{G}{\rho}} \qquad\qquad (7.23)$$

where G is the shear modulus.

Elastically anisotropic solids (most foams and all woods are anisotropic) have sound velocities which depend on direction. If $E_\parallel$ is the modulus along the direction of the rise or grain and $E_\perp$ is that perpendicular to it, the ratio of the two velocities, as given by Eqn. (7.21), is

$$\frac{v_\parallel}{v_\perp} = \sqrt{\frac{E_\parallel}{E_\perp}}.$$

These concepts are routinely used to estimate elastic constants of cellular solids. For foams with relative densities near unity, this works well, but as the relative density falls, wave propagation is increasingly influenced by the elastic response of the gas within the cells, and the multiple reflections at the cell walls.

Data for sound velocities in foams are very limited. There exists a body of data for woods which is informative. Available information is assembled in Table 7.4. The calculated sound velocity (second-last column) is that predicted by Eqn. (7.21). It is compared with the measured sound velocity (last column). For high relative densities the two are close; but at low relative density (cork) the estimate is too low. This is partly because it neglects the contribution of the air contained in the cells, giving as a lower limit the velocity 340 m/s. The anisotropy of sound velocity in woods is apparent.

Data for longitudinal sound velocity in spruce are plotted as a function of density in Fig. 7.11. Young's modulus for woods, measured parallel to the grain, scales approximately as ρ^*/ρ_s (Chapter 10, Eqn. (10.6)) so the longitudinal sound velocity is expected to be almost independent of density. For spruce, this

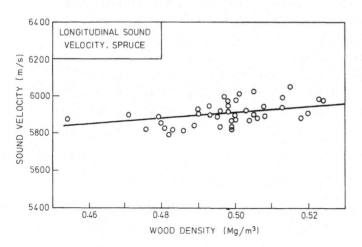

Figure 7.11 The longitudinal sound velocity in spruce wood, as a function of density. Data from Kollman and Krech (1960).

Table 7.4 Sound velocities in cellular solids[*]

Material or wood species	Density ρ (Mg/m^3)	Young's modulus E^* (GPa)	Sound velocity, calculated v^* (m/s)	Sound velocity, measured v^* (m/s)
PS foam, low density	0.02	0.005	500	—
PS foam, high density	0.78	1.6	1430	—
(Solid PS)	(1.04)	(3.6)	(2350)	(2280–2400)
Cork	0.25	0.035	380	430–530
Spruce, ‖ to grain	0.47	10.8	4800	5820–5980
Spruce, ⊥ to grain	0.47	0.54	1070	—
Fir, ‖ to grain	0.45	10.8	4900	5250–5500
Fir, ⊥ to grain	0.45	0.48	1030	1700–1900
Beech, ‖ to grain	0.73	15.7	4640	3400–3500
Beech, ⊥ to grain	0.73	1.5	1433	1770–1980
Oak, ‖ to grain	0.69	12.7	4300	3381–4310
Oak, ⊥ to grain	0.69	1.0	1200	1790–2040

[*]Data from Kollman and Côté (1968) and McDonald (1978).

appears to be true, but data for other woods which cover a wider range of density, suggest that it rises slowly with increasing density. This is because the microfibrillar angle in the cell walls changes with the phase of growth of the tree, and thus with density. By contrast, the sound velocity in equiaxed foams falls steeply with density, because the moduli scale approximately as $(\rho^*/\rho_s)^2$, giving $v \approx \rho^*/\rho_s$ (see Section 5.3(a)). This gives low-density foams a low sound velocity, often not much greater than that of air.

The velocities of Table 7.4 correspond to wavelengths, λ, which are large compared with the cell size in even the largest-celled foams. Using Eqn. (7.21) and the expression developed earlier for Young's modulus of an isotropic foam (Eqn. (5.6a)), we find

$$\lambda = \frac{v}{f} = \frac{1}{f}\sqrt{\frac{E^*}{\rho^*}} = \frac{1}{f}\sqrt{\frac{E_s}{\rho_s} \cdot \frac{\rho_s}{\rho^*}} \tag{7.24}$$

giving wavelengths in the acoustic range ($f = 20$–$20\,000$ Hz) of between 50 mm and 5 m.

(b) Sound management

The mechanism for reducing sound intensity within a given enclosed space (a room, for instance) depends where it comes from. If it is generated within the room, one seeks to *absorb* the sound. If it is airborne and comes from outside, one seeks to *insulate* the space to keep the sound out. And if it is transmitted through the frame of the structure itself, one seeks to *isolate* the structure from the source of vibration. Cellular, porous solids are good for absorbing sound; and – in combination with other materials – they can help in isolation. They are very poor at providing insulation against sound.

Soft porous materials absorb incident, airborne sound waves, converting them into heat. Sound power, even for very loud noise, is small, so the temperature rise is negligible. Porous or highly flexible materials such as low-density polymeric and ceramic foams (or similar materials such as plaster and fibreglass) absorb well; so do woven polymers in the form of carpets and curtains. Several mechanisms of absorption are at work here. First, there is the viscous loss as air is pumped into and out of the open, porous structures – sealing the surface (with a paint-film, for instance) greatly reduces the absorption. Second, there is the intrinsic damping in the material. It measures the fractional loss of energy of a wave, per cycle, as it propagates within the material itself. Intrinsic damping in most metals and ceramics is low, typically 10^{-6}–10^{-2}; in polymers and foams made from them it is high – in the range 10^{-2}–0.2. The proportion of sound absorbed by a surface is called the *sound-absorption coefficient*. A material with a coefficient of 0.8 absorbs 80% of the sound which is incident on it, and material of coefficient 0.03, absorbs only 3% of the sound, reflecting 97%. The sound-absorption coefficients for a range of materials are plotted as a function of frequency in Fig. 7.12.

So foams make good sound-absorbers. They are not good at sound insulation, that is, mitigation of airborne noise from outside the room. The degree of insulation is proportional to the mass of the wall, floor or roof through which sound has to pass. This is known as the *mass-law*: the heavier the material, the better it insulates. The lightweight walls of modern buildings, designed for good thermal insulation, do not in general provide good sound insulation. Foams, for obvious reasons, are not a good choice here; instead, the practice is to add mass to the wall or floor with an extra layer of brick or concrete, or a lead cladding.

Impact noise is transmitted directly into the structure or fabric of a building. Elastic materials, and particularly a steel frame if there is one, can transmit vibrations throughout the building. Unlike airborne sound, noise of this sort is not attenuated by additional mass. Since it is transmitted by the continuous solid part of the structure, it can be reduced by interrupting the sound path by floating the floor or building foundation on a resilient material, and here cellular materials can be useful. The impact sound of footsteps can be suppressed by using

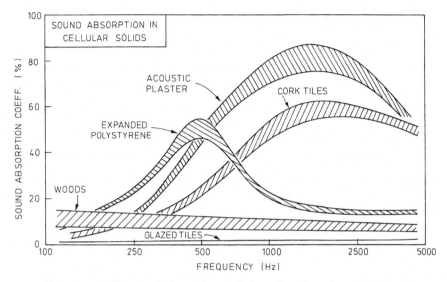

Figure 7.12 The sound-absorption coefficient plotted as a function of frequency for cellular solids, contrasted with that for glazed tiles. Data from Parkin *et al.* (1979) and Beranek (1960).

cork, porous rubber or plastic tiles. On a larger scale, buildings are isolated by setting the entire structure on resilient pads. A composite of rubber filled with cork particles is a good choice. The low shear modulus of the rubber isolates the building from shear waves, and the compressibility of the cork adds a high impedance to compressive waves.

(c) Sound wave impedance and radiation of sound energy

If a sound-transmitting material is interfaced with a second one with different properties, part of the sound-wave is transmitted across the interface, and part is reflected back into the first material. The transmission and reflection factors are determined by the relative impedances of the two materials. The impedance is defined by

$$Z = \rho v = \sqrt{\rho E} \qquad\qquad (7.25)$$

where E is the appropriate modulus, and ρ the density. The reflection and transmission coefficients between material 1 and material 2 are given by

$$R = \frac{Z_1 - Z_2}{Z_1 + Z_2} \quad \text{and} \quad T = 1 - R = \frac{2Z_2}{Z_1 + Z_2}.$$

Thus if the two impedances are also equal, most of the sound is transmitted, but if they differ greatly, most is reflected. This is the origin of the 'mass-law' cited earlier: a heavy wall gives a large impedance mis-match with air, so that most of the sound is reflected from it and does not penetrate to the neighbouring room.

Typical values of the acoustic impedance of foams are compared with those of conventional solids in Table 7.5. Its value, for cellular solids, is much lower than that for fully dense materials.

In the design of sound boards (the front plate of a violin, the sound board of a harpsicord, the panel of a loudspeaker), the intensity of sound radiation is an important design parameter. Fletcher and Rossing (1991) and Meyer (1995) demonstrate that the intensity, I, is proportional to the surface velocity, and that for a given driving function, this scales with modulus and density as:

$$I \propto \frac{v_{\mathrm{b}}}{\rho} \propto \sqrt{\frac{E}{\rho^3}}. \tag{7.26}$$

A high value of the combination of properties $\sqrt{E/\rho^3}$, called the 'radiation factor', is used by instrument makers to select materials for sound boards. When

Table 7.5 Acoustic impedances and radiation factors[*]

Material	Density ρ (Mg/m^3)	Young's modulus E^* (GPa)	Acoustic impedance, calculated $\sqrt{\rho E}$ (GN s/m^3)	Radiation factor, calculated $\sqrt{E/\rho^3}$ (m^4/kg s)
PS foam, low density	0.03	0.01	0.01	19.2
PS foam, high density	0.78	1.6	1.1	2.0
Solid PS	1.04	3.6	1.9	1.8
Nylon 6/6	1.14	2.4	1.7	1.3
Spruce, ‖ to grain	0.38	10.8		
		$\bar{E} = 4.1$	1.2	8.6
Spruce, ⊥ to grain	0.38	0.54		
Maple, ‖ to grain	0.45	10.8		
		$\bar{E} = 5.1$	1.7	5.4
Maple, ⊥ to grain	0.45	0.48		
Aluminium	2.7	69	13.6	1.87
Steel	7.9	210	40.7	0.65
Silver	10.4	75	27.9	0.26
Glass	2.24	46	10.1	2.0

[*]Data from American Institute of Physics, Handbook (1972) and Meyer (1995).

the material is elastically anisotropic (as is wood), E is replaced by $\bar{E} = (E_{\parallel} \cdot E_{\perp})^{1/2}$. The last column of Table 7.5 gives some representative values for the radiation factor. Spruce, widely used for the front plates of violins, has a particularly high value – nearly twice that of maple which is used for the back plate, the function of which is to reflect, not radiate. Solids generally have low values of $\sqrt{E/\rho^3}$, but low-density foams do not. They could find application as sound radiators in acoustic systems.

7.5   Conclusions

The thermal properties of foams are fairly well understood. The melting or softening point, the thermal expansion coefficient and the specific heat are, in most cases, the same as those of the solid material from which the foam is made. The thermal conductivity and diffusivity, however, are not: they are almost always much smaller. This is because up to 90% of the heat is transferred by conduction through the gas within the cells, which itself has a low conductivity. The remaining heat transfer occurs by conduction through the solid and by radiation. Convection is suppressed in cells with a diameter of less than about 10 mm; the cells in most foams are much smaller than this. There is an optimum foam density which minimizes the thermal conductivity; for polymer foams it is around $\rho^*/\rho_s = 0.05$. Above this density, conduction through the solid increases heat flow; below, radiation passes easily through the cell walls and heat flow again increases. For a given density of foam, the contribution of radiation decreases as the cells become smaller because of repeated reflection at the larger number of cell walls. The thermal-shock resistance of a foam is improved by reducing its density, as the cellular structure accommodates thermal strain more easily.

These properties are widely exploited in low-technology applications: in the food industry, in insulation of housing, and in relatively low-performance refractories, for instance. Optimization of properties in these applications is not essential, though it could lead to economies. Increasingly, however, foams are contributing to high technology: in cryogenic equipment, in insulation of fuel tanks in space vehicles, and in thermal shielding of re-entry capsules. And in these cases, it is essential to exploit their thermal and thermomechanical properties to the maximum.

The electrical properties of foams derive from those of the solid of which they are made, modified by a function of the relative density. Thus, as the relative density decreases towards zero, the dielectric constant ϵ decreases linearly from that of the solid, ϵ_s, towards the value 1 (that of air), and the electric loss-factor, D_ϵ decreases from D_{es} to zero. This allows the construction of radar-transparent housings ('radomes') for navigational radar with high resolution and range. The electrical conductivity of conducting foams, too, is modified by foaming, falling towards zero as the relative density decreases.

Foamed acoustic tiles are a familiar remedy for a too-noisy room, but they only help if the noise is airborne and generated within the room itself. This is because open-cell foams absorb sound well, but they do not prevent its transmission from a neighbouring room or space. Beyond this, certain cellular solids – particularly certain woods – have acoustic properties which make them specially attractive for the sound boards of musical instruments, something already well-known to Stradivari (1644–1737).

References

American Institute of Physics (1972), Handbook, 3rd edition, editor D. E. Gray, McGraw-Hill, NY, USA.

Bauman, G. (1982) Proceedings of the SPI Annual Conference.

Baxter, S. and Jones, T. T. (1972) *Plast. Polym.*, April, p. 69.

Beranek, L. L. (Editor) (1960) *Noise Reduction* McGraw-Hill, New York. pp. 349–395.

Cunningham, A., Sparrow, D. J., Rosbotham, I. D. and de Nazelle, G. M. R. du Cauze (1989) A fundamental study of the thermal conductivity ageing of rigid PUR foam blown with HFA-141b/CO_2 and HFA-123/CO_2 mixtures. Proceedings of the SPI 32nd Annual Conference, San Francisco, p. 56–60.

DIAB (1995), Data Sheets for Divinycell PVC Foams, Centre d'Activities Stroch, 61 rue Coquillep, 45200 Montagis, France.

Doherty, D. J., Hurd, R. and Lester, G. R. (1962) *The Physical Properties of Rigid Polyurethane Foams*, Chem. Ind. London, p. 1340.

Fletcher, N. H. and Rossing, T. D. (1991) *The Physics of Musical Instruments* Springer-Verlag, Berlin, Germany.

Glicksman, L. R. (1994) Heat transfer in foams, Chapter 5 in *Low Density Cellular Plastics* eds. Hilyard, N. C. and Cunningham, A., Chapman and Hall, London.

Gorring, R. L. and Churchill, S. W. (1961) *Chem. Eng. Prog.*, **57** (7), 53.

Griffin, J. D. and Skochdopole, R. E. (1964) *Engineering Design for Plastics*, ed. E. Baer. Van Nostrand Reinhold, London, Chapter 15.

Guenther, F. O. (1962) *SPE Trans.*, **2**, 243.

Haskins, J. F. and Hertz, J. (1962) *Adv. Cryogen. Eng.*, **7**, 353.

Holman, J. P. (1981) *Heat Transfer*, 5th edn. McGraw-Hill, New York.

King, W. J. (1932) *Mech. Eng.*, **54**, 347.

Knox, R. E. (1962) *ASHRAE J.*, **4** (10), 43.

Kollman, F. F. P. and Côté, W. A. Jr (1968) *Principles of Wood Science and Technology* Vols 1 and 2, Springer-Verlag, Berlin, Germany.

Kollman, F. F. P. and Krech, H. (1960), *Holz als Rod-und Werstoff*, **18**, p. 41–51.

Lauriks, W. (1994) Acoustic characteristics of low density foams, in *Low Density Cellular Plastics*, eds. N. C. Hilyard and A. Cunningham, Chapman and Hall, London, Ch. 10.

McDonald, K. A. (1978) USDA Forestry Service. Forest Products Laboratory Research Paper FPL-311, US Department of Agriculture Forest Service, Madison, Wisconsin, U.S.A.

McFarland, M. (1992) Investigations of the environmental acceptability of fluorocarbon alternatives to chlorofluorocarbons. *Proc. Nat. Acad. Sci. USA*, **89**, 807–11.

McIntire, O. R. and Kennedy, R. N. (1948) *Chem. Eng. Prog.*, **9**, 727.

Mepura (1995), Data Sheets, *Metallpulvergesellschaft m.b.h.*, Ranshofen, A-5282 Braunau-Ranshofen, Austria.

Meyer, H. G. (1995) *Catgut Acoust. Soc. J.*, **2**, No 7, 9–12.

Parkin, P. H., Humphreys, J. R. and Cowell, J. R.
    (1979) *Acoustics, Noise and Buildings* 4th
    edition, Faber, London, pp. 279–83.

Patten, G. A. and Skochdopole, R. E. (1962) *Mod.
    Plast.*, **39**, 149.

Reitz, D. W., Schuetz, M. A. and Glicksman, L. R.
    (1983) Proceedings of the Society of Plastics
    Industry 6th International Technical/Marketing
    Conference, San Diego, California, 31 Oct–4
    Nov., pp. 332–40.

Schuetz, M. A. and Glicksman, L. R. (1983)
    Proceedings of the Society of Plastics Industry
    6th International Technical/Marketing
    Conference. San Diego, California, pp. 341–7.

Skaar, C. (1948), N.Y.S. College of Forestry,
    Syracuse University Tech. Pub. No. 69.

Skochdopole, R. E. (1961) *Chem. Eng. Prog.*, **57**, 55.

Stamm (1929) *Ind. Eng. Chem., Amal.* ed. **1**, 94–7.

Toohy, R. P. (1961) *Chem. Eng. Prog.*, **57**, 52.

Traeger, R. K. (1967) *J. Cell. Plast.*, **3**, 405.

Chapter 8

Energy absorption in cellular materials

8.1 Introduction and synopsis

Packaging surrounds most things we buy or do. Food is packaged, parcels through the post are packaged, and within a car or aeroplane, we ourselves are carefully packaged. It is hard to say how much is spent on it, or the worth of the goods damaged due to inadequate packaging, but the sums involved are certainly considerable, and the potential return on any improvement is large.

The essence of protective packaging is the ability to convert kinetic energy into energy of some other sort – usually, heat – via plasticity, viscosity, visco-elasticity or friction; and this must be done whilst keeping the peak force (and thus the deceleration or acceleration) on the packaged object below the threshold which will cause damage or injury. And there is more to it than that. The direction of impact may not be predictable; then the package must offer omni-directional protection, that is, it must absorb impact from any side. Since the package must be carried with the object it protects, light weight is important. And – since much packaging is discarded – it must (almost always) be cheap.

Foams are especially good at this. The energy-absorbing capacity of a foam is compared with that of the solid of which it is made in Fig. 8.1. For the same energy-absorption, the foam always generates a lower peak force. Energy is absorbed as the cell walls bend plastically, or buckle, or fracture (depending on the material of which the foam is made), but the stress is limited by the long, flat plateau of the stress–strain curve (Figs. 4.2 and 5.1). By choosing the right cell-wall material and relative density, the foam can be tailored to give the best com-

309

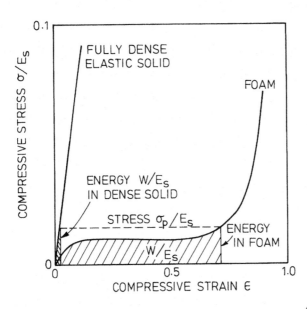

Figure 8.1 (a) Stress–strain curves for an elastic solid and a foam made from the same solid, showing the energy per unit volume absorbed at a peak stress σ_p. (b) Energy absorbed per unit volume against peak stress generated by an impact for both foams and the solid from which the foams are made. The foams always absorb more energy than the solid for a given maximum peak stress, σ_p. Both axes are normalized by the solid modulus, for reasons explained later.

(a)

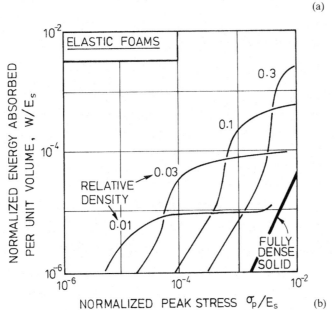

(b)

bination of properties for a given package. The energy absorbed by equiaxed foams (unlike that absorbed by honeycombs) is independent of the direction of impact; and most foams are light and cheap.

The chapter starts with analysis of energy-absorption mechanisms in foams, including the effects of pore fluid. A number of ways of characterising the energy-absorption have been proposed. A broad and useful way of making the comparison is described in Chapter 13; but it does not include the effect of

strain-rate. Methods which allow more detailed comparison have been proposed: among them are the Janssen factor, J, the Cushion factor, C, and the
energy-absorption diagram. All three are described below. Among them, the
last has the greatest generality and utility. Its uses are illustrated here by examples from case studies.

The design of effective packaging and crash padding involves many other considerations which lie beyond the scope of this book or are touched on only briefly.
They include the shape of the packaged object (important because it influences
load-transfer during impact); the capacity of the foam to dissipate elastic energy
(which controls rebound); and the time for which the packaged object can tolerate a given deceleration (which influences the choice of foam thickness, density
and geometry). There is a considerable literature on the use of foams for cushioning and packaging. The interested reader might wish to consult the book by Mustin (1968), that edited by Hilyard (1982), the series of papers by Lockett,
Cousins and Dawson (Cousins, 1976a,b) and Lockett et al. (1981), the papers
by Green et al. (1969), by Rusch (1970, 1971), by Lee and Williams (1971), by
Melvin and Roberts (1971) and by Schwaber and his co-workers (Meinecke and
Schwaber, 1970; Meinecke et al., 1971; Schwaber and Meinecke, 1971; Schwaber, 1973). A recent review of a number of aspects of energy absorption in foam
is presented by Mills in the book edited by Hilyard and Cunningham (1994).

Before lauching into detail, one point should be made. Much packaging
guards against the rare, unpredictable event: the delicate object (a bottle of
wine, let us say) sent by mail. Here the nature of the impact and the energy to
be absorbed can only be guessed, and guidance in selecting the best candidates
from among available, cheap, foams is the challenge. For this class of problem,
the way forward is the simple, empirical selection procedure outlined in Chapter 13. There are, however, a range of packaging problems – particularly
those involved with human safety – in which optimised selection is essential:
in an automobile accident, prevention of injury at 20 mph is a significant
improvement over that at 18 mph. The methods detailed below allow such optimised choice.

8.2   Energy-absorption mechanisms

When a foam is loaded, work is done by the forces applied to it. The work per unit
volume in deforming the foam to a strain ϵ is simply the area under the stress–
strain curve up to the strain ϵ (Fig. 8.2). Very little energy is absorbed in the
short, linear-elastic regime. As the figure shows, it is the long plateau of the
stress–strain curve, arising from cell collapse by buckling, yielding or crushing,
which allows large energy/absorption at near-constant load.

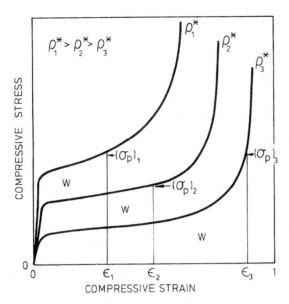

Figure 8.2 The peak stresses generated in foams of three densities in absorbing the same energy W are given by $(\sigma_p)_1$, $(\sigma_p)_2$ and $(\sigma_p)_3$. The lowest-density foam 'bottoms out' before absorbing the energy W, generating a high peak stress. The highest-density foam also generates a high peak stress before absorbing the energy W. Between these two extremes there is an optimum density, which absorbs the energy, W, at the lowest peak stress, $(\sigma_p)_2$.

The level of this plateau depends on the material and density of the foam, and on the rate of straining. Here we need to be clear about what we mean by 'low', 'intermediate' and 'high' strain-rates and impact-velocities. Ordinary laboratory tensile or compression-testing equipment provide *low* strain-rates: the range is roughly $10^{-8} - 10^{-2}$/s. Converted to an impact-velocity on a compressible foam 100 mm thick, this gives 10^{-9}–10^{-3} m/s. Almost all packaging is required to cope with higher impact-velocities than this, so data obtained in this way are suspect.

Free fall from a height of 1 metre gives an impact-velocity of just over 4 m/s, with associated strain-rates (in a 100 mm cushion) of around 40/s. These *intermediate* strain-rates are reproduced in the laboratory with high-speed servo-hydraulic testing equipment and drop-hammer tests. Auto design requires protection against impacts of up to 40/s; and ballistic impacts can be 50 times faster than this, corresponding to strain-rates as high as 10^4 m/s. Dynamic loading devices, gas guns and explosive loading equipment allow the *high-rate* range to be investigated. The upper end of this range lies above the sound velocity in all foams, and here inertial effects drive the plateau level upwards rapidly. Dynamic loading of foams was described briefly in Section 6.2, but it is rarely relevant for the kind of packaging applications in which foams excel.

A number of mechanisms are at work in absorbing energy (Schwaber, 1973). Some are related to the elastic, plastic or brittle deformation of the cell walls, and to the compression or flow of the fluid within the cells. The relevant mechanisms for a particular foam depend on the behaviour of the cell-wall material and on whether or not the cells are open or closed.

The plateau stress for elastomeric foams (used in cushions and in soft padding) is determined by the elastic buckling of the cells (Section 5.3). Since this is a form of elastic deformation, much of the external work stored during loading is released again when the foam is unloaded. But because elastomeric materials show damping, or hysteresis, not all of the external work is recovered; a fraction is dissipated as heat. The fraction, η, is known as the loss coefficient; typical values for it are given in Table 8.1.

Plastic and brittle foams are different. For these, the work done in the plateau region is completely dissipated as plastic work, or as work of fracture, or in friction between broken fragments of cell walls. These materials are particularly effective in high-performance packaging applications, giving large, controlled absorption of energy with none of the rebound that can be as damaging as the initial impact itself.

Other mechanisms, related to the deformation of the fluid within the cells, add to these. In open-cell foams, the pore fluid is expelled as the foam is compressed, giving rise to viscous dissipation, analysed later. It is helpful to know that it is strongly dependent on strain-rate, and that, unless the cells are small and the cell fluid is very viscous, it becomes important only at high rates (10^3/s or more). In closed-cell foams, the cell fluid is compressed as the foam deforms, storing energy which is largely recovered when the foam is unloaded. Unlike viscous dissipation, this storage mechanism is almost independent of strain-rate.

Natural cellular materials can absorb energy in yet other ways. Wood, bone, and leaves (all discussed in later chapters) have cell walls which are themselves composites. When these materials are deformed the fibres pull out and unravel in complicated ways which dissipate a great deal of energy; man-made materials which use them are now being developed. It is an area with scope for further exploitation (see, for example, Gordon and Jeronimides, 1974).

We now review the understanding of energy absorption in foams, developing simple equations for each mechanism. These are used later to construct diagrams which summarize the energy-absorbing capacity of foams.

Table 8.1 Loss coefficients for polymers*

Material	Frequency range (Hz)	Temperature (°C)	Loss coefficient η
PMMA	0.001–1000	25	0.01–0.05
Polyester	1–10 000	43	0.02–0.03
Polystyrene	0.01–100	–	0.01–0.02
Polypropylene	10–100	20	0.1–0.2

*Data from Lazan (1968).

(a) Plateau absorption in open-cell foams

The work done in compressing an open-cell foam is absorbed by the elastic buckling, plastic yielding or brittle crushing of the cell walls. Each of these collapse mechanisms proceeds at nearly constant load until the foam is almost completely crushed; then the load rises sharply. The energy absorbed per unit volume, up to a strain, ϵ, is

$$W = \int_0^\epsilon \sigma(\epsilon)\, d\epsilon. \tag{8.1}$$

When the plateau is flat, the plateau stress σ^* is almost constant and W is given approximately by:

$$W \approx \sigma^* \epsilon. \tag{8.2}$$

The energy, W, and stress, σ^*, depend on strain-rate, $\dot{\epsilon}$, and on temperature, T, through the intrinsic strain-rate and temperature-dependence of the cell wall material, described in Chapters 3 and 6. But since W is proportional to σ^*, Eqn. (8.2) still holds, and $\dot{\epsilon}$ and T do not appear explicitly.

(b) Viscous dissipation in open-cell foams

When an open-cell foam is compressed work is done in expelling the viscous fluid within the pores (Kosten and Zwikker, 1939; Gent and Rusch, 1966a,b; Hilyard and Kanakkannatt, 1970). The contribution of the fluid to the strength σ^* of an open-cell foam was analysed in Chapter 6; it is (Eqn. (6.27))

$$\sigma_g^* = \frac{C\mu\dot{\epsilon}}{(1-\epsilon)}\left(\frac{L}{l}\right)^2. \tag{8.3}$$

Here μ is the dynamic viscosity of the fluid, L the sample dimension, l the cell size and C a constant close to unity. The work done in compressing the fluid in an open-cell foam to a strain ϵ_f at *constant* strain-rate is then

$$W_g = \int_0^{\epsilon_f} \sigma_g^*\, d\dot{\epsilon} = C\mu\dot{\epsilon}\left(\frac{L}{l}\right)^2 \ln\left(\frac{1}{1-\epsilon_f}\right). \tag{8.4}$$

More usually we want the energy absorbed in bringing a moving object to rest, so that the strain-rate starts at $\dot{\epsilon}_i$ but decreases to 0 at strain ϵ_f. Assuming that the strain-rate decreases linearly, then:

$$\dot{\epsilon} = \frac{\dot{\epsilon}_i(\epsilon_f - \epsilon)}{\epsilon_f} \tag{8.5}$$

and

$$W_g = C\mu\dot{\epsilon}_i\left(\frac{L}{l}\right)^2\left(1 + \left(\frac{1-\epsilon_f}{\epsilon_f}\right)\ln(1-\epsilon_f)\right). \tag{8.6}$$

In both cases the energy dissipated increases as the viscosity of the fluid increases, and as the cell size decreases.

(c) Compression of the pore fluid in closed-cell foams

When a closed-cell foam is compressed the fluid within the cells is compressed also. Almost always the fluid is a gas; then a restoring force, proportional to the volume change in the gas, opposes further compression. The pressure in the cells after an axial strain ϵ (Chapter 5, Eqn. (5.8)) is:

$$p = p_0 \left\{ \frac{(1 - \rho^*/\rho_s)}{1 - \epsilon(1 - 2\nu^*) - \rho^*/\rho_s} \right\} \tag{8.7}$$

where p_0 is the initial pressure (usually atmospheric). The work done in compressing the gas to a strain, ϵ_f, is then

$$W_c = \int_0^{\epsilon_f} (p - p_0)\mathrm{d}\epsilon$$

$$= p_0 \left\{ \left(\frac{1 - \rho^*/\rho_s}{1 - 2\nu^*} \right) \ln \left(\frac{1 - \rho^*/\rho_s}{1 - \rho^*/\rho_s - (1 - 2\nu^*)\epsilon_f} \right) - \epsilon_f \right\}. \tag{8.8}$$

At all but the highest rates of strain ($\dot{\epsilon} > 100/\mathrm{s}$) the compression is essentially isothermal (see Chapter 6) because the thermal mass of the polymer forming the cell walls greatly exceeds that of the gas contained within the cells. This means that the compression of the pore gas gives an additional elastic contribution to the loading curve and to the stored energy which is independent of strain-rate.

8.3 Methods of characterizing energy-absorption in foams

As mentioned already, the aim in packaging is to absorb the kinetic energy of the packaged object while keeping the force on it below some limit. There is an optimum foam density for a given package. Figure 8.2 illustrates how, if the density is too low, the foam 'bottoms out' with a sharp increase in force before all the energy has been absorbed, while if it is too dense, the force exceeds the critical value before enough energy has been dissipated. Roughly speaking, the 'ideal' foam is one with a plateau-stress just below the critical damaging level, and with an area under the stress–strain curve up to the strain at which densification begins, ϵ_D, which is just equal to the kinetic energy to be absorbed per unit volume of packaging material.

A number of schemes have been suggested for characterizing energy absorption by foams. For an initial, broad, selection from among available foams, the plateau stress/densification strain diagram described in Chapter 13 is best. Here we are concerned with more exact characterization of individual foams. There are several methods. Some, like the Janssen factor, J, and the cushion factor, C (described below) measure the 'efficiency' of a foam – that is, how close it comes to being ideal; but they lack generality: a new diagram is needed for each foam, and for each impact energy. An alternative approach is that of the energy-absorption diagram which allows the optimum foam to be designed for a given packaging application, using information on the allowable peak stress, the total energy to be absorbed, and the strain-rate.

(a) The Janssen factor, J

The deceleration, a, of an object of mass, m, packaged in foam is given by Newton's law:

$$a = \frac{F}{m}. \tag{8.9}$$

The force, F, on the packaged object is the compressive stress of the foam times the area of contact between it and the foam. One way of estimating the efficiency of a real foam in absorbing a given impact energy is to compare the peak deceleration it produces, a_p, with that caused by an ideal foam, a_i, for a given impact energy (Fig. 8.3(a)). The ideal foam absorbs energy at a constant deceleration a_i; it is calculated by equating the kinetic energy to the work done by the constant force in the foam, acting through a displacement equal to its thickness, t, so that

$$\tfrac{1}{2}mv^2 = ma_i t \tag{8.10}$$

from which

$$a_i = \frac{v^2}{2t}. \tag{8.11}$$

The effectiveness of the foam at absorbing impact energy is measured by the ratio

$$J = \frac{a_p}{a_i} \tag{8.12}$$

and is called the Janssen factor, J. The Janssen factor for a foam depends on the energy of the impact; it is high at both low and high energies and reaches a minimum at some intermediate energy; the shape of a typical plot is shown in Fig. 8.3(a). It is constructed, for a given foam, by measuring the peak acceleration in the foam during pendulum impact tests; examples can be found in the papers by Woolam (1968) and Hilyard and Djiauw (1971). The J factor is useful for comparing the efficiency of different foams in absorbing energy. But because it is an empirical measure it requires a large amount of data collection (each thickness of foam, as well as each density, requires a new curve) and it does not relate the energy-absorbing capacity to the mechanisms by which the foam deforms.

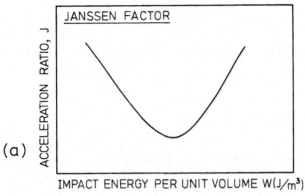

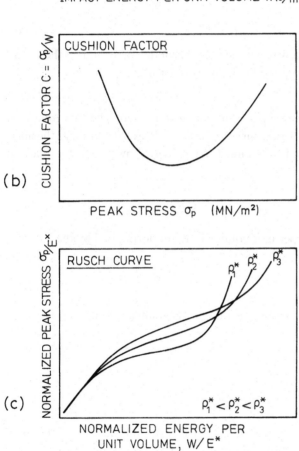

Figure 8.3 Three diagrams which are used to characterize energy absorption in foams: (a) the Janssen factor, J; (b) the cushion factor, C; and (c) the curves constructed from the stress–strain equations of Rusch (1970, 1971).

(b) The Cushion factor

Since a large amount of uniaxial stress–strain data for foams already exists, it is helpful to relate the efficiency of a foam in absorbing energy to its uniaxial stress–strain behaviour. The simplest way to do this is to plot the energy absorbed up to a given stress, σ_p, against the stress itself. A variation of the method is to normalize this energy by the peak stress, σ_p, and plot peak stress divided by energy absorbed up to that stress against the stress. Such a plot gives the cushion factor, C, for a foam (Gordon, 1974). The shape of a typical plot is shown in Fig. 8.3(b).

(c) Rusch's curve

Rusch (1970, 1971) has improved on this method. He first notes that the shape of the stress–strain curve for a foam can be defined by an empirical shape factor, $\psi(\epsilon)$, defined by

$$\sigma = E^* \psi(\epsilon) \epsilon \qquad (8.13a)$$

where σ is the stress, E^* (as always) is the modulus of the foam, and ϵ is the strain. The shape factor is found empirically to have the form:

$$\psi(\epsilon) = me^{-n} + re^s \qquad (8.13b)$$

where m, n, r and s are constants of that particular foam. The two equations define a shape for the stress–strain curve of the foam. Rusch defines K as the maximum deceleration produced by an ideal foam divided by that of the foam in question $(= 1/J)$:

$$K = \frac{v^2}{2ta_p} \qquad (8.14)$$

and I is the impact energy per unit volume of foam normalised by the Young's modulus of the foam:

$$I = \frac{W}{E^*}. \qquad (8.15)$$

The ratio I/K gives the peak stress generated in the foam normalized by the foam modulus

$$\frac{I}{K} = \frac{mv^2}{2AtE^*} \cdot \frac{2ta_p}{v^2} = \frac{ma_p}{AE^*}$$

$$= \frac{\sigma}{E^*}. \qquad (8.16)$$

Both I and K can be related to the uniaxial stress–strain behaviour of the foam through the shape factor $\psi(\epsilon)$. The optimum foam for absorbing a given amount of energy with a maximum allowable peak stress can then be determined by plotting I/K, the peak stress normalized by the foam modulus, against I, the energy per unit volume of the impact normalized by the foam modulus. A typical plot is shown in Fig. 8.3(c). Rusch (1971) also analyses the efficiency of composite foam systems, with two foams placed either in series or in parallel. He finds that each foam behaves independently but at a reduced efficiency, so that composite systems are never optimum. Rusch's approach has greater generality than methods based on the J factor; but depending as it does on an empirical function to describe the shape of the stress–strain curve, it lacks any mechanistic basis.

8.4   Energy-absorption diagrams

A different approach, which allows empiricism to be combined with physical modelling, and which has attractive generality as a way of optimizing the choice of foam, is offered by *energy-absorption diagrams* (Maiti *et al.*, 1984). The procedure for constructing them from experimental stress–strain curves is shown in Fig. 8.4. The contribution of modelling is left to the next section.

(a) Construction of the diagrams

Consider the sequence of operations shown in Fig. 8.4. Samples of a given foam with a range of densities are tested in compression, at a fixed strain-rate, $\dot{\epsilon}_1$, and temperature T_1, to give a family of stress–strain curves. The area under each curve up to the stress σ_p (stepping upward in σ_p) is measured; this area is the energy absorbed per unit volume, W. The value of W is plotted against σ_p for each curve, normalizing both by the modulus of the solid, E_s, measured at a standard strain-rate and temperature (the diagrams of this chapter use 10^{-3}/s and $20\,°C$).

The best foam for a given package is the one that absorbs the most energy up to the maximum permitted package stress σ_p. Each foam density has a σ_p for which it is the best choice. It is given by the shoulder on the energy curve of the middle diagram of Fig. 8.4, because here the curve for that foam lies above that for any other. The heavy line shows the envelope of these points. It describes a relationship between W and σ_p for the *optimum* foam density, for loading at a strain-rate, $\dot{\epsilon}_1$, and at a temperature, T_1.

This line is replotted, on the same axes, in the bottom diagram. The individual curves are no longer shown, but the optimum density, ρ^*/ρ_s, read from the middle diagram, is marked on the line. The tests are now repeated at a series of different strain-rates, $\dot{\epsilon}_2$, $\dot{\epsilon}_3$, etc., and the data from these are treated in the same way. This allows a family of optimum energy absorption curves to be built up, as shown. The points corresponding to equal densities are connected by lines to give an intersecting family. A change of temperature can be treated in the same way.

Two examples will help to make a number of further points. Families of stress–strain curves for two elastomeric foams (polyurethane and polyethylene) and for one plastic foam (polymethylacrylimid) were shown in Fig. 5.3. The temperature ($20\,°C$) and the strain-rate ($= 10^{-3}$/s) of the tests were constant. The

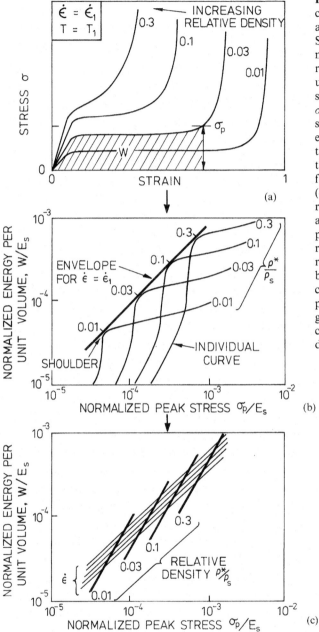

Figure 8.4 The construction of energy-absorption diagrams. (a) Stress–strain curves are measured at a single strain-rate, $\dot{\epsilon}_1$. (b) The area W under each curve up to the stress σ_p is plotted against σ_p, both normalized by the solid modulus E_s. The envelope which just touches each curve defines the optimum choice of foam at the strain-rate $\dot{\epsilon}_1$. (c) The envelope is replotted on the same axes, and marked with density points. The procedure is repeated at other strain-rates (or temperatures) to build up the family of curves. Finally the density points are connected to give a family of intersecting contours of constant density.

areas under each, up to the stress σ_p, are plotted in Figs. 8.5 and 8.6. Both axes are normalized by E_s, the modulus of the solid polymer; it is this which allows the polyurethane and the polyethylene to be plotted together. The envelope of optimum energy absorption is shown: it is almost a straight line. The tests were

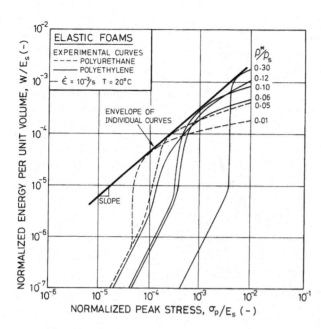

Figure 8.5 Energy-absorption curves for two elastomeric foams, constructed by measuring the area under the stress-strain curves of Figs. 5.3(a) and 5.3(b). The envelope line relates W, σ_p and relative density for the optimum choice of foam at the strain-rate 10^{-2}/s. It has a mean slope of 0.9.

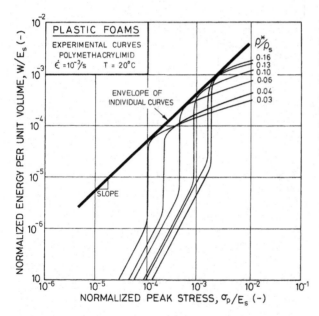

Figure 8.6 Energy-absorption curves for a plastic foam, constructed by measuring the area under the stress-strain curves of Fig. 5.3(c). The envelope line relates W, σ_p and relative density for the optimum choice of foam at the strain-rate 10^{-2}/s.

repeated for a range of strain-rates; one example for a single density, is shown in Fig. 8.7. The additional data give a set of optimum energy envelopes, one for each strain-rate, as shown in Figs. 8.8 and 8.9. The optimum energy lines are intersected by the family of constant density curves.

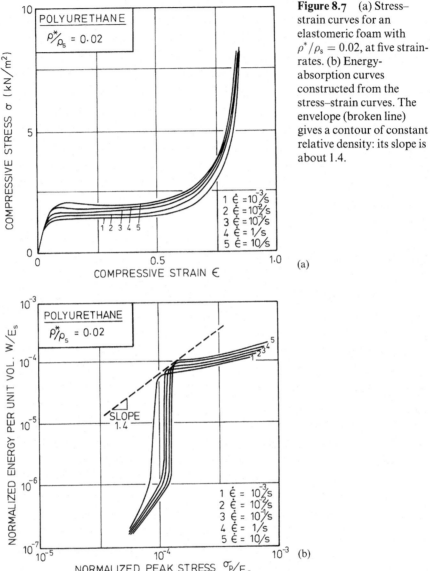

Figure 8.7 (a) Stress–strain curves for an elastomeric foam with $\rho^*/\rho_s = 0.02$, at five strain-rates. (b) Energy-absorption curves constructed from the stress–strain curves. The envelope (broken line) gives a contour of constant relative density: its slope is about 1.4.

The figures summarize the energy absorption capacity of these classes† of foams at 20 °C (further tests would give further sets of curves for other temperatures). The effect of strain-rate in these foams is small, but in others it can be larger. For any given allowable peak stress, the diagrams allow the best foam density to be selected, and show the maximum energy per unit volume, W, which can be absorbed without exceeding that peak stress, σ_p. This forms the

† Kurauchi *et al.* (1984) have studied energy absorption in brittle (glass) foams. These, like plastic foams, show large energy absorption at constant stress. The energy absorption can be calculated from the plateau stress in the manner outlined for plastic foams in this section.

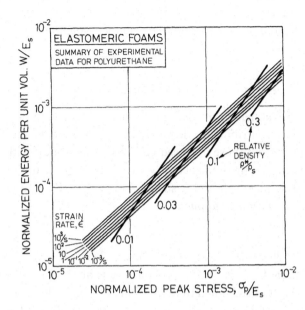

Figure 8.8 An energy-absorption diagram for an elastomeric foam. Although it was constructed using data for polyurethane, it is broadly typical of all elastomeric foams.

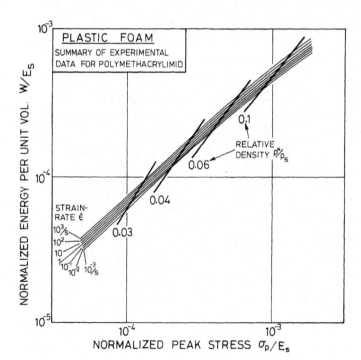

Figure 8.9 An energy-absorption diagram for a plastic foam. It was constructed using data for polymethacrylimid, but is broadly typical of plastic foams with the same value of σ_{ys}/E_s, as polymethacrylimid ($\sigma_{ys}/E_s = 1/30$).

basis of a procedure for selecting foams, detailed below. But first we examine how modelling gives further understanding of the diagrams and their potential.

(b) Modelling of energy-absorption diagrams

Additional insight is given by modelling the individual energy-absorption processes and examining how each contributes to the total. The dominant contributions in open-cell foams differ from those in closed-cell foams, so we deal with each separately. Modelling allows the influence of viscous absorption, and of compression of cell gas, to be included; and it suggests the use of W/E_s and σ_p/E_s as axes to give diagrams with wide application.

Open-cell elastomeric foams

When an open-cell elastomeric foam is compressed, energy is absorbed in the bending and buckling of the cell walls, and in the expulsion of the pore fluid. We idealize the stress–strain curve in the way shown in the inset of Fig. 8.10. The linear-elastic portion of the curve is followed by a horizontal plateau at a stress, σ_{el}^*, truncated by a vertical rise at the densification strain, ϵ_D (Eqn. (5.22)):

$$\epsilon_D = 1 - 1.4\rho^*/\rho_s. \tag{8.17}$$

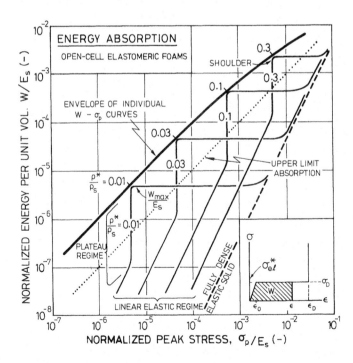

Figure 8.10 The modelling of the energy-absorption diagram for open-cell elastomeric foams. The envelope line has a slope near 1 at low stresses, falling to 0.85 at high. Viscous dissipation, caused by the expulsion of pore fluid, displaces the envelope to the right. It is important only at high strain-rates.

In the linear-elastic region the energy stored per unit volume in loading the foam up to a stress σ_p is

$$W = \frac{1}{2} \frac{\sigma_p^2}{E^*}.$$

Using Eqn. (5.6a) for E^* gives

$$\frac{W}{E_s} = \frac{1}{2} \left(\frac{\sigma_p}{E_s}\right)^2 \frac{1}{(\rho^*/\rho_s)^2}. \tag{8.18}$$

The normalized energy per unit volume, W/E_s, is a function of σ_p/E_s and the relative density, ρ^*/ρ_s, only. It is shown on Fig. 8.10, which has logarithmic scales, as a family of lines of slope 2, one for each value of ρ^*/ρ_s, labelled 'linear-elastic regime'.

The important part of the stress–strain curve, from the point of view of packaging, is the plateau. In this regime, energy is absorbed at constant stress σ_{el}^*, so that

$$dW = \sigma_{el}^* d\epsilon.$$

Using Eqn. (5.18a) for σ_{el}^* and integrating over strain

$$\frac{W}{E_s} = 0.05 \left(\frac{\rho^*}{\rho_s}\right)^2 (\epsilon - \epsilon_0) \tag{8.19}$$

where ϵ_0 is the strain at the end of the linear-elastic regime. This plots as a family of vertical lines on Fig. 8.10. The plateau ends abruptly at the densification strain, at which point (in our idealization) the energy-absorption curves become horizontal. The value of this maximum useful energy absorption per unit volume, W_{max}, is given by replacing ϵ in Eqn. (8.19) by the densification strain, Eqn. (8.17), giving (neglecting ϵ_0):

$$\frac{W_{max}}{E_s} = 0.05 \left(\frac{\rho^*}{\rho_s}\right)^2 \left(1 - 1.4 \frac{\rho^*}{\rho_s}\right). \tag{8.20}$$

The stress continues to rise until the foam is compressed to a solid. At this point the energy curve joins that for a fully dense elastomer (broken line) for which

$$\frac{W}{E_s} = \frac{1}{2} \left(\frac{\sigma_p}{E_s}\right)^2.$$

As already explained, the optimum choice of foam is the one with a shoulder which lies at that value of $\sigma_p = \sigma_D$. The curve of optimum energy absorption is then the envelope of the individual $W - \sigma_p$ curves, as shown by the heavy, solid line in Fig. 8.10. Its equation is given by Eqn. (8.20), with ρ^*/ρ_s replaced by the plateau stress for that density, that is, by

$$\frac{\rho^*}{\rho_s} = \left(\frac{20\sigma_D}{E_s}\right)^{1/2} \tag{8.21}$$

giving

$$\frac{W_{\max}}{E_s} = \frac{\sigma_D}{E_s}\left(1 - 6.26\left(\frac{\sigma_D}{E_s}\right)^{1/2}\right). \qquad (8.22)$$

This line has a slope of 1 at low stresses, falling to about 7/8 at higher stresses.

Although approximate, this development of the form of the energy diagram is illuminating. The individual energy curves have the same shape as the experimental ones (Fig. 8.5), giving physical meaning to each segment. The optimum energy line (full line) has the same slope – slightly less than 1 – as that observed in real foams. More important, the equations developed above show that W/E_s depends on σ_p/E_s and ρ^*/ρ_s only; that is, the one diagram describes *all* elastomeric foams, of all densities, no matter what they are made of (the material properties are contained in the normalizing parameter E_s). Individual foam materials will, in practice, differ a little, for reasons discussed later. But the diagram has value as an approximate, universal plot; and it allows some further features of energy absorption to be illustrated.

The first is the relative efficiency with which foams absorb energy. The full, heavy envelope line describing foams on Fig. 8.10 converges towards the broken line describing fully dense solids at high stresses. But for the range of energy absorption covered by the diagram, the peak stress of the foam is less, by a factor of 10^{-4}–10^{-1}, than that caused by the dense solid. That is one reason why foams are so widely used in packaging.

The second concerns the dependence of energy absorption on strain-rate. The modulus E_s of a polymer increases with strain-rate, particularly when the temperature is near its glass temperature (see Chapter 3). Both the energy absorption and the peak stress are proportional to E_s, so normalizing them by E_s at the appropriate strain-rate would remove this strain-rate dependence – that is why the simple modelling of this section leads to a single, unique envelope line for all $\dot{\epsilon}$ and T. The experimental data used to construct Fig. 8.8 were normalized by a value of E_s at a single fixed strain-rate (10^{-3}/s). This, and a small additional dependence derived from the way in which E_s is measured (it is the slope of the stress–strain curve at small strains – the cell walls of the collapsing foam are subjected to large strains), causes the separation of the lines into a family, one for each strain-rate.

There is another potential energy-absorbing process in open-cell foams. Cell fluid is expelled when the foam is compressed. The viscous work involved in doing this was calculated earlier: for a constant strain-rate it is given by Eqn. (8.4). It must be added to the absorption by cell-wall deformation (Eqn. (8.20)), giving at the peak (when $\epsilon = \epsilon_D = 1 - 1.4\rho^*/\rho_s$),

$$\frac{W}{E_s} = 0.05\left(\frac{\rho^*}{\rho_s}\right)^2\left(1 - 1.4\frac{\rho^*}{\rho_s}\right) + \frac{C\mu\dot{\epsilon}}{E_s}\left(\frac{L}{l}\right)^2\ln\left(\frac{1}{1.4\rho^*/\rho_s}\right)$$

and

$$\frac{\sigma_p}{E_s} = 0.05\left(\frac{\rho^*}{\rho_s}\right)^2 + \frac{C\mu\dot{\epsilon}}{E_s}\left(\frac{L}{l}\right)^2\left(\frac{1}{1.4\rho^*/\rho_s}\right).$$

$$(8.23)$$

When the second term in each of these equations is much smaller than the first, viscous dissipation is negligible. But when the second is comparable with, or larger than, the first – as it will be if the strain-rate $\dot{\epsilon}$ is large enough or if the fluid viscosity is high enough – then this new contribution becomes important. In the limit of very high strain-rates, the first term in each equation can be neglected; then, substituting the second equation into the first,

$$\frac{W}{E_s} = 1.4\frac{\sigma_p}{E_s}\frac{\rho^*}{\rho_s}\ln\left(\frac{1}{1.4\rho^*/\rho_s}\right).$$

The term involving ρ^*/ρ_s has a value of about 0.1 for the normal range of ρ^*/ρ_s, so that, for pure viscous dissipation

$$\frac{W}{E_s} \approx 0.1\frac{\sigma_p}{E_s}.$$

This is plotted at a dotted line on Fig. 8.10. As the strain-rate is raised past the value at which the two terms in Eqn. (8.23) are equal, the envelope line, which is almost independent of $\dot{\epsilon}$ at low $\dot{\epsilon}$, moves to the right towards the upper limit given by the dotted line.

That, then, is the physical basis of the energy-absorption diagram in open-cell foams. Closed-cell foams differ in several ways, described next.

Closed-cell elastomeric foams

When a closed-cell elastomeric foam is compressed, energy is absorbed by the bending, buckling and stretching of the cell walls, and in the compression of the fluid – usually a gas – contained within the cells. Compression of the gas gives a stress–strain curve which rises with strain. When the foam density is low, the relative contribution of the gas is large, and may completely dominate the behaviour (as it does in packaging materials made by trapping air bubbles between polyethylene membranes). When the density is high, or the foam is made of a stiff material (that is, E_s is large) the relative contribution of gas compression is much less. So, for a given foam material, there is a transition from gas-dominated behaviour to cell-wall-dominated behaviour at a characteristic density. We idealize the stress–strain curve as shown in the inset of Fig. 8.11. The linear-elastic regime has a slope given by Eqn. (5.13a): for the present purposes we shall take the fraction $1 - \phi$ of material in the cell faces to be small, and assume that the initial gas pressure, p_0, is atmospheric, so that p_0/E_s is small also. Then, with adequate precision,

$$\frac{E^*}{E_s} \approx \left(\frac{\rho^*}{\rho_s}\right)^2$$

storing energy up to the strain ϵ_0 (where $\epsilon_0 < \epsilon_D$) of

$$\frac{W}{E_s} = \frac{1}{2}\left(\frac{\sigma_p}{E_s}\right)^2 \frac{1}{(\rho^*/\rho_s)^2}$$

as before. The plateau, however, has a positive slope, given by Eqn. (5.20):

$$\frac{\sigma^*}{E_s} = 0.05\left(\frac{\rho^*}{\rho_s}\right)^2 + \frac{p_0}{E_s}\left\{\frac{\epsilon}{1 - \epsilon - \rho^*/\rho_s}\right\} \tag{8.24}$$

storing energy per unit volume up to the strain, ϵ (Eqn. (8.8) with $v^* = 0$) of:

$$\frac{W}{E_s} = 0.05\left(\frac{\rho^*}{\rho_s}\right)^2 \epsilon + \frac{p_0}{E_s}\left\{(1 - \rho^*/\rho_s)\ln\left(\frac{1 - \rho^*/\rho_s}{1 - \rho^*/\rho_s - \epsilon}\right) - \epsilon\right\}. \tag{8.25}$$

The plateau ends, and the stress rises sharply, at the densification strain ϵ_D given by Eqn. (8.17).

The resulting $W - \sigma_p$ curves are plotted on Fig. 8.11, taking p_0/E_s as 6.7×10^{-4}, corresponding to $p_0 = 1$ atmosphere and $E_s = 150\,\mathrm{MN/m^2}$, for polyethylene. When the density is high, the first term in the last two equations dominates, and the curves have the same general shape as those of open-cell foams. But when the density is lower, the second pair of terms dominates, and energy absorption is dominated by compression of the gas within the cells. Then

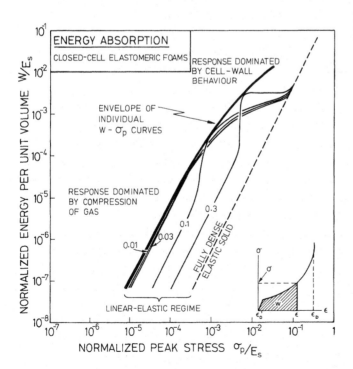

Figure 8.11 The modelling of the energy-absorption diagram for closed-cell elastomeric foams. At high densities, cell-wall deformation is the dominant energy-absorbing process. At low densities, compression of the gas in the cells becomes dominant. ($p_0/E_s = 6.7 \times 10^{-4}$, corresponding to $p_0 = 1$ atmosphere and $E_s = 150\,\mathrm{MN/m^2}$, for polyethylene).

the individual energy-absorption curves become independent of density (they simply reflect the compression of the gas) and the envelope tends, at low stress, to a slope of 2. The foam, in this limit, has become a pneumatic damper.

Plastic foams

When a plastic foam is compressed, work is done in bending and stretching the cell walls. The fluid within the cells plays a less important role than it did for elastomeric foams, because the plastic foams are so much stiffer and stronger. We shall ignore the pore fluid in the following treatment (its effect can be included, if desired, in the way illustrated in the last two sections). With these provisos, the stress–strain curve can be idealized in the way shown in the inset of Fig. 8.12. The development parallels that for open-celled elastomeric foams, with the plateau controlled, now, by plastic collapse.

In the linear-elastic regime energy is stored in the bending of the cell walls, giving the contribution to energy absorption described by Eqn. (8.18). It is shown in Fig. 8.12 as the family of lines labelled 'linear-elastic regime'.

As far as energy absorption is concerned, the plateau is again the important part of the stress–strain curve. In this regime, energy is absorbed at a constant stress, σ_{pl}^*, so that:

$$dW = \sigma_{pl}^* \, d\epsilon.$$

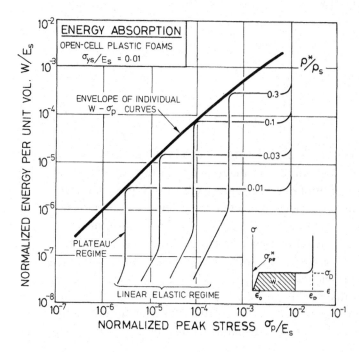

Figure 8.12 The modelling of the energy-absorption diagram for plastic foams $(\sigma_{ys}/E_s = 0.01)$.

Using Eqn. (5.27a) for σ_{pl}^* and integrating over strain gives:

$$\frac{W}{E_s} = 0.3 \frac{\sigma_{ys}}{E_s} \left(\frac{\rho^*}{\rho_s}\right)^{3/2} (\epsilon - \epsilon_0) \qquad (8.26)$$

where, as before, ϵ_0 is the strain at the end of the linear-elastic regime. The equation appears as a family of vertical lines on Fig. 8.12, more closely spaced than in Fig. 8.10 because of the less rapid dependence of σ_{pl}^* on density.

The plateau of the stress–strain curve ends abruptly (in our idealization) at the densification strain ϵ_D given by Eqn. (8.17). At this strain the energy-absorption curve becomes horizontal. The maximum useful energy absorption per unit volume, W_{max}, is found by inserting ϵ_D into Eqn. (8.26) giving (neglecting ϵ_0):

$$\frac{W_{max}}{E_s} \approx 0.3 \frac{\sigma_{ys}}{E_s} \left(\frac{\rho^*}{\rho_s}\right)^{3/2} (1 - 1.4\rho^*/\rho_s). \qquad (8.27)$$

The stress continues to rise until the foam is compressed to a solid, when the energy curve joins that for the plastic compression of the fully dense solid (taken here to be $\sigma_{ys} = E_s/100$).

The optimum choice of foam, as before, is the one which absorbs the most energy without the stress rising above the chosen limit σ_D: it is the foam with the shoulder at the value σ_D. The curve of optimum energy absorption (shown as a heavy line on Fig. 8.12) is the envelope that touches each $W - \sigma_p$ curve at its shoulder. Its equation is given by Eqn. (8.27) with ρ^*/ρ_s replaced by the plateau stress for that density, that is, by:

$$\left(\frac{\rho^*}{\rho_s}\right) = \left(\frac{3.3\sigma_p}{\sigma_{ys}}\right)^{2/3}$$

giving

$$\frac{W_{max}}{E_s} = \frac{\sigma_D}{E_s} \left\{1 - 3.1 \left(\frac{\sigma_D}{E_s}\right)^{2/3}\right\}. \qquad (8.28)$$

This line has a slope of 1 at low stresses, falling to about 5/6 at higher stresses.

This approximate derivation of the energy diagram for plastic foams explains and quantifies the broad features of the experimental curves of Fig. 8.6 remarkably well. The individual curves have the predicted shape, and the optimum line has the slope that the theory leads us to expect. The modelling also makes it clear that the diagram is less general than that for elastic foams: energy absorption in plastic foams depends not only on E_s but on the ratio σ_{ys}/E_s as well, so a given diagram relates to a single, given value for σ_{ys}/E_s. All foams made from materials with the same value of this ratio are described by a single diagram, and since the ratio for many polymers is close to $1/100$, a diagram for this value has broad application. Figure 8.12 is constructed with this value of σ_{ys}/E_s.

8.5   The design and selection of foams for packaging

There are two scenarios. First, the big picture: which, from all available foams, are the best candidates for a given application? This is explored in Chapter 13. Here the focus has a higher resolution: how can we optimize the selection of material, foam density, and thickness to maximize the efficiency of package? The answer is of particular importance where human safety is a concern.

Energy-absorption diagrams give a systematic way of addressing this. Some of their uses are obvious. Given a foam of known material and density, the diagram immediately identifies the peak stress σ_p/E_s and the energy absorbed per unit volume, W/E_s, at which it is best used. The area of contact, A, between the foam and the packaged object can then be chosen to give a peak force ($F = \sigma_p A$) which is non-damaging, and the thickness, t, of the package can be adjusted so that the kinetic energy of the object, U, is completely absorbed ($U = WAt$).

The diagrams do more than this. Consider first the problem of choosing the best foam density and thickness for a given package. The foam material (and hence E_s at some standard strain-rate – $10^{-3}/s$, for instance) is given. The mass, m, of the packaged object, its contact area with the package, its kinetic energy U and the maximum force, F, for acceleration, a, it can tolerate ($F = ma$) are specified. The procedure, illustrated by Fig. 8.13, is as follows.

1.   A vertical line is drawn corresponding to the peak stress level, σ_p/E_s.

2.   An arbitrary thickness, t_1, is chosen and, using this, an approximate strain-rate, $\dot{\epsilon}_1$, is calculated from the velocity, v, of the impact:

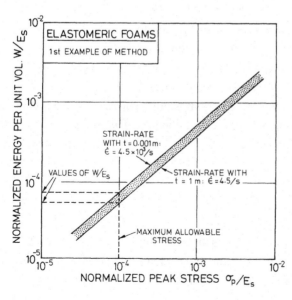

Figure 8.13 Example 1 of the use of the energy-absorption diagram. Only two absorption lines are shown, corresponding to the two strain-rates which enter the first iteration of the problem. The final choice lies between them.

$$\dot{\epsilon}_1 \approx \frac{v}{t_1} = \frac{1}{t_1}\left(\frac{2U}{m}\right)^{1/2}.$$

3. The corresponding value of $\log(W/E_s)$ is determined from the intersection of the stress line and the $\dot{\epsilon} = \dot{\epsilon}_1$ line (Fig. 8.13) and W is calculated from the known value of E_s.

4. A new thickness, t_2, is calculated using this value of W, the known energy to be absorbed, and the known area of contact of the package with the foam, A.

5. A new strain-rate $\dot{\epsilon}_2$ is calculated using t_2.

6. The corresponding value of $\log(W/E_s)$ is determined from which a new thickness can be calculated.

The procedure is repeated and converges rapidly. The numerical example outlined in Table 8.2 illustrates the method. An object weighing 0.5 kg must be packaged so that, when dropped on to a hard floor from a height of 1 m, it is not subjected to a deceleration of greater than $10g$ (a very low deceleration). The two initial choices of foam thickness of 1000 mm and 1 mm both converge quickly to the optimum value of 150 mm, and lead to a choice of foam with a relative density of a little less than 0.01 (using Fig. 8.8). The low density and large foam thickness result from the demanding specification that the deceleration of the packaged object must not exceed $10g$.

Now consider a second, more general, example in which the foam material is not specified (Table 8.3). We are asked to find the best foam material and density to package an object of mass 2.5 kg. The contact area, A (0.025 m^2), and foam thickness, t (20 mm), the drop height, h (1 m) and the maximum allowable deceleration ($100g$) are given. The procedure is illustrated in Fig. 8.14.

1. Calculate the energy to be absorbed per unit volume of foam, W, and the allowable peak stress, σ_p (they are given in Table 8.3).

2. Choose an arbitrary value of E_s (100 MN/m^2, for example) and plot W/E_s and σ_p/E_s on to the diagram, giving point A.

3. Construct a line of slope 1 through this point (broken line). Moving along this line simply changes E_s for constant values of W and σ_p.

4. Calculate the strain-rate from

$$\dot{\epsilon} = \frac{v}{t} = \frac{1}{t}\left(\frac{2U}{m}\right)^{1/2}.$$

Select the point where the broken line intersects the energy-absorption line at this strain-rate (roughly 10^2/s in this example), giving the point B.

5. Read off the values of σ_p/E_s and W/E_s. Divide these into the values of σ_p and W to extract E_s, thus identifying the best choice of foam material (in

Table 8.2   Example 1: selection of foams

Specification of the problem

Mass of the package object, $m = 0.5\,kg$
Area of contact between foam and object, $A = 0.01\,m^2$
Velocity of package on impact (drop height $h = 1\,m$), $v = 4.5\,m/s$
Energy to be absorbed, $U = mv^2/2 = 5\,J$
Maximum allowable package force (based on deceleration of 10g), $F = ma = 50\,N$
Maximum allowable peak stress, $\sigma_p = F/A = 5\,kN/m^2$
Solid modulus in foam (flexible polyurethane), $E_s = 50\,MN/m^2$
Maximum allowable normalized peak stress, $\sigma_p/E_s = 10^{-4}$

Iterative procedure		
1st Iteration	$t_1 \gg t$	$t_1 \ll t$
Initial choice of t_1	1 m	0.001 m
Resulting strain-rate, $\dot\epsilon = v/t_1$	$4.5\,s^{-1}$	$4.5 \times 10^3\,s^{-1}$
Resulting (W/E_s) at $\sigma_p/E_s = 10^{-4}$	5.25×10^{-5}	7.4×10^{-5}
Energy absorbed per unit volume, W	$2620\,J/m^3$	$3700\,J/m^3$
2nd Iteration		
Revised t_2 (from $U = WAt$)	0.19 m	0.14 m
Revised $\dot\epsilon = v/t_2$	$24\,s^{-1}$	$32\,s^{-1}$
Revised (W/E_s)	6.6×10^{-5}	6.7×10^{-5}
Revised W	$3300\,J/m^3$	$3350\,J/m^3$
3rd Iteration		
Revised t_3 (from $U = WAt$)	0.15 m	0.15 m
Optimum density, ρ^*/ρ_s (Fig. 8.8)	A little below 0.01	

---

this example, the result is $E_s \approx 28\,MN/m^2$, so that a low-modulus, flexible polyurethane is the best choice). Replotting point B on to the more detailed diagram (Fig. 8.8) shows that the desired relative density is 0.1.

If the initial design point A lies above all the energy contours and the line of slope 1 through it does not intersect them, the specification cannot be achieved; then either A or t must be increased. If it lies well below the contours, the packaging is over-generous, and the values of A or t can be reduced while still meeting the specification.

    We conclude this chapter with two further, more detailed, examples of the energy-absorption diagram method.

Table 8.3   Example 2: selection of foams

Specification of the problem

Mass of the package object, $m = 2.5\,\text{kg}$

Area of contact between foam and object, $A = 0.025\,\text{m}^2$

Thickness of foam, $t = 20\,\text{mm}$

Drop height, $h = 1\,\text{m}$

Velocity of impact $v = (2gh)^{1/2} = 4.5\,\text{m/s}$

Strain-rate $\dot{\epsilon} = v/t = 225/\text{s}$

Energy to be absorbed $U = mgh = 25\,\text{J}$

Energy to be absorbed per unit volume of foam $W = U/At = 5 \times 10^4\,\text{J/m}^3$

Maximum allowable force (based on decleration of $100g$) = 2500 N

Maximum allowable peak stress $\sigma_\text{p} = F/A = 10^5\,\text{N/m}^2$

Trial design point A, using $E_\text{s} = 100\,\text{MN/m}^2$

Normalized energy $W/E_\text{s} = 5 \times 10^{-4}$

Normalized peak stress $\sigma_\text{p}/E_\text{s} = 10^{-3}$

Final design point B, read from diagram

Normalized energy $W/E_\text{s} = 1.8 \times 10^{-3}$

Normalized stress $\sigma_\text{p}/E_\text{s} = 3.7 \times 10^{-3}$

Resulting derived value of $E_\text{s} = 28\,\text{MN/m}^2$

Desired foam density ≈ 0.1

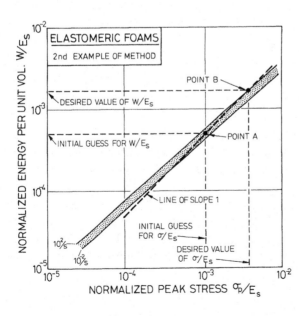

Figure 8.14   Example 2 of the use of the energy-absorption diagram. The initial choice of modulus gives point A. The line of slope 1 through this point intersects the required strain-rate contour at B. This is the design point and gives the required value for E_s and ρ^*/ρ_s.

8.6   Case studies in the selection of foams for packaging

In most applications, it appears that some thought has gone into the selection and design of foams for energy absorption, but that there are difficulties with the methods employed. The use of J curves requires extensive impact testing of foams to compare performance; they cannot be constructed from available uni-axial stress–strain behaviour. The empirical energy-absorption diagrams developed by Rusch (1970, 1971) can be constructed from uniaxial data but they ignore the effect of strain-rate and temperature. There are problems of specification, too: in estimating the maximum peak stress that a component can tolerate without damage; in deciding the number and size of the impacts that the component may encounter; in trying to apply the results of tests using flat indenters to impact situations where a component may be spherical or cylindrical, or irregularly shaped (Lee and Williams, 1971); and in satisfying size, weight, or cost constraints for the package.

Energy diagrams of the kind derived in this chapter provide a consistent and systematic method for foam selection and design for given impact situations. Their advantage over other methods lies not so much in their greater accuracy (since the difficulties above still apply) but in the fact that they bring together foam and strain-rate variables in one diagram and provide a summary of data from which optimum density, thickness and section can be read quickly and with adequate precision. We conclude this chapter with two illustrative case studies of their use.

Case study 1: design of a car head-rest

Legislation in most countries now requires that the front seats of cars be equipped with head-rests. Their primary function is to support and cushion the head, protecting the occupant from injury in a rear-end collision. Ideally, the head-rest should be able to absorb the kinetic energy of the head without exerting a force which exceeds that which would cause damage to the brain or skull. The maximum tolerable acceleration depends on the time over which it is applied: Lockett *et al.* (1981) suggest that accelerations which exceed $100g$ over 5 milliseconds or $50g$ over 100 milliseconds are dangerous to life.

We first analyse a head-rest from a contemporary family car and then consider how the design might be improved. The head-rest padding is made of a flexible polyester foam with a relative density close to 0.06. The compressive stress–strain curve, and the energy-absorption diagram, at a strain-rate of 10^{-3}/s, are shown in Fig. 8.15. Tests at a higher strain-rate showed a slight rate-dependence, very like that of polyurethane.

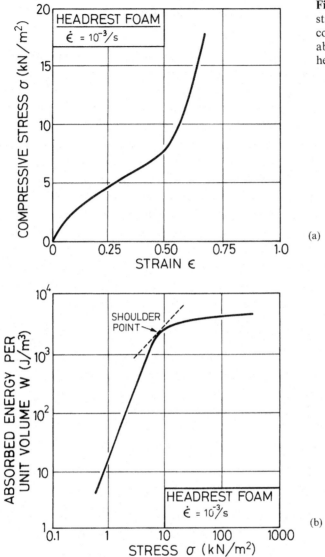

Figure 8.15 A stress–strain curve and the corresponding energy-absorption curve for the head-rest foam.

(a)

(b)

The data required for the analysis of the problem are summarized in Table 8.4. The foam padding has a thickness, t, and contacts the back of the head over an area, A. In a rear-end collision at a velocity, v, the kinetic energy which is imparted to the head (of mass m) is $0.5mv^2$. The energy to be absorbed, per unit volume of foam, is:

$$W = 0.5mv^2/At \quad [\text{J/m}^3]. \tag{8.29}$$

The maximum permissible force the head can sustain without damage is that corresponding to an acceleration of $a = 50g$, giving,

$$F_{max} = ma = 1250\,\text{N}.$$

Table 8.4 Data for car head-rest design

Foam: Polyester, relative density 0.06	
Mass of head	$m = 2.5\,\text{kg}$
Maximum permissible deceleration	$a = 50g = 500\,\text{m/s}^2$
Area of contact between head and rest	$A = 0.01\,\text{m}^2$
Thickness of padding	$t = 0.17\,\text{m}$
Maximum permissible force	$F_{max} = ma = 1250\,\text{N}$
Maximum permissible stress	$\sigma_p = F/A = 125\,\text{kN/m}^2$
Energy to be absorbed per unit volume	$W = 0.5mv^2/At = 735v^2\,\text{J/m}^3$
Peak strain-rate	$\dot{\epsilon} = v/t\,\text{s}^{-1}$
Temperature	$T = -10 \text{ to} +30\,°\text{C}$

The peak allowable stress is then:

$$\sigma_{max} = \frac{ma}{A} = 125\,\text{kN/m}^2. \tag{8.30}$$

From Fig. 8.15(b) we find that the maximum energy absorbed, per unit volume, W, up to a stress of $125\,\text{kN/m}^2$, is $5 \times 10^3\,\text{J/m}^3$. Hence, from Eqn. (8.29), the maximum collision velocity is:

$$v = 2.6\,\text{m/s} = 5.8\,\text{m.p.h.}$$

The head-rest is thus ineffectual for all but the mildest collisions. The choice of a low-density polyester reflects design more for comfort than for safety.

The design can be modified without altering the dimensions to give a larger margin of safety. Figure 8.16 shows the energy-absorption diagram for elastomeric foams, with two additional scales (top and right) of stress and absorbed energy for the polyester of which the head-rest is made, for which $E_s = 15\,\text{MN/m}^2$. The shoulder point for the head-rest foam, taken from Fig. 8.15(b), is shown. It lies a little below the line for $\dot{\epsilon} = 10^{-3}$/s, but close to the contour for $\rho^*/\rho_s = 0.06$. The discrepancies reflect experimental error, and the fact that the foam has some closed cells.

The maximum permissible stress was calculated above, as $125\,\text{kN/m}^2$. This is plotted on the diagram, and shows that a much higher density of polyester foam $(\rho^*/\rho_s \approx 0.2)$ is allowable, giving an energy absorption W/E_s approaching 2.6×10^{-3}, almost trebling the safe impact speed to

$$v = 7.3\,\text{m/s} = 16\,\text{m.p.h.}$$

A foam of relative density 0.2 is not, however, a good choice. It results in a head-rest that is too heavy and, at this density, the material begins to behave more like a solid than a foam. The obvious step is to select an alternative material.

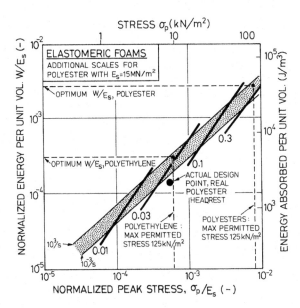

Figure 8.16 The choice of the optimum foam for the head-rest is made using the energy-absorption diagram as explained in the text.

Table 8.5 Optimum choice of foam for car head-rest

	Polyester	Polyethylene
Solid modulus, E_s (MN/m^2)	15	200
Normalized peak stress, σ_p/E_s	8.3×10^{-3}	6.3×10^{-4}
Optimum density, ρ^*/ρ_s	0.2	0.03
Optimum W/E_s	2.6×10^{-3}	3.2×10^{-4}
Optimum energy absorbed, W (J/m^3)	3.9×10^4	6.4×10^4
Maximum safe impact velocity, v (m/s)	7.3	9.3

Consider, as an example, the alternative choice of an open-cell low-density polyethylene, with a solid modulus E_s of 200 MN/m^2 (Table 8.5). The normalized peak stress is now

$$\frac{\sigma_p}{E_s} = 6.3 \times 10^{-4}. \tag{8.31}$$

It is marked on Fig. 8.16, which shows that at the expected strain-rate

$$\dot{\epsilon} = \frac{v}{t} = 10^2/\text{s} \tag{8.32}$$

the energy absorbed is

$$\frac{W}{E_s} = 3.2 \times 10^{-4}. \tag{8.33}$$

This allows a further increase in impact velocity, calculated from Eqn. (8.29), to

$$v = 9.3\,\text{m/s} \approx 21\,\text{m.p.h.}$$

This is an important improvement, and illustrates how the diagrams offer a logical approach to foam selection. This particular choice might, of course, face consumer resistance because the material feels less soft and comfortable than the polyester, but this can be overcome by facing the stiffer polyethylene with a skin of soft polyester. The head then encounters the polyethylene only under accident conditions.

Case study 2: packaging for a microcomputer

Delicate electronic devices like microcomputers are commonly packaged for delivery in moulded expanded polystyrene. Its function is to protect the computer against accelerations or decelerations which might damage its components or the connections between them. The effectiveness of the packaging might be measured by the height from which the system could be dropped on to a hard floor without damage. A reasonable expectation for this height is 1 m: that is roughly the distance it would fall if dropped when carried in the hands. We first analyse the packaging of a real microcomputer, and then examine whether the design of the package could be improved.

The relevant details of the package and its content are summarized in Table 8.6 and Fig. 8.17. The slab-shaped computer is supported on ribs moulded on to the inner walls of the package. It will be assumed that impact occurs in the least favourable way – that is, along the shortest side of the package. The calculation is complicated by the section profile of the impacting edge.

Figure 8.18 shows the stress–strain curve and energy-absorption diagram for a sample cut from the package. As the package crushes, the columns compress first; when a displacement of 11 mm is reached, the deformation spreads across the whole section.

Consider first crushing of the columns only. Then, using the data in Table 8.6, the peak stress which the package can exert on the contents is:

$$\sigma_p = \frac{F_{max}}{A} = 0.41\,\text{MN/m}^2. \tag{8.34}$$

From Fig. 8.18 we find that this corresponds to a point on the shoulder of the curve (and thus is close to the optimum) at which the energy absorbed per unit volume is $W = 1.5 \times 10^5\,\text{J/m}^3$. Then the maximum drop height, h, is given by equating the potential energy of the dropped computer to the energy which the columns can absorb:

$$mgh_1 = WA_1 t_1 \tag{8.35}$$

Table 8.6 Data for the microcomputer package

Foam: Expanded polystyrene (relative density 0.1)

Mass of microcomputer	$m = 3.65 \text{ kg}$
Maximum permissible deceleration	$a = 60g = 600 \text{ m/s}^2$
Energy to be absorbed in a drop of 1 m	$U = 36.5 \text{ J}$
Impact velocity in a drop of 1 m	$v = 4.5 \text{ m/s}$
Area of contact between microcomputer and foam	
(a) six columns only	$A_1 = 5.3 \times 10^{-3} \text{ m}^2$
(b) full edge section	$A_2 = 22.7 \times 10^{-3} \text{ m}^2$
Thickness of packaging material	
(a) columns	$t_1 = 0.011 \text{ m}$
(b) full edge section	$t_1 = 0.021 \text{ m}$
Depth of package	$b = 0.063 \text{ m}$
Maximum permissible force	$F_{max} = ma = 2190 \text{ N}$
Maximum permissible package stress	$\sigma_p = F_{max}/A$
Energy to be absorbed per unit volume	$W = mgh/At \text{ (J/m}^3)$
Peak strain-rate	$\dot{\epsilon} = v/t = 100/s$
Temperature	$T = -10 \text{ to } 30\,^\circ\text{C}$

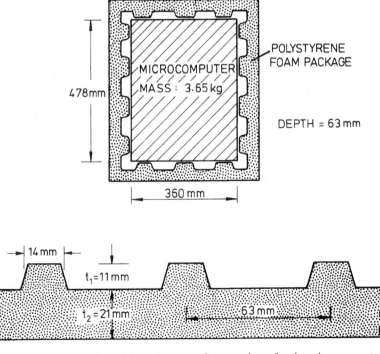

Figure 8.17 The design of the polystyrene foam package for the microcomputer. The upper diagram is schematic; there are six ribs along the short side. The lower drawing is to scale.

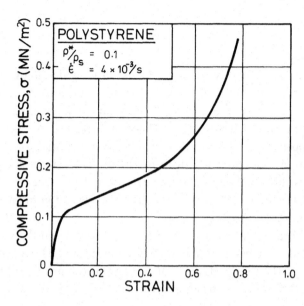

Figure 8.18 A stress–strain curve and an energy-absorption curve for the polystyrene of the microcomputer package. The choice of foam is near-optimum.

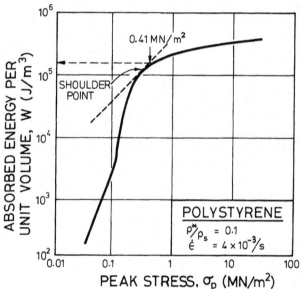

from which

$$h_1 = 0.24 \, \text{m}.$$

The ribs are an important feature of the design of the package. If the full section crushes, the maximum stress falls to $F_{max}/A_2 = 0.096 \, \text{MN/m}^2$, the maximum energy absorbed drops, dramatically, to $2.4 \times 10^3 \, \text{J/m}^3$ and the maximum drop height is given by:

$$mgh_2 = WA_2t_2 \tag{8.36}$$

from which

$$h_2 = 0.032\,\text{m}.$$

But a drop height of 0.24 m is hardly adequate. Consider the redesign of the package to increase the critical drop height. One obvious approach is to increase the height of the ribs: if these are extended to a height of 20 mm (making the package wall 41 mm thick at the ribs) then the safe drop height is increased to 0.44 m. A more far-reaching approach is to reconsider the choice of material for the package. Consider, then, a redesign for a drop height of 1 m, with a safety factor of $S = 1.5$ on stress, and using a low-density polyethylene foam ($E_s = 200\,\text{MN/m}^2$). This foam has the advantage that it recovers after impact, and can protect against multiple impacts.

Assume that the load-bearing face is flat (there are no ribs) and that the side length and width are the same as before, but that the foam density and thickness can be chosen. The energy to be absorbed is $U = 36.5\,\text{J}$. The maximum allowable stress is $F_{\text{max}}/(A_2 S)$ where S is the safety factor, giving

$$\sigma_p = 0.064\,\text{MN/m}^2$$

or

$$\sigma_p/E_s = 3.2 \times 10^{-4}$$

and the velocity on impact is $\sqrt{2gh} = 4.5\,\text{m/s}$. If the foam thickness is of order 40 mm, the strain-rate is roughly $10^2/\text{s}$.

From the energy-absorption diagram for elastomeric foams (Fig. 8.8) we find that the optimum foam has a relative density

$$\left(\frac{\rho^*}{\rho_s}\right)_{\text{optimum}} = 0.02$$

and that the maximum energy that can be absorbed per unit volume is

$$W/E_s = 2 \times 10^{-4}$$

or

$$W = 4 \times 10^4\,\text{J/m}^3.$$

The foam thickness is then obtained from

$$U = W A_2 t$$

giving

$$t = 0.04\,\text{m}.$$

So for a polyethylene foam package the solution is a 40 mm thick foam of relative density 0.02. If this density were not one of the standard ones available, then the next lowest density would be chosen and the thickness recalculated; the maximum stress reached would then be slightly lower. Note that the redesign has given a safer package, with lower density, able to withstand repeated shocks. It may, of course, cost more; and in packaging problems cost can be the decisive factor. But where performance outweighs cost, the method illustrated here has merit.

8.7   Conclusions

There is an optimum foam material and density for any given packaging application. The optimum gives a package which completely absorbs the kinetic energy of the packaged object while limiting the force to a level below the threshold for damage. Initial screening to identify possible candidates is described in Chapter 13. Optimum selection requires the methods described here – the J-factor, the C-factor, Rusch curves or the energy-absorption diagrams. The most promising appears to be the use of energy-absorption diagrams, which condense the data for energy-absorption as a function of peak-stress, strain-rate, density and foam type into a single master diagram. The features of the diagram can be related to the physical mechanisms by which foams absorb energy. But further experimental and theoretical work is needed to develop diagrams of sufficient accuracy for general use.

References

Cousins, R. R. (1976a) *J. Appl. Polymer Sci.*, **20**, 2893.

Cousins, R. R. (1976b) Design Guide to the Use of Foams for Crash Padding. NPL Report DMA 237, London.

Gent, A. N. and Rusch, K. C. (1966a) *J. Cell. Plast.*, **2**, 46.

Gent, A. N. and Rusch, K. C. (1966b) *Rubber Chem. Technol.*, **39**, 389.

Gordon, G. A. (1974) *Testing and Approval, Impact Strength and Energy Absorption*, PIRA.

Gordon, J. E. and Jeronimides, G. (1974) *Nature*, **232**, 116.

Green, S. J., Schierloh, F. L., Perkins, R. D. and Babcock, S. G. (1969) *Exp. Mech.*, March, p. 103.

Hilyard, N. C. (ed.) (1982) *Mechanics of Cellular Plastics*, Applied Science Publishers, London.

Hilyard, N. C. and Cunningham, A. (1994) *Low Density Cellular Plastics, Physical Basis of Behaviour*, Chapman and Hall, London, UK.

Hilyard, N. C. and Djiauw, L. K. (1971) *J. Cell. Plast.*, **7**, 33.

Hilyard, N. C. and Kanakkanatt, S. V. (1970) *J. Cell. Plast.*, **6**, 87.

Kosten, C. W. and Zwikker, A. (1939) *Rubber Chem. Technol.*, **12**, 105.

Kurauchi, T., Sata, N., Kamigaito, O. and Komatasu, N. (1984) *J. Mat. Sci.*, **19**, 871.

Lazan, B. J. (1968) *Damping of Materials and Members in Structural Mechanics*, Chapter 8. Pergamon Press, Oxford.

Lee, W. M. and Williams, B. M. (1971) *J. Cell. Plast.*, **7**, 72.

Lockett, F. J., Cousins, R. R. and Dawson, D. (1981) *Plast. Rubber Proc. Appl.*, **1**, 25.

Maiti, S. K., Gibson, L. J. and Ashby, M. F. (1984) *Acta Met.*, **32**, 1963.

Meinecke, E. A. and Schwaber, D. M. (1970) *J. Appl. Polymer Sci.*, **14**, 2239.

Meinecke, E. A., Schwaber, D. M. and Chiang, R. R. (1971) *J. Elastoplast.*, **3**, 19.

Melvin, J. W. and Roberts, V. L. (1971) *J. Cell. Plast.*, **7**, 97.

Mills, N. J. (1994) Chapter 9 in *Low Density Cellular Plastics, Physical Basis of Behaviour*, eds. Hilyard N. C. and Cunningham, A., Chapman and Hall, London.

Mustin, G. S. (1968) *Theory and Practice of Cushion Design*. US Government Printing Office, Washington, DC.

Rusch, K. C. (1970) *J. Appl. Polymer Sci.*, **14**, 1263 and 1433.

Rusch, K. C. (1971) *J. Cell. Plast.*, **7**, 78.

Schwaber, D. M. (1973) *Polymer-Plast. Technolg. Eng.*, **2**, 231.

Schwaber, D. M. and Meinecke, E. A. (1971) *J. Appl. Polymer Sci.*, **15**, 2381.

Woolam, W. E. (1968) *J. Cell. Plast.*, **4**, 79.

Chapter 9

The design of sandwich panels with foam cores

9.1   Introduction and synopsis

Structural members made up of two stiff, strong skins separated by a lightweight core are known as sandwich panels. The separation of the skins by the core increases the moment of inertia of the panel with little increase in weight, producing an efficient structure for resisting bending and buckling loads. Because of this, sandwich panels are often used in applications where weight-saving is critical: in aircraft, in portable structures, and in sports equipment. Figure 9.1 shows the rotor blade from a helicopter and a flooring panel from a Boeing 747; Fig. 9.2 shows a prefabricated housing wall panel; and Fig. 9.3 shows sections from a downhill ski and from the hull of a racing yacht. In these examples, the skins or face materials are, typically, aluminium or fibre-reinforced composites; the cores are aluminium or paper–resin honeycombs, polymeric foams or balsa wood, all of which have a cellular structure.

Nature, too, makes use of sandwich designs. Sections through the skull of a human and the wing of a bird both clearly show a low-density core separating the solid faces (Fig. 9.4). Plants, too, can have a sandwich structure: that of the iris leaf, described in more detail below, is shown in Fig. 9.22(b) while that of a stalk was shown earlier, in Fig. 2.6(h). In all of these examples, the faces and core are made from the same material with the faces almost fully dense and the core a foam.

The mechanical behaviour of a sandwich panel depends on the properties of the face and core materials and on its geometry. In most applications the panel

(a)

Figure 9.1 Examples of sandwich panel design in aircraft components. (a) A helicopter rotor blade made with fibreglass faces and a paper-phenolic honeycomb core. (b) Aircraft flooring panels made using carbon fibre composite faces and a paper-phenolic core.

(b)

Figure 9.2 A section of a prefabricated housing wall panel. The plywood and gypsum faces are separated by a polyurethane foam core.

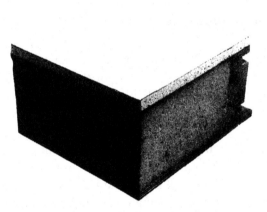

must have some required minimum stiffness, it must not fail under some maximum service loading and it must be as light as possible. Its design can be formulated as an optimization problem: the goal is the panel with minimum weight which meets the constraints on stiffness and strength. The optimization can be carried out with respect to the core and skin thicknesses, with respect to the core and skin materials and (since foams have continuously variable density) with respect to the core density.

It is a problem which has received considerable attention (see, for example, Allen, 1969; Zenkert, 1995). At the simplest level, the skin and core materials and the core density are given, and the task is that of selecting the optimum skin and core thicknesses. Allen (1969) gives solutions for a beam subject to a stiffness constraint; Huang and Alspaugh (1974) and Ueng and Liu (1979) include addi-

(a)

Figure 9.3 Examples of sandwich panel design in modern sporting equipment. (a) A section from a downhill ski made with aluminium skins separated by a polyurethane foam core. (b) A section from the desk of a sailboat made with fibreglass skins and a balsa wood core.

(b)

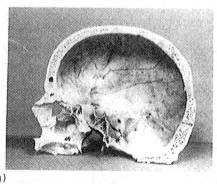

(a)

Figure 9.4 Examples of sandwich construction in nature. (a) A section from a human skull, showing two layers of dense, compact bone separated by a layer of spongy, cancellous bone (after Hodgson, 1973, Fig. 1b, courtesy of Plenum Press). (b) A section through a bird's wing (after Thompson, 1961, Fig. 103, courtesy of Cambridge University Press).

(b)

tional constraints on strength. A more complete optimization is possible if the core density, as well as the skin and core thicknesses, are treated as variables. Wittrick (1945), Ackers (1945) and Kuenzi (1965) have analysed this problem for panels subjected to various structural constraints (on stiffness, or strength or buckling resistance) assuming that the core properties all varied *linearly* with the core density. Although this is adequate for some cores (for example, wood loaded along the grain) it is not, in general, true for foams, or even for woods or

honeycombs. We saw in Chapter 5, for instance, that the moduli, strength and
fracture resistance of foams all vary as a power (between 1.5 and 2) of the density;
and for woods loaded across the grain the same is true (Chapter 10). In this chapter we develop results for optimum sandwich-panel design using these more accurate relationships.

Briefly, Section 9.2 describes how the stiffness of a sandwich beam or plate is
calculated, and then shows how the weight can be minimized for a stiffness constraint. Section 9.3 presents equations for the strength of sandwich beams for
several possible modes of failure, and tabulates the results. Section 9.4 discusses
design to meet both stiffness and strength constraints simultaneously. The chapter ends with two illustrative case studies, one analysing the design of a downhill
ski, the other describing the mechanics of the iris leaf.

9.2   The stiffness of sandwich structures and its optimatization

(a) Theory of stiffening and optimizing stiffness/weight

Figure 9.5(a) shows a sandwich beam. Its response to load is relatively easy to
analyse, so it gives a good way of introducing the main ideas and methods.
These can be applied (though with greater mathematical complexity) to other

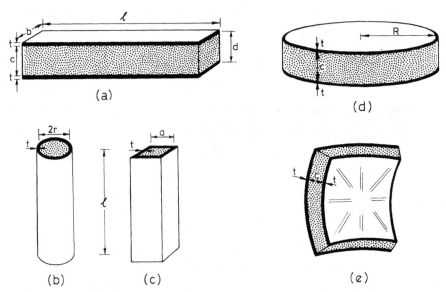

Figure 9.5   The geometry of sandwich structures: (a) a rectangular beam, (b) a
circular column, (c) a square column, (d) a circular plate, (e) a shell element.

sandwich structures, including the columns, plates and shells shown in Fig. 9.5(b)–(e).

The geometric and material parameters which enter the analyses are listed in Table 9.1. (In this chapter the subscript 'f' refers to the faces, the subscript 'c' to the foamed core and the subscript 's' to the solid from which the core is made.)

Consider a sandwich beam under a load P in three-point bending (Fig. 9.6). The span is l, the width, b, the core thickness, c, and the face thickness, t. The central deflection is δ. The density and Young's modulus of the faces are ρ_f and E_f, respectively. Those of the core are ρ_c^* and E_c^*. The shear modulus of the core is G_c^*. The core moduli vary continuously with the foam density, ρ_c^* (Eqn. 5.6a,b); at their simplest, they are given by:

$$E_c^* = C_1 E_s \left(\frac{\rho_c^*}{\rho_s}\right)^2 \tag{9.1}$$

Table 9.1 Sandwich panel parameters

Geometrical parameters

a Width of the core in a square sandwich tube [m]
b Width of a rectangular sandwich beam [m]
c Core thickness in a rectangular sandwich beam, plate or shell [m]
d Distance between centroids of faces in a sandwich panel [m]
l Span of a sandwich beam or height of a sandwich column [m]
R Radius of a circular sandwich plate [m]
r Radius of the core in a circular sandwich tube [m]
t Face thickness of a sandwich panel [m]

Material properties

	Face	Foamed core	Solid unfoamed core
Density (kg/m^3)	ρ_f	ρ_c^*	ρ_s
Young's modulus (GN/m^2)	E_f	E_c^*	E_s
Shear modulus (GN/m^2)	—	G_c^*	—
Yield strength (MN/m^2)	σ_{yf}	σ_{yc}^*	σ_{ys}

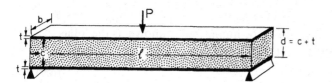

Figure 9.6 A sandwich beam loaded in three-point bending.

and

$$G_c^* = C_2 E_s \left(\frac{\rho_c^*}{\rho_s}\right)^2 \tag{9.2}$$

where C_1 (≈ 1) and C_2 (≈ 0.4) are constants of proportionality.

The stiffness is calculated from the *equivalent flexural rigidity*, $(EI)_{eq}$, and the *equivalent shear rigidity*, $(AG)_{eq}$, of the beam (Allen, 1969). The equivalent flexural rigidity, found from the parallel axis theorem, gives, for the rectangular beam of Fig. 9.6:

$$(EI)_{eq} = \frac{E_f b t^3}{6} + \frac{E_c b c^3}{12} + \frac{E_f b t d^2}{2}. \tag{9.3a}$$

The first and second terms describe the bending stiffnesses of the faces and of the core about their centroids: together they give the stiffness of the beam if the faces were not bonded to the core. In optimal sandwich design, both are small compared to the third term, which describes the bending stiffness of the faces about the centroid of the beam itself. In practice, the faces are always much thinner than the core so that $d \approx c$. To a good approximation,

$$(EI)_{eq} = \frac{E_f b t c^2}{2}. \tag{9.3b}$$

The equivalent shear ridigity is (Allen, 1969)

$$(AG)_{eq} = \frac{b d^2 G_c^*}{c} \tag{9.4a}$$

which, for thin faces ($d \approx c$) becomes

$$(AG)_{eq} = b c G_c^*. \tag{9.4b}$$

When the load, P, is applied to the beam, it deflects. The deflection δ is the sum of the bending and shear components:

$$\delta = \delta_b + \delta_s = \frac{P l^3}{B_1 (EI)_{eq}} + \frac{P l}{B_2 (AG)_{eq}} \tag{9.5}$$

where B_1 and B_2 are constants which depend on the geometry of loading; Table 9.2 lists values for common geometries. Using Eqns. (9.3b) and (9.4b), the compliance of the beam is

$$\frac{\delta}{P} = \frac{2 l^3}{B_1 E_f b t c^2} + \frac{l}{B_2 b c G_c^*}. \tag{9.6}$$

We wish to minimize the weight of the beam for a given bending stiffness, P/δ. In optimization theory the equation to be minimized is known as the 'objective function'. In this case it is the weight, W

$$W = 2 \rho_f g b l t + \rho_c^* g b l c \tag{9.7}$$

Table 9.2 Constants for bending and failure of beams

Mode of loading (all beams of length l)	B_1	B_2	B_3	B_4
	$\delta_b = \dfrac{Pl^3}{B_1(EI)_{eq}}$	$\delta_s = \dfrac{Pl}{B_2(AG)_{eq}}$	$M = \dfrac{Pl}{B_3}$	$Q = \dfrac{P}{B_4}$
Cantilever, end load, P	3	1	1	1
Cantilever, uniformly distributed load, †$q = P/l$	8	2	2	1
Three-point bend, central load, P	48	4	4	2
Three-point bend, uniformly distributed load, $q = P/l$	$\dfrac{384}{5}$	8	8	2
Ends built in, central load, P	192	4	8	2
Ends built in, uniformly distributed load, $q = P/l$	384	8	12	2

†q is a distributed load per unit length.

where g is the acceleration due to gravity. The span, l, the width, b, and the stiffness, P/δ, are fixed; the free variables are t, c and ρ_c^*.

If the core density is fixed, the optimization is easy. The procedure is to solve Eqn. (9.6) for t, substitute this into the objective function (9.7) which is then minimized with respect to the only other free variable, c, by setting dW/dc equal to zero; this gives the optimum core thickness, c_{opt}; substituting this back into the constraining Eqn. (9.6) gives the optimum face thickness, t_{opt}.

More insight is given by examining the problem graphically. For simplicity, the core density is held constant and the free variables are the face and core thicknesses. The objective function and the constraints are plotted in Fig. 9.7, which is constructed as follows. Inverting the objective function (9.7) gives a relationship between the two variables of the problem, t/l and c/l:

$$\left(\frac{t}{l}\right) = \frac{W}{2bl^2\rho_f g} - \frac{\rho_c^*}{2\rho_f}\left(\frac{c}{l}\right).$$

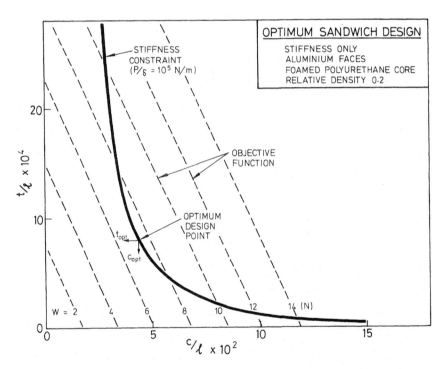

Figure 9.7 The graphical method of optimization. The family of broken lines shows the objective function (Eqn. 9.7). The solid curve shows the stiffness constraint (Eqn. 9.6): points which lie to the right of this line satisfy the constraint. The optimum design point defines the structure with the minimum weight. (Aluminium skins with $\rho_f = 2700\,\text{kg/m}^3$, and $E_f = 70\,\text{GN/m}^2$; foamed polyurethane core with $\rho_c^*/\rho_s = 0.2$, $\rho_s = 1200\,\text{kg/m}^3$, $E_s = 1.6\,\text{GN/m}^2$, and $G_c^* = 45\,\text{MN/m}^2$. The stiffness constraint is $P/\delta = 10^5\,\text{N/m}$. Beam width $b = 50\,\text{mm}$, length $l = 1\,\text{m}$, central load in three-point bend).

It plots as a family of straight, parallel lines, one for each value of W (since the material properties ρ_c^* and ρ_f, and the dimensions b and l are fixed). The constraint (9.6) gives a second relationship between t/l and c/l. Inverting it gives

$$\frac{t}{l} = \frac{2B_2}{B_1} \frac{G_c^*}{E_f(c/l)} \left\{ \frac{1}{B_2(\delta/P)bG_c^*(c/l) - 1} \right\}.$$

It is shown as a full line (Fig. 9.7), cutting through the contours of constant W. The optimum design point is found by inspection (it is marked on the figure); from it, the optimum dimensions t/l and c/l and the optimum weight W can be read off. Changing the design stiffness, P/δ (the constraint), moves the full line, and therefore changes the best choice of t/l and c/l. They can, of course, also be found by the analytical procedure outlined above.

A more complete optimization is possible if the density of the core is also treated as a free variable. Substituting for G_c^* (Eqn. 9.2) in the constraining Equation (9.6) and solving for ρ_c^* gives

$$\rho_c^* = \left\{ \frac{B_1}{C_2 B_2} \frac{E_f}{E_s} \left(\frac{ltc}{B_1 btc^2 E_f(\delta/P) - 2l^3} \right) \right\}^{1/2} \rho_s. \tag{9.8}$$

This is now substituted into the objective function (9.7) which is minimized† by setting

$$\frac{\partial W}{\partial c} = \frac{\partial W}{\partial t} = 0$$

to give optimum values of c and t. These, when put into Eqn. (9.8), give the optimum core density. The procedure is summarized in the flow chart of Table 9.3. Results for a rectangular beam and a circular plate are listed in the appendix to this chapter. A number of useful generalizations emerge. The ratio of the weight of the two faces to that of the core is a constant for any particular type of structural sandwich independent of the stiffness, span, loading conditions or materials used. In addition, the ratio of the bending component of deflection to the shear component is also independent of stiffness, span, loading conditions or materials used for a given type of sandwich construction. These ratios are listed in Table 9.4.

(b) Experiments to measure sandwich stiffness

For every ten publications on the theory of sandwich structures there is less than one which describes experiments which might test the theory. The most complete

---

†An extension of the graphical method is also viable. The objective function (Eqn. (9.7)) is plotted as in Fig. 9.7 for one value of ρ_c^*. The constraining equation in the form of Eqn. (9.8) gives a second family of lines, for that value of ρ_c^*. A series of graphs, one for each value of ρ_c^*, can be plotted and the values of ρ_c^*, t/l, and c/l which minimize the weight read off.

Table 9.3   Optimum design of a sandwich panel subject to a stiffness constraint

---

Formulate objective function for the weight of the beam

$$W = 2\rho_f g b l t + \rho_c^* g b l c \qquad (Eqn.\ (9.7))$$

↓

Formulate the stiffness constraint

$$\delta/P = \frac{2l^3}{B_1 E_f b t c^2} + \frac{l}{B_2 b c G_c^*} \qquad (Eqn.(9.6))$$

↓

Solve stiffness constraint for one variable (e.g. ρ_c^*)

$$\rho_c^* = \left\{ \frac{B_1}{C_2 B_2} \frac{E_f}{E_s} \left(\frac{ltc}{B_1 b t c^2 E_f (\delta/P) - 2l^3} \right) \right\}^{1/2} \rho_s \qquad (Eqn.(9.8))$$

and substitute this into the objective function

↓

Form $\partial W/\partial c = 0$ and $\partial W/\partial t = 0$

and solve to give

$$c_{opt} = f(\delta/P, material\ properties, beam\ geometry)$$
$$t_{opt} = f(\delta/P, material\ properties, beam\ geometry)$$

Substitute these into Eqn. (9.8) for core density to give optimum density, ρ_{copt}^*.

↓

Final Result:

Optimum	*As function of*
Core thickness, c_{opt}	Design stiffness, δ/P
Face thickness, t_{opt}	Material properties, $\rho_f, \rho_s, E_f, E_s, C_2$
Core density, ρ_{copt}^*	Loading geometry, B_1, B_2, l, b

Table 9.4   Optimization analysis for sandwich panels subject to a stiffness constraint

Geometry	W_f/W_c	δ_b/δ	δ_s/δ
Rectangular beam	1/4	1/3	2/3
Circular plate (distributed load over entire plate)	1/4	1/3	2/3
Circular plate (distributed load over radius r)	1/4	1/3	2/3

experimental study of optimal sandwich design for stiffness is that of Gibson (1984). It gives considerable confidence in the calculations outlined above, confirming almost all of its predictions. In this study, aluminium faces were bonded to foamed polyurethane cores to give rectangular beams, all with the same span, l, but with differing t, c and ρ_c^*. The beams were all designed to have the same stiffness, but because of the different dimensions and core densities, they had different stiffness per unit weight, $P/\delta W$. This quantity was measured in three-point bending and is shown, plotted against t, c and ρ_c^* in Fig. 9.8. The stiffness of each beam can be calculated from Eqn. (9.6) provided G_c^* is known as a function of density. This was measured separately, and is shown in Fig. 9.9; it is well described by

$$G_c^* = 0.7 E_s \left(\frac{\rho_c^*}{\rho_s}\right)^2.$$

$$(9.9)$$

The calculated stiffness, divided by the weight (Eqn. (9.7)) is shown as a full line on Fig. 9.8, which shows how $P/\delta W$ varies with a pair of the parameters c, t and ρ_c^* while the third was kept constant at the optimum value. Each shows a well-defined maximum closely followed by the experimental measurements. The values corresponding to the peaks in the experimental and theoretical curves are listed in Table 9.5. They confirm that the predicted values of t_{opt}, c_{opt} and ρ_{opt}^* do lead to a minimum-weight design.

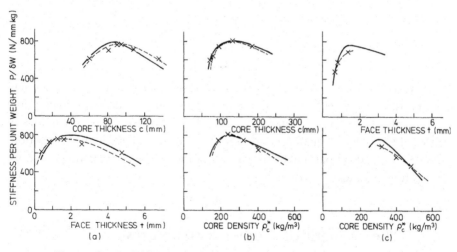

Figure 9.8   (a) Stiffness per unit weight, $P/\delta W$, for rectangular sandwich beams with the optimum core density ($\rho_c^* = 320\,\mathrm{kg/m^3}$) against core thickness, c and face thickness, t. (b) The same, for beams with the optimum face thickness ($t = 1.3\,\mathrm{mm}$) against core thickness, c and core density, ρ_c^*. (c) The same, for beams with the optimum core thickness ($c = 93\,\mathrm{mm}$) against face thickness, t, and core density, ρ_c^*. Each beam was designed to have a stiffness of 1070 N/mm.

Table 9.5   Optimum face and core thicknesses and core density†

	c_{opt} (mm)	t_{opt} (mm)	ρ^*_{copt} (kg/m^3)
Measured	95–130	1.3–1.6	240–320
Predicted	109	1.4	280

†From Gibson (1984), for rectangular beams with aluminium faces and rigid polyurethane cores ($\rho_f = 2.7\,\text{Mg/m}^3$, $\rho_s = 1.2\,\text{Mg/m}^3$, $E_f = 70\,\text{GN/m}^2$, $E_s = 1.6\,\text{GN/m}^2$, $P/\delta = 1\,\text{MN/m}$).

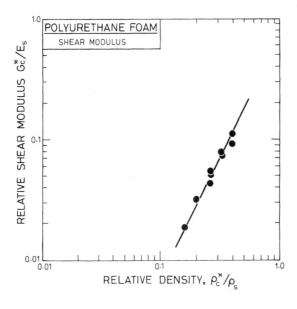

Figure 9.9   The relative shear modulus G^*_c/E_s for the foams used in the experimental study of optimum beam and plate design. The modulus is well described by $G^*_c/E_s = 0.7(\rho^*_c/\rho_s)^2$.

Similar tests on circular sandwich plates were equally satisfactory (Demsetz and Gibson, 1987). Here, too, aluminium faces were bonded to polyurethane cores, and the face and core thicknesses and the core density were systematically varied. The results adequately confirm the procedure outlined in the last section, and demonstrate, in particular, the success of the theory in predicting optimum design for the plates.

9.3   The strength of sandwich structures

The obvious attraction of sandwich structures is that they are light and stiff. But stiffness alone is not enough. The beam or panel must also have strength: it must carry the design loads without failing. At least five different failure modes (described below) are possible; a given sandwich will fail by the one which occurs at the lowest load. And as the geometry and loading change the failure mode can change, too. So it is not enough to design against one mode; all must be considered, and the dominant mode – the one which determines failure – identified and evaluated.

In this section we analyse the failure modes and develop a way of identifying the one which is dominant. In the next we return to the problem of optimum design; it is achieved by arranging that two or more failure modes occur at the same load.

(a) Theory of sandwich strength

The failure modes of a sandwich beam or panel are shown in Fig. 9.10. It may fail by the yielding or fracture of the faces (Kuenzi, 1965; Ueng and Liu, 1979; Froud, 1980; Ciba-Giegy, 1980; Triantafillou and Gibson, 1987a,b). The compression face may 'wrinkle' – a local buckling of the skin into the core (Kuenzi, 1965; Allen, 1969) – or it may 'dimple' – a local buckling of the compression face on a honeycomb core, with a wavelength equal to twice the cell size (Kuenzi, 1965; Ciba-Geigy, 1980). The core, too, can fail, usually in shear (Ciba-Geigy, 1980; Hall and Robson, 1984) though compressive or tensile failure is also possible. Then there is the bond between the face and core: it can fail; and since resin

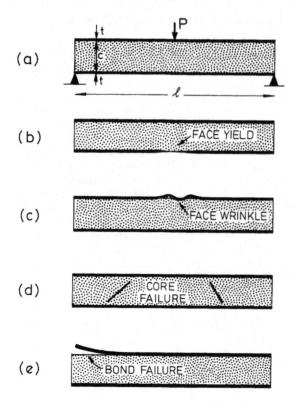

Figure 9.10 Failure modes for sandwich panels: (a) the loading geometry, (b) face yielding, (c) face wrinkling, (d) core failure, (e) decohesion of the interfacial bond.

adhesives are usually brittle, debonding is by brittle fracture (Hong and Jeong, 1985). Finally, there is the possibility of indentation of the faces and core at the loading point; but because this can be suppressed by distributing the load over an area about equal to the core section, it is easily avoided. Similar failure modes occur in end-loaded sandwich columns: both face yielding and wrinkling have been reported (Ackers, 1945; Wittrick, 1945; Allen, 1969; Wrzecioniarz, 1983). In addition, columns can fail by overall buckling (Ackers, 1945; Wittrick, 1945; Kuenzi, 1965; Allen, 1969; Ciba-Geigy, 1980).

The analysis of these failure modes requires expressions for the normal stresses, σ_f and σ_c, and the shear stresses, τ_f and τ_c, acting in the faces and the core. The maximum normal stresses are related to the applied moment by (Allen, 1969):

$$\sigma_f = \frac{MyE_f}{(EI)_{eq}} = \frac{M}{btc} \qquad (9.10a)$$

$$\sigma_c = \frac{MyE_c^*}{(EI)_{eq}} = \frac{M}{btc}\frac{E_c^*}{E_f} \qquad (9.10b)$$

where M is the applied moment, y is the distance from the neutral axis, and $(EI)_{eq}$ is given by Eqn. (9.3b). They are shown in Fig. 9.11. The shear stress varies through the face and core in a parabolic fashion; but if the faces are much stiffer and thinner than the core (as, in practice, they always are) the shear stress can be thought of as linear through the face and constant through the core at the value:

$$\tau_c = \frac{Q}{bc}. \qquad (9.11)$$

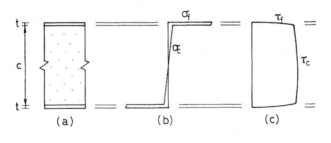

(a)   (b)   (c)

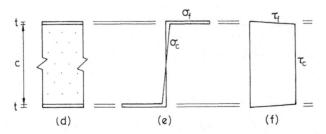

(d)   (e)   (f)

Figure 9.11 Stress distributions in a rectangular sandwich beam: (a) beam cross-section, (b) normal stress distribution, (c) shear stress distribution, (d) beam cross-section, (e) approximate normal stress distribution, (f) approximate shear stress distribution.

The maximum normal stress occurs at the section where the bending moment is maximum; this can be related to the applied load, P, by a constant, B_3, which depends on loading geometry. The maximum moment is:

$$M = \frac{Pl}{B_3} \tag{9.12}$$

giving:

$$\sigma_f = \frac{Pl}{B_3 btc} \tag{9.13a}$$

and

$$\sigma_c = \frac{Pl}{B_3 btc} \frac{E_c^*}{E_f}. \tag{9.13b}$$

The mean shear stress in the face and the core is:

$$2\tau_f = \tau_c = \frac{P}{B_4 bc} \tag{9.14}$$

where B_4 is a further constant relating the maximum shear force in the beam, Q, to the applied load, P (Table 9.2). Finally, the principal stresses, σ_1 and σ_2, in the face and core, and the maximum shear stress, τ_{max}, are given by (Timoshenko and Goodier, 1970):

$$\sigma_1 = \frac{\sigma}{2} \left(1 + \left(1 + \left(\frac{2\tau}{\sigma} \right)^2 \right)^{1/2} \right) \tag{9.15a}$$

$$\sigma_2 = \frac{\sigma}{2} \left(1 - \left(1 + \left(\frac{2\tau}{\sigma} \right)^2 \right)^{1/2} \right) \tag{9.15b}$$

and

$$\tau_{max} = \frac{\sigma}{2} \left(1 + \left(\frac{2\tau}{\sigma} \right)^2 \right)^{1/2} \tag{9.15c}$$

where σ is the normal stress (Eqns. (9.13)) and τ is the shear stress (Eqn. (9.14)) in either the face or core. In the faces the ratio of τ_f/σ_f ($(= B_3/B_4) \cdot t/l$) is small, so that $\sigma_1 = \sigma_f$, $\sigma_2 = 0$ and $\tau_{max} = \tau_f$. Conversely, in the core the ratio of

$$\tau_c/\sigma_c = \frac{B_3}{B_4} \frac{t}{l} \frac{E_f}{E_c^*}$$

is usually, although not always, large. When this is so, $\sigma_1 = \tau_c$, $\sigma_2 = -\tau_c$ and $\tau_{max} = \tau_c$.

Considering each failure mode in turn, we find the following:

(i) *Face yielding* occurs when the normal stress in the face equals the strength of the face material, σ_{yf}, or when (Eqn. (9.13a)):

$$\sigma_f = \frac{Pl}{B_3 btc} = \sigma_{yf} \tag{9.16}$$

where B_3 is the constant relating the applied load, P, to the maximum moment in the beam (Table 9.2).

(ii) *Face wrinkling* occurs when the normal stress in the compressive face of the beam reaches the local instability stress. Allen (1969) gives the result; wrinkling occurs when the compressive stress in the face is:

$$\sigma_f = \frac{Pl}{B_3 btc} = \frac{3E_f^{1/3} E_c^{*2/3}}{(12(3 - v_c^*)^2 (1 + v_c^*)^2)^{1/3}}. \tag{9.17}$$

Taking v_c^* (Poisson's ratio for the core) to be $1/3$, and using the result $E_c^* = (\rho_c^*/\rho_s)^2 E_s$ (Eqn. (5.6a)) gives

$$\frac{Pl}{B_3 btc} = 0.57 E_f^{1/3} E_s^{2/3} \left(\frac{\rho_c^*}{\rho_s}\right)^{4/3} \tag{9.18}$$

where B_3 is given in Table 9.2.

(iii) *Core failure* occurs in a foam with a plastic-yield point when the principal stresses satisfy the yield criterion (Eqn. (6.46)). If the shear stress in the core (Eqn. (9.14)) is large compared to the normal stress (Eqn. (9.13b)), failure occurs when the shear stress, τ_c, equals the yield strength of the foam in shear, τ_c^*:

$$\tau_c = \tau_c^*. \tag{9.19}$$

The yield strength of the foam in shear depends on density in the same way as the uniaxial strength (since plastic bending is the dominant failure mode in both). Core failure is then given by:

$$\frac{P}{B_4 bc} = C_{11} \left(\frac{\rho_c^*}{\rho_s}\right)^{3/2} \sigma_{ys} \tag{9.20}$$

where C_{11} is a constant of proportionality. The foam continues to yield until final tensile fracture produces the diagonal cracks shown in Fig. 9.10(d).

A brittle foam core, with an initial crack length of $2a$, fails when the brittle failure criterion is satisfied (Eqns. (6.50) and (6.52)). If the normal stresses in the core are small compared to the shear stresses, and if the crack length is more than about four cell diameters, then failure occurs when the maximum principal stress equals the tensile fracture strength of the foam (see Fig. 6.28), or when:

$$\sigma_1 = \tau_c = \sigma_f^* = \frac{K_{IC}^*}{\sqrt{\pi a}}. \tag{9.21}$$

Using Eqn. (9.14) for the shear stress in the core and Eqn. (5.47) for the fracture toughness gives:

$$\frac{P}{B_4 bc} = C_8 \left(\frac{\rho_c^*}{\rho_s}\right)^{3/2} \sigma_{fs} \sqrt{\frac{l^*}{a}} \tag{9.22}$$

where l^* is used here to denote the cell size.

(iv) *Failure of the adhesive bond* between the skin and the core is the most difficult of the mechanisms to analyse. It is important: the loss of military aircraft has been traced to this failure mode in wings of sandwich construction, and skis (which are sandwiches – see Section 9.5) sometimes delaminate by bond failure. Epoxy adhesives are generally stronger than the core itself, so that when the bond is perfect, delamination is seldom a problem. But if the interface between the skin and the core contains a crack-like defect, this may propagate when the sandwich is loaded. A full analysis is complicated by the vastly different moduli of skins, adhesive and core.

A lower limit for the failure load, valid when there is a large crack at the interface between the face and core, can be made as follows. The energy, U, stored in a beam which is subjected to a moment, M, is $\frac{1}{2} M\theta$ where θ is the angle of bend. Using $M = (EI)_{eq}/R$ and replacing the radius of curvature R by l/θ gives

$$U = \frac{1}{2} \frac{M^2 l}{(EI)_{eq}}.$$

Consider a through crack of length $2a$, contained in the interface between the skin and the core. Its area is $2ba$. If the crack propagated over the entire interface, all the energy U would be released. We will assume that the energy released increases linearly with crack area. Then, if the crack extends by δa, the change in area is $2b\delta a$ and the change in energy is

$$\delta U = -\frac{2b\delta a}{lb} U = -\delta a \frac{M^2}{(EI)_{eq}}.$$

The strain energy release rate, G, is defined as $-\delta U/2b\delta a$ giving:

$$G = \frac{M^2}{2b(EI)_{eq}}.$$

If this exceeds the toughness of the adhesive, G_c, the crack will propagate and the beam will fail. Then (using $M = Pl/B_3$) the failure load is

$$\frac{Pl}{B_2 btc} = \left(\frac{G_c E_f}{t}\right)^{1/2}. \tag{9.23}$$

If G_c of the core is less than that of the adhesive, the crack will propagate through the core (along the interface); in this case G_c of the core should be used. This is a *lower bound* on P: if a large crack pre-exists it will propagate at this load. The load is low: taking $G_c = 1 \, \text{J/m}^2$, $E_f = 70 \, \text{GN/m}^2$ (aluminium) and $t = 1 \, \text{mm}$ gives

$$\left(\frac{G_c E_f}{t}\right)^{1/2} \approx 8 \, \text{MN/m}^2$$

which is much less than the stress required for face-yield, for instance. If, in addition to a bending moment, M, the beam resists a shear force, Q, the shear strain energy† in the core must be included in the energy of the beam, U. Experiments confirm that large interface cracks ($a \gg c$) propagate readily at loads which are lower than those for the other failure modes (Triantafillou and Gibson, 1988). But if the interface is perfect a crack must first nucleate; and this requires far higher loads. A proper analysis for this case is not yet available. Experiments, described below, show that debonding is not design-limiting, provided care is taken to ensure that the bond is defect-free.

(v) *Core indentation*, mentioned earlier, is a problem only when loads are very localized. The problem can always be overcome by ensuring that the load is distributed over an area of, at least,

$$A = \frac{P}{\sigma_c^*} \tag{9.24}$$

where σ_c^* is the compressive strength of the core. We will not consider it further.

(b) Failure-mode maps

Table 9.6 lists the equations for the failure load for each of the mechanisms. Each equation contains three sets of variables: those relating to the loading configuration (B_3 and B_4, Table 9.2); those relating to the material properties of the face and the solid from which the core is foamed (ρ_f, E_f, σ_{yf}, ρ_s, E_s and σ_{ys}) and those relating to the beam design (t/l, c/l, b/l and ρ_c^*/ρ_s). We wish to determine how the failure mode depends on the beam design.

The *dominant* mechanism, for a given design, is the one giving failure at the lowest load. A *transition* in failure mechanism takes place when two mechanisms have the same failure load. This information can be displayed as a diagram or map (Fig. 9.12). The axes are the design parameters of the beam. The diagram is divided into fields, within which one failure mechanism is dominant. The fields are separated by field boundaries (heavy lines), which are the loci of design points for which two mechanisms have the same failure load. The field boundaries are found by equating the equations in Table 9.6, taken in pairs. When this is done it is found that the beam width, b, and core thickness, c, cancel out (because all mechanisms depend on them in the same way). This leaves just two design parameters: t/l and ρ_c^*/ρ_s. They are used as the axes for the diagram.

†More recent analysis by Triantafillou and Gibson (1988) shows that the shear strain energy is significant, giving a slightly different criterion for debonding which again shows that debonding only occurs if there are large interfacial cracks ($a \gg c$).

Table 9.6 Failure-mode equations for rectangular sandwich beams

Failure mode	Failure load	Equation number
Face yielding (from Eqn. (9.16))	$P = B_3 bc \left(\dfrac{t}{l}\right) \sigma_{yf}$	(F1)
Face wrinkling (from Eqn. (9.18))	$P = 0.57 B_3 bc \left(\dfrac{t}{l}\right) E_f^{1/3} E_s^{2/3} \left(\dfrac{\rho_c^*}{\rho_s}\right)^{4/3}$	(F2)
Core shear (from Eqn. (9.20))	$P = C_{11} B_4 bc \sigma_{ys} \left(\dfrac{\rho_c^*}{\rho_s}\right)^{3/2}$	(F3)
Core fracture (from Eqn. (9.22))	$P = C_8 B_4 bc \sigma_{fs} \left(\dfrac{\rho_c^*}{\rho_s}\right)^{3/2} \sqrt{\dfrac{l^*}{a}}$	(F4)
Bond failure (from Eqn. (9.23))	$P = B_3 bc \left(\dfrac{t}{l}\right) \sqrt{\dfrac{G_c E_f}{t}}$	(F5)

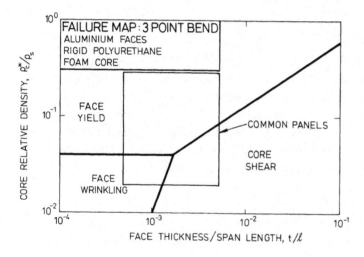

Figure 9.12 A failure mode map for a rectangular sandwich beam with aluminium faces and a polyurethane foam core loaded in three-point bending. This one diagram with axes of relative density, and face thickness to span ratio, t/l, shows the failure modes for all possible designs of beams with aluminium faces and polyurethane foam cores in three-point bending. The range of t/l and ρ_c^*/ρ_s for common sandwich panels is indicated by the box.

Consider the transition between face yielding and wrinkling. Equating failure Eqns. F1 and F2 from Table 9.6 gives the equation of the field boundary:

$$\frac{\rho_c^*}{\rho_s} = \left(\frac{\sigma_{yf}}{0.57 E_f^{1/3} E_s^{2/3}}\right)^{3/4}. \tag{9.25}$$

It is independent of t/l, so it appears as a horizontal line in Fig. 9.12. Similar equations are found for the other transitions by equating pairs of equations from Table 9.6 (the important ones are listed in Table 9.7); in general, they depend on both ρ_c^*/ρ_s and t/l. Figure 9.12 is the typical result: it describes all

Table 9.7 Transition equations for failure-mode maps†

Failure mode	Transition equation	Equation number
Face yield – face wrinkling	$$\frac{\rho_c^*}{\rho_s} = \left(\frac{\sigma_{yf}}{0.57 E_f^{1/3} E_s^{2/3}}\right)^{3/4}$$	(T1)
Face yield – core shear	$$\frac{t}{l} = \frac{C_{11}B_4}{B_3}\left(\frac{\rho_c^*}{\rho_s}\right)^{3/2}\left(\frac{\sigma_{ys}}{\sigma_{yf}}\right)$$	(T2)
Face wrinkling –core shear	$$\frac{t}{l} = \frac{C_{11}B_4}{0.57 B_3}\left(\frac{\sigma_{ys}}{E_f^{1/3} E_s^{2/3}}\right)\left(\frac{\rho_c^*}{\rho_s}\right)^{1/6}$$	(T3)

†The transition equations are based on the assumption that normal stresses in the core are insignificant in the core shear failure mode. More general forms of the equations are given by Triantafillou and Gibson (1987a).

sandwich beams with aluminium skins and foamed polyurethane cores, loaded in three-point bending. The face thickness to span ratio, t/l, varies from $1/10\,000$ to $1/10$; most practical sandwich beams have values between $1/2000$ and $1/200$. The relative density of the core ρ_c^*/ρ_s, varies from 0.01 to 1; practical sandwich beams have values between 0.02 and 0.3. Three failure modes dominate the map: face yielding, face wrinkling and core shear. At low values of t/l, the face fails by wrinkling if the core density is low (and hence cannot resist the buckling of the face) and by yielding if the core density is high. As the face thickness increases, there is a transition from failure of the face to that of the core. For this loading configuration and combination of face and solid core materials, the core fails by yielding in shear. Bond failure does not appear because the bond is assumed to be perfect. The mode of loading changes the map a little (because B_3 and B_4 change): that for an end-loaded cantilever, for instance, has a slightly larger field of core shear, but the general disposition of the fields is the same.

At this level of detail the map of Fig. 9.12 has generality: it describes all sandwich beams with aluminium faces and foamed polyurethane cores loaded in three-point bending. More detail can be added, but generality is lost. If the core thickness, c, and beam width, b, are given, for example, it is possible to calculate the failure load of the beam at each point on the diagram. This information (which is useful in examining the design of a beam for a specific application) can be plotted as a set of contours on to the map, which then summarizes both failure mechanism and failure load. An example is shown later (Fig. 9.14).

Failure-mode maps can also be constructed for sandwich plates and columns. The procedure is the same: equations for the normal and shear stresses in the face and core are equated to the strength of the face, or its wrinkling load, or the strength of the core. These failure equations, taken in pairs, give the field boundaries of the failure map. In practice the transition equations cannot be expressed in closed form, and the map has to be constructed by numerical methods. Triantafillou and Gibson (1987a) give an example.

(c) Experiments to measure sandwich strength

Most investigations of sandwich strength have focused on a single failure mode – face yield, for instance. One, that of Triantafillou and Gibson (1987a), studied the spectrum of failure modes discussed in the last section, in a way which allowed a critical test of the theory. This study used sandwich beams with the same aluminium faces and foamed polurethane cores used to characterize the optimum stiffness (Section 9.2). The beams had differing values of t/l and ρ_c^*/ρ_s, and were loaded in three-point bending. 'Failure' was defined as the first pronounced deviation from linearity in the loading curve. Most of the beams failed by either face yielding, face wrinkling or core shear; a few failed by debonding or local indentation. The results are plotted on the failure-mode maps in Figs. 9.13 and 9.14. There is close agreement between the observed and expected mechanisms of failure, and between the calculated and measured failure loads.

A very few (three out of 124) of the beams failed by debonding, almost certainly because the adhesive layer contained a fabrication defect – a region that was imperfectly bonded. Since the beams tested cover most of the practical

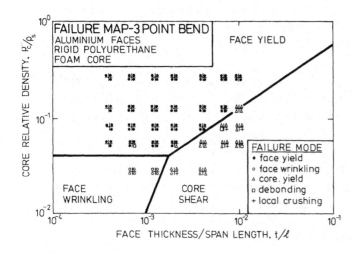

Figure 9.13 The failure mode map for sandwich beams with aluminium faces and rigid polyurethane foam cores loaded in three-point bending. The mode of failure of each beam tested is shown on the diagram.

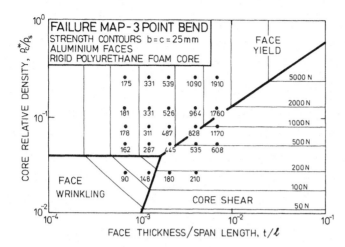

Figure 9.14 The failure mode map for sandwich beams with aluminium faces and rigid polyurethane foam cores loaded in three-point bending. The strength contours correspond to beams with $b = 25$ mm and $c = 25$ mm. The points indicate the average failure load for the test beams with $b = c = 25$ mm.

range of t/l and ρ_c^*/ρ_s, debonding is unlikely to be a problem if steps are taken to avoid interfacial defects.

9.4   Optimization of sandwich design: stiffness, strength and weight

The optimization of sandwich structures for strength has received some attention, though there is no single definitive analysis. Kuenzi (1965), Ueng and Liu (1979) and Froud (1980) all analysed the optimization of beams with respect to one or, at best, two of the many possible failure mechanisms. And optimization on strength does not guarantee adequate stiffness: that gives an additional constraint, considered by Huang and Alspaugh (1974) and Ueng and Liu (1979). It is also included in the work of Ackers (1945), Wittrick (1945) and Kuenzi (1965), who examined the optimization of end-loaded sandwich columns – a very similar problem, with overall Euler buckling as an additional (stiffness-controlled) failure mode. But none of these studies includes all possible failure modes in the optimization.

In this section we develop ways of optimizing sandwich structures for both strength and stiffness. We start with the simplest common problem: the design of a sandwich beam for which the skin and core materials are given. The two variables are then the skin thickness, t, and the core thickness, c.

When the skin and core thicknesses, t and c, are the only design variables, the graphical method is the best approach. It is illustrated in Fig. 9.15. The axes are the same as those of Fig. 9.7, and so, too, are the family of broken lines describing the objective function (Eqn. (9.7)) and the full, heavy line describing the stiffness constraint for a given P/δ (Eqn. (9.6)). Superimposed on these are lines (for a given required strength, P) for

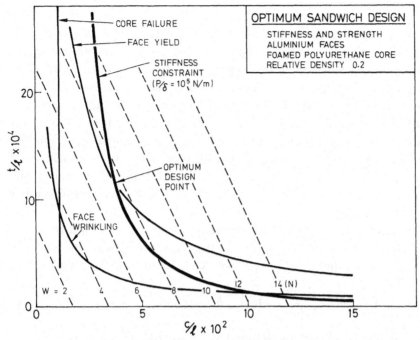

Figure 9.15   The graphical method for optimization for stiffness and strength. The family of straight broken lines shows the objective function (Eqn. 9.7). The heavy solid curve shows the stiffness constraint (Eqn. 9.6) for a given value of stiffness, P/δ. The other three full lines show failure by face yield, face wrinkling and core shear for a given value of the load, P. The optimum design point (marked) satisfies all four constraints at minimum weight. (Data as for Fig. 9.7, with face yield strength of $87 \, MN/m^2$, core yield strength of $3.4 \, MN/m^2$ and requirement that the beam support a load of 750 N. Beam width 50 mm and length 1 m.)

(a)   failure by face yield (Eqn. (F1), Table 9.6);

(b)   failure by face wrinkling (Eqn. (F2), Table 9.6); and

(c)   failure by core shear (Eqn. (F3), Table 9.6).

The design limiting constraints are immediately obvious: for the values of P/δ and P chosen here they are stiffness and face yield. The optimum design point is marked; from it the optimum face thickness $(t/l)_{opt}$ and core thickness $(c/l)_{opt}$ and the weight W_{opt}, can be read off. If the design criteria P/δ and P are changed, the relative positions of the lines change and the optimum design point moves. The graphical method is attractive because it displays the whole picture, showing not only the design-limiting constraints and the expected failure mode, but how close to all other modes of failure the design lies. It can be extended to allow optimization with respect to core density ρ_c^* as well as t and c (using an adaptation of the approach described in Section 9.2), but it is somewhat cumbersome. Alternative analytical methods are cumbersome, too, but are sometimes useful. We outline these next.

The failure-mode maps and the experiments described in the previous section showed that failure is usually caused by face yield, face wrinkling or core shear. Equations describing these are listed in Table 9.6. The analytical approach to optimization is based on the observation that, in minimum weight design *based on strength alone*, the face and the core must fail at the same load; if they do not the panel is overdesigned. The map of Fig. 9.14 shows how the strength contours for face yield are independent of the core density, so that (in this field) ρ_c^* can be reduced to the value which corresponds to the field boundary without loss of strength. Similarly, strength contours for core shear are independent of face thickness t/l, so t, too, can be reduced to the value at the field boundary without loss of strength. Optimum design, then, lies on the field boundaries (Triantaffilou and Gibson, 1987b).

Failure by face yielding and by core shear are described by Eqns. (F1) and (F3) in Table 9.6. We solve the first for face thickness, t, and the second for core density, ρ_c^*, and substitute these into the objective function (9.7), giving

$$W = 2bl\rho_f g\left(\frac{P}{B_3 b\sigma_{yf}}\right)\left(\frac{l}{c}\right) + bcl\rho_s g\left(\frac{P}{C_{11}B_4 bc\sigma_{ys}}\right)^{2/3}. \tag{9.26}$$

The optimum core thickness is found by setting $\partial W/\partial c = 0$, giving

$$\left(\frac{c}{l}\right)_{opt} = \left\{\frac{6\rho_f}{\rho_s}\frac{P}{B_3 bl\sigma_{yf}}\left(\frac{C_{11}B_4 bl\sigma_{ys}}{P}\right)^{2/3}\right\}^{3/4}.$$

The optimum values of t/l and ρ_c^*/ρ_s which minimize the weight of the beam are found by substituting this back into the inverted forms of Eqns. (F1) and (F3). For this solution to be valid the core density must be greater than that given by the boundary between face yielding and face wrinkling on the failure-mode map, Fig. 9.12; if it is not, the beam will fail by face wrinkling instead. In practice this optimization gives values of (c/l) which are unreasonably large, such that the core thickness, c, is of the same order as the span of the beam, l. An additional constraint on the maximum permissible core thickness is then introduced; the optimum design is found using the strength contours on the failure-mode map for the maximum allowable core thickness.

The other combination – failure by face wrinkling and core shear – is approached in a similar way. The equation for wrinkling (Eqn. (F2), Table 9.6) is solved for t, and that for core shear (Eqn. (F3)) is solved, as before, for ρ_c^*. Both are substituted into the objective function, giving

$$W = 2\rho_f g\frac{Pl^2}{0.57B_3 C_1^{2/3}E_f^{1/3}E_s^{2/3}(\rho_c^*/\rho_s)^{4/3}c} + \left(\frac{P}{B_4 C_{11}\sigma_{ys}}\right)^{2/3}(bc)^{1/3}l\rho_s g. \tag{9.27}$$

This is optimized by setting $\partial W/\partial c = 0$, giving

$$\left(\frac{c}{l}\right)_{\text{opt}} = \left\{ 3\frac{\rho_{\text{f}}}{\rho_{\text{s}}}\left(\frac{P}{blE'}\right)\left(\frac{C_{11}B_4 bl\sigma_{\text{ys}}}{P}\right)^{14/9} \right\}^{9/4}$$

with

$$E' = 0.57B_3 C_1^{2/3} E_{\text{f}}^{1/3} E_{\text{s}}^{2/3}.$$

As before, the optimum t/l and $\rho_{\text{c}}^*/\rho_{\text{s}}$ are found by substituting this back into the inverted forms of Eqns. (F2) and (F3). The core density must be less than that given by the boundary between simultaneous face wrinkling and yielding for this to be a valid solution; otherwise the beam will fail by face-yielding instead (see the failure-mode map, Fig. 9.12).[†]

Some interesting results follow (Triantafillou and Gibson, 1987b). When face yield and core shear occur together, the ratio of the weight of the faces to that of the core is 1:3. When, instead, face wrinkling and core shear occur together, the ratio is 3:1. The core density given by either optimization analysis may not produce the assumed failure modes. If this is the case, and if the assumed mode of failure was by simultaneous face yielding and core shearing, then the core density must be increased to give this mode of failure. Similarly, if the assumed failure was by simultaneous face wrinkling and core shearing, the core density must be decreased. This implies that the optimum design lies at the intersection of all three transition lines on the failure-mode map, and that face yielding, face wrinkling and core shearing all occur simultaneously. The three beam design variables are then uniquely determined by the three failure equations for face yielding, face wrinkling and core shear.

The method can obviously be extended. It has been applied to the design of sandwich plates by Triantafillou and Gibson (1987b). It can be adapted for face or core materials which fail in a brittle manner (fibre-reinforced skins, with foamed-glass core, for instance). The obvious difficulty with it is its complexity, and the way that the overall picture is obscured by the mathematics: it is not easy to establish that all possible failure modes have been considered, and that the right combination (out of a large number of possible combinations) has been chosen for the final optimization.

In a problem with as many variables as this the best route to a solution is, in the end, likely to be a graphical one. That described earlier readily encompasses all failure modes, and stiffness constraints as well. It can be extended to allow the core density to be a variable in one of several ways. The simplest (probably the best, because it retains the clear physical picture) is to build the objective function and all the constraints into an interactive computer code which plots diagrams like Fig. 9.15; the operator then steps through values of $\rho_{\text{c}}^*/\rho_{\text{s}}$, recording the mini-

[†]More recent analysis, analogous to that above, shows that simultaneous face yielding and face wrinkling can sometimes be the critical failure mode.

mum weight at each step, and selecting the one that gives the design point of lowest weight. The optimum t/l, c/l and ρ_c^*/ρ_s are then read from the appropriate plot.

Finally, there is the question of which materials to use for the face and the core. For a sandwich beam of a given stiffness, the weight is minimized by substituting the equations for the optimum values of the face thickness, t, the core thickness, c, and the core density, ρ_c^*, (Appendix 9A) into the weight equation (9.7):

$$W = 3.18bl^2 \rho_f^{1/5} \rho_s^{4/5} \left[\frac{1}{B_1 B_2^2 C_2^2} \frac{1}{E_f E_s^2} \left(\frac{P}{b\delta} \right)^3 \right]^{1/5}.$$

The weight is minimized by selection of the face material with the maximum value of E_f/ρ_f and the core material with the maximum value of $E_s^{1/2}/\rho_s$. Bending a sandwich beam induces normal stresses in the faces and shear stresses in the foam core (which, in turn, produce bending of the solid cell-wall material): the two parameters correspond, as expected, to the performance indices which minimize the weight of a component of a given stiffness subject to axial and bending loads, respectively (Ashby, 1992). The calculation for a sandwhich plate gives the identical result.

Sandwich panels sometimes have the same face and core materials, as in structural foams, for instance. Writing $\rho_f = \rho_s = \rho$ and $E_f = E_s = E$ we find that the weight of the panel is minimized by selection of the material with the maximum value of $E^{3/5}/\rho$. Using this criterion, the best polymers for structural foams are polystyrene, epoxy and polyester.

9.5   Case studies in sandwich design

In this section we analyse two real (and so more complicated) sandwich structures: a ski, and the leaf of an iris. Both must meet a constraint on stiffness, while keeping weight low. And both must satisfy further constraints on their strength.

(a) Optimum design of downhill skis

The characteristics of a ski

The stiffness of a ski is critical in giving the skier the right 'feel'. A ski that is too flexible is difficult to control, while one that is too stiff can leave the skier suspended, as on a plank, between bumps. Because of this, skis are designed primarily for stiffness. Thirty years ago skis were made from a single piece of wood, shaped to give a suitable overall bending stiffness and with an upturned tip to cut through the snow. Subsequently designers found that they could reduce the weight by making a laminated ski: a light wood (such as pine or spruce) was bonded between facing sheets of a denser, stiffer wood (often ash or hickory). This early design was evolved into the sophisticated sandwich structure of the modern downhill ski. Faces made of either aluminium of fibre-reinforced polymer are separated by a core made of aluminium honeycomb or rigid polyurethane foam; the core and face thickness are tailored to give a progressive change in stiffness along the length of the ski. The design gives the combination of properties that 'feel' right, while reducing the weight. But is the design optimal? Could the sandwich be modified (using the methods developed in this chapter) to reduce the weight further?

Figure 9.16 shows a 200 cm Rossignol Equipe 3G downhill ski. It is a sandwich structure: a rigid foamed polyurethane core separates aluminium skins. The profile of the ski curves upward under the skier's boot, and its thickness and width vary along its length to control the longitudinal stiffness. A typical cross-section is shown in Fig. 9.17. The main structural features are the aluminium faces and the rigid polyurethane foam core. The top and bottom of the ski are both coated with polymer layers: the top for decoration; the bottom to reduce

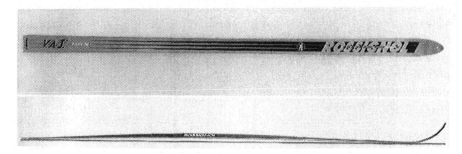

Figure 9.16 Plan and profile of a modern downhill ski. It has a sandwich structure with aluminium faces bonded to a foamed polyurethane core.

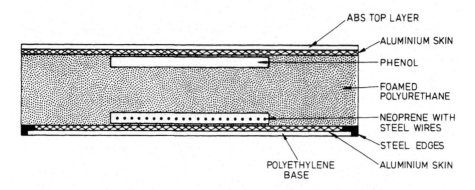

Figure 9.17 A cross-section through the ski about half-way along its length.

the sliding friction. Steel edges incorporated into the base give better control but
have no important effect on stiffness. A short strip of phenol is moulded into the
foam core to allow the binding to be screwed in, and a 300 mm strip of neoprene
rubber reinforced with steel wires is built into the lower half of the foam core to
increase damping. The steel wires (which are there to introduce shear strain in
the rubber) increase the stiffness of the rubber to a value close to that of the rigid
foam core so that, in modelling the stiffness of the ski, only the aluminium faces
and rigid foam core play a significant role.

The properties of the face and core materials are listed in Table 9.8. Their
thicknesses are shown in Fig. 9.18. The aluminium and polymer facing layers
have a constant thickness; subtracting the sum of these from the total thickness
gives the core thickness. The measured bending stiffness of the ski, loaded as
shown in Fig. 9.18, is 3.5 N/mm.

The bending stiffness of the ski can be calculated by the methods of Section
9.2. Figure 9.19 shows how the core thickness, c, the flexural rigidity, $(EI)_{eq}$, the
normalized moment, M/Pl, and $M/(EI)_{eq}$ vary along the length of one half of

Table 9.8   Properties of ski face and core materials

Property	Aluminium	Solid polyurethane	Polyurethane foam
Density (Mg/m^3)	2.7	1.2	0.53
Young's modulus (GN/m^2)	70	1.94	0.38
Shear modulus (GN/m^2)	—	—	0.14

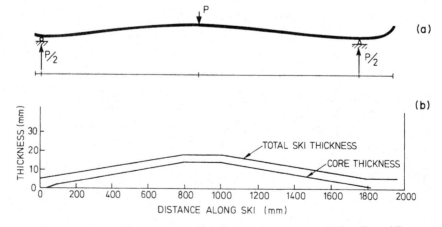

Figure 9.18   (a) The geometry of loading used to measure the bending stiffness.
(b) The thickness along the length of the ski.

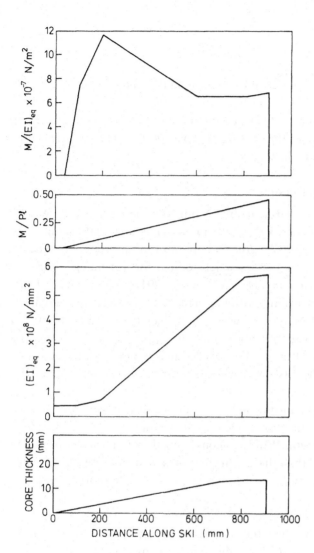

Figure 9.19 The development of an $M/(EI)_{eq}$ diagram for calculating bending deflection.

the ski. Using the moment – area method to calculate the bending deflection of the ski gives

$$\delta_b = 0.28P$$

where δ_b is in mm and P is in N. The shear deflection of the ski can be found from the average equivalent shear rigidity, giving

$$\delta_s = Pl/(AG)_{eq}$$
$$= 0.0045P.$$

The resulting stiffness for the ski is then the sum of these two components, or:

$$\delta = \delta_b + \delta_s$$
$$= 0.29P$$

giving a stiffness of

$$P/\delta = 3.5\,\text{N/mm}$$

which agrees with the measured value, confirming that the ski does behave as a sandwich beam.

The mass of the ski is dominated by the aluminium skins and foam core. In the Rossignol design the mass of the load-bearing part of the ski (the 1.70 m central span, neglecting the tip and the tail) is 1.3 kg.

Optimum weight design for skis

Could the ski be reshaped to give the same bending stiffness, but a lower weight? There are several possibilities. First (for simplicity) assume that the thickness of the ski is constant along its length. The optimization procedure outlined in Section 9.2 then gives the minimum weight design for a bending stiffness of 3.5 N/mm: it is $c = 70$ mm, $t = 0.095$ mm and $\rho_c^* = 29\,\text{kg/m}^3$. The mass of such a ski, 72 mm wide and 1.70 m long, is 0.31 kg, a reduction of 75% from the current design. This is achieved by increasing the core thickness from 15 to 70 mm, decreasing the face thickness from about 1 mm to 0.1 mm and by decreasing the core density from 525 to $29\,\text{kg/m}^3$. But this design is impractical: the thick core is unwieldy and its low density gives inadequate strength, a problem ignored in the optimization so far.

A better method is to fix the maximum acceptable value of the core thickness beneath the binding, and then to profile the core so that the value of $M/(EI)_{\text{eq}}$ is constant along the length. With this constraint the optimization procedure can be carried out to find the values of the face thickness and core density which minimize weight for the required bending stiffness (3.5 N/mm). Figure 9.20 shows the moment diagram for half the ski (which can be thought of as canti-levered from its mid-point); it increases linearly along its length. The efficient design for the ski is one in which the bending rigidity $(EI)_{\text{eq}}$ also varies linearly along the length, so that the $M/(EI)_{\text{eq}}$ diagram is a constant. Since

$$M = Px$$

and

$$(EI)_{\text{eq}} = \frac{E_{\text{f}} b t c^2}{2}$$

the profile of the core thickness should vary as the square root of x, the distance along the length of the ski. If the half-length is 870 mm and the core thickness is zero at the front and back of the ski and is 15 mm in the middle then a constant $M/(EI)_{\text{eq}}$ requires a core thickness profile of:

$$c = \frac{15}{(870)^{1/2}} x^{1/2} = 0.51 x^{1/2}. \tag{9.28}$$

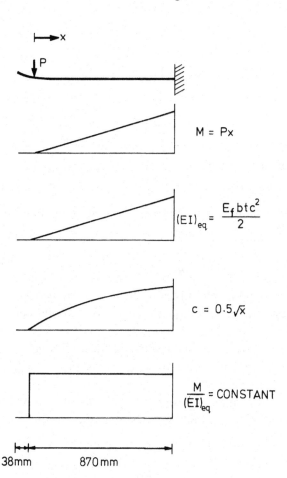

Figure 9.20 An 870 mm cantilever beam with a constant $M/(EI)_{eq}$.

$M = Px$

$(EI)_{eq} = \dfrac{E_f btc^2}{2}$

$c = 0.5\sqrt{x}$

$\dfrac{M}{(EI)_{eq}} = \text{CONSTANT}$

38 mm   870 mm

The average thickness of the core is 2/3 of the maximum value, or in this case, 10 mm.

To proceed with the optimization we need an expression for the stiffness, P/δ, of the ski. It is found by the methods of Section 9.2. The deflection arises partly from bending, determined by $(EI)_{eq}$ and partly from shear, determined by $(AG)_{eq}$. Evaluating and summing the two gives (Gibson and Kellogg, 1986):

$$\frac{\delta}{P} = \frac{2l^3}{B_1 E_f bt(c_{max} + t)^2} + \frac{l}{B_2 C_2 bc_{max}(\rho_c^*/\rho_s)^2 E_s}. \tag{9.29}$$

This expression can be solved for the core density, ρ_c^*:

$$\frac{\rho_c^*}{\rho_s} = \left\{ \frac{A}{\dfrac{\delta}{P} - \dfrac{B}{t(c_{max} + t)^2}} \right\}^{1/2} \tag{9.30}$$

where

$$A = \frac{l}{C_2 B_2 bc_{max} E_s}$$

and

$$B = \frac{2l^3}{B_1 E_f b}$$

and substituted into the weight equation,

$$W = 2\rho_f gbtl + \tfrac{2}{3}\rho_c^* gbc_{max} l. \qquad (9.31)$$

Setting $dW/dt = 0$ gives an equation for the optimum face thickness which will minimize the mass of the ski:

$$\frac{6}{A^{1/2}B}\frac{\rho_f}{\rho_s} - \frac{c_{max}(c_{max} + 3t)}{t^{1/2}((\delta/P)t(c_{max} + t)^2 - B)^{3/2}} = 0. \qquad (9.32)$$

The optimum thickness, t, is found graphically or by an iterative numerical technique, and its value can be substituted into Eqn. (9.30) to find the optimum core density.

The results of solving this equation for the ski, with a stiffness 3.5 N/mm and a maximum allowable core thickness of 15 mm, are summarized in Table 9.9. We find that the optimum face thickness is 1.03 mm and the optimum core density is 163 kg/m³. The mass of the ski using this design is 0.88 kg, considerably more than the initial optimum (at 0.31 kg) but still much less than the current mass of 1.27 kg.

(b) The structure and mechanics of the iris leaf

Plants live by photosynthesis, capturing solar energy with their leaves. For this to take place, the leaf must be stiff enough to support its own weight without drooping excessively. Some leaves (such as ivy or maple) achieve their stiffness by means of a dendritic network of thick woody veins. Others, like the iris and most grasses, derive it from a structure which is not unlike that of the ski: a pair of stiff, parallel skins bonded to a spongy core. Figure 9.21 shows exam-

Table 9.9   Optimum ski design (parabolic core profile)

Core	$c = 0.51x^{1/2}$
	$c_{max} = 15$ mm
Face	$t = 1.03$ mm
Stiffness	$P/\delta = 1.75$ N/mm
Weight	$W = 8.6$ N (or 0.88 kg)

Figure 9.21 Macrophotographs of (a) ivy and (b) maple, (c) grass and (d) iris leaves. The first two have a network of thick veins to support the leaf; the second two have outer layers of parallel longitudinal ribs separated by a low-density core.

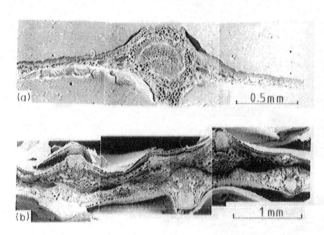

Figure 9.22 Micrographs of transverse sections through (a) an ivy leaf and (b) an iris leaf. In the ivy leaf the veins run through the centre of the leaf, while in the iris the ribs run along the top and bottom surface of the leaf.

ples of both arrangements; Fig. 9.22 shows magnified views of transverse sections of both. It is the second – the sandwich-like leaf of the iris – which interests us here.

The iris leaf

Figure 9.22 shows the iris leaf. It has skins reinforced by stiff ribs, made up of dense sclerenchyma cells; the bulk of the leaf, supporting and separating the skins, is made of lower-density parenchyma cells. The width of the leaf is roughly constant for about two-thirds of its length; the final third tapers to a pointed tip. The thickness varies along the length and across the width from a maximum of 6 mm at the centre of the base to about 0.5 mm at the tip and edges.

Figure 9.23 shows longitudinal sections of the outer skin and the core of the leaf. The outer skin is made up of dense ribs connected by a single layer of cells of roughly square transverse section. Jointly they act like a fibre-reinforced sheet. The cells in the interior of the leaf differ, in that they are roughly equiaxed

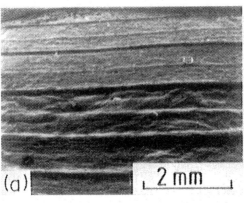

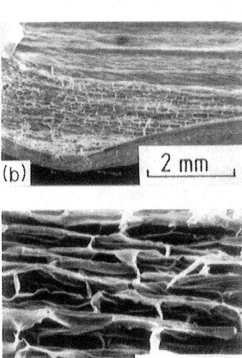

Figure 9.23 Scanning electron micrographs of longitudinal sections of (a) the outer face and (b) and (c) the inner core of an iris leaf.

in the transverse section (Fig. 9.22(b)) and somewhat elongated in the longitudinal section (Fig. 9.23(c)). They have thin cell walls and form the low-density, foam-like 'core' of the leaf. A schematic drawing of a transverse section of the leaf is shown in Fig. 9.24. Typical dimensions are listed in Table 9.10. Most remain roughly constant along the length of the leaf; only the core thickness changes, tapering from about 6 mm at the base to 0.5 mm at the tip.

Table 9.10 Dimensions of the iris leaf

	Mean (mm)	Std. dev. (mm)
At thin end of leaf		
Depth of layer of 'face' cells, f	0.03	0.0043
Thickness of square 'face' cell wall, t_f	0.0014	—
Length of square 'face' cells, l_f	0.04	0.0058
Depth of 'core' layer, c	0.5	—
Thickness of a 'core' cell, t_c	0.0014	—
Length of 'core' cells, l_c	0.05	0.023
Diameter of rib, d	0.13	0.04
Spacing of rib, s	1.2	0.46
Volume fraction of solid in rib, v_f	0.8	—
At mid-length of leaf		
Depth of layer of 'face' cells, f	0.03	0.0054
Thickness of square 'face' cell wall, t_f	0.0014	—
Length of square 'face' cells, l_f	0.03	0.005
Depth of 'core' layer, c	3.0	—
Thickness of 'core' cell wall, t_c	0.0014	—
Length of 'core' cells, l_c	0.07	0.025
Diameter of rib, d	0.19	0.058
Spacing of rib, s	0.92	0.32
Volume of fraction of solid in rib, v_r	0.8	—

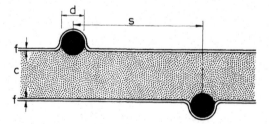

Figure 9.24 A schematic drawing of a transverse section of an iris leaf showing the parameters characterizing its structure.

To calculate the bending stiffness in terms of the theory of Section 9.2 the longitudinal elastic properties of the 'face' and 'core' must be known. The two Young's moduli, E_f and E_c^*, can be estimated using a simple rule-of-mixtures if Young's modulus for the cell-wall material, E_s, and the volume fractions of solid in the face and core are known.

The modulus of the cell walls

The extensive measurements of Young's modulus for the cell walls of plants are summarized in Table 9.11. It varies from about 0.12 GN/m^2 for *Acetabularia crenulata* to 2.2 GN/m^2 for *Nitella translucens*. The table includes data for iris (Gibson *et al.*, 1988), inferred from measurements of the longitudinal stiffness of the leaf.

Table 9.11   Young's modulus of plant cell walls

Plant material	Condition	E (GN/m^2)	Reference
Cell wall			
Penicellus dumetosus	Wet	0.16–0.22	(a)
Acetabularia crenulata	Wet	0.12–0.14	(b)
Nitella translucens	Wet	1.8–2.2	(a)
Nitella opaca	Wet	0.5–4	(c)
Nitella flexilis	Wet	0.4–0.7	(d)
Potato tuber	Wet	0.5	(e)
Iris	Wet	4 ± 1.6	(f)
Leaf components			
Leaf fibre	Dry	12–23	(g)
Leaf fibre	Wet	23	(h)
Leaf bundle	Wet	0.84	(h)
Cell-wall components			
Cellulose	Theory	130	(i)
Lignin	Theory	2	(i)
Fibres with high content of cellulose			
Flax		110	(j)
Ramie		60	(j)
Hemp	Wet	35	(k)
	Dry	70	(k)
Sisal fibre	Dry	10–21	(g)
Cotton hair	Dry	8.5	(g)
	Wet	2.9	(g)

Sources of data: (a) Sellen (1979); (b) Haughton and Sellen (1969); (c) Probine and Preston (1962); (d) Kamiya *et al.* (1963); (e) Nilsson *et al.* (1958); (f) Gibson *et al.* (1988); (g) Preston (1974); (h) Vincent (1982); (i) Bodig and Jayne (1982); (j) Treitel (1946); (k) Wainwright *et al.* (1976).

Several factors affect the cell-wall modulus: the most significant are the direction of loading (because the cell wall is made up of aligned cellulose microfibrils in a matrix of lignin and hemicelluloses) and the moisture content of the cell wall. The rate or the duration of loading has a small effect on the modulus; Sellen (1979), for example, measures a decrease in the relaxation modulus of about 20% when the duration of loading is increased by four orders of magnitude.

There is disagreement on whether the stiffness of the bulk leaf is affected by the turgor pressure (the osmotic pressure within the cell). Falk et al. (1958) claim that the stiffness depends on turgor pressure throughout its normal range, but Kamiya et al. (1963) find that the longitudinal stiffness of the cells increases with turgor pressure only up to pressures of $0.4\,\mathrm{MN/m^2}$; above, the stiffness is constant. We shall assume that for freshly picked leaves the turgor pressure is roughly constant at a relatively high level (around $0.6\,\mathrm{MN/m^2}$) and that (following Kamiya et al.) the longitudinal stiffness of the leaf is independent of pressure.

The water content of the leaf affects the measured longitudinal stiffness. Vincent (1983) reports that the longitudinal stiffness of two grasses was constant at water contents of between 100 and 400% of the dry weight, but rose sharply as the water content was reduced from 100% to 10%. Fresh leaves are always in a range in which the stiffness is independent of water content.

Analysis of the iris leaf as a sandwich beam

If the iris leaf derives its resistance to drooping from its sandwich-like construction it should be possible to calculate its stiffness, at least approximately, from the equations developed in Section 9.2. This has been attempted by Gibson et al. (1988), with fair success. They measured the flexural stiffness of freshly picked iris leaves by cutting cantilever beams parallel to the long dimensions of the leaf and bending them by hanging small weights from the free end. The bending stiffness was calculated from the load–deflection curve. The results are given in Table 9.12.

Gibson et al. (1988) estimate the flexural stiffness of the iris leaf by adapting the method of Section 9.2. The modulus of the faces is calculated by a rule-of-mixtures, summing the contributions from the ribs and from the face cells (the volume fractions and densities of which are given in Table 9.10) and using a solid modulus, E_s, of $4.4\,\mathrm{GN/m^2}$; the resulting face modulus is $E_f = 1.6\,\mathrm{GN/m^2}$. The shear modulus of the core is calculated from its relative density (again using the cell dimensions in Table 9.10) giving a value of $G_c^* = 0.11\,\mathrm{GN/m^2}$. As one would expect in a sandwich panel, the stiffness of the face is much greater than that of the core.

Calculation of the flexural stiffness is complicated by the varying thickness of the core across the section. Gibson et al. (1988) deal with this by dividing the transverse section into sub-units, each of constant thickness, and summing their

Table 9.12 Beam bending results

Specimen	1	2	3	4
Measured beam stiffness, P/δ (N/mm)	0.66	0.54	0.41	0.25
Beam length, l (mm)	35	35	35	35
Thickness, t_1 (mm)	4.85	3.53	2.71	1.73
Face thickness, f (mm)	0.03	0.03	0.03	0.03
Maximum core thickness, c (mm)	4.63	3.31	2.49	1.51
Width, b (mm)	18	18	18	18
Flexural rigidity, D (Nm^2)	0.019	0.0090	0.0046	0.002
Bending compliance, $(\delta/P)_b$ (m/N)	0.00075	0.0016	0.0031	0.0068
Shear compliance, $(\delta/P)_s$ (m/N)	5.7×10^{-6}	7.3×10^{-6}	9.2×10^{-6}	1.2×10^{-5}
Calculated beam stiffness, P/δ (N/mm)	1.32	0.62	0.32	0.15

contributions to the overall stiffness. The flexural stiffness of the leaf was then found from Eqn. (9.6). The results are given in the bottom row of Table 9.12 and should be compared with the measured values in the top row. The agreement is as good as could be expected of the approximations made in estimating the moduli of the face and core, and in modelling the irregular cross-section of specimens. The results clearly support the view that the iris leaf behaves mechanically like a sandwich structure.

9.6   Conclusions

Sandwich structures – stiff skins bonded to a light, foam-like core – have exceptional stiffness and strength per unit weight. They appear frequently in nature, and are increasingly manufactured by man for applications in which minimizing weight is important. The understanding of honeycombs and of foams, developed in Chapters 4 and 5, provides the tools required for optimum sandwich design. Designs for stiffness and strength, at minimum weight, are analysed in this chapter, where procedures and results for common geometries are developed. The results are successfully applied to two very different problems – the optimum design of a downhill ski, and the origin of the stiffness of iris leaves, both of which derive their stiffness from their sandwich-like construction.

Appendix 9A: Results for stiffness-optimized sandwich structures

Here we list results for the optimum skin thickness, core thickness and core density for beams and plates, subject to a constraint on stiffness. It is important to ensure that strength constraints (Section 9.3) are also met.

(a) Rectangular beam under a concentrated load, P

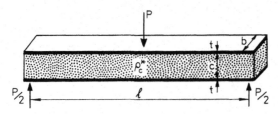

The optimum core thickness, skin thickness and core density are:

$$\left(\frac{c}{l}\right)_{opt} = 4.3\left\{\frac{C_2 B_2}{B_1^2}\left(\frac{\rho_f}{\rho_s}\right)^2\frac{E_s}{E_f^2}\left(\frac{P}{b\delta}\right)\right\}^{1/5}$$

$$\left(\frac{t}{l}\right)_{opt} = 0.32\left\{\frac{1}{B_1 B_2^2 C_2^2}\left(\frac{\rho_s}{\rho_f}\right)^4\frac{1}{E_f E_s^2}\left(\frac{P}{b\delta}\right)^3\right\}^{1/5}$$

$$\left(\frac{\rho_c^*}{\rho_s}\right)_{opt} = 0.59\left\{\frac{B_1}{B_2^3 C_2^3}\left(\frac{\rho_s}{\rho_f}\right)\frac{E_f}{E_s^3}\left(\frac{P}{b\delta}\right)^2\right\}^{1/5}$$

δ = central beam deflection.

(b) Circular plate under a uniformly distributed load q per unit area

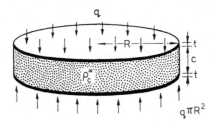

The optimum core thickness, skin thickness and core density are:

$$\left(\frac{c}{R}\right)_{opt} = \frac{1}{2}\left\{48C_2(5 + v_f)^2(1 - v_f)^2\left(\frac{\rho_f}{\rho_s}\right)^2\frac{E_s}{E_f^2}\left(\frac{qR}{w}\right)\right\}^{1/5}$$

$$\left(\frac{t}{l}\right)_{opt} = \frac{1}{16}\left\{\frac{27(5 + v_f)(1 - v_f)}{8C_2^2}\left(\frac{\rho_s}{\rho_f}\right)^4\frac{1}{E_f E_s^2}\left(\frac{qR}{w}\right)^3\right\}^{1/5}$$

$$\left(\frac{\rho_c^*}{\rho_s}\right)_{opt} = \frac{1}{2}\left\{\frac{9}{4(5 + v_f)(1 - v_f)C_2^3}\left(\frac{\rho_s}{\rho_f}\right)\frac{E_f}{E_s^3}\left(\frac{qR}{w}\right)^2\right\}^{1/5}$$

w = central plate deflection.

(c) Circular plate under a distributed load q per unit area over a small central area of radius r

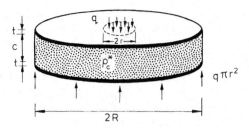

The optimum core thickness, skin thickness and core density are:

$$\left(\frac{c}{r}\right)_{opt} = \left\{ \frac{24C_2(1 - v_f^2)^2 g_1^2}{g_2} \left(\frac{\rho_f}{\rho_s}\right)^2 \frac{E_s}{E_f^2} \left(\frac{qr}{w}\right) \right\}^{1/5}$$

$$\left(\frac{t}{r}\right)_{opt} = \frac{1}{16} \left\{ \frac{27}{2} \frac{(1 - v_f^2) g_1 g_2^2}{C_2^2} \left(\frac{\rho_s}{\rho_f}\right)^2 \frac{1}{E_f E_s^2} \left(\frac{qr}{w}\right)^3 \right\}^{1/5}$$

$$\left(\frac{\rho_c^*}{\rho_s}\right)_{opt} = \frac{1}{4} \left\{ \frac{18 g_2^3}{C_2^3(1 - v_f^2) g_1} \frac{\rho_s}{\rho_f} \frac{E_f}{E_s^3} \left(\frac{qr}{w}\right)^2 \right\}^{1/5}$$

where

$$g_1 = \left(\frac{3 + v_f}{1 + v_f}\right)\left(\frac{R}{r}\right)^2 + \ln\left(\frac{r}{R}\right) - \frac{7 + 3v_f}{4(1 + v_f)}$$

$$g_2 = 1 + 2\ln\left(\frac{R}{r}\right)$$

w = central plate deflection.

References

Ackers, P. (1945) U.K. Aeronautical Research Council, Farnborough, Hants. U.K. Reports and Memoranda 2015.

Allen, H. G. (1969) *Analysis and Design of Structural Sandwich Panels.* Pergamon Press, Oxford.

Ashby, M. F. (1992) *Material Selection in Mechanical Design.* Pergamon Press, Oxford, UK.

Bodig, J. and Jayne, B. A. (1982) *Mechanics of Wood and Wood Composites.* Van Nostrand Reinhold, New York.

Ciba-Geigy (1980) *Aeroweb Honeycomb Sandwich Design.* Instruction sheet No. AGC 33a (Part 2) Bonded Structures Division, Duxford, Cambridge.

Demsetz, L. A. and Gibson, L. J. (1987) *Mat. Sci. Eng.,* **85**, 33.

Falk, S., Hertz, H. and Virgin, H. I. (1958) *Physiol. Plant,* **11**, 802.

Froud, G. R. (1980) *Composites,* **11**, 133.

Gibson, L. J. (1984) *Mat. Sci. Eng.,* **67**, 125.

Gibson, L. J. and Kellogg, K. (1986) Dept. of Civil Engineering, M.I.T. Report No. R86-16.

Gibson, L. J., Ashby, M. F. and Easterling, K. E. (1988) *J. Mat. Sci.*, **23**, 3041–48.

Hall, D. J. and Robson, B. L. (1984) *Composites*, **15**, 266.

Haughton, P. M. and Sellen, D. B. (1969) *J. Exp. Bot.*, **20**, 516.

Hodgson, V. R. (1973) in *Human Impact Response: Measurement and Simulation* (ed. King, W. F. and Mertz, H. J.) Plenum Press, New York.

Hong, C. S. and Jeong, K. Y. (1985) *Eng. Fract. Mech.*, **21**, 285.

Huang, S. N. and Alspaugh, D. W. (1974) *AIAA J.*, **12**, 1617.

Kamiya, N., Tazawa, M. and Takata, T. (1963) *Protoplasma*, **57**, 501.

Kuenzi, E. W. (1965) *Minimum Weight Structural Sandwich*. US Forest Service Research Note FPL-086. Forest Products Laboratory, Madison WI.

Nilsson, S. B., Hertz, C. H. and Falk, S. (1958) *Physiol. Plant*, **11**, 818.

Preston, R. D. (1974) *The Physical Biology of Plant Cell Walls*. Chapman and Hall, London.

Probine, M. C. and Preston, R. D. (1962) *J. Exp. Bot.*, **13**, 111.

Sellen, D. B. (1979) in Symposia of the Society for Experimental Biology: The Mechanical Properties of Biological Materials, Leeds University, Leeds, 4–6 September.

Thompson, D. W. (1961) *On Growth and Form*, abridged edn (ed. J. T. Bonner). Cambridge University Press, Cambridge.

Timoshenko, S. P. and Goodier, J. N. (1970) *Theory of Elasticity*. 3rd edn, McGraw-Hill, New York, p. 22.

Treitel, O. (1946) *J. Colloid Sci.*, **1**, 327.

Triantafillou, T. C. and Gibson, L. J. (1987a) *Mat. Sci. Eng.*, **95**, 37.

Triantafillou, T. C. and Gibson, L. J. (1987b) *Mat. Sci. Eng.*, **95**, 55.

Triantafillou, T. C. and Gibson, L. J. (1989) *Materials and Structures*, **22**, 64–9.

Ueng, C. E. S. and Liu, T. L. (1979) *Proc. ASCE Engineering Mechanics Division 3rd Speciality Conf.* University of Texas at Austin, 17–19 September, p. 41.

Vincent, J. F. V. (1982) *J. Mat. Sci.*, **17**, 856.

Vincent, J. F. V. (1983) *Grass Forage Sci.*, **38**, 107.

Wainwright, S. A., Gibbs, W. D., Currey, J. D. and Gosline, J. M. (1976) *Mechanical Design in Organisms*. Princeton University Press, Princeton, NJ.

Wittrick, W. H. (1945) UK Aeronautical Research Council, Farnborough, Hants, U.K. Reports and Memoranda 2016.

Wrzecioniarz, P. A. (1983) *J. Eng. Mech.*, **109**, 1460.

Zenkert, D. (1995) *An Introduction of Sandwich Construction*. Engineering Materials Advisory Services Ltd., Solihull UK.

Chapter 10

Wood

10.1 Introduction and synopsis

Wood is the most ancient, but still the most widely used, structural material in the world – indeed – the word 'material' itself derives from the Latin *materies, materia*: the trunk of a tree. The use of wood in buildings, ships and furniture is as old as the pyramids – wooden artefacts at least 5000 years old have been found in them. During the sixteenth century the demand in Europe for stout oaks for shipbuilding was so great that the population of suitable trees was depleted; by the seventeenth century ships' timbers had to be imported into England from the New World.† By 1800 much of Europe had been deforested by the exponential growth in the consumption of wood, a problem that programmes of reforestation have only partly overcome. Today the world production of wood is roughly the same as that of iron and steel: roughly 10^9 tonnes per year. This production finds many uses, in everything from musical instruments to pit props. Table 10.1 lists some of these, with the species of wood best suited for each. Much of the total production is used structurally: for beams, joists, flooring and supports which bear load. Then the properties which interest the designer are the moduli, the crushing strength and the toughness. These properties vary enormously from one wood to another: oak is more than 10 times stiffer, stronger and tougher than balsa. And wood can be very anisotropic, too: some species

†In 1665 Samuel Pepys, then Clerk to the Navy Board in London, recorded in his diary that he was arranging 'a great contract for New England Masts' for the English Navy which was about to go to war with the Dutch (Pepys, 1660–9).

Table 10.1 Uses for various types of wood[†]

Species	Uses	Properties
Hardwoods		
Apple	Mallets, cogs, shuttles, golf clubs	Hard, strong and tough
Ash	Handles, oars, baseball bats, frames for cars and trucks	High shock resistance
Bamboo	Frames for building, canoes, early aircraft	Flexible, strong
Balsa	Floats, rafts, insulation, cushioning, models, cores for sandwich panels	Low density, good heat and sound insulation
Beech	Flooring, furniture, veneer chairs, wooden spoons, brushes, barrels	Heavy, strong, stiff, resistant to shock, suitable for steam bending
Birch	Plywood, flooring and furniture	Tough, works well but decays easily
Boxwood	Used since biblical times for musical instruments, athletic goods, furniture	High shock resistance, uniform texture, works well
Cedar (hard)	Furniture, boat building, panelling, cigar boxes	Extremely durable
Cherry	Furniture and cabinetry	Dense with attractive texture
Ebony	Musical instruments, canes, doorknobs, billiard cues	Hard and brittle, difficult to work, but attractive appearance
Elm	Boats, docks, coffins, furniture	Moderate strength but good in waterlogged conditions
Horse chestnut	Boxes and crates, handles	A plain wood of low strength
Lignum vitae	Gears and bearings	Very high density and hardness, wear-resistant
Lime	Boat building, carving, piano parts	Strong and easily worked
Mahogany	Furniture, doors, panelling, boat building, plywood	Light and moderately durable
Maple	Flooring, bowling alleys, piano actions, violin making	Heavy, strong, stiff, resistant to shock
Red oak	Lumber, veneer, flooring, furniture	Heavy, strong, stiff

Table 10.1 *(cont.)*

Species	Uses	Properties
White oak	Whisky casks, veneer, boat building, furniture	Heavy strong, stiff, good resistance to decay
Sassafras	Used for dugout canoes by the American Indians	Durable when exposed to conditions of decay
Teak	Boats and ships, furniture, benchtops	Does not cause rust or corrosion when in contact with a metal
Walnut	Furniture, architectural woodwork, cabinets, rifle butts and stocks	Heavy, stiff, strong, straight-grained, easily worked, stable
Willow	Artificial limbs, crickets bats, flooring	Absorbs energy without splintering, works well
Softwoods		
Cedar (soft)	Pencils, blinds, shingles	Noted resistance to decay
Fir, Douglas	Building and construction, plywood	High stiffness and strength
Fir, Western	Building and construction shrinkage	Good strength, low shrinkage
Pine, Eastern	Containers and packaging, furniture	Moderate stiffness and strength
Pine, red	Building and construction (particularly log cabins)	Moderate stiffness and strength
Red cedar, Eastern	Fence posts, furniture	Moderate stiffness and strength; low shrinkage; resistant to decay
Red cedar, Western	Building and construction shingles, exterior siding, boat building	Low in strength and stiffness but very resistant to decay
Spruce	Building and construction, pulpwood, furniture, aircraft construction, violin making	Moderate stiffness, high strength

†References: *Wood Handbook* (1974); *The International Book of Wood* (1976)

are more than 50 times stiffer when loaded along the grain than when loaded across it.

In this chapter we review the structure and mechanical properties of wood and show how its stiffness, strength and toughness can be explained by extensions of the ideas developed in earlier chapters. We show that the properties depend primarily, like those of all cellular solids, on the properties of the cell wall, on the relative density, and on the shape of the cells. Other factors (age and moisture content, for example) play a secondary, though still important, role.

10.2   **The structure of wood**

If a sample of wood is cut at a sufficient distance from the centre of the tree that the curvature of the growth rings can be neglected, its properties are orthotropic. It has three orthogonal planes of symmetry: the radial, the tangential and the axial (Fig. 10.1). The stiffness and strength are greatest in the axial direction, that is, parallel to the trunk of the tree; in the radial and tangential directions they are less by a factor of $1/2$–$1/20$, depending on the species.

These differences all relate to the structure of the wood. At one scale (that of millimetres), wood is a cellular solid: cells walls, often with the shape of hexagonal prisms, enclose pore space. The relative density, ρ^*/ρ_s (the density of the wood divided by that of the cell-wall material) can be as low as 0.05 for balsa, and as high as 0.80 for lignum vitae. Three features characterize the microstructure (Dinwoodie, 1981; Bodig and Jayne, 1982):

(a)   the highly elongated cells which make up the bulk of the wood, called *tracheids* in softwoods and *fibres* in hardwoods;

(b)   the *rays*, made up of radial arrays of smaller, more rectangular, parenchyma cells; and

(c)   the *sap channels*, which are enlarged cells with thin walls and large pore spaces which conduct fluids up the tree.

There are, of course, structural differences between softwoods and hardwoods. The rays in softwoods are narrow and extend only a few cells in the axial direction, while those in hardwoods are wider and extend hundreds of cells in the axial direction. In softwoods the sap channels make up less than 3% of the

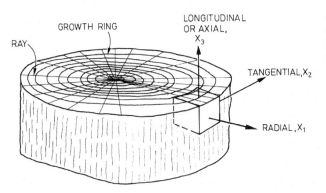

Figure 10.1   A section through the trunk of a tree showing the axial, radial and tangential directions.

wood volume while in hardwoods they can account for as much as 55%. The growth rings in softwoods are made up of alternating circumferential bands of thick- and thin-walled tracheids, while those in ring-porous hardwoods are characterized by bands of large- and small-diameter sap channels. Diffuse–porous hardwoods have a uniform distribution of sap channels of the same size; they do not exhibit any characteristic growth rings on a microscopic scale. All of these features can be seen in Fig. 10.2, which shows the structures of a number of common softwoods and hardwoods. The volume fractions and dimensions of the cells in wood are listed in Table 10.2.

At a finer scale (that of microns) wood is a fibre-reinforced composite. The cell walls are made up of fibres of crystalline cellulose embedded in a matrix of amorphous hemicellulose and lignin, rather like the glass-fibre-in-polymer composite

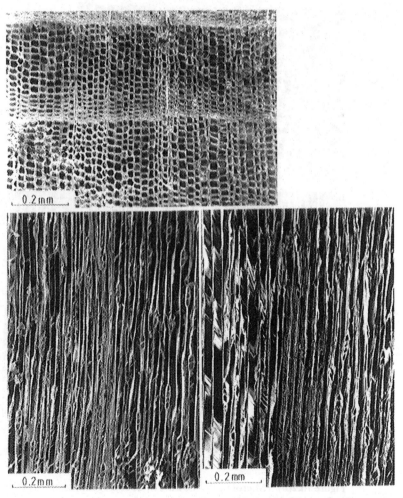

Figure 10.2(a)   Micrographs showing three orthogonal sections of cedar.

Table 10.2   Volume fractions and dimensions of wood cells†

	Softwoods		Hardwoods		
	Tracheid	Ray cells	Fibre	Vessel	Ray cells
Volume fraction (%)	85–95	5–12	37–70	6–55	10–32
Axial dimension (mm)	2.5–7.0		0.6–2.3	0.2–1.3	
Tangential dimension (μm)	25–80		10–30	20–500	
Radial dimension (μm)	17–60		10–30	20–350	
Cell-wall thickness, t (μm)	2–7		1–11	—	

†Bodig and Jayne (1982), p. 14.

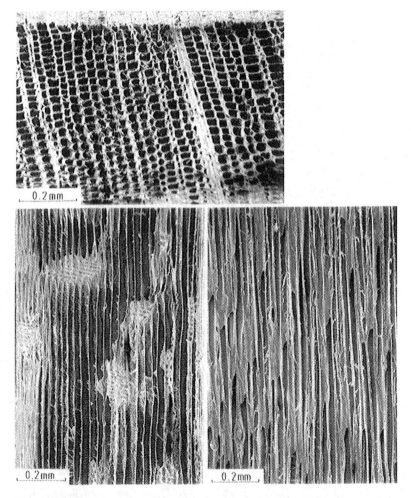

Figure 10.2(b)   Micrographs showing three orthogonal sections of Columbian pine.

used to make the hollow shaft of a fibreglass tennis racquet. The lay-up of the cellulose fibres in the wall is complicated but important because it accounts for part of the great anisotropy of wood – the difference in properties along and across the grain (Mark, 1967). It is helpful (though a simplification) to think of the cell walls as helically wound, like the shaft of the racquet, with the fibre direction nearer the cell axis than across it. This gives the cell wall a modulus and a strength which are large parallel to the axis of the cell, and smaller, by a factor of about 3, across it. But this accounts for only a part of the large anisotropy of wood. The rest is related to the cell shape: elongated cells are stiffer and stronger when loaded along the long axis of the cell than when loaded across it. Details are given in the next section.

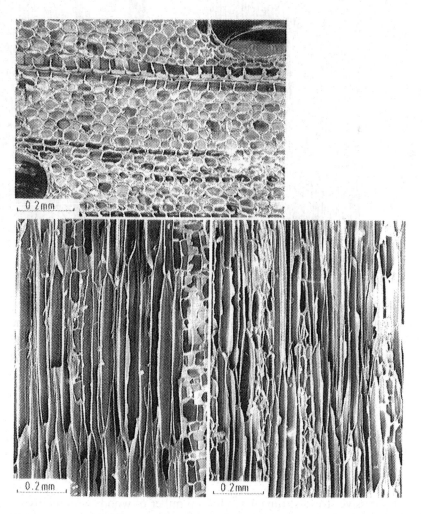

Figure 10.2(c)   Micrographs showing three orthogonal sections of balsa.

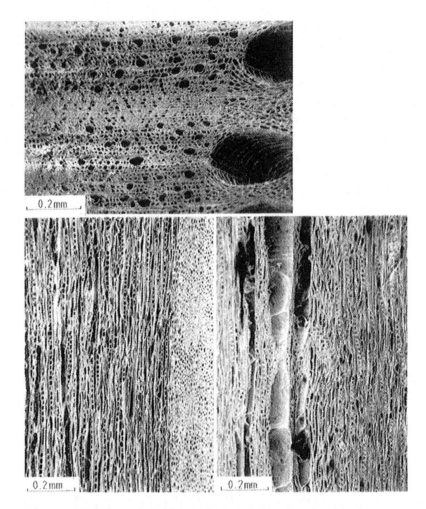

Figure 10.2(d)   Micrographs showing three orthogonal sections of oak.

Although woods differ enormously in their density and mechanical properties, the properties of the cell wall are, as a rough approximation, the same for all woods. Woods as different as balsa and beech have cell walls with a density near $1500 \, kg/m^3$ and moduli and strengths, which, according to Cave (1968, 1969) have values close to those listed in Table 10.3.

10.3   The mechanical properties of wood

(a) The stress–strain curve

Compressive stress–strain curves for a number of woods with relative densities ranging from 0.05 to 0.5, loaded in the tangential and axial directions, are

Table 10.3 Cell wall properties for wood†

Property	Literature value	Value inferred from data plots
Density, ρ_s (kg/m^3)	1500 (a)	—
Axial Young's modulus, E_s (GN/m^2)	35 (b)	35
Transverse Young's modulus (GN/m^2)	10 (b)	19
Shear modulus, from A–R loading (GN/m^2)	—	2.6
Shear modulus, from R–T loading (GN/m^2)	—	2.6
Axial yield strength, σ_{ys} (MN/m^2)	350 (c)	120
Transverse yield strength (MN/m^2)	135 (c)	50
Shear yield strength (MN/m^2)	—	30
Toughness, peeling mode, G_{cs}^p (J/m^2)	—	350
Toughness, breaking mode, G_{cs}^b (J/m^2)	—	1650
Fracture toughness, peeling mode, $(K_{IC}^p)_s$ (MN/m$^{3/2}$)	—	1.9
Fracture toughness, breaking mode, $(K_{IC}^b)_s$ (MN/m$^{3/2}$)	—	4.1

†*References*: (a) Dinwoodie (1981), p. 33; (b) Cave (1968); (c) Cave (1969).

shown in Fig. 10.3. That for loading in the radial direction is very like that for tangential loading but usually starts with a small yield drop (see, for example, Figs. 10.4). The strength of wood is affected by the age and moisture content, and depends also on the temperature and strain-rate at which testing is carried out: the data analysed in this chapter refer to well-seasoned wood, with a constant moisture content of about 12%, tested at 18 °C ± 2 °C and at a strain-rate of close to 10^{-3}/s.

The general observations are as follows. At small strains (less than about 0.02) the behaviour is linear-elastic in all three directions. Young's modulus in the axial direction is much larger than that in the tangential and radial directions, which are roughly equal. Beyond the linear-elastic regime, the stress–strain curves for loading in all three directions show a stress plateau extending to strains between 0.2 and 0.8 depending on the density of the wood. At the end of the plateau the stress rises steeply. Compression in the tangential direction (Fig. 10.3(a)) gives a smooth stress–strain curve which rises gently throughout the plateau. Compression along the radial direction is distinguished by a small yield drop at the end of the linear-elastic regime, followed by a slightly irregular or wavy stress plateau. The tangential and radial yield stresses are about equal. That in the axial direction is much higher and it is followed by a sharply serrated

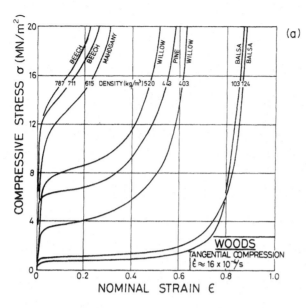

(a)

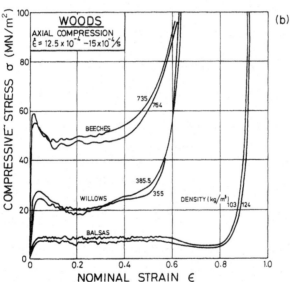

(b)

Figure 10.3 Compressive stress–strain curves for several species of wood loaded in the (a) tangential and (b) axial directions. The curves for radial loading are similar to those for tangential loading.

plateau (Fig. 10.3(b)). As the density of the wood increases the moduli and plateau strengths increase.

The compression curves of balsa wood in all three directions are compared in Fig. 10.4(a). Insight into the mechanisms of deformations is given by observing the cells directly in a deformation stage mounted in a scanning electron microscope (Easterling *et al.*, 1982). Compression in the tangential direction (Fig. 10.4(b)) causes uniform bending and collapse by plastic yielding of the cell walls, as shown in the micrographs in Figs. 10.4(b) and 10.5 and (schemati-

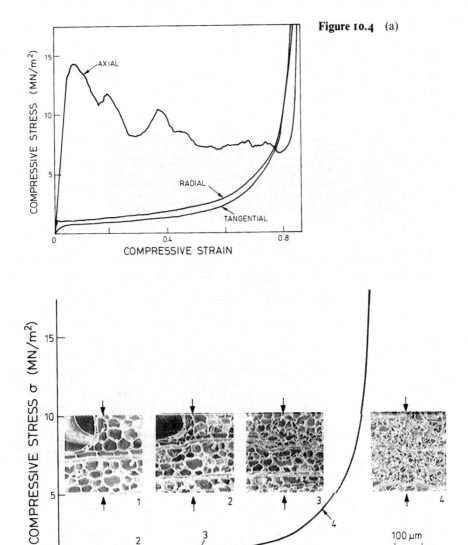

Figure 10.4 (a)

Figure 10.4 (b)

cally) in Fig. 10.6. In species of wood which have more pronounced growth rings than balsa the growth rings act as stiff reinforcement; yielding corresponds to their plastic buckling, on a macroscopic scale, as shown for the Douglas fir in Fig. 10.7.

Compression of balsa in the radial direction is shown in Fig. 10.4(c). At first the cell walls bend uniformly (as in tangential loading). Plastic collapse of the

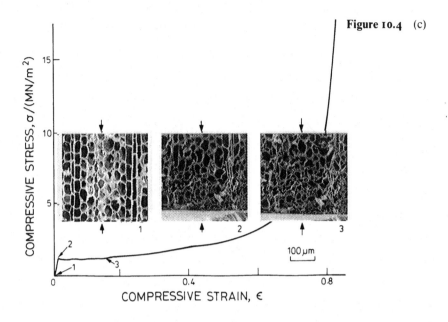

Figure 10.4   (c)

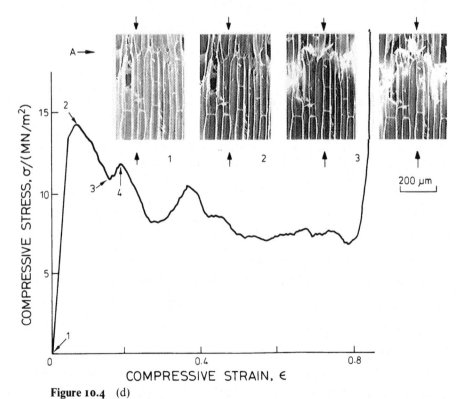

Figure 10.4   (d)

Figure 10.4   (a) Compressive stress-strain curves for balsa: (b) tangential compression, with micrographs showing the deformation of the cells; (c) radial compression, showing cell deformation; (d) axial compression showing cell deformation (from Easterling *et al.*, 1982).

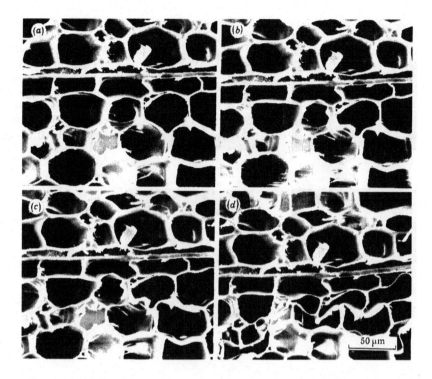

Figure 10.5 Micrographs showing the cell deformation in balsa at higher magnification. Loading is in the tangential direction.

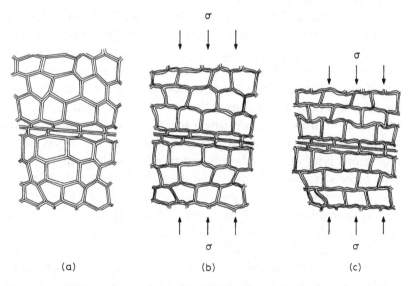

Figure 10.6 Schematic drawings of the deformation of wood cells under loading in the tangential direction: (a) no applied load. (b) cell-wall bending in the linear-elastic range. (c) uniform cell collapse by plastic yielding.

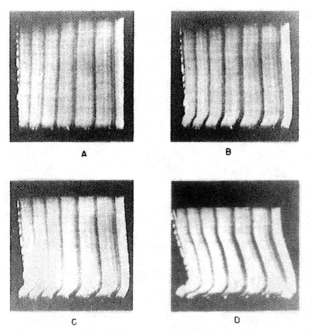

Figure 10.7 Failure in a specimen of Douglas fir loaded in compression in the tangential direction: (a) undeformed specimen, (b) first failure at 8% compression, (c) 10% compression, (d) 20% compression (from Bodig and Jayne, 1982, Fig. 7.14 courtesy of Krieger Publishing).

cells is non-uniform, starting at the surface of the loading platen and propagating inward along the length of the specimen. The same feature has been observed in other species of wood, such as the Douglas fir shown in Fig. 10.8. The deformation is shown schematically in Fig. 10.9.

The mechanism of compressive deformation in the axial direction is fundamentally different (Fig. 10.4(d)). There is little evidence of cell-wall bending in the linear-elastic regime; instead, the cells compress uniaxially. At a sufficiently high strain the cells collapse by the yielding and fracturing of planes of material where the ends of adjacent layers of cells join together (section A–A in Fig. 10.4(d)). In woods of low density the cell walls are forced between each other like the teeth of two combs meshing together. The stress drops until the next layer of cell ends is intercepted, one cell length away. The stress increases until this second layer of cell-ends fails and then drops off again as the cells mesh. This process is repeated, giving the dramatically serrated plateau, until the wood has almost completely densified at a strain of about 0.8. In woods of higher density the mechanism can be different: cells undergo local, plastic buckling, as shown for spruce in Fig. 10.10. Both fracturing and buckling of the cell wall are shown schematically in Fig. 10.11.

Data for the linear-elastic moduli and compressive and shear strengths for a number of species of wood are plotted in Figs. 10.12–10.15. The lower axis shows the density normalized by that of the cell wall ($1500 \, \text{kg/m}^3$). That on the left-hand side shows the modulus normalized by Cave's (1968) value for the

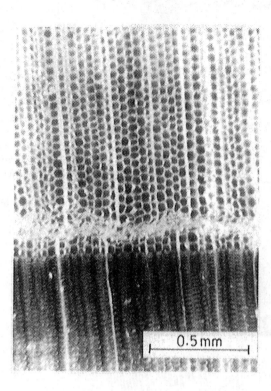

Figure 10.8 A micrograph of a specimen of Douglas fir loaded in the radial direction showing the non-uniform collapse of the cells (from Bodig and Jayne, 1982, Fig. 7.13, courtesy of Krieger Publishing).

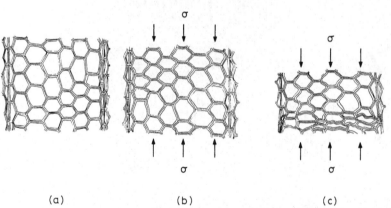

(a)   (b)   (c)

Figure 10.9 Schematic drawings of the deformation of wood cells under loading in the radial direction: (a) no applied load (b) cell-wall bending in the linear-elastic range (c) non-uniform cell collapse by plastic yielding.

axial modulus of the cell wall ($35\,\text{GN/m}^2$) or the strength normalized by Cave's (1969) value for the axial strength of the cell wall in tension ($350\,\text{MN/m}^2$). The unnormalized density and modulus or strength are shown on the remaining two sides. Data for the Poisson's ratios of wood are listed in Table 10.4.

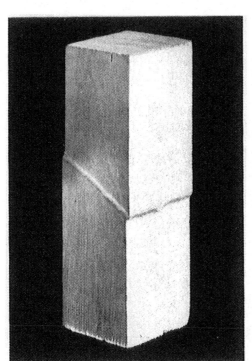

(a)

Figure 10.10
Micrographs showing the
local plastic buckling of the
cell walls in Norway spruce
loaded in compression in
the axial direction: (a) axial
collapse of a wood
specimen showing the
crease formed by the
yielding of the cell walls;
(b) collapsed layer in cells;
(c) buckled cell walls (from
Dinwoodie, 1981, Fig.
5.14, © Crown copyright,
reproduced with
permission of the Building
Research Establishment,
and Bariska and Kucera,
(1982), Figs. 1,3, courtesy
of Springer-Verlag).

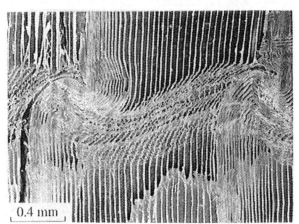

(b)

0.4 mm

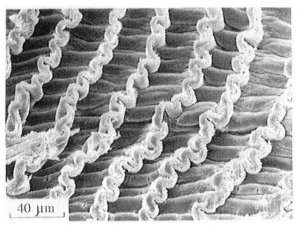

(c)

40 μm

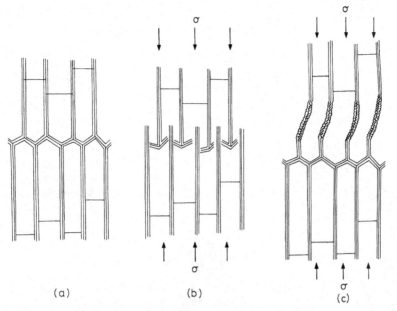

Figure 10.11 Schematic drawings of the deformation in wood cells under loading in the axial direction: (a) no applied load; (b) cell collapse by end cap fracture; (c) cell collapse by local buckling of the cell walls.

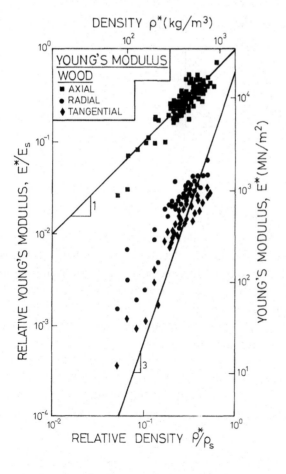

Figure 10.12 The Young's moduli of wood plotted against density. One pair of axes is normalized by the cell-wall Young's modulus in the axial direction ($35 \, GN/m^2$) and by the cell wall density ($1500 \, kg/m^3$). The other pair of axes corresponds to the raw data. Data from: Goodman and Bodig (1970); Bodig and Goodman (1973); *Wood Handbook* (1974); Dinwoodie(1981); Bodig and Jayne (1982); and Easterling *et al.* (1982).

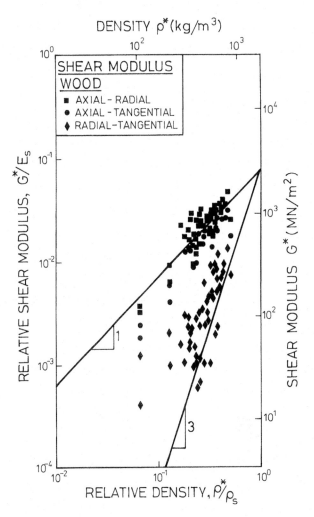

Figure 10.13 The shear moduli of wood plotted against density. One pair of axes is normalized as in Fig. 10.12. The other pair corresponds to the raw data. Data from: Goodman and Bodig (1970); Bodig and Goodman (1973); Dinwoodie (1981); and Bodig and Jayne (1982).

(b) Fracture and toughness

Traditionally, the resistance of wood to tensile fracture has been measured by the work of fracture (the area under the load–deflection curve), either in static or impact bending tests (Dinwoodie, 1981). More recently, the methods of fracture mechanics have been applied to wood (Wu, 1963; Schniewind and Pozniak, 1971; Walsh, 1971; Schniewind and Centano, 1973; Johnson, 1973; Mindess *et al.*, 1975; Williams and Birch, 1976; Jeronimidis, 1980; Barrett, 1981; Nadeau *et al.*, 1982; Ashby *et al.*, 1985; Bentur and Mindess, 1986).

In wood, eight systems of crack propagation can be identified. They are illustrated in Fig. 10.16. Each system is identified by a pair of letters, the first indicating the direction normal to the crack plane, the second describing the direction of crack propagation. The LT and LR systems correspond to crack propagation

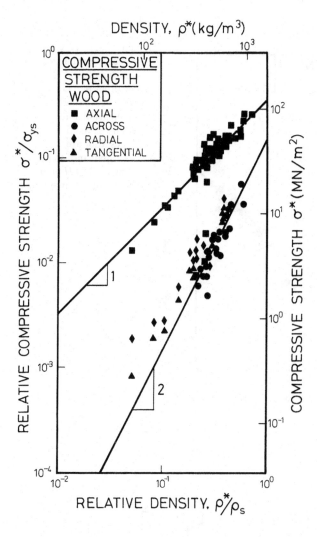

Figure 10.14 The compressive strength of wood plotted against density. One pair of axes is normalized by the yield strength of the cell wall in the axial direction ($350\,\mathrm{MN/m^2}$) and by the cell wall density ($1500\,\mathrm{kg/m^2}$). The other pair corresponds to the raw data. Data from: Goodman and Bodig (1971); *Wood Handbook* (1974); Dinwoodie (1981); and Easterling *et al.* (1982).

across the grain, the others to crack propagation along the grain. The growth rings, which contain a density gradient, introduce further loss of microscopic symmetry. The + sign means that the crack propagates away from the centre of the tree; the − sign that it propagates towards the centre.

Wood is not, strictly speaking, a linear-elastic solid. Even at room temperature, and loaded rapidly, it is viscoelastic. Because of this, the standard tests of linear-elastic fracture mechanics do not always give consistent results. Notched three-point bend and double edge-notch specimens, for instance, give values of fracture toughness that differ, for some woods, by up to 30% (Ashby *et al.*, 1985). None the less, the results of a given test are broadly reproducible and the conditions for valid fracture toughness tests are approached nearly enough for the data to be useful for mechanical design.

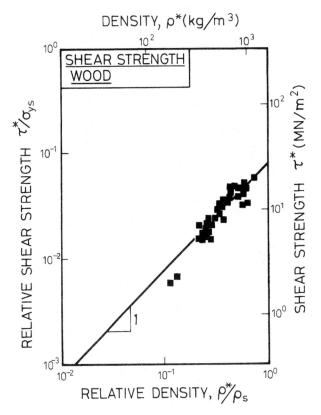

Figure 10.15 The shear strength of wood plotted against density. One pair of axes is normalized in the same way as in Fig. 10.14. The other pair correspond to the raw data. Data are for shearing in the axial–radial and axial–tangential planes; they are from: Goodman and Bodig (1971); *Wood Handbook* (1974); and Dinwoodie (1981).

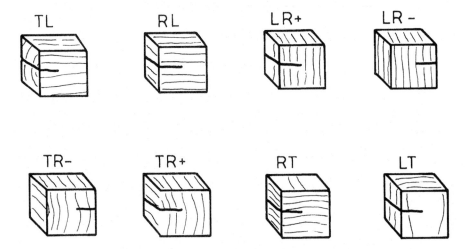

Figure 10.16 The eight modes of crack propagation in wood. The distinction between the + and − directions arises because of the asymmetric structure of the growth rings.

Table 10.4 Poisson's ratios for wood

Species	Density (kg/m^3)	v^*_{RT}	v^*_{RA}	v^*_{TR}	v^*_{TA}	v^*_{AR}	v^*_{AT}	Reference
Hardwoods								
Balsa	200	0.66	0.02	0.24	0.01	0.23	0.49	(a)
Aspen	300	—	—	0.50	—	0.49	0.37	(b)
Yellow poplar	380	0.70	0.03	0.33	0.02	0.32	0.39	(c)
Khaya	440	0.60	0.03	0.26	0.03	0.30	0.64	(a)
Sweetgum	530	0.68	0.04	0.30	0.02	0.33	0.40	(c)
Oak	580	—	—	0.26	—	0.29	0.48	(b)
Walnut	590	0.72	0.05	0.37	0.04	0.49	0.63	(a)
Birch	620	0.78	0.03	0.38	0.02	0.49	0.43	(a)
Ash	670	0.71	0.05	0.36	0.03	0.46	0.51	(a)
Beech	750	0.75	0.07	0.36	0.04	0.45	0.51	(a)
Softwoods								
Engleman spruce	350	—	—	0.22	—	0.44	0.50	(b)
Norway spruce	390	0.51	0.03	0.31	0.03	0.38	0.51	(a)
Sitka spruce	390	0.43	0.03	0.25	0.02	0.37	0.47	(a)
Scotch pine	550	0.68	0.04	0.31	0.02	0.42	0.51	(a)
Douglas fir	430	—	—	0.56	—	0.28	0.50	(b)
Douglas fir	590	0.63	0.03	0.40	0.02	0.43	0.37	(a)

†References: (a) Dinwoodie (1981); (b) Goodman and Bodig (1970); (c) *Wood Handbook* (1974).

Data for the fracture toughness of wood are plotted against relative density in Fig. 10.17. Values for K^*_{IC} for crack propagation normal to the grain (K^{*n}_{IC}) are roughly 10 times greater than those for crack propagation along the grain (K^{*a}_{IC}), regardless of the relative density of the wood. Some researchers detected significant differences between the different directions which lie parallel to the grain – the TL or RL orientations, for instance (Schniewind and Pozniak, 1971; Williams and Birch, 1976) – but these are small (Schniewind and Centano, 1973; Ashby *et al.*, 1985), and largely masked by the sample-to-sample variations inevitable in a natural material like wood. Large changes in temperature (−195–20 °C), in moisture content (1–72%) and in the rate of loading (10^{-4} to 10^2 mm/s) have a significant effect on the modulus, strength and fracture toughness of woods (Jeronimidis, 1980). The small differences between the conditions of the tests of Fig. 10.17 (18–22 °C; 7–12% moisture content; and loading rates around 10^{-2} mm/s) do not introduce significant variation

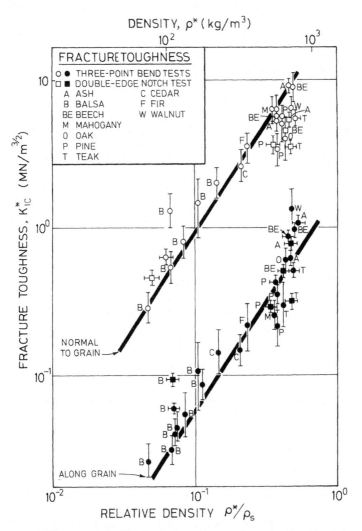

Figure 10.17   The fracture toughness of wood plotted against relative density. Data from: Wu (1963); Schniewind and Pozniak (1971); Walsh (1971); Johnson (1973); Schniewind and Centano (1973); Williams and Birch (1976); Jeronimidis (1980); and Ashby *et al.* (1985).

into the data (Dinwoodie, 1981; Silvester, 1967). The obvious and important implication of Fig. 10.17 is that, for wood in a standard state, the fracture toughness depends principally on its density.

Crack propagation through wood can be studied by loading suitable specimens in the scanning electron microscope. The observations from an extensive study of knot-free samples of six different woods, loaded in tension normal to the crack plane, were as follows (Ashby *et al.*, 1985).

(a) The crack is stable until a critical load is reached. As this load is approached the crack first advances stably by one or a few cell diameters and then becomes unstable and propagates rapidly over many hundred cell diameters. The initial crack extension is almost always parallel to the grain, even when the starter crack lies across the grain (i.e. in the LT or LR orientations).

(b) Crack advance in low-density woods ($\rho^*/\rho_s < 0.2$) and in the thin-walled early wood of higher-density woods is commonly by *cell-wall breaking* (Koran, 1966; Bodig and Jayne, 1982; Ashby *et al.*, 1985). It is shown in the micrograph and schematic drawing of Fig. 10.18. When cracks propagate in the RT orientation, the crack advance is almost entirely by cell-

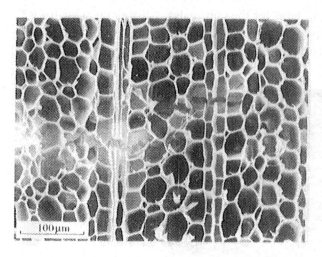

(a)

Figure 10.18 (a) A micrograph showing crack advance by cell-wall breaking (balsa, RL loading): (b) a schematic drawing of crack advance by cell-wall breaking (from Ashby *et al.*, 1985).

100 µm

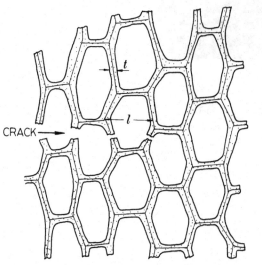

CRACK →

(b)

wall breaking (Bodig and Jayne, 1982; Ashby *et al.*, 1985). Johnson (1973) reports the same thing for the RL orientation.

(c)   Crack advance in woods of higher density ($\rho^*/\rho_s > 0.2$) involves both *cell-wall breaking* and *cell-wall peeling*: the pulling apart of two halves of the cell wall which debond along the central lamella (Bodig and Jayne, 1982; Ashby *et al.* 1985). It is shown in the micrograph and schematic drawing of Fig. 10.19. When cracks propagate in the TR orientation, cell-wall peeling predominates. A crack running at an angle between the RT and TR orientations tends to deviate towards TR and adopt a peeling mode, suggesting that the toughness in this mode is lower than that for cell-wall breaking. Cracks in the TR orientation seek rays and propagate along them (Koran, 1966; Johnson, 1973; Schniewind and Pozniak, 1971). When they do, they advance mainly by cell-wall peeling even in low-density woods. Figures 10.20(a) and 10.20(b) show, for the same balsa, cell-wall breaking for RT propagation and cell-wall peeling (along a ray) for TR propagation. Figures 10.20(c) and 10.20(d) show the same two fracture mechanisms in a softwood.

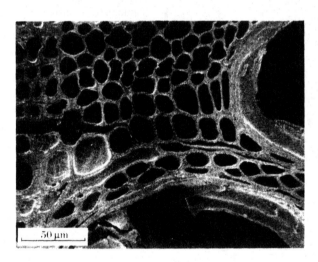

Figure 10.19   (a) A micrograph showing crack advance by cell-wall peeling (ash, TR loading): (b) a schematic drawing of crack advance by cell-wall peeling (from Ashby *et al.*, 1985).

50 μm

(a)

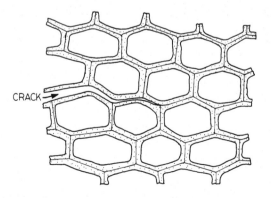

CRACK→

(b)

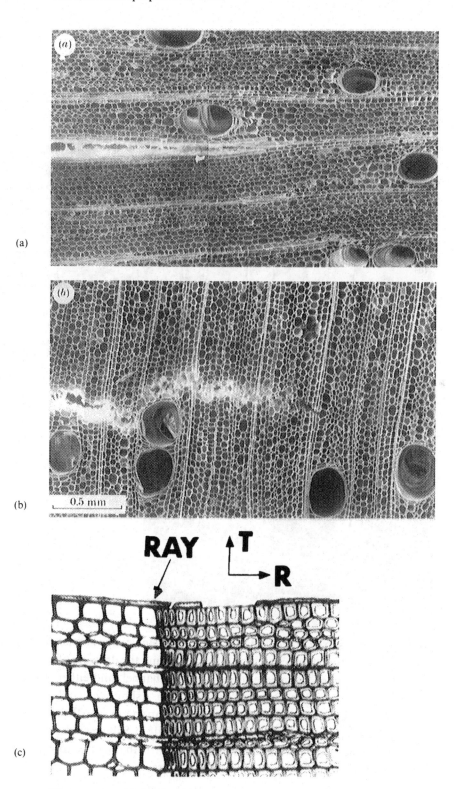

Figure 10.20 (a)–(c) *Caption on p. 412*

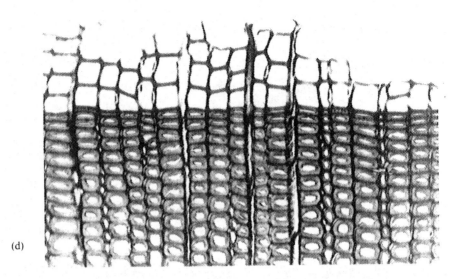

(d)

Figure 10.20   Crack propagation in balsa: (a) cell-wall peeling in TR mode, (b) cell wall breaking in RT mode. Crack propagation in a softwood: (c) cell-wall peeling in TR mode and (d) cell-wall breaking in RT mode (from Ashby *et al.*, 1985 and Koran, 1966, courtesy of the Pulp and Paper Research Institute of Canada).

Figure 10.21   (a)

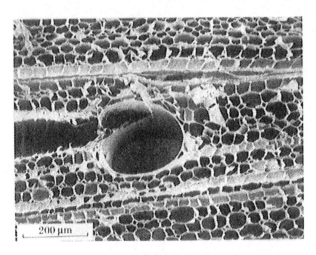

200 µm

(d)   Clusters of sap channels and, less commonly, single sap channels, can act as crack arrestors as shown in the micrograph and the drawing of Fig. 10.21. The crack tends to deviate towards a sap channel, and either enter it or run partly around its periphery and then stop. Ashby *et al.* (1985) observed that a crack often jumped from one layer of sap channels to the next, arresting at each layer, confirming an inference made by Schniewind and Pozniak (1971) when they saw discontinuous crack growth on a macroscopic scale in their wood specimens loaded in the TR direction.

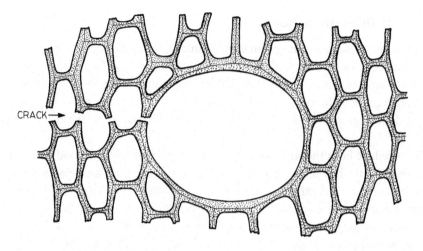

Figure 10.21 (b)

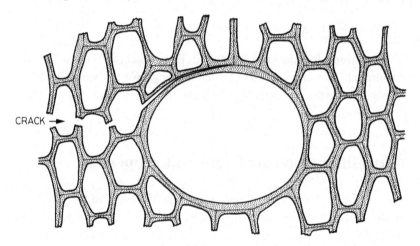

Figure 10.21 (c)

Figure 10.21 Crack arrest at a sap channel: (a) a micrograph showing a crack breaking into a sap channel, (b) a schematic drawing of a crack breaking into a sap channel, and (c) a schematic drawing of a crack splitting the wall of a sap channel. Loading in the TR direction (from Ashby *et al.*, 1985).

Additional observations of the mechanisms of cell-wall breaking for loading across the grain (LT and LR systems of crack propagation) have been made by Jeronimidis (1980). He finds that the cell-wall fracture is associated with 'pseudo-buckling' caused by the shear stresses induced in a helically wound fibre composite loaded in axial tension.

(c) Influence of knots

It is easy, in focusing on the cellular structure of wood, to forget that it may contain defects on a larger scale which affect its properties. Any wood sample with a volume of more than a few cubic centimetres contains knots where branches were accommodated into the trunk of the tree. The grain around a knot is distorted, and the knot itself may be poorly bonded to the rest of the wood.

Knots reduce both the stiffness and the strength of wood (Dinwoodie, 1981). More important, they reduce the fracture strength. The knot and the distorted grain around it are a centre of weakness, and can behave like an incipient crack so that, when the wood is loaded, failure starts from the knot. The behaviour is sometimes characterized in terms of the knot-area ratio R_k (defined as the total area of knots divided by that of the cross-section); then, very roughly, the fracture strength σ_f^* falls as:

$$\sigma_f^* = \sigma_f^{*0}(1 - 1.2R_k)$$

where σ_f^{*0} is the strength of the knot-free wood (Dinwoodie, 1981). But this is not the best approach. A more complete understanding awaits the application of fracture mechanics to the problem. We shall not attempt it here.

10.4 Modelling wood structure and properties

The cellular structure of wood is shown, somewhat idealized, in Fig. 10.22. The *tracheids* and *fibres* which make up the bulk of the cells in softwoods and hardwoods, respectively, can be thought of as a regular array of long, hexagonal prisms with occasional transverse membranes. They are traversed by *rays*: radial bands of shorter, more rectangular cells. Circular *sap channels* run up the axis of the tree. Table 10.2 lists the volume fractions and dimensions for each type of cell. We seek to model the mechanical behaviour of this structure. The problem is further complicated by the differences between softwoods, ring–porous hardwoods and diffuse–porous hardwoods. The rays and growth rings of softwoods and hardwoods differ: in softwoods the rays are narrow and only a few cells long, while in hardwoods they are much wider and longer; and the growth rings in softwoods are made up of alternating circumferential bands of thick- and thin-walled tracheids, while those in ring-porous hardwoods have bands of large- and small-diameter sap channels, or vessels. The distribution of sap channels in diffuse–porous hardwoods is almost uniform, so that they do not have obvious growth rings. In devising mechanical models for wood these differences are introduced only in the most approximate way.

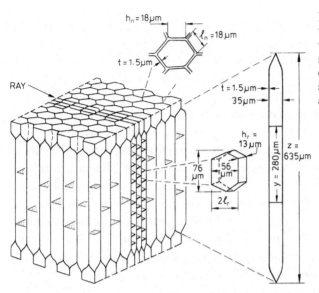

Figure 10.22 Model for the cellular structure of wood. The cell dimensions shown are for balsa; typical cell dimensions for softwoods and hardwoods are given in Table 10.2.

(a) Mechanical models for wood structure

The modelling of wood properties starts with the modelling of the cell wall. Its structure has been described already: it consists of strong fibres of cellulose, laid up in stacked plies at angles between 5° and 60° to the axial direction and embedded in a matrix which is predominantly hemicellulose and lignin. Its resemblance to a fibre-reinforced laminate has prompted attempts to model its properties using composite theory, among which that of Mark (1967) is the most complete and successful. Starting with properties of the fibres and the matrix, Mark was able to calculate the stiffness and strength of a single ply, and to combine the properties of plies, laid up at angles which simulated those in the cell wall, to give the complete stiffness tensor and the axial and transverse strengths, of the cell wall.

There are, inevitably, many points in a calculation of this sort at which approximations or assumptions must be made. Its value lies in the convincing way it demonstrates that the cell wall acts as a fibre-reinforced composite, and has properties that can be understood in simple physical terms; but the precision of its predictions depends in a sensitive way on the details of the fibre lay-up, and these are not adequately characterized. We learn from it that the cell wall itself is anisotropic, with a lay-up that makes it stiffest and strongest in the axial direction. To progress further, we use experimentally measured values for the stiffness and strengths, provided by the direct measurements of Cave (1968, 1969), given in Table 10.3. We find that the analysis of wood properties is not entirely consistent with these values, and arrive at a slightly different set, justified in the following sections.

With this starting point we can develop models for the stiffness and strength of wood itself. Attempts to model the elastic moduli of wood by analysing the deformation of an idealized cell date back at least 60 years. Price (1928), in a remarkably detailed paper, modelled the elongated wood cells as a parallel array of cylindrical tubes, like a box full of drinking straws. Loaded axially the tubes are extended or compressed; loaded transversely they distort from a circular to an oval section predominantly by bending. Price analysed the distortion of a single tube, and found that the transverse Young's moduli should vary as the cube of the density while the axial Young's modulus should vary linearly with density – precisely the same as the results for a hexagonal honeycomb derived in Chapter 4. Recognizing that the ray cells, having a higher density than the tracheids, add stiffness in the radial direction, Price suggested that the radial Young's modulus should be somewhat greater than the tangential one. He also calculated Poisson's ratios for each loading direction: as discussed below, his results are in broad agreement with observations. The model was developed further by Srinavasan (1942), who added a second set of tubes, at right angles to the first, to simulate the rays.

The tracheid and fibre cells which make up the bulk of most woods are more nearly hexagonal prisms than cylinders. In the rest of this chapter we will analyse the properties of woods, thinking of the cells as a parallel array of close-packed hexagonal prisms, stiffened (when appropriate) by rays, and caps and transverse membranes. This seems the most realistic choice of geometry, and it allows us to draw extensively on the results already developed in Chapter 4. The results, in most cases, are almost identical with those of Price's tube model (though the analysis for the hexagons is simpler); the results will be compared where appropriate.

To progress further, we require a relationship between the cell wall thickness, t, the edge-length, l, and the relative density. For simple hexagonal prisms the result is (Chapter 2, Eqn. (2.14b)):

$$\frac{\rho^*}{\rho_s} = \frac{2}{\sqrt{3}} \frac{t}{l} \left\{ 1 - \frac{t}{2\sqrt{3}l} \right\}. \qquad (10.1)$$

The first term in the brackets describes the contribution of the cell walls; the second accounts for the corners. For low-density woods the equation is adequately approximated by:

$$\frac{\rho^*}{\rho_s} = \frac{2}{\sqrt{3}} \frac{t}{l}. \qquad (10.2)$$

We can now proceed to the calculation of wood properties, using the results of Chapter 4. The complexity of the real structure – the rays, the end caps on the hexagonal cells, the transverse membranes and so forth – means that it is not always possible to calculate the absolute magnitude of the property with accuracy. But the analysis gives the value approximately, and (more important) it reveals the way in which the property depends on the density of the wood and on the direction of loading, giving insight into the magnitude and origin of anisotropy in wood. A simple calibration procedure then gives approximate analytical equations for wood properties.

(b) Linear–elastic moduli

When wood is compressed in the tangential direction the cell walls bend (Figs. 10.4(b), (10.5) and (10.6)). Thinking of the cells as a two-dimensional hexagonal array or honeycomb, the modulus in the tangential direction, E_T^*, is (Chapter 4, Eqn. (4.7) with Eqn. (10.2)):

$$\frac{E_1^*}{E_s} \propto \left(\frac{t}{l}\right)^3 = C_1 \left(\frac{\rho^*}{\rho_s}\right)^3 . \tag{10.3}$$

Here, E_s is the Young's modulus of the cell wall in the axial direction ($35\,\mathrm{GN/m^2}$; Table 10.3). The anisotropy of the cell wall itself introduces complications, which we deal with by normalizing in every equation by the axial property of the cell wall (as here), and incorporating the difference between the axial and transverse cell-wall property in the constant C_1. The values of these constants (which measure the ratio of axial to transverse cell-wall properties) can be found from the data plots of Figs. 10.12 to 10.15. Values are given below, and in Table 10.5.

The rays, the end caps of the cells and the transverse membrane which subdivide the cells all tend to increase the tangential stiffness. The rays act in two ways: they are denser, and so stiffer, than the bulk of the cells in the tree; and, because of their rectangular cross-section, they probably constrain the lateral spreading of the adjacent hexagonal cells and thus stiffen them. The transverse walls and end caps stretch laterally when the wood is compressed; because their in-plane tensile stiffness is high, they, too, constrain the bending deformation of the neighbouring portion of the cell. This in-plane deformation of the cell faces gives an additional contribution to the overall stiffness which varies linearly with density (Chapter 5, Eqn. (5.11)). It is perhaps for this reason that the data plotted in Fig. 10.12 lie somewhat above the line, a plot of Eqn. (10.3), with a slope of 3. The transverse cell-wall modulus can be found from the intercept at $\rho^*/\rho_s = 1$; it is $19\,\mathrm{GN/m^2}$, rather larger than the value obtained by Cave (1968). C_1 is equal to 0.54.

At first sight one might expect the moduli in the radial and tangential directions to be equal. But the rays, which merely constrain lateral spreading in tangential loading, act as reinforcing plates when the loading is radial (Figs 10.4(c), (10.8) and (10.9)). The rays have a higher density, and therefore modulus, than the rest of the wood – in balsa they are nearly twice as dense. Then using a simple weighted average, the radial modulus, E_R^*, is:

Table 10.5 Approximate equations for wood properties in terms of density

Property	Direction of loading		
	Tangential	Radial	Axial
Young's modulus	$\dfrac{E_T^*}{E_s} = C_1\left(\dfrac{\rho^*}{\rho_s}\right)^3 = 0.54\left(\dfrac{\rho^*}{\rho_s}\right)^3$	$\dfrac{E_R^*}{E_s} = 1.5C_1\left(\dfrac{\rho^*}{\rho_s}\right)^3 = 0.8\left(\dfrac{\rho^*}{\rho_s}\right)^3$	$\dfrac{E_A^*}{E_s} = C_2\left(\dfrac{\rho^*}{\rho_s}\right) = \left(\dfrac{\rho^*}{\rho_s}\right)$
Shear modulus		$\dfrac{G_{RT}^*}{E_s} = C_3\left(\dfrac{\rho^*}{\rho_s}\right)^3 = 0.074\left(\dfrac{\rho^*}{\rho_s}\right)^3$	$\dfrac{G_{AR}^*}{E_s} = \dfrac{G_{AT}^*}{E_s} = C_4\left(\dfrac{\rho^*}{\rho_s}\right) = 0.074\left(\dfrac{\rho^*}{\rho_s}\right)$
Poisson's ratio	$\nu_{TR}^* = 1$ $\nu_{TA}^* = 0$	$\nu_{RT}^* = 1$ $\nu_{RA}^* = 0$	$\nu_{AR}^* = \nu_s$ $\nu_{AT}^* = \nu_s$
Crushing strength	$\dfrac{\sigma_T^*}{\sigma_{ys}} = C_5\left(\dfrac{\rho^*}{\rho_s}\right)^2 = 0.14\left(\dfrac{\rho^*}{\rho_s}\right)^2$	$\dfrac{\sigma_R^*}{\sigma_{ys}} = 1.4C_5\left(\dfrac{\rho^*}{\rho_s}\right)^2 = 0.20\left(\dfrac{\rho^*}{\rho_s}\right)^2$	$\dfrac{\sigma_A^*}{\sigma_{ys}} = C_6\left(\dfrac{\rho^*}{\rho_s}\right) = 0.34\left(\dfrac{\rho^*}{\rho_s}\right)$
Shear strength		$\dfrac{\tau_{RT}^*}{\sigma_{ys}} = C_7\left(\dfrac{\rho^*}{\rho_s}\right)^2$	$\dfrac{\tau_{AR}^*}{\sigma_{ys}} = \dfrac{\tau_{AT}^*}{\sigma_{ys}} = C_8\left(\dfrac{\rho^*}{\rho_s}\right) = 0.086\left(\dfrac{\rho^*}{\rho_s}\right)$
Fracture toughness $(\mathrm{MN/m^{3/2}})$		$K_{IC}^{*a} = 1.8\left(\dfrac{\rho^*}{\rho_s}\right)^{3/2}$	$K_{IC}^{*n} = 20\left(\dfrac{\rho^*}{\rho_s}\right)^{3/2}$

The equations are described in the text of Section 10.4. The normalization, throughout, has used the following data for cell-wall properties: ρ_s = cell-wall density = $1500\,\mathrm{kg/m^3}$; E_s = axial cell-wall Young's modulus = $35\,\mathrm{GN/m^2}$; σ_{ys} = axial cell-wall yield strength = $350\,\mathrm{MN/m^2}$.

$$E_R^* = V_R R^3 E_T^* + (1 - V_R)E_T^* \qquad (10.4)$$

where V_R, the volume fraction of ray cells, is about 10% in softwoods and 20% in hardwoods, and R, the density of the ray cells relative to that of the tracheids or fibres, is, typically, between 1.1 and 2. For balsa, $V_R = 0.14$ and $R = 2$, giving $E_R^* \approx 2E_T^*$. For higher densities R approaches 1 and E_R^* becomes almost equal to E_T^*, so typically:

$$E_R^* \approx 1.5E_T^*. \qquad (10.5)$$

To summarize, the radial modulus has roughly the same dependence on density as the tangential modulus, but is usually a little larger, the difference increasing with decreasing density.

The axial modulus of wood, E_A^*, is much larger than either of the other two (Figs. 10.3 and 10.12). This is because the cell walls are loaded axially, instead of in bending. It is like the loading of a honeycomb in the X_3 direction, analysed in Chapter 4. The modulus is simply (Eqn. (4.91)):

$$\frac{E_A^*}{E_s} = C_2\left(\frac{\rho^*}{\rho_s}\right) \qquad (10.6)$$

where E_s is again the axial modulus of the cell wall and $C_2 \approx 1$. It is identical with the result given by Price (1928).

The upper full line on Fig. 10.12 shows the axial modulus as predicted by Eqn. (10.6). Agreement is good. The figure shows that the axial modulus is always larger than the other two, partly because the microfibrils of cellulose in the cell wall lie more nearly along that direction, making the cell wall stiffer against axial deformation than against bending, and partly because hexagonal prismatic cells are intrinsically stiff along the prism axis for the reasons analysed in Chapter 4. When cell-wall bending determines the modulus (with small contributions from axial stiffening of the transverse membranes in the cells) Young's modulus varies roughly as $(\rho^*/\rho_s)^3$. But when uniaxial extension or compression determines the modulus, it varies as ρ^*/ρ_s. As a result, the anisotropy in the Young's moduli of woods varies in a systematic way with density, decreasing from almost 100 for low-density balsa to about 2 for very dense woods.

The shear moduli of wood can also be estimated using the ideas of Chapter 4. In-plane shearing in the tangential–radial plane bends the cell walls; following the argument given in Chapter 4 we find (Eqn. (4.18) with Eqn. (10.2)):

$$\frac{G_{RT}^*}{E_s} = C_3\left(\frac{\rho^*}{\rho_s}\right)^3 \qquad (10.7)$$

where C_3 is the ratio of the shear modulus of the cell wall to its axial Young's modulus, E_s. The rays and transverse membranes again constrain the deformation. It is difficult to estimate their contribution to the shear modulus, but the data plotted in Fig. 10.13 suggest it is similar to that for the Young's moduli. Like E_T^* and E_R^*, the data for G_{RT}^* lie somewhat above the line of slope 3. The cell-wall modulus relevant to in-plane shearing is given by the intercept at $\rho^*/\rho_s = 1$; it is $2.6\,\mathrm{GN/m^2}$, giving a value of $C_3 = 0.074$.

For shear loading in the axial–tangential or axial–radial planes, the result (Eqn. (4.100) with (10.2)) is:

$$\frac{G^*_{AT}}{E_s} = \frac{G^*_{AR}}{E_s} = C_4\left(\frac{\rho^*}{\rho_s}\right) \tag{10.8}$$

where, as before, E_s is the Young's modulus of the cell wall in the axial direction and C_4 is a factor to adjust for the relevant modulus for out-of-plane shearing. The rays and transverse cell membranes have little effect on these shear moduli, since they offer no resistance to this particular deformation. Data for the axial–tangential and axial–radial shear moduli of a number of woods are plotted in Fig. 10.13. The two shear moduli are roughly equal, and they vary about linearly with density in agreement with Eqn. (10.8). The relevant out-of-plane cell-wall modulus is $2.6\,\mathrm{GN/m^2}$, giving $C_4 = 0.074$.

Poisson's ratios, too, can be estimated using the ideas from Chapter 4. We define v_{ij} to be the negative ratio of strain in the j direction to that in the i direction, when loading is applied in the i direction. An ideal array of regular hexagonal–prismatic cells has the following Poisson's ratios:

$$
\begin{aligned}
v^*_{AR} &= v_s & v^*_{TR} &= 1 & v^*_{TA} &= 0 \\
v^*_{AT} &= v_s & v^*_{RT} &= 1 & v^*_{RA} &= 0.
\end{aligned}
\tag{10.9}
$$

Poisson's ratio for any pair of directions is independent of relative density. Similar results were derived by Price (1928) from his model of parallel tubes.

Data for the Poisson's ratios for woods are listed in Table 10.4. They do not match the ideal values exactly, but some general observations can be made. The axial strain resulting from loading in either the tangential or radial direction is roughly 0.03 times that in the loading direction, in good agreement with the predicted value of 0. The in-plane Poisson's ratios, v^*_{RT} and v^*_{TR}, are about 0.65 and 0.35 (instead of 1), probably because of the constraining effect of the rays and transverse cell membranes. As Price observed, this constraint is more significant for loading in the tangential than in the radial direction; the data seem to bear out this view. Loading in the axial direction produces roughly equal Poisson's ratios of about 0.4 in the tangential and radial directions, close to the expected value for the cell-wall material itself.

(c) Compressive strength

The plateau of the stress–strain curves of Fig. 10.3 corresponds to the progressive crushing of the wood. The mechanism by which this happens depends on the direction of loading: in the tangential direction it is by the plastic bending of the cell walls (Fig. 10.4(b)), or by plastic buckling preceded by plastic bending (Fig. 10.7). Plastic collapse occurs at the stress (Eqn. (4.28)):

$$\frac{\sigma_T^*}{\sigma_{ys}} \propto \left(\frac{t}{l}\right)^2 \tag{10.10}$$

where σ_{ys} is the yield strength of the cell wall. Cave (1969) reports that the axial strength of the cell wall is greater than its transverse strength, because the cellulose fibres are oriented more nearly towards the axial direction. To avoid confusion we normalize throughout by the axial cell-wall strength in tension, using Cave's value of $\sigma_{ys} = 350\,\mathrm{MN/m^2}$. Replacing t/l by ρ^*/ρ_s (Eqn. (10.2)) gives:

$$\frac{\sigma_T^*}{\sigma_{ys}} = C_5 \left(\frac{\rho^*}{\rho_s}\right)^2 \tag{10.11}$$

where C_5 is a constant which measures the ratio of the transverse to the axial strength of the cell wall.

The stress–strain curve for compression in the radial direction resembles that for tangential compression, but it starts with a yield drop (Fig. 10.4(c)). Microscopic examination shows that the deformation is non-uniform and occurs by the progressive collapse of cells from the surface inwards, although the unit step is still the plastic collapse of a cell. For this direction of loading the rays act as thin, plate-like reinforcement (Figs. 10.8 and 10.9); being denser than the rest of the wood, they are also stronger. Proceeding as with the modulus to form the weighted mean of the strengths of the fibre and ray cells, and using Eqn. (10.11), gives:

$$\sigma_R^* = V_R R^2 \sigma_T^* + (1 - V_R)\sigma_T^* \tag{10.12}$$

where V_R is the volume fraction of rays and R is the ratio of the density of the ray to that of the rest of the wood (as before). Using data for balsa ($V_R = 0.14$, $R = 2$) gives $\sigma_R^* = 1.4\sigma_T^*$, but the difference decreases as the density increases, and for most woods it is small. The radial strength, then, is a little larger than that in the tangential direction, and both vary approximately as the second power of the density.

Axial collapse is the most complicated. The micrographs show that it can occur in two ways. In low-density wood the pyramidal end caps of the fibres can bend, yield and then fracture (Fig. 10.4(d)). In woods of higher density the long, axial cell walls yield and then undergo local plastic buckling (Fig. 10.10). The first mechanism is analysed by equating the work done in moving the applied load through a displacement, δ, to the plastic work done in the corresponding deformation of the faces of the end cap. The result is given by Easterling et al. (1982): collapse and crushing occur when

$$\frac{\sigma_A^*}{\sigma_{ys}} = C_6 \left(\frac{\rho^*}{\rho_s}\right). \tag{10.13}$$

It has been suggested (Grossman and Wold, 1971) that axial crushing is triggered by elastic buckling of the cell walls. But the evidence suggests that the buckling shown in Fig. 10.10 is preceded by simple plastic yielding (Dinwoodie, 1981). Kink bands first form in tracheids or fibres where the cells are displaced tangentially to accommodate rays, at stresses below the ultimate compressive strength; their number and width increase with stress. Dinwoodie notes that the kink bands do not lie on a plane normal to the compression axis (thought of as vertical); instead, the kink plane is horizontal when viewed in the radial direction and at an angle of between 45° and 60° to the vertical when viewed in the tangential direction, giving the characteristic 'crease' which forms in specimens compressed in the axial direction (Fig. 10.10(a)). At the maximum compressive stress, yielding has occurred throughout the plane of the crease; additional deformation rotates the cells, triggering the buckling of the cell walls shown in Figs. 10.10(b) and 10.10(c).

The picture, then, is that the complicated mechanisms of axial crushing of wood are all preceded, and triggered, by axial plastic yielding in compression. When the cell walls are very thin this may involve the elaborate folding of the cell walls described in Section 4.5; but when they are as thick as those of most woods, yield is by simple axial compression. Then (Eqn. (4.81))

$$\frac{\sigma_A^*}{\sigma_{ys}} = C_6\left(\frac{\rho^*}{\rho_s}\right) \qquad (10.14)$$

for all mechanisms.

Data for the crushing strengths of woods in the tangential, radial and axial directions are plotted against density, on log scales, in Fig. 10.14. The full lines are the strengths predicted by Eqns. (10.11) and (10.14). As with the moduli, the strength in the axial direction is always greater than that in the other two. The difference is again largely due to the elongated shape of the cells, which requires that, under axial loading, the cell wall is compressed, but under tangential or radial loading it bends. This is why the axial strength varies as ρ^*/ρ_s while the other two vary as $(\rho^*/\rho_s)^2$; and it is why the anisotropy in the crushing strength decreases from about 10 for low-density balsa to about 2 for very dense woods. The cell-wall yield strenths in transverse and axial compression, found from the intercepts at $\rho^*/\rho_s = 1$, are 50 and 120 MN/m² , giving $C_5 = 0.14$ and $C_6 = 0.34$. These are roughly 1/3 the strengths given by Cave (1969).[†]

The shear failure of wood is analysed in a similar way. Shear in the plane normal to the axial direction (the 'radial–tangential' plane) requires plastic bending of the cell walls themselves. Then (Section 4.3, Eqn. (4.29))

[†]The difference between Cave's measurement of axial strength (350 MN/m²) and the one we infer from Fig. 10.14 (120 MN/m²) probably arises from the fact that Cave measured a tensile strength, whereas the value we infer is for a compressive strength. It is known that composites are usually less strong in compression.

$$\frac{\tau_{TR}^*}{\sigma_{ys}} \propto \left(\frac{t}{l}\right)^2 = C_7 \left(\frac{\rho^*}{\rho_s}\right)^2 \tag{10.15}$$

where σ_{ys} is the axial yield strength of the cell wall and C_7 is a constant which measures the ratio of the strength of the cell wall in bending to its axial strength.

Shear in a plane containing the axial direction (the 'axial–radial' and the 'axial–tangential' planes in wood terminology) loads the cell walls in shear. Plastic collapse in this mode of loading was analysed in Section 4.5, with the result:

$$\frac{\tau_{AT}^*}{\sigma_{ys}} = \frac{\tau_{AR}^*}{\sigma_{ys}} \propto \left(\frac{t}{l}\right) = C_8 \left(\frac{\rho^*}{\rho_s}\right) \tag{10.16}$$

where C_8 measures the ratio of the shear strength of the cell wall to its axial strength.

Data for the shear strength of woods are shown in Fig. 10.15. The out-of-plane shear strength (for loading in the axial–tangential or axial–radial planes) is linearly related to the density of the wood in agreement with Eqn. (10.16). The cell-wall yield strength relevant to out-of-plane shearing is $30\,MN/m^2$, giving $C_8 = 0.086$. The data for the in-plane shear strength are too limited to draw any conclusions about the dependence on density; they have been omitted from the plot.

(d) Fracture toughness

When wood containing a crack which lies in the plane of the grain is subjected to mode I, or crack-opening, loads, the crack advances in its own plane. The load–deflection curve is linear to failure, which occurs by fast, unstable fracture. Data for the fracture toughness, K_{IC}^{*a}, for this sort of crack propagation are plotted as a function of relative density in Fig. 10.17.

If, instead, the crack lies normal to the grain, the problem is more complicated. The load–deflection curve is linear up to the load at which the crack first extends. This initial extension is stable and, almost always, parallel to the grain (and thus perpendicular to the starter crack). Thereafter the crack extends in a step-wise manner, partly along the grain and partly across it (Fig. 10.23), giving a load–deflection curve which passes through a maximum and then falls. It is then necessary to distinguish two values of K_{IC}^*: that for initial crack extension, and that for failure. Data for the smaller of the two – that for initial extension, K_{IC}^{*n} – are plotted as a function of relative density in Fig. 10.17.

The figure shows that the fracture toughness for cracking normal to the grain, K_{IC}^{*n}, is about 10 times greater than that for cracking along the grain, K_{IC}^{*a}. To understand this we must first examine the stress state around the tip of a mode I crack in an orthotropic material (Sih et al., 1965). When the crack is loaded (Fig. 10.24), tensile stresses appear on the plane TB ahead of its tip. But there are also tensile stresses on the plane TA: at a given distance from the tip they are

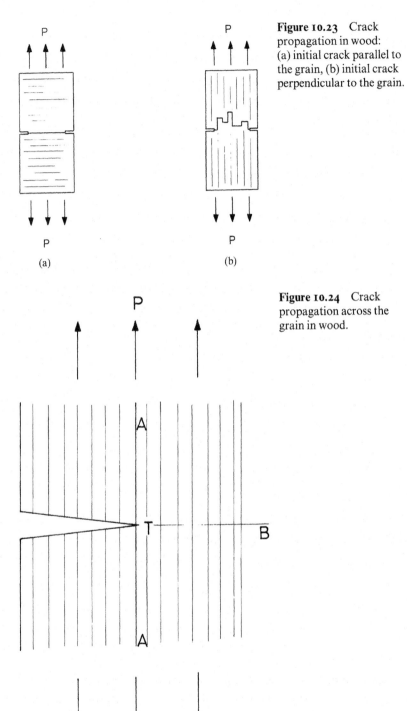

Figure 10.23 Crack propagation in wood: (a) initial crack parallel to the grain, (b) initial crack perpendicular to the grain.

Figure 10.24 Crack propagation across the grain in wood.

less than those on the plane TB by a factor, F, which depends on the degree of anisotropy in the material. For woods this factor is between 5 and 12 (Ashby et al., 1985). But the fracture toughness on plane TA is much lower than that on plane TB, so that the crack starts to propagate on TA when the load is larger, by the factor, F, than that required to propagate a simple mode I crack parallel to the grain: that is why K_{IC}^{*n} is a constant factor (roughly 10) times larger than K_{IC}^{*a}, independent of the density of the wood. As the crack propagates along the grain it seeks out weak fibres, or defects in the wood (knots for instance) and at these points it breaks across the grain, giving the zig-zag path sketched in Fig. 10.23.

Figure 10.17 shows that the fracture toughness depends principally on the relative density of the wood. For crack propagation along the grain the data are described by:

$$K_{IC}^{*a} = 1.8 \left(\frac{\rho^*}{\rho_s} \right)^{3/2} \text{MN/m}^{3/2}. \tag{10.17}$$

This can be understood as follows. When cell walls peel apart along the central lamella, as in Fig. 10.19, the fracture process is like that of the peeling apart of an adhesion joint. Let the energy absorbed per unit area of peeling be G_{cs}^p; we expect this to be about constant for all woods (at a given moisture content) since the composition and structure of the cell walls varies very little between species. During fracture this energy is supplied by the release of elastic energy from the surrounding wood, plus any work done by the applied loads. Using standard results the energy release rate, G_1, is, to a sufficient approximation, given by:

$$G_I = \frac{K_I^2}{E_R^*} \tag{10.18}$$

where E_R^* is Young's modulus of the wood across the grain and K_I is the stress intensity. Equating this to G_{cs}^p and using the result that the transverse modulus is related to the relative density and Young's modulus, E_s, for the cell-wall material by Eqns. (10.3) and (10.5) gives, for the peeling mode:

$$K_{IC}^{*a} = (C_1 E_s G_{cs}^p)^{1/2} \left(\frac{\rho^*}{\rho_s} \right)^{3/2}. \tag{10.19}$$

The quantity $(C_1 E_s G_{cs}^p)$ is simply the fracture toughness of the cell wall in the peeling mode, $(K_{IC}^p)_s$, so that:

$$K_{IC}^{*a} = (K_{IC}^p)_s \left(\frac{\rho^*}{\rho_s} \right)^{3/2}. \tag{10.20}$$

When the cell walls break, as in Figs. 10.18 and 10.20, the energy is absorbed, G_{cs}^b per unit area, in breaking a cell wall. Fibres of cellulose in the cell wall must be broken or pulled out when the cell wall breaks, so we expect this energy to be larger than that for the peeling mode, G_{cs}^p, which does not involve pull-out. The area fraction occupied by cell walls in the crack plane of Fig. 10.18 is approximately t/l, or (using Eqn. (10.2)) roughly (ρ^*/ρ_s). Thus, the energy absorbed per unit area of crack is $(\rho^*/\rho_s)G_{cs}^b$. Equating this to the energy-release rate (Eqn. (10.18)) and using Eqns. (10.3) and (10.5) for E_R^* gives, for the fracture toughness in the breaking mode:

$$K_{IC}^{*a} = (C_1 E_s G_{cs}^b)^{1/2}\left(\frac{\rho^*}{\rho_s}\right)^2 \tag{10.21}$$

or

$$K_{IC}^{*a} = (K_{IC}^b)_s \left(\frac{\rho^*}{\rho_s}\right)^2 . \quad \text{(breaking mode)} \tag{10.22}$$

Thus the fracture toughness for a crack propagating along the grain in the breaking mode varies as a higher power of the density than that for one propagating in the peeling mode. Remembering that $G_{cs}^b > G_{cs}^p$ and that $\rho^*/\rho_s < 1$, we find an interesting result: cracks propagating along the grain should do so by cell-wall breaking when the relative density is low, and by cell-wall peeling when it is high. There is experimental evidence suggesting that this is the case (see Section 10.3); the transition occurs at a relative density of about 0.2. Equating Eqns. (10.17) and (10.19) gives a value for the peeling toughness, G_{cs}^p:

$$G_{cs}^p \approx 350 \, \text{J/m}^2.$$

A value for the breaking toughness G_{cs}^b can be inferred from the transition from breaking to peeling at a relative density of about 0.2. Equating Eqns. (10.19) and (10.21) at this density gives:

$$G_{cs}^b \approx 1650 \, \text{J/m}^2.$$

10.5   Conclusions

Wood is complicated stuff. At a high magnification it has the structure of a fibre-reinforced composite. At a lesser magnification it is a cellular material – one with very anisotropic cells. In most woods the composition and lay-up of the fibre-reinforced cell walls seems to be similar, so that the primary differences between woods relate to the differences in their cellular structure. Above all, it is the relative thickness of the cell walls, and thus the relative density, ρ^*/ρ_s, which determines the mechanical properties of wood. Age, moisture content, strain-rate and temperature all have an effect, of course, but the small variations in these when dry, seasoned, knot-free wood is tested under laboratory conditions mean that relative density is the most important variable.

Young's moduli for woods vary over a range of 1000. The tangential and radial moduli, E_T^* and E_R^*, for loading across the grain, are determined by cell-wall bending, and vary roughly as the cube of the relative density. By contrast, the modulus for loading along the grain is determined by cell-wall extension or compression and varies linearly with relative density. This difference in dependence on density causes the elastic anisotropy of wood to increase as the density decreases, so that the ratio of E_A^* or E_T^* can be as high as 75 for low-density balsa. The shear moduli show the same large differences. The modulus for shear in the radial–tangential plane varies as the cube of the density; the other two shear moduli vary linearly, for the same reasons as before: the one bends cell walls, the other two cause stretching. Poisson's ratios, too, are very anisotropic, being almost zero for some directions and more than 0.5 for others. Figures 10.12 and 10.13, and Table 10.4, summarize information about the moduli. From them, the axial and transverse modulus of the cell wall itself can be inferred. Values are given in Table 10.3.

The compressive strengths of woods vary by a factor of about 100. The compressive strength across the grain is determined by the plastic bending of the cell walls and varies as the square of the density. The compressive strength along the grain is determined by the uniaxial yielding of the cell walls or the end caps of the cells, and varies linearly with density. Here, too, the anisotropy increases as density decreases. The shear strength follows a similar pattern. Information about the strength of woods is summarized in Figs. 10.14 and 10.15, from which the axial and transverse strengths of the cell wall can be inferred. Their values are given in Table 10.3.

To a first approximation, the fracture toughness of dry, knot-free wood depends mainly on density, too. The fracture toughness for wood containing a mode I crack in a plane containing the grain, K_{IC}^{*a}, is less, by a factor of about 10, than that for wood containing a mode I crack in a plane normal to the grain, K_{IC}^{*n}. Unlike the moduli and the compressive strengths, the anisotropy in the fracture toughness for crack initiation in wood is independent of density: the ratio of K_{IC}^{*n}/K_{IC}^{*a} is roughly constant at about 10. This can be understood when the details of the crack-tip field in a very anisotropic solid are considered. It appears that cracking is initiated when the local stress at a characteristic distance from the crack tip reaches a critical value. Cracking always starts on a plane parallel to the grain regardless of the orientation of the starter crack. When the starter crack is parallel to the grain the subsequent propagation is unstable; when it is normal to the grain propagation is stable. Figure 10.17 summarizes the observations. From the data it contains, values for the intrinsic toughness of a cell wall can be inferred. Values are given in Table 10.3

At a more precise level the fracture toughness depends on the geometry of the wood structure. The boundaries between ray cells and tracheids or fibres are planes of weakness which give rise to a low-energy, peeling mode of failure. Sap channels, and particularly rows of sap channels, act as crack arresters. Temperature, moisture content and strain-rate all have some influence on fracture toughness. Further work is required to characterize these effects.

It is sometimes convenient to have simple equations describing the way in which wood properties depend on density. The best such simple descriptions, corresponding to the straight lines on Figs. 10.12–10.15 and 10.17 are collected in Table 10.5.

References

Ashby, M. F., Easterling, K. E., Harrysson, R. and Maiti, S. K. (1985) *Proc. Roy. Soc.*, **A398**, 261.

Bariska, M. and Kurcera, L. A. (1982) *Wood Science and Technology*.

Barrett, J. D. (1981) *Phil. Trans. Roy. Soc.*, **A299**, 217.

Bentur, A. and Mindess, S. (1986) *J. Mat. Sci.*, **21**, 559.

Bodig, J. and Goodman, J. R. (1973) *Wood Sci.*, **5**, 249.

Bodig, J. and Jayne, B. A. (1982) *Mechanics of Wood and Wood Composites*. Van Nostrand Reinhold, New York.

Cave, I. D. (1968) *Wood Sci. Tech.*, **2**, 268.

Cave, I. D. (1969) *Wood Sci. Tech.*, **3**, 40.

Dinwoodie, J. M. (1981) *Timber, Its Nature and Behaviour*. Van Nostrand Reinhold, New York.

Easterling, K. E., Harrysson, R., Gibson, L. J. and Ashby, M. F. (1982) *Proc. Roy. Soc.*, **A383**, 31.

Goodman, J. R. and Bodig, J. (1970) *J. Struct. Div., ASCE*, **ST11**, 2301.

Goodman, J. R. and Bodig, J. (1971) *Wood Sci.*, **4**, 83.

Grossman, P. U. A. and Wold, M. B. (1971) *Wood Sci. Tech.*, **5**, 147.

Jeronimidis, G. (1980) *Proc. Roy. Soc.*, **B208**, 447.

Johnson, J. A. (1973) *Wood Sci.*, **6**, 151.

Koran, Z. (1966) Pulp and Paper Research Institute of Canada Technical Report, No. 472.

Mark, R. E. (1967) *Cell Wall Mechanics of Tracheids*. Yale University Press.

Mindess, S., Nadeau, J. S. and Barrett, J. D. (1975) *Wood Sci.*, **8**, 389.

Nadeau, J. S., Bennett, R. and Fuller, E. R. (1982) *J. Mat. Sci.*, **17**, 2831.

Pepys, S. (1660–9) Diary; published as *The Diary of Samuel Pepys* (ed. R. Latham and W. Matthews) (1983). Bell and Hyman, London.

Price, A. T. (1928) *Phil. Trans.*, **A228**, 1.

Schniewind, A. P. and Centano, J. C. (1973) *Wood Fiber*, **5**, 152.

Schniewind, A. P. and Pozniak, R. A. (1971) *Eng. Fract. Mech.*, **2**, 223.

Sih, G. C., Paris, P. C. and Irwin, G. R. (1965) *Int. J. Fract. Mech.*, **1**, 189.

Silvester, F. D. (1967) *Mechanical Properties of Timber*. Pergamon Press, Oxford.

Srinavasan, P. S. (1942) *J. Indian Inst. Sci.*, **23B**, 222.

Walsh, P. F. (1971) *Cleavage Fracture in Timber*. Div. Forest Prod., Tech. Paper No. 65, CSIRO, Melbourne, Australia.

Williams, J. G. and Birch, M. W. (1976) in *Cracks and Fracture*, ASTM Stand. No. 601, p. 125.

Wood Handbook (1974) U.S. Dept. of Agriculture, Agricultural Handbook No. 72, Forest Products Lab., Madison, WI.

Wood, The International Book of (1976) Mitchell Beazley, London.

Wu, E. M. (1963) 'Applications of Fracture Mechanics to Orthotropic Plates'. Theor. Appl. Mech. Report No. 1418, University of Illinois, Urbana, IL.

Chapter 11

Cancellous bone

11.1   **Introduction and synopsis**

Superficially, bones look fairly solid. But looks are deceptive. Most bones are an elaborate construction, made up of an outer shell of dense *compact* bone, enclosing a core of porous cellular, *cancellous*, or *trabecular* bone (trabecula means 'little beam' in Latin). Examples are shown in Fig. 11.1: cross-sections of a femur, a tibia, and a vertebra. In some instances (as at joints between vertebrae or at the ends of the long bones) this configuration minimizes the weight of bone while still providing a large bearing area, a design which reduces the bearing stresses at the joint. In others (as in the vault of the skull or the iliac crest) it forms a low weight sandwich shell like those analysed in Chapter 9. In either case the presence of the cancellous bone reduces the weight while still meeting its primary mechanical function.

An understanding of the mechanical behaviour of cancellous bone has relevance for several biomedical applications. In elderly patients with osteoporosis the mass of bone in the body decreases over time to such an extent that fractures can occur under loads that, in healthy people, would be considered normal. Such fractures are common in the vertebrae, hip and wrist, and are due in part to a reduction in the amount of cancellous bone in these areas. The degree of bone loss in a patient can be measured using non-invasive techniques, so an understanding of the relationship between bone density and strength helps in predicting when the risk of a fracture has become high. It helps, too, in the design of artificial hips. Most of the bone replaced by an artificial hip is cancellous; an

429

improved understanding of the structure–property relationships for cancellous bone allows the design of artificial hips with properties which more closely match those of the bone they replace. The mismatch of properties between current artificial hips and the surrounding bone is thought to be one reason that they work loose, an unpleasant development from the patient's point of view

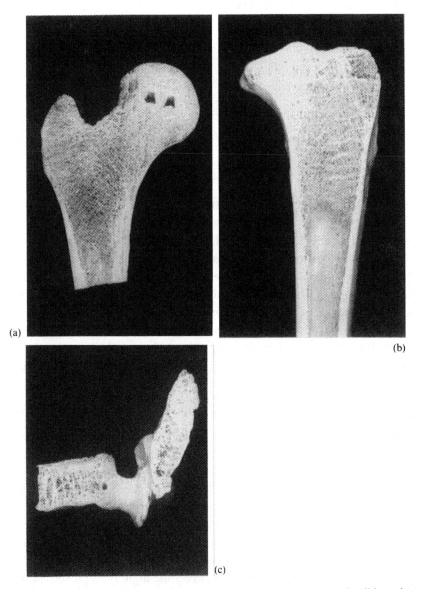

(a)

(b)

(c)

Figure 11.1   Cross-sectional views of: (a) the head of a femur. (b) the tibia and (c) a lumbar vertebra. In each case there is an outer shell of almost fully dense compact bone surrounding a core of porous, low-density, cancellous bone.

because replacement is difficult. Cancellous bone may also play a role in osteoar-
thritis, which is thought to be related to a breakdown in the lubrication process
at joints. The distribution of forces acting across a joint is directly related to the
mechanical properties of the underlying cancellous bone, so changes in its structure (and hence properties) may change the distribution of forces and cause
damage to the lubrication system.

The cellular structure of cancellous bone is shown in Fig. 11.2. It is made up of
an interconnected network of rods or plates. A network of rods produces low-
density, open cells, while one of the plates gives higher-density, virtually closed
cells. In practice the relative density of cancellous bone varies from 0.05 to 0.7
(technically, any bone with a relative density less than 0.7 is classified as 'cancellous').

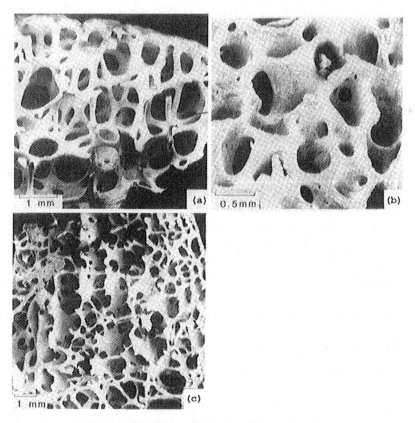

Figure 11.2 Scanning electron micrographs showing the cellular structure of
cancellous bone. (a) A specimen taken from the femoral head, showing a low-
density, open-cell, rod-like structure. (b) A specimen from the femoral head,
showing a higher-density, roughly prismatic cell structure. (c) A specimen from the
femoral condyle, of intermediate density, showing a stress-oriented, parallel plate
structure, with rods normal to the plates (from Gibson, 1985).

The mechanical behaviour of cancellous bone is typical of a cellular material. The compressive stress–strain curve, for instance, has the three distinct regimes characteristic of all cellular solids and (as will be shown below) the modulus and strength, and probably the toughness, all vary with density in the way expected of a foam. In this chapter we review the structure and mechanical behaviour of cancellous bone, model its behaviour using an extension of the ideas developed in Chapters 4 and 5, and then compare the results of the model with data for its properties. The variability of bone is enormous, so our ability to predict properties is less good than in the case of man-made foams or wood; but the models help explain the obvious trends in properties with density.

11.2   The structure of cancellous bone

Low-magnification micrographs (Fig. 11.2) reveal the cellular structure of cancellous bone (Dyson et al., 1970; Whitehouse et al., 1971a,b; Whitehouse and Dyson, 1974; and Whitehouse, 1974, 1975). At the lowest densities the cells are open and like a network of rods. As the density increases the rods progressively spread and flatten, become more plate-like, and finally fuse to give almost closed cells.

It has long been known that bone grows in response to stress (Wolff, 1869; Thompson, 1961; Currey, 1984), although the mechanism is not yet completely understood. Bone is piezoelectric (that is, it generates an electric potential when stressed) and it has been suggested that in some way this is responsible for stress-induced growth. Comparisons between the disposition of trabeculae in cancellous bone and the directions of principal stress in similarly loaded solid members strongly suggest that trabeculae develop along the principal stress trajectories in the loaded bone (Thompson, 1961; Currey, 1984). Such comparisons are striking: an example, taken from the head of the human femur is shown in Fig. 11.3. Other evidence supporting this idea derives from finite element studies of the stresses acting in the human patella (Hayes and Snyder, 1981) and from in-vivo measurements of the strains in the cortex of the calcaneus of sheep (Lanyon, 1974).

It is generally accepted that the density of cancellous bone depends on the magnitude of the loads it experiences. Whitehouse and his co-workers, in their study of the human femur, show micrographs from various parts in the femoral head along with a density contour map of the same femur. Comparison of the structure with the density map shows that low-density, open-cell, rod-like structures develop where the stress is low, while denser, almost closed, plate-like structures are found in regions of higher stress.

The structural anisotropy of cancellous bone seems to depend on the ratios of the principal stresses, as one might expect if stress stimulates growth. In the

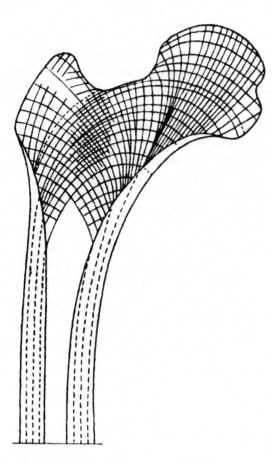

Figure 11.3 A schematic drawing showing the principal stress trajectories in the head of the human femur. The pattern of the trajectories is similar to that of the cancellous bone shown in Fig. 11.1(a). (From Thompson, 1961, Fig. 101 courtesy of Cambridge University Press.)

femoral head, for instance, the principal stresses are roughly equal, and the bone grows as a foam with almost equiaxed cells (Fig. 11.2(a)). But in the femoral condyle at the knee the loading in one direction is much greater than that in the other two, and then the bone forms in parallel plates aligned in the direction of the larger load; normal to these plates, it forms thin rods which act as spacers (Fig. 11.2(c)).

The structure of cancellous bone changes over time due to reductions in bone mass with aging. For instance, there is roughly a 50% reduction in the relative density of vertebral trabecular bone between 20 and 80 years of age (Moskilde, 1989). Only part of the density reduction is due to thinning of the cell walls: the remainder is due to enlargement of circular perforations within the cell wall and to the complete loss of some cell walls (Parfitt, 1992).

At a finer level of structure, bone is a composite of a fibrous, organic matrix (proteins, largely collagen) filled with inorganic calcium compounds (crystalline hydroxyapatite, $Ca_{10}(PO_4)_6(OH)_2$ and amorphous calcium phosphate, $CaPO_3$);

it is these which give bone its stiffness. The compositions of compact and cancellous bone are almost the same. In both, the organic matrix makes up about
35% of the wet weight of bone, the calcium compounds 45%, and water the
remainder (Gong *et al.*, 1964; Carter and Spengler, 1978). The densities, too,
are similar: that of compact bone lies in the range 1800–2000 kg/m^3 (Carter
and Hayes, 1977; Behiri and Bonfield, 1984) while that of the individual trabecula taken from specimens of cancellous bone averages 1820 kg/m^3 (Galante
et al., 1970). The mechanical properties of compact bone have been measured
in many studies; they are reviewed by Currey (1970 and 1984), Reilly and Burstein (1974) and Carter and Spengler (1978). Stress–strain curves for wet compact bone are shown in Fig. 11.4. In the longitudinal direction (that is,
parallel to the length of the bone) it is linearly elastic up to strains of about
0.7%; beyond, it yields plastically up to strains of about 3%, both in tension
and compression. In the transverse direction wet compact bone is less stiff,
strong and ductile, and in tension it fails at strains of only 0.6%. The anisotropy
derives from the structure; the hydroxyapatite is plate- or fibre-like, and its partial alignment in the longitudinal direction makes this the stiffer and stronger
axis of the material. When bone is dried the elastic moduli increase, and the

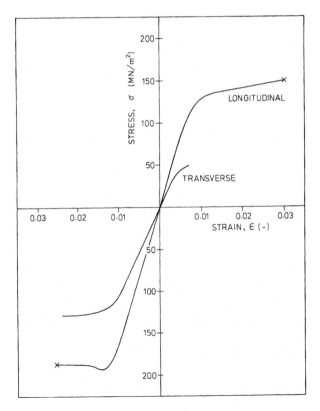

Figure 11.4   Two stress–
strain curves for wet
compact bone loaded in the
longitudinal and transverse
directions (based on curves
and data from Reilly and
Burstein, 1975 and Currey,
1984).

strength and strain to failure decrease. Completely dry bone does not yield at all; it is linear-elastic to fracture. Increasing the strain-rate at which bone is loaded produces similar effects. Data for the properties of wet compact bone are summarized in Table 11.1.

The mechanical properties of the individual trabeculae, making up the cell walls of trabecular bone, have long been assumed to be the same as those of cortical bone (McElhaney *et al.*, 1970a,b; Townsend *et al.*, 1975a,b; Carter and

Table 11.1   The properties of wet compact bone[†]

	Human	Bovine
Density, ρ_s (kg/m^3)	1800–2000	2060
Young's modulus, E_s (GN/m^2)		
longitudinal	17.0	22.6
radial	11.5	10.2
tangential	11.5	10.2
Shear modulus, G_s (GN/m^2)		
longitudinal–radial	3.3	3.6
longitudinal–tangential	3.3	3.6
radial–tangential		
Poisson's ratio, ν_s		
ν_{sLR}	—	0.36
ν_{sLT}	0.41	0.36
ν_{sTR}	0.41	0.51
Compressive strength, σ_{ys} (MN/m^2)		
along	193	254
normal	133	146
Tensile yield strength, σ_{ys}^T (MN/m^2)		
along	148	144
normal	49	46
Toughness, G_{cs} (J/m^2)		
along	—	1690
normal	—	4330
Fracture toughness, K_{ICs} (MN/m$^{3/2}$)		
along	—	3.5
normal	—	6.1

†Data for all mechanical properties are from static tests on wet compact bone. A more complete summary of the properties of compact bone is available in Currey (1984), from which this table was compiled.

Hayes, 1977, and Gibson, 1985). Recently, there have been several attempts to determine the Young's modulus of the trabeculae using a number of different techniques: direct mechanical testing of a single trabecula in tension, bending or buckling (Townsend *et al.*, 1975; Runkle and Pugh, 1975; Ryan and Williams, 1989; and Choi *et al.*, 1990); ultrasonic wave propagation in trabecular bone specimens (Ashman and Rho, 1988); and finite element analysis (Pugh *et al.*, 1973b; Williams and Lewis, 1982; Mente and Lewis, 1987; and van Rietbergen *et al.*, 1995). The results are summarized in Table 11.2.

Direct tension or bending tests are difficult: unmachined specimens have varying cross-sectional areas and are curved along their length while machined specimens may have significant surface defects introduced by the machining. In addition, the deformations to be measured in tensile or bend tests on a single trabecula are typically small. Any additional displacement introduced as an artifact of the testing technique (e.g. slippage or stress concentrations at loading or support points) results in a lower measured modulus than the true value. A lower

Table 11.2 Young's modulus of trabeculae

Reference	Type of bone†	Method	E_s (GPa)
Mechanical tests			
Ryan and Williams (1989)	B F	tension unmachined	0.76 (0.39)
Kuhn *et al.* (1987)	H T	3-point bend machined	3.17 (1.5)
	H cortical		3.8
Choi *et al.* (1990)	H PT	3-point bend machined	4.59 (1.60)
	H cortical		5.44 (1.25)
Townsend *et al.* (1975)	H PT	buckling unmachined	11.4 (wet)
			14.1 (dry)
Runkle and Pugh (1975)	H DF	buckling	8.69 (3.17) dry
Ultrasound tests			
Ashman and Rho (1988)	B F	ultrasound on	10.9 (1.6)
	H F	trabecular bone	13.0 (1.5)
Finite element analysis			
Pugh *et al.* (1973b)	H DF	2D FEM	$E_s < E_{compact}$
Williams and Lewis (1982)	H PT	2D FEM	1.30
Mente and Lewis (1987)	H F	2D FEM	5.3 (2.6) dry
Rietbergen *et al.* (1995)	H PT	3D FEM	2.23–10.1

†All specimens are of individual trabeculae unless stated otherwise. B = bovine; H = human; F = femur; T = tibia; P = proximal; D = distal.

bound for E_s can be estimated by considering that at a relative density of 0.3 the Young's modulus of porous trabecular bone is between 0.8 and 2.7 GPa (Fig. 11.7). Linear extrapolation to the fully dense solid gives a lower bound for E_s of 2.7–9.0 GPa. Ryan and Williams' (1989) tensile value of 0.76 GPa, obtained from excised trabeculae, is clearly too low, probably as a result of geometrical irregularities in their specimens. Kuhn et al. (1987) and Choi et al., (1990) attempted to overcome this difficulty by machining three-point bend microspecimens of both trabecular tissue (specimen size $\sim 100\mu \times 100\mu \times 1550\mu$): they obtained mean values of E_s of 3.17–4.59 GPa. Using the same technique on similar sized specimens of cortical bone they measured its Young's modulus to be 3.8–5.44 GPa, much lower than the accepted value of 16–17 GPa (Currey, 1984 and Choi et al., 1990). Taking the ratio of trabecular tissue to cortical bone moduli measured on the microspecimens and multiplying by the accepted value of 17 GPa gives $E_s = 14$ GPa. Buckling tests on a single trabecula gives values of E_s of 8.69–11.4 GPa for wet bone (Townsend et al., 1975; Runkle and Pugh, 1975).

Ultrasonic testing of trabecular bone specimens eliminates the problems associated with direct mechanical tests. The Young's modulus of the individual trabeculae can be found by measuring the speed of a high frequency ultrasonic wave on a specimen of porous trabecular bone, eliminating the need for testing of individual trabeculae. Since the Young's modulus is calculated from the wave speed and density, deformations do not have to be measured. The method is described in more detail by Ashman and Rho (1988) who find values of E_s of 10.9 and 13.0 for bovine and human bone, respectively.

Finite element methods can also be used to estimate the Young's modulus of the individual trabeculae. The finite element grid is designed to represent the trabecular architecture of a specimen of bone of which the modulus is known, either by a mechanical test or by the use of empirical modulus–density relationships. The value for the modulus of the individual trabeculae used in the finite element analysis is then chosen so that the calculated modulus of the trabecular bone specimen matches the measured or estimated modulus. Computational limits restricted the initial finite element studies to a two-dimensional analysis which gave values of $E_s = 1.30$ GPa for wet bone (Williams and Lewis, 1982) and $E_s = 5.3$ GPa for dry bone (Mente and Lewis, 1987). Both values appear to be too low: the first is lower than the lowest value extrapolated from data for trabecular bone (2.7 GPa) and the second is about half of the value obtained using the ultrasonic technique. The lower than expected values are consistent with the fact that the Young's modulus of a two-dimensional honeycomb (with either uniform hexagonal cells or Voronoi cells) decreases more rapidly with decreasing density than that of a three-dimensional foam (Eqns. (4.12), (5.3), and Silva et al., 1995). Recently, using serial reconstruction and more efficient computational techniques, van Rietbergen et al., (1995) have performed a three-dimensional finite element analysis of a 5 mm cube of trabecular bone. The modulus of the specimen was estimated using Hodgkinson and Currey's (1992) empirical equations relating modulus and density. Their results suggest values of 2.23–10.1 GPa, with a mean of 5.91 GPa. Van Rietbergen and his co-workers note that the empirical equations were fitted to data which were suspect because interfacial friction at the specimen ends gave low values of E^*. Odgaard and Linde (1991) and Keaveny et al. (1993) suggest that such artifacts give measured moduli 20–70% less than the true modulus. A 50% error would give a mean value of E_s of about 12 GPa, close to that measured ultrasonically. This is the value of E_s we use here.

There have been no measurements of the strength of the individual trabeculae, σ_{ys}, reported. Estimating the strength is further complicated by the fact that the tensile and compressive strengths of compact bone differ (Table 11.1). Here we assume that the reduction in the strength of the trabeculae below that of compact bone is the same as that for the modulus:

$$\sigma_{ys} = \frac{E_s}{E_{cortical}} \sigma_{cortical}.$$

The trabecular properties used in this chapter are summarized in Table 11.3.

11.3   The mechanical properties of cancellous bone

The compressive stress–strain curve of cancellous bone is typical of that of a cellular solid. Figure 11.5 shows the three familiar regimes of behaviour described in Chapter 5. The small strain, linear-elastic response of low-density, near-isotropic cancellous bone derives from the elastic bending of the cell walls (Pugh et al., 1973a; Stone et al., 1983; Gibson, 1985). Axial or membrane stresses may be present, too (Pugh et al., 1973b), but the deformations resulting from these are small compared to those from bending. Stress-oriented bone can be different: in it, the tube- or plate-like cell walls are aligned along the direction of largest principal stress, and loads applied in this direction extend or compress the walls (Townsend et al., 1975a,b; Williams and Lewis, 1982) while transverse loads give rise to bend-

Table 11.3 Solid cell-wall properties for trabecular bone

Property	Value
Young's modulus	12 GPa
Compressive strength	136 MPa
Tensile strength	105 MPa

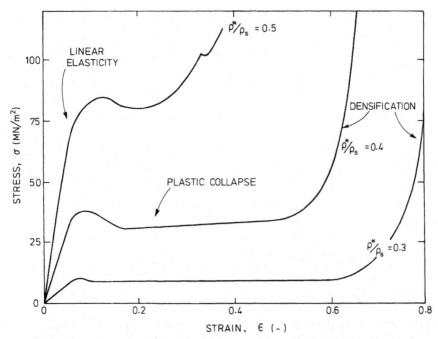

Figure 11.5 Compressive stress–strain curves for several relative densities of wet cancellous bone (after Hayes and Carter, 1976, Fig. 1a, with kind permission of John Wiley & Sons).

ing deformations in the connecting rods between the plates (Williams and Lewis, 1982). (Wood, too, is like this: loaded along the grain the cell walls are compressed or extended axially; across the grain they are bent (Chapter 10).)

The linear-elastic regime ends when the cells begin to collapse. The rod- or plate-like cell walls in low-density bone have a high slenderness ratio (the ratio of the length of a column to its thickness) and fail by elastic buckling, both in wet and dry conditions. At higher densities the slenderness ratio is lower and buckling is more difficult; then it is found that wet specimens microcrack while dry ones fracture in a brittle manner (Townsend *et al.*, 1975a; Gibson, 1985). Elastic buckling and shear fracture have also been suggested as possible failure modes (Pugh *et al.*, 1973b; Behrens *et al.*, 1974) and there is some suggestion of both in micrographs of failed samples (Hayes and Carter, 1976). Progressive compressive collapse gives the long, horizontal plateau of the stress–strain curve which continues until opposing cell walls meet and touch, causing the stress to rise steeply (Hayes and Carter, 1976). It is instructive to compare Fig. 11.5, which shows how the modulus and strength of bone increase with density, with Fig. 5.3(c), which shows the same thing for a rigid polymer foam.

A tensile stress–strain curve for wet cancellous bone is shown in Fig. 11.6. The initial linear-elastic portion of the curve results from the elastic bending or extension of the trabeculae, as in compression. At strains of about 1% the stress-strain

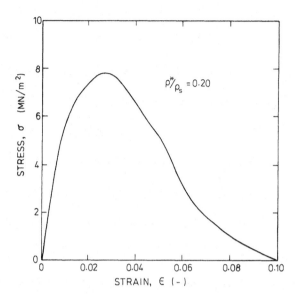

Figure 11.6 Tensile stress–strain curves for wet cancellous bone (after Carter *et al.*, 1980, Fig. 3 courtesy of Munksgaard International Publishers Ltd.).

curve becomes non-linear as the trabeculae start to deform irreversibly and crack. Beyond the peak, the stress–strain curve falls gradually as the trabeculae progressively fail by tearing and fracturing (Carter *et al.*, 1980). Early studies suggested that the tensile and compressive strengths of cancellous bone were about equal (Carter *et al.*, 1980; Bensusan *et al.*, 1983; and Neil *et al.*, 1983). More recent work indicates that cancellous bone is weaker in tension that in compression (Stone *et al.*, 1983 and Kaplan *et al.*, 1985) and that the difference in strengths increases with relative density (Keaveny *et al.*, 1994).

Data for cancellous bone of unspecified trabecular orientation are plotted against density in Figs. 11.7, 11.8 and 11.9. The first of the three figures shows Young's modulus, and includes data for both tension and compression, for equiaxed and stress-oriented bone, and for all directions of loading. The second and third figures show compressive and tensile strengths for a similarly wide selection of bone samples. The spread in the data is large, for several reasons. Unless the cells are perfectly equiaxed, cancellous bone is anisotropic; a great part of the scatter arises from differences in the bone structure and in the directions of loading of the samples. Some may arise from variations in the properties of the bone making up the cell walls as a result of small differences in its porosity or inorganic content. And the data shown derive from tests carried out at strain-rates varying over five orders of magnitude: it is known that both Young's modulus and strength depend on strain-rate (Carter and Hayes, 1977). Finally, the moisture content is important: some of the scatter in the strength data, particularly, can be traced to different levels of dryness of the bone.

More limited data for Young's modulus and the compressive strength of stress-oriented cancellous bone are plotted against relative density in Figs.

11.10 and 11.11. Here greater selectivity has been used: open and closed symbols distinguish the direction of loading.

As one would expect for a cellular solid, the density has a profound influence in determining the stiffness and strength of cancellous bone. Other factors are important, too. The figures show how the stress-induced orientation of the trabeculae gives large anisotropies in properties: the longitudinal to transverse stiffness of cancellous bone in the human tibia, for instance, can differ by a factor of as much as 10 (Williams and Lewis, 1982). And the mechanical properties of the trabeculae depend on moisture and strain-rate. In a living creature, of course, the cells contain marrow, but this does not affect the behaviour of cancellous bone at strain-rates below about 10/s (Carter and Hayes, 1977), an observation which is consistent with the analysis of the influence of pore fluids given in Chapter 6.

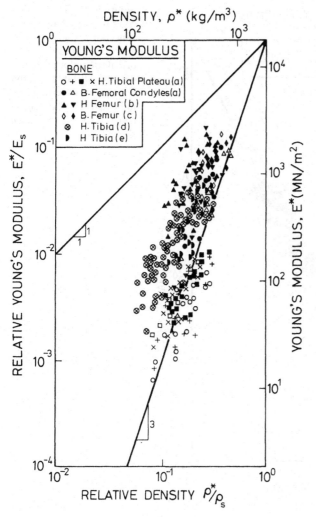

Figure 11.7 Young's moduli of cancellous bone of unspecified trabecular orientation plotted against density. The data fall between lines of slope 1 and 3. Data from: (a) Carter and Hayes, 1977; (b) Carter *et al.*, 1980; (c) Bensusan *et al.*, 1983; (d) Hvid *et al.* (1989); (e) Linde *et al.* (1991). H = human, B = bovine.

The fracture toughness of bone, too, is an important property from a biomechanics point of view. A low toughness makes the bone more liable to fracture – and it is widely thought that bone grows brittle with age. Measuring fracture toughness is not easy because samples with dimensions which meet the requirements of a valid test can be cut only from large bones. Attempts to do so for dense bone give values in the range 600–3000 J/m^2 for the toughness, G_c, corresponding to values of fracture toughness, K_{IC}, in the range 2.5–6 MN/m$^{3/2}$ (Melvin and Evans, 1973; Margel-Robertson, 1973; Bonfield and Datta, 1974, 1976; Wright and Hayes, 1977; Bonfield *et al.*, 1978; Behiri and Bonfield, 1980, 1984). The considerable spread in the data can be traced to differences in bone density and moisture content, in the strain-rate of the test and in the orientation of the crack. There are no data for the fracture toughness of cancellous bone.

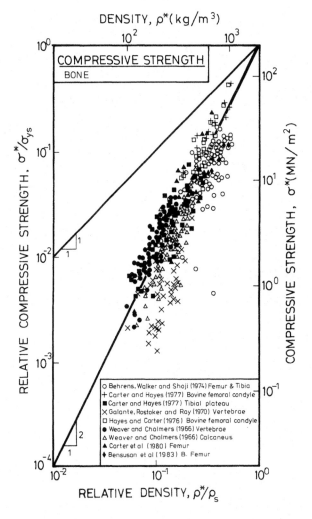

Figure 11.8 The compressive strength of cancellous bone of unspecified trabecular orientation plotted against density. All data for human bone unless otherwise indicated. The data fall roughly around the line of slope 2.

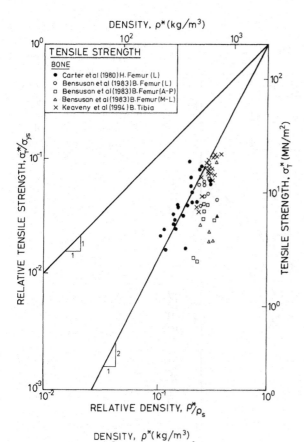

Figure 11.9   The tensile strength of cancellous bone of unspecified trabecular orientation plotted against density. The long axis of the specimens coincided with the bone axis. Specimens were loaded in the longitudinal direction (L), the anterior–posterior direction (A–P) and the medial–lateral direction (M–L). The data for longitudinal loading fall on a line of slope 2. H = human, B = bovine.

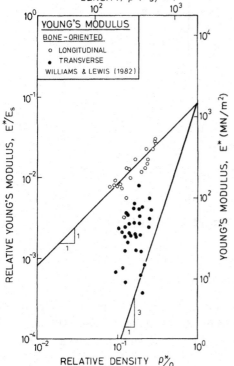

Figure 11.10   Young's modulus of cancellous bone with a prismatic structure plotted against density. The data for loading along the prism axis fall on a line of slope 1; those for loading across the prism show a slope near 3.

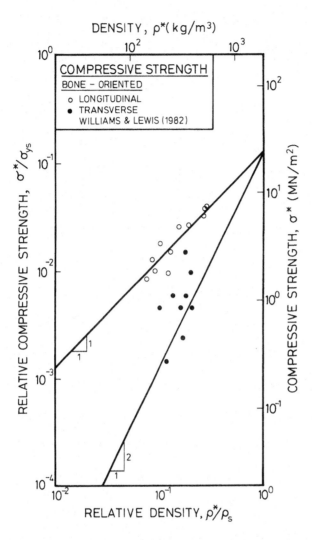

Figure 11.11 The compressive strength of cancellous bone with prismatic structure plotted against density. The data for loading along the cylinders lie on a line of slope 1; those for loading across the cylinders lie near a line of slope 2.

In the next section the structure and deformation mechanisms of cancellous bone are modelled using the ideas developed in Chapters 4 and 5, giving some insight into the origins of its properties.

11.4   Modelling the structure and properties of cancellous bone

(a) Idealizations of the structure

The foam-like structure of cancellous bone has suggested various idealizations. McElhaney *et al.* (1970b) think of it as a block, divided into a three-dimensional array of cubic elements, each of which is solid or void. The elements can be arranged in various patterns, and their relative numbers altered, to give a porous block with the density, and approximating the structure, of bone. Beaupre and Hayes (1985) introduce porosity in another way, thinking instead of stacked cubical unit cells, each containing a spherical void. The diameter of the sphere is larger than the edge-length of the cell, so that the voids interconnect; then, depending on the sphere diameter, the solid which remains in a cell is either in the form of rods (the cell edges) or perforated plates (cell edges plus faces).

These idealizations, though ingenious in their attempt to be general, tend to obscure the structural features which determine stiffness and strength. The origins of these properties are more clearly seen by considering the two classes of model shown in Fig. 11.12. Equiaxed cells are described by the first class: at low densities the cells are like a network of rods (Fig. 11.12(a)) at higher densities they are like a framework of perforated plates (Fig. 11.12(b)). Although these resemble the structures proposed by Beaupre and Hayes, there is an important difference. The stacking of the cells in the models of Fig. 11.12(a) and 11.12(b) is staggered, whereas that in the model of Beaupre and Hayes is not. It is this feature which allows bending in the cell walls, as has been observed in several studies (Pugh et al., 1973b; Stone et al., 1983). The second class of model describes stress-oriented bone. Some cancellous bone – probably that which develops in response to uniaxial loads – is like an array of prismatic tubes (Fig. 11.12(c)), with occasional cross-members. In other cases – probably in response to biaxial loading – it grows as an array of parallel plates, oriented so as to contain the directions of largest principal stress, and separated by slender cross-members which act as spacers (Fig. 11.12(d)). Intermediate cases are found which combine features of two or more of these structures.

The attraction of the explicit models of Fig. 11.12 is that the mechanics of their deformation can be analysed, using the methods of Chapters 4 and 5. This we now do.

(b) Mechanical modelling of cancellous bone

It is clear from the data plots of Figs. 11.7–11.10 that the mechanical properties of cancellous bone depend strongly on its density. There is much scatter, part caused by the natural variability of the solid bone itself, and part caused by the considerable differences between the structures described in the last section. Some order can be introduced by considering the properties of each in turn.

Experimental evidence supports the view that the linear-elastic behaviour of equiaxed cancellous bone occurs by bending of the cell walls (Pugh et al., 1973a,b; Stone et al., 1983). The elastic response of equiaxed, open cells was analysed in Section 5.3, where it was shown that Young's modulus is (Eqn. (5.6a)),

$$\frac{E^*}{E_s} = C_1 \left(\frac{\rho^*}{\rho_s}\right)^2 . \tag{11.1}$$

The perforated plate model (Fig. 11.12(b)) responds to load by bending, too. If the perforation was not present, the modulus would be that of a closed-cell foam, including membrane stresses (Eqn. (5.11)). With the perforation, the modulus is reduced. And in the limit of a large perforation (relative to the cell-edge length) the model resembles the low-density equiaxed structure of Fig. 11.12(a). The relative modulus of the perforated plate model, then varies with relative density raised to a power between 1 and 2, depending on the relative diameter of the perforation.

Prismatic cells (Fig. 11.12(c)) are different. Loaded across the prism axis (the 'transverse' direction), the bone behaves like a honeycomb, loaded in-plane (Section 4.3). The modulus varies as the cube of the density (Eqn. (4.7) with Eqn. (4.1b)):

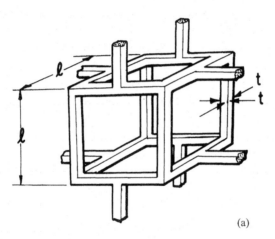

Figure 11.12 Models for the structure of cancellous bone: (a) the low-density equiaxed structure, (b) the higher-density equiaxed structure, (c) the stress-oriented prismatic structure, and (d) the stress-oriented parallel plate structure.

(a)

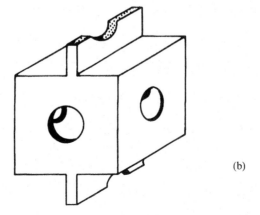

(b)

$$\frac{E_T^*}{E_s} = C_2 \left(\frac{\rho^*}{\rho_s}\right)^3. \tag{11.2}$$

Loaded along the prism axis (the 'longitudinal' direction) the dependence is linear (Eqn. (4.91)):

$$\frac{E_A^*}{E_s} = C_3 \left(\frac{\rho^*}{\rho_s}\right). \tag{11.3}$$

The plate-like structure (Fig. 11.12(d)) has similar properties, though with a different symmetry. Stress applied normal to the plates causes bending in the plates (and in the rods if they are loaded eccentrically) giving, roughly:

$$\frac{E_T^*}{E_s} = C_4 \left(\frac{\rho^*}{\rho_s}\right)^3.$$

Stress applied in either direction parallel to the plates loads the plates in their own plane, leading to the usual linear dependence on density:

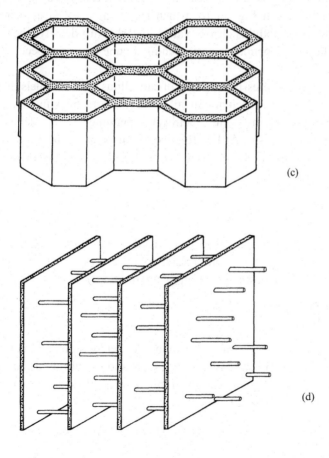

(c)

(d)

$$\frac{E_L^*}{E_s} = C_5 \left(\frac{\rho^*}{\rho_s} \right)$$

assuming that the volume fraction of solid in the plates is very much greater than that in the rods.

These results give a partial explanation of the experimental data for Young's modulus. Figure 11.7 shows tests on bone chosen without regard for its structure or for the direction of loading. The models suggest that the data should lie between the line with slope 1 (a linear dependence) and that with slope 3 (a cubic dependence), constructed to pass through the point corresponding to the modulus of fully dense bone. Most do, although a significant fraction of the samples have moduli even lower than the lowest prediction of the models, perhaps because the moduli were measured at low strain-rates when creep contributes to the strain. A statistical analysis of pooled data from Carter and Hayes (1977), Williams and Lewis (1982) and Bensusan *et al.* (1983) indicates that the modulus of trabecular bone was better described by $A\rho^2$ than either $B\rho^3$ or $(C\rho^2 + D\rho^3)$, where A, B, C and D are constants (Rice *et al.*, 1988). The anisotropy in modulus is more clearly revealed in Fig. 11.10, which shows data for stress-oriented bone. The longitudinal modulus clearly varies linearly with density; the transverse modulus shows a dependence approaching a cube law, as expected from Eqns. (11.2) and (11.3).†

The analysis of the compressive strength of cancellous bone parallels that for foams given in Chapter 5. Trabeculae can fail either by elastic buckling or by progressive microfracture (giving some ductility) depending on the slenderness ratio (Townsend *et al.*, 1975a). At low relative densities elastic buckling dominates and the compressive collapse stress for equiaxed cells is proportional to the square of the density (Eqn. (5.18a)):

$$\frac{\sigma^*}{E_s} = C_6 \left(\frac{\rho^*}{\rho_s} \right)^2. \tag{11.4}$$

At higher densities the trabeculae have lower slenderness ratios and fail by progressive microfracturing, giving a compressive collapse stress which should vary as the density raised to the power $3/2$ (Eqn. (5.27) or (5.33)):

$$\frac{\sigma^*}{\sigma_{ys}} = C_7 \left(\frac{\rho^*}{\rho_s} \right)^{3/2}. \tag{11.5}$$

As with the modulus, the perforated plate model behaves in a manner intermediate to that of a closed-cell foam (Eqns. (5.28) and (5.37b)) and that of an open-cell foam (Eqn. (5.18a)), with the result that its strength depends on relative density raised to a power between 1 and $3/2$.

†The absolute values of the moduli measured by Williams and Lewis (1982) are lower, by a factor of about 10, than those reported by other investigators, for reasons which are unclear.

Stress-oriented bone has strengths which differ in the longitudinal and transverse directions. The plateau strength for loading in the transverse direction, like that of honeycombs and of wood, is expected to vary as the second power of density (Eqn. (4.28) with (4.1b)):

$$\frac{\sigma^*}{\sigma_{ys}} = C_8 \left(\frac{\rho^*}{\rho_s}\right)^2 \tag{11.6}$$

while the longitudinal strength, reflecting the axial loading of the walls, should vary linearly (Eqn. (4.114)):

$$\frac{\sigma^*}{\sigma_{ys}} = C_9 \left(\frac{\rho^*}{\rho_s}\right). \tag{11.7}$$

As with the moduli, these results give a partial explanation of observed strengths of cancellous bone. Figure 11.8 shows data chosen without regard for its structure or the direction of loading. The models suggest that the data should be bracketed by lines of slope 1 and 2. In fact almost all the points cluster around the line of slope 2, and the same is true of the tensile strengths (Fig. 11.9). The square dependence on density is also supported by the statistical analysis of Rice *et al.* (1988). The anisotropy is more clearly revealed in the tests of Williams and Lewis (1982) on stress-oriented bone (Fig. 11.11). Here the longitudinal strength has the expected linear-dependence on density (Eqn. (11.7) and the transverse strength follows the predicted quadratic dependence fairly closely (Eqn. (11.6)).†

The fracture toughness of equiaxed foams varies as the density raised to the power 3/2 (Eqn. 5.47). Cancellous bone should show a similar dependence.

11.5   Conclusions

Cancellous bone has a cellular structure. The shape and density of the cells depend on the loads which, in the living animal, it has to support. If the loads are about equal in all three principal directions the bone tends to form roughly equiaxed cells. But if one load is much larger than the others the cell walls tend to align and thicken in the direction which will best support it. The relative density of the cells depends on the magnitude of the loads. That of lightly loaded bone is low, structured in a rod-like network of open cells; as the loads increase the cell walls thicken and spread until they resemble perforated plates. When the loads are predominantly in one direction the cells form with a structure which can be thought of as an aggregate of parallel prismatic cells, or of parallel

---

†As with their modulus data, the absolute magnitudes of the strength data reported by Williams and Lewis (1982) are a factor of 10 lower than those of other studies. The reasons are unclear.

plates, aligned with the direction of maximum principal stress. The loading on bone depends on its location in the body, and because of this the density and structure of cancellous bone varies greatly, and the structure is sometimes equiaxed and sometimes strongly oriented.

The mechanical behaviour of cancellous bone is typical of that of a cellular solid, and can be analysed by the methods of Chapters 4 and 5. The roughly equiaxed cells deform at first by bending or extension of the trabeculae (the cell walls or edges), giving linear-elastic behaviour and a Young's modulus that varies with density raised to a power which lies between 1 and 3, depending on cell shape and orientation. The trabeculae then either buckle or progressively micro-fracture (depending on their slenderness ratio) giving the characteristic plateau of the compressive stress–strain curve. The plateau-stress varies with density to a power of between 1 and 2, again depending on cell shape and orientation. The plateau ends when opposing cell walls touch.

Data for moduli and strengths of cancellous bone with relative densities between 0.08 and 0.3 are broadly consistent with the predictions of the models for bone structure and properties, though the detailed dependence of E^* and σ^* on density is not fully understood. Further work remains to be done in character-izing the microstructural anisotropy in cancellous bone, and relating this to the anisotropy in mechanical behaviour. Work is needed, too, to elucidate the way in which fracture toughness varies with structure and density (experience with foams suggests a dependence on density to a power of between 1 and 2).

Such work could have important consequences. An improved understanding of the relationship between the structure and properties of cancellous bone has many biomedical applications. It helps in designing prosthetic hips and knees with mechanical properties which match more closely those of the bone they are replacing. It helps, too, in diagnosing when patients with osteoporosis are at risk of bone fractures caused by the loss of bone mass associated with that dis-ease. And it could lead to an improved understanding of the mechanics of joints which is of relevance to the understanding of oesteoarthritis.

References

Ashman, R. B. and Rho, J. Y. (1988) Elastic modulus of trabecular bone material. *J. Biomech.*, **21**, 177–81.

Beaupre, G. S. and Hayes, W. C. (1985) *J. Biomech. Eng.*, **107**, 249.

Behiri, J. C. and Bonfield, W. (1980) *J. Mat. Sci.*, **15**, 1841.

Behiri, J. C. and Bonfield, W. (1984) *J. Biomech.*, **17**, 25.

Behrens, J. C., Walker, P. S. and Shoji, H. (1974) *J. Biomech.*, **7**, 201.

Bensusan, J. S., Davy, D. T., Heiple, K. G. and Verdin, P. J. (1983) *19th Annual Orthopaedic Research Society Meeting*, **8**, 132.

Bonfield, W. and Datta, P. K. (1974) *J. Mat. Sci.*, **9**, 1609.

Bonfield, W. and Datta, P. K. (1976) *J. Biomech.*, **9**, 121.

Bonfield, W., Grynpas, M. D. and Young, R. J. (1978) *J. Biomech.*, **11**, 473.

Carter, D. R. and Hayes, W. C. (1977) *Bone Joint Surg.*, **59A**, 954.

Carter, D. R. and Spengler, D. M. (1978) *Clin. Orthopaed. Rel. Res.*, **135**, 192.

Carter, D. R., Schwab, G. H. and Spengler, D. M. (1980) *Acta Orthopaed. Scand.*, **51**, 733.

Choi, K., Kuhn, J. L., Ciarelli, M. J. and Goldstein, S. A. (1990) The elastic moduli of human subchondral trabecular and cortical bone tissue and the size dependence of cortical bone modulus. *J. Biomech.*, **23**, 1103–13.

Currey, J. D. (1970) *Clin. Orthopaed. Rel. Res.*, **73**, 210.

Currey, J. D. (1984) *The Mechanical Adaptations of Bones*. Princeton University Press, Princeton, NJ.

Dyson, E. D., Jackson, C. K. and Whitehouse, W. J. (1970) *Nature*, **225**, 957.

Galante, J., Rostoker, W. and Ray, R. D. (1970) *Calc. Tiss. Res.*, **5**, 236.

Gibson, L. J. (1985) *J. Biomech.*, **18**, 317.

Gong, J. K., Arnold, J. S. and Cahn, S. H. (1964) *Anat. Rec.*, **149**, 325.

Hayes, W. C. and Carter, D. R. (1976) *J. Biomed. Mat. Res.*, Symposium, **7**, 537.

Hayes, W. C. and Synder, B. (1981) in Cowin, S. C. (ed.) *Mechanical Properties of Bone*, AMD, vol. 45, American Society of Mechanical Engineers.

Hodgkinson, R. and Currey, J. D. (1992) Young's modulus, density and material properties in cancellous bone over a large density range. *J. Mater. Sci. Mater. Med.*, **3**, 377–81.

Hvid, I., Bentzen, S. M., Linde, F., Mosekilde, L. and Pongsoipetch, B. (1989) X-ray quantitative computed tomography: the relations to physical properties of proximal tibial trabecular bone specimens. *J. Biomech.*, **22**, 837–44.

Kaplan, S. J., Hayes, W. C. and Stone, J. L. (1985) Tensile strength of bovine trabecular bone. *J. Biomech.*, **18**, 723–7.

Keaveny, T. M., Borchers, R. E., Gibson, L. J. and Hayes, W. C. (1993) Theoretical analysis of the experimental artifact in trabecular bone compressive modulus. *J. Biomech.*, **26**, 599–607.

Keaveny, T. M., Wachtel, E. F., Ford, C. M. and Hayes, W. C. (1994) Differences between the tensile and compressive strengths of bovine tibial trabecular bone depend on modulus. *J. Biomech.*, **27**, 1137–46.

Kuhn, J. L., Goldstein, S. A., Choi, K. W., Landon, M., Herzig, M. A. and Matthews, L. S. (1987) The mechanical properties of single trabeculae. *Trans. 33rd Orthopaedic Research Society*, **12**, 48.

Lanyon, L. E. (1974) *J. Bone Joint Surg.*, **56B**, 160.

Linde, F., Norgaard, P., Hvid, I., Odgaard, A. and Soballe, K. (1991) Mechanical properties of trabecular bone: dependency on strain rate. *J. Biomech.*, **24**, 803–9.

McElhaney, J. H., Fogle, J. L., Melvin, J. W., Haynes, R. R., Roberts, V. L. and Alem, N. M. (1970a) *J. Biomech.*, **3**, 495.

McElhaney, J. H., Alem, N. and Roberts, V. (1970b) ASME publication 70-WA/BHF-2.

Margel-Robertson, D. (1973) Ph.D. thesis, Stanford University.

Melvin, J. W. and Evans, F. G. (1973) *ASME Biomaterials Symposium*. ASME, New York.

Mente, P. L. and Lewis, J. L. (1987) Young's modulus of trabecular bone tissue. *Trans. 33rd Orthopaedic Research Society*, **12**, 49.

Moskilde, L. (1989) Sex differences in age-related loss of vertebral trabecular bone mass and structure – biomechanical consequences. *Bone*, **10**, 425–32.

Neil, J. L., Demos, T. C., Stone, J. L. and Hayes, W. C. (1983) *29th Annual Orthopaedic Research Society Meeting*, **8**, 344.

Odgaard, A. and Linde, F. (1991) The underestimation of Young's modulus in compressive testing of cancellous bone specimens. *J. Biomech.*, **24**, 691–8.

Parfitt, A. M. (1992) Implications of architecture for the pathogenesis and prevention of vertebral fracture. *Bone*, **13**, S41–S47.

Pugh, J. W., Rose, R. M. and Radin, E. L. (1973a) *J. Biomech.*, **6**, 475.

Pugh, J. W., Rose, R. M. and Radin, E. L. (1973b) *J. Biomech.*, **6**, 657.

Reilly, D. T. and Burstein, A. H. (1974) *J. Bone Joint Surg.*, **56A**, 1001.

Reilly, D. T. and Burstein, A. H. (1975) *J. Biomech,.* **8**, 393.

Rice, J. C., Cowin, S. C. and Bowman, J. A. (1988) On the dependence of the elasticity and strength of cancellous bone on apparent density. *J. Biomech.*, **21**, 155–68.

van Rietbergen, B., Weinans, H., Huiskes, R. and
    Odgaard, A. (1995) A new method to determine
    trabecular bone elastic properties and loading
    using micromechanical finite element models. *J.
    Biomech.*, **28**, 69–81.

Runkle, J. C. and Pugh, J. W. (1975) The
    micromechanics of cancellous bone – II.
    Determination of the elastic modulus of
    individual trabeculae by a buckling analysis.
    Bull. Hosp. Jt. Dis., **36**, 2–10.

Ryan, S. D. and Williams, J. L. (1989) Tensile
    testing of rodlike trabeculae excised from
    bovine femoral bone. *J. Biomech.*, **22**, 351–5.

Silva, M. J., Hayes, W. C. and Gibson, L. J. (1995)
    The effects of non-periodic microstructure on
    the elastic properties of two-dimensional
    cellular solids. *Int. J. Mech. Sci.* **37**, 1161–77.

Stone. J. L., Beaupre, G. S. and Hayes, W. C. (1983)
    J. Biomech., **16**, 743.

Thompson, D. W. (1961) *On Growth and Form*.
    Cambridge University Press, Cambridge.

Townsend, P. R., Rose, R. M. and Radin, E. L.
    (1975a) *J. Biomech.*, **8**, 199.

Townsend, P. R., Raux, P., Rose, R. M., Miegel, R.
    E. and Radin, E. L. (1975b) *J. Biomech.*, **8**, 363.

Weaver, J. K. and Chalmers, J. (1966) *J. Bone Joint
    Surg.*, **48A**, 289.

Whitehouse, W. J. (1974) *J. Microsc.*, **101**, 153.

Whitehouse, W. J. (1975) *J. Pathol.*, **116**, 213.

Whitehouse, W. J. and Dyson, E. D. (1974) *J. Anat.*,
    118, 417.

Whitehouse, W. J., Dyson, E. D. and Jackson, C. K.
    (1971a) *Brit. J. Radiol.*, **44**, 367.

Whitehouse, W. J., Dyson, E. D. and Jackson, C. K.
    (1971b) *J. Anat.*, **108**, 481.

Williams, J. L. and Lewis, J. L. (1982) *J. Biomech.
    Eng.*, **104**, 50.

Wolff, J. (1869) *Zentralblatt für die medizinische
    Wissenschaft*, **VI**, 223.

Wright, T. M. and Hayes, W. C. (1977) *J. Biomech.*,
    10, 419.

Chapter 12

Cork

12.1   Introduction and synopsis

Cork has a remarkable combination of properties. It is a light yet resilient; it is an outstanding insulator for heat and sound; it has a high coefficient of friction; and it is impervious to liquids, chemically stable and fire-resistant. Such is the demand that production now exceeds half a million tonnes a year (and 1 tonne of cork has the volume of 56 tonnes of steel).

In pre-Christian times cork was used (as we still use it today) for fishing floats and soles of shoes. When Rome was besieged by the Gauls in 400 BC, messengers crossing the Tiber clung to cork for buoyancy (Plutarch, AD 100). And ever since man has cared about wine, he has cared about cork to keep it sealed in flasks and bottles. 'Corticum abstrictum pice demovebit amphorae'† sang Horace (27 BC) to celebrate his miraculous escape from death from a falling tree. But it was in the Benedictine Abbey at Hautvilliers where, in the seventeenth century, the technology of stopping wine bottles with clean, unsealed cork was perfected. Its elasticity and chemical stability mean that it seals the bottle without contaminating the wine, even when it must mature for many years. No better material is known, even today.

Commercial cork is the bark of an oak (*Quercus suber*) that grows in Portugal, Spain, Algeria and California. Pliny describes it thus (Pliny, AD 77): 'The Cork-Oak is a small tree; its only useful product is its bark which is extremely thick

---

†Pull the cork, set in pitch, from the bottle.

and which, when cut, grows again.' Modern botanists add that the cork cells (phellem) grow from the cortex cells via an intermediate structure known as cork cambium (phellogen). Their walls are covered with thin layers of an unsaturated fatty acid (suberin) and waxes which make them impervious to air and water, and resistant to attack by many acids (Esau, 1965; Zimmerman and Brown, 1971; Eames and MacDaniels, 1951). All trees have a thin layer of cork in their bark. *Quercus suber* is unique in that, at maturity, the cork forms a layer several centimetres thick around the trunk of the tree. Its function in nature is to insulate the tree from heat and loss of moisture, and perhaps to protect it from damage by animals (suberin tastes unpleasant). We use it today for thermal insulation in refrigerators and rocket boosters, acoustic insulation in submarines and recording studios, as a seal between mating surfaces in woodwind instruments and internal combustion engines and as an energy-absorbing medium in flooring, shoes and packaging. Its use has widened further since 1892, when a Mr John Smith of New York patented a process for making cork aggregate by the simple hot-pressing of cork particles: the suberin provides the necessary bonding.

In this chapter we describe the structure of cork, review data for the moduli and collapse stresses, and examine the way in which the theory of Chapter 4 can be used to explain them. We conclude with a survey of the ways in which the special properties of cork are exploited in several applications.

12.2   The structure of cork

Cork occupies a special place in the histories of microscopy and of plant anatomy. When, around 1660, Robert Hooke perfected his microscope, one of the first materials he examined was cork. What he saw led him to identify the basic unit of plant and biological structure, which he termed 'the cell' (Hooke, 1664). His careful drawings of cork cells (Fig. 12.1) show their roughly hexagonal shape in one section and their box-like shape in the other. Hooke noted that the cells were stacked in long rows, with very thin walls, like the wax cells of the honeycomb. Subsequent descriptions of cork-cell geometry add very little to this. Esau (1965), for example, describes cork cells as 'approximately prismatic in shape – often somewhat elongated parallel to the long axis of the stem'. Lewis (1928) concluded that their shape lay 'somewhere between orthic and prismatic tetrakaidecahedrons', Eames and MacDaniels (1951) simply described them as 'polygonal', but their drawing shows the same shape that Hooke described. These descriptions conflict, and none is quite correct. Figure 12.2 shows the three faces of a cube of cork (Gibson *et al.*, 1981). In one section the cells are roughly hexagonal; in the other two they are shaped like little bricks, stacked as

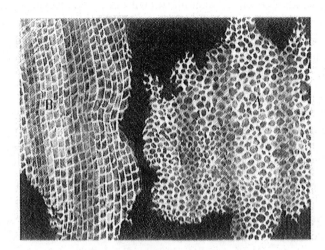

Figure 12.1 Radial (a) and tangential (b) section of cork, as seen by Robert Hooke through his microscope in 1664.

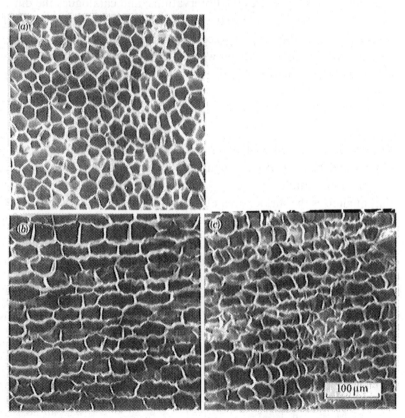

Figure 12.2 Scanning electron micrographs of (a) radial, (b) axial and (c) tangential sections of cork.

one would stack them in building a wall. The similarity with Hooke's drawing is obvious.

From micrographs such as these the cell shape can be deduced. Roughly speaking, the cells are closed hexagonal prisms (Fig. 12.3) stacked in rows so that the hexagonal faces register and are shared by two cells; but the rows are staggered so that the membranes forming the hexagonal faces are not continuous across rows. Figure 12.4 shows how the cells lie with respect to the trunk of the tree. The axes of the hexagons lie parallel to the radial (X_3) direction. A cut normal to the radial direction shows the hexagonal cross-section; any cut containing the radial direction shows the rectangular section, stacked like bricks in a wall because of the staggering of the rows. At higher magnifications the scanning microscope reveals details that Hooke could not see, because their scale is comparable with the wavelength of light. Six of the eight walls of each cell are corrugated (Fig. 12.5). Each cell has two or three complete corrugations, so that it is shaped like a concertina, or bellows.

Figure 12.6 summarizes the observations, and catalogues the dimensions of the cork cells. The cell walls have a uniform thickness (about 1 μm) of height h (about 45 μm) and hexagonal face-edge l (about 20 μm). The density ρ^* of the cork is related to that of the cell-wall material ρ_s and the cell-wall dimensions by (Table 2.2, closed hexagonal prisms):

$$\frac{\rho^*}{\rho_s} = \frac{t}{l}\left(\frac{l}{h} + \frac{2}{\sqrt{3}}\right). \tag{12.1}$$

The density of the cell-wall material is close to 1150 kg/m^3 (Gibson *et al.*, 1981). The mean density of cork is roughly 170 kg/m^3, giving a relative density of 0.15.

The aspect ratio of the cells, h/l, is about 2; this is larger than the value (1.7) that minimizes the surface area of a close-packed array of hexagons. The radial section of the structure does not always show hexagonal sections: five-, six-,

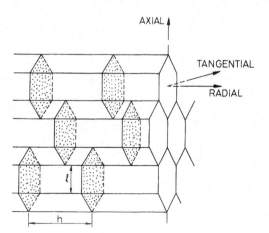

Figure 12.3 The shape of cork cells, deduced from the micrographs shown in Fig. 12.2. They are hexagonal prisms, having 8 faces, 18 edges and 12 vertices. There are, of course, imperfections, involving cells with different numbers of faces, edges, etc., and the cell walls are not straight (as here) but corrugated.

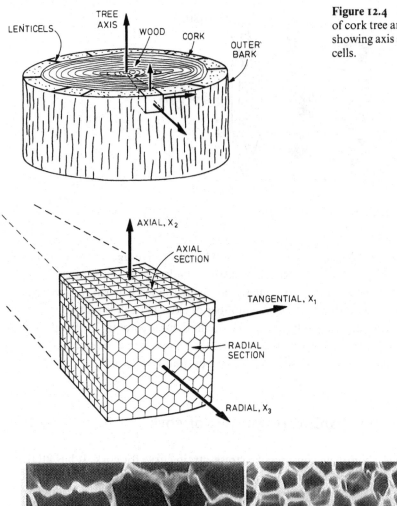

Figure 12.4 Diagram of cork tree and cork, showing axis system and cells.

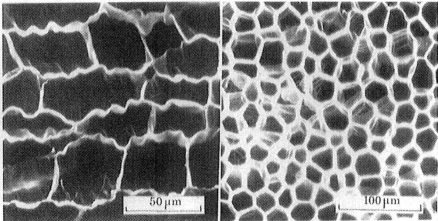

Figure 12.5 Scanning electron micrographs of cork cells, showing corrugations.

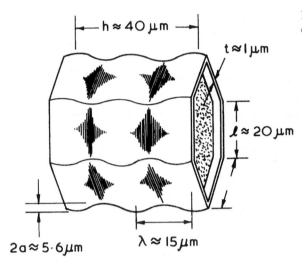

Figure 12.6   A corrugated cell, showing dimensions.

seven- and eight-sided figures are all observed. But the average number or sides per cell in the radial section is six (Lewis, 1928, finds 5.978). This, of course, is an example of the operation of Euler's law which asserts, when applied to a three-connected net, that the average number of sides per cell is six (Chapter 2, Eqn. (2.4)). The cells themselves are very small; there are about 20 000 of them in a cubic millimetre. They are much smaller than those in normal foamed plastics, and comparable with those in 'microporous' foams (Chapter 6).

12.3   The mechanical properties of cork

Figure 12.7 is a complete compressive stress–strain curve for cork. It has all the characteristics we expect of a cellular solid (Chapters 4 and 5). It is linear-elastic up to about 7% strain, at which point elastic collapse gives an almost horizontal

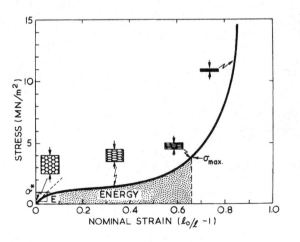

Figure 12.7   Stress–strain curve for cork.

plateau which extends to about 70% strain when complete collapse of the cells causes the curve to rise steeply. Typical mean values for Young's moduli, the shear moduli and Poisson's ratios are recorded in Table 12.1. The Young's modulus along the prism axis is roughly one and a half times that along the other two directions; additional tests by Fortes and his co-workers confirm this result (Fortes and Nogueira, 1989; Rosa and Fortes, 1991). The moduli (and the other properties) have circular symmetry about the prism axis. In the plane normal to this axis cork is roughly isotropic, as might be expected from its structure. The table lists the stress (σ_{el}^*) and the strain at the start of the plateau in compression, and the fracture stress (σ_f^*) and fracture strain in tension. Tensile fracture along

Table 12.1 The mechanical properties of cork†

Young's modulus

tangential	E_1^*	$\left.\right\}= 13 \pm 5 \text{ MN/m}^2$
axial	E_2^*	
radial	E_3^*	$= 20 \pm 7 \text{ MN/m}^2$

Shear modulus

in 1–2 plane	G_{12}^*	$= 4.3 \pm 1.5 \text{ MN/m}^2$
in 1–3 plane	G_{13}^*	$\left.\right\}= 2.5 \pm 1.0 \text{ MN/m}^2$
in 2–3 plane	G_{23}^*	

Poisson's ratio

$v_{12}^* = v_{21}^* = 0.25^a\text{–}0.50$

$v_{13}^* = v_{31}^* = v_{23}^* = v_{32}^* = 0\text{–}0.10^a$

Collapse stress (and strain)

tangential	$(\sigma_{el}^*)_1$	$\left.\right\}= 0.7 \pm 0.2 \text{ MN/m}^2, \text{ 6% strain}$
axial	$(\sigma_{el}^*)_2$	
radial	$(\sigma_{el}^*)_3$	$= 0.8 + 0.29 \text{ MN/m}^2, \text{ 4% strain}$

Fracture stress (and strain)

tangential	$(\sigma_f^*)_1$	$\left.\right\}= 1.1 \pm 0.2 \text{ MN/m}^2, \text{ 9% strain}$
axial	$(\sigma_f^*)_2$	
radial	$(\sigma_f^*)_3$	$= 1.0 + 0.2 \text{ MN/m}^2, \text{ 5% strain}$

Fracture toughness $\quad K_{IC}^* \quad = 60\text{–}130 \text{ MPa m}^{1/2b}$

Loss coefficient (at 0.01 Hz)

tangential	η_1^*	$\left.\right\}= 0.3 \text{ at 20% strain}$
axial	η_2^*	
radial	η_3^*	$= 0.1 \text{ at 1% strain}$

†Data from Gibson *et al.* (1981), except (a) Fortes and Nogueira (1989) and (b) Rosa and Fortes (1991).

the prism axis occurs at 5% strain, but in the other two directions the strain is larger – about 9%. The fracture toughness, K^*_{ICij}, depends on both the direction of the normal to the crack plane, i, and the direction of crack propagation, j. It is roughly 115–130 kPa$\sqrt{\text{m}}$ for systems with the radial direction normal to the crack plane and 60–100 kPa$\sqrt{\text{m}}$ for all other systems (Rosa and Fortes, 1991). The final item in the table is the loss coefficient:

$$\eta^* = \frac{D}{2\pi U}$$

where D is the energy dissipated in a complete tension–compression cycle and U is the maximum energy stored during the cycle. It is roughly constant from 0.01 to 4 kHz, with a broad peak at 2 kHz (Fernandez, 1978) and increases from 0.1 at low strain amplitudes to 0.3 at high. It is this which gives cork good damping and sound-absorbing properties, and a high coefficient of friction. Additional data for creep and rate effects have been reported by Rosa and Fortes (1988a,b).

When cork deforms, the cell walls bend and buckle. The behaviour when the axis of deformation lies along the prism axis differs from that when it lies across the prisms. Both tensile and compressive deformation across the prism axis first bend the cell walls. In compression, at higher strains, the cell walls buckle (Fig. 12.8) giving large recoverable strains of order 1.

Tensile deformation along the prism axis unfolds the corrugations (Fig. 12.9), straightening the prism walls. About 5% extension is possible in this way; by then the walls have become straight, and further tension at first stretches and then breaks them, causing the cork to fail. Compressive deformation, on the other hand, folds the corrugations. The folding is unstable; once it reaches about 10% a layer of cells collapses completely, suffering a large compressive strain (Fig. 12.10). Further compression makes the boundary of this layer propagate; cells collapse at the boundary, which moves through the cork like a Lüders band through steel or a drawing band through polyethylene.

The in-plane mechanical properties for loading normal to the prism axis can be understood in terms of the models developed in Chapter 4 for a two-dimensional array of hexagonal cells. Results developed there for the in-plane properties explain the values of the moduli we have called E^*_1, E^*_2, G^*_{12}, v^*_{12}, and v^*_{21}, and the elastic collapse stresses $(\sigma^*_{el})_1$ and $(\sigma^*_{el})_2$. It is convenient to recall the results, obtained by substituting Eqn. (12.1) into Eqns. (4.12), (4.18), (4.13) and (4.21):

$$E^*_1 = E^*_2 = 0.5E_s \left(\frac{\rho^*}{\rho_s}\right)^3 \tag{12.2a}$$

$$G^*_{12} = G^*_{21} = 0.13E_s \left(\frac{\rho^*}{\rho_s}\right)^3 \tag{12.2b}$$

$$v^*_{12} = v^*_{21} = 1.0 \tag{12.2c}$$

$$(\sigma^*_{el})_2 = 0.05 \left(\frac{\rho^*}{\rho_s}\right)^3. \tag{12.2d}$$

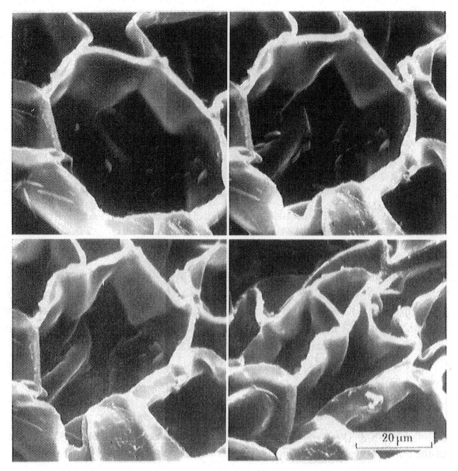

Figure 12.8   Micrographs showing the bending and buckling of cell walls as cork is compressed across the prism axis. The load is applied from top left to bottom right on the micrographs.

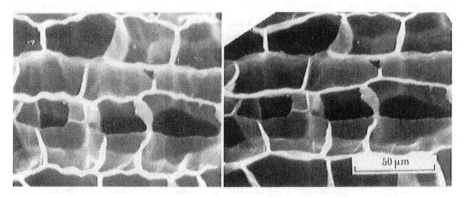

Figure 12.9   Micrographs showing the progressive straightening of cell walls as cork is pulled in the radial direction (tensile axis parallel to the prism axis).

Figure 12.10   The catastrophic collapse of cork cells compressed in the radial direction (along the prism axis).

Here E_s and ρ_s are the modulus and density of the solid of which the cell walls are made and ρ^* is the overall density of the cork. The collapse strain is given by $\epsilon^* = \sigma^*_{el}/E^* = 0.1$. Below this strain the structure is linear-elastic; above it is non-linear but still elastic. Buckling allows deformation to continue until the cell walls touch at a nominal strain ϵ_D (given by Eqn. (4.75)) of about

$$\epsilon_D = 1 - 1.4\left(\frac{\rho^*}{\rho_s}\right).$$

If a honeycomb of regular prismatic cells like that of Fig. 12.3 is compressed parallel to the prism axis, the modulus is determined by the axial compression of the material in the cell walls. This leads to the obvious result (Eqn. (4.91)):

$$\frac{E_3^*}{E_s} = \frac{\rho^*}{\rho_s}. \tag{12.3}$$

This equation properly predicts the axial modulus of wood (Chapter 10) but it over-estimates, by a factor of 50 or more, that of cork. The discrepancy arises because the cell walls have corrugations which fold or unfold like the bellows of a concertina when the cork is deformed (Fig. 12.9). The axial stiffness of a corrugated cell wall of thickness, t, and corrugation amplitude, a, is derived by Gibson *et al.* (1981). It is:

$$\frac{E_3^*}{E_s} = 0.7\left(\frac{\rho^*}{\rho_s}\right)\left\{\frac{1}{1 + 6\left\{\frac{a}{t}\right\}^2}\right\}. \tag{12.4}$$

This deformation has another interesting feature. Axial compression produces no lateral expansion, because the cells simply fold up. We therefore expect:

$$v_{13}^* = v_{31}^* = v_{23}^* = v_{32}^* = 0. \tag{12.5}$$

The properties ρ_s and E_s of the cell walls of cork are discussed by Gibson *et al.* (1981). The best estimates of their values are:

$$\rho_s = 1150 \, \text{kg/m}^3$$
$$E_s = 9 \, \text{GN/m}^2. \tag{12.6}$$

This information, together with the dimensions of the cells given earlier, give the moduli and collapse stresses for cork given in Table 12.2. Agreement is remarkably good. In particular, the understanding of the cork structure explains the isotropy in the plane normal to the radial (X_3) direction and the factor of 1.5 difference between Young's modulus in the radial direction and in the other two; and it explains the striking anisotropy in the values of Poisson's ratios and the magnitude of the elastic collapse loads. The biggest discrepancy is in the value of Poisson's ratio, v_{12}^*, and is probably due to a variation in cell shape and orientation, and to the constraining effect of the membranes which form the hexagonal end-faces of the cells.

12.4 Uses of cork

For at least 2000 years cork has been used for 'floats for fishing nets, and bungs for bottles, and also to make the soles of woman's winter shoes' (Pliny, AD 77). Few materials have such a long history, or have survived so well the competition from man-made substitutes. We now examine briefly how the special structure of cork has suited it so well to its uses.

Table 12.2 Comparison between calculated and measured properties of cork[†]

	Calculated	Measured
Moduli		
E_1^*, E_2^* (MN/m^2)	15	13 ± 5
E_3^* (MN/m^2)	20	20 ± 7
G_{12}^*, G_{21}^* (MN/m^2)	4	4.3 ± 1.5
$G_{13}^*, G_{31}^*, G_{23}^*, G_{32}^*$ (MN/m^2)	—	2.5 ± 1
$v_{12}^* = v_{21}^*$	1.0	0.25^{a}–0.50
$v_{13}^* = v_{31}^* = v_{23}^* = v_{32}^*$	0	0–0.10[a]
Compressive collapse stress		
$(\sigma_{\text{el}}^*)_1, (\sigma_{\text{el}}^*)_2$ (MN/m^2)	1.5	0.7 ± 0.2
$(\sigma_{\text{el}}^*)_3$ (MN/m^2)	1.5	0.8 ± 0.2

[†]Data from Gibson *et al.* (1981), except for (a) Fortes and Nogueira (1989).

(a) Bungs for bottles and gaskets for woodwind instruments

Connoisseurs of wine agree that there is no substitute for cork. Plastic corks are hard to insert and remove, they do not always make a very good seal, and they may contaminate the wine. Cork corks have none of these problems. The excellence of the seal is a result of the elastic properties of the cork. It has a low Young's modulus (E^*) but, much more important, it also has a low bulk modulus ($K^* = \frac{1}{3}E^*$, Fortes *et al.*, 1989). Solid rubber and solid polymers above their glass transition temperature have a low E but a large K, and it is this which makes them hard to force into a bottle, and gives a poor seal when they are inserted.

One might expect that the best seal would be obtained by cutting the axis of the cork parallel to the prism axis of the cork cells; then the circular symmetry of the cork and of its properties are used to best advantage. This idea is correct: the best seal is obtained by cork cut in this way. But natural cork contains *lenticels*: tubular channels that connect the outer surface of the bark to the inner surface, allowing oxygen into, and carbon dioxide out from, the new cells that grow there. A glance at Fig. 12.4 shows that the lenticels lie parallel to the prism axis, and that a cork cut parallel to this axis will therefore leak. That is why most commercial corks are cut with the prism axis (and the lenticels) at right angles to the axis of the bung.

A way out of this problem is shown in Fig. 12.11. The base of the cork, where sealing is most critical, is made of two discs cut with the prism axis (and lenticels) parallel to the axis of the bung itself. The leakage problem is overcome by laminating the two discs together so that the lenticels do not connect. Then the cork,

Figure 12.11 A section through a champagne cork and through a normal cork.

when forced into the bottle, is compressed (radially) in the plane in which it is isotropic, and it therefore exerts a uniform pressure on the inside of the neck; and the axial load needed to push the cork into the bottle produces no radial expansion (which would hinder insertion) because this Poisson's ratio (v_{31}^*) is zero.

Cork makes good gaskets for the same reason that it makes good bungs: it accommodates large elastic distortion and volume change, and its closed cells are impervious to water and to oil. Thin sheets of cork are used, for instance, for the joints of woodwind and brass instruments. The sheet is always cut with prism axis (and lenticels) normal to its plane. The sheet is then isotropic in its plane, and this may be the reason for cutting it so. But it seems more likely that it is cut like this because the Poisson ratio for compression down the prism axis is zero. Then, when the joints of the instrument are mated, there is no tendency for the sheet to spread in its plane, and wrinkle.

(b) Friction for shoes and floor coverings

Manufacturers who sell cork flooring sometimes make remarkable claims that it retains its friction even when polished or wet.

Friction between a shoe and a cork floor has two origins. One is *adhesion*: atomic bonds form between the two contacting surfaces, and work must be done to break and re-form them if the shoe slides. Between a hard shoe and a tiled or stone floor this is the only source of friction and, since it is a surface effect, it is completely destroyed by a film of polish or soap. The other source of friction is due to *anelastic loss*. When a rough shoe slides on a cork floor the bumps on the shoe deform the cork. If the cork were perfectly elastic no net work would be done: the work done in deforming the cork ahead of a bump is recovered as the bump moves on. But if the cork has a high loss coefficient (as it does) then it is like riding a bicycle through sand: the work done in deforming the material ahead of the bump is not recovered, and a large coefficient of friction appears. This anelastic loss is the main source of friction when rough surfaces slide on cork; and since it depends on processes taking place below the surface, it is not affected by films of polish or soap. Exactly the same thing happens when a cylinder or sphere rolls on cork, which therefore shows a high coefficient of rolling friction.

(c) Energy absorption and packaging

Many of the uses of cork depend on its capacity to absorb energy. Cork is attractive for soles of shoes and flooring because, as well as having frictional properties, it is resilient under foot, absorbing the shocks of walking. It makes good packaging because it compresses on impact, limiting the stresses to which the contents of the package are exposed. It is used as handles of tools to insulate the hand from the impact loads applied to the tool. In each of these applications it is essential that the stresses generated by the impact are kept low, but that considerable energy is absorbed.

Cellular materials are particularly good at this (Chapter 8). The stress–strain curve for cork (Fig. 12.7) shows that the collapse stress of the cells is low, so that the peak stress during impact is limited. But large compressive strains are possible, absorbing a great deal of energy as the cell walls progressively collapse. In this regard its structure and properties resemble polystyrene foam, which has replaced cork (because it is cheap) in many packaging applications, discussed in detail in Chapter 8.

(d) Insulation

The cork tree, it is thought, surrounds itself with cork to prevent loss of water in hotter climes. The properties involved – low thermal conductivity and low permeability to water – make it an excellent material for the insulation of cold, damp habitations. Caves fall into this category: the hermit caves of southern Portugal, for example, are liberally lined with cork. For the same reasons, crates and boxes are sometimes lined with cork. And the cork tip of a cigarette must appeal to the smoker because it insulates (a little) and prevents the tobacco becoming moist.

Heat flow through cellular materials is discussed in Chapter 7. Flow by conduction depends only on the amount of solid in the foam (ρ^*/ρ_s) and so it does not depend on the cell size. Flow by convection does depend on cell size because convection currents in large cells carry heat from one side of the cell to the other. But when cells are less than about 10 mm in size, convection does not contribute significantly. Flow by radiation, too, depends on the cell size: the smaller the cells, the more times the heat has to be absorbed and re-radiated, and the lower is the rate of flow.

So the small cells are an important feature of cork. They are very much smaller than those in ordinary foamed plastics and it is this, apparently, that imparts exceptional insulating properties to the material.

(e) Indentation and bulletin boards

Cellular materials densify when they are indented: the requirement that volume is conserved, so important in understanding indentation problems for fully dense solids, no longer applies. So when a sharp object like a drawing pin is stuck into cork, the deformation is highly localized. A layer of cork cells, occupying a thickness of only about one quarter of the diameter of the indenter, collapses, suffering a large strain. The volume of the indenter is taken up by the collapse of the cells so that no long-range deformation is necessary. For this reason the force needed to push the indenter in is small. And since the deformation is (non-linear)-elastic, the hole closes up when the pin is removed.

12.5 Conclusions

Cork is the epitome of an elastomeric cellular solid. Its low density ($\rho^*/\rho_s = 0.15$) and closed cells, and its chemical stability and resilience, give it special properties which have been exploited by man for at least 2000 years. These properties derive from its cellular structure, and can be understood in terms of the models of Chapter 4, with modifications to include the curious shape of the cells in cork. This shape gives the cork anisotropic elastic properties; and even these can be exploited to advantage in its applications.

References

Eames, A. J. and MacDaniels, L. H. (1951) *An Introduction to Plant Anatomy*. McGraw-Hill, London.

Esau, K. E. (1965) *Plant Anatomy*, p. 340. Wiley, New York.

Fernandez, L. V. (1978) *Inst. Nac. Invest. Agrar. (Spain)*, Cuad. No. 6, p. 7.

Fortes, M. A., Fernandes, J. J., Serralheiro, I. and Rosa, M. E. (1989) *J. Test. Eval.*, **17**, No. 1, 67–71.

Fortes, M. A. and Nogueira, M. T. (1989) *Mats. Sci. and Eng.*, **A122**, 227–32.

Gibson, L. J., Easterling, K. E. and Ashby, M. F. (1981) *Proc. Roy. Soc.*, **A377**, 99.

Hooke, R. (1664) *Micrographica*, pp. 112. Royal Society, London.

Horace, Q. (27 BC) *Odes*, book III, ode 8, line 10.

Lewis, P. T. (1928) *Science*, **68**, 635.

Pliny, C. (AD 77) *Natural History*, vol. 16, section 34.

Plutarch (AD 100) *Life of Camillus, Parallel Lives*, vol. II,. ch. xxv, p. 154.

Rosa, M. E. and Fortes, M. A. (1988a) *J. Mat. Sci.*, **23**, 35–42.

Rosa, M. E. and Fortes, M. A. (1988b) *J. Mat. Sci.*, **23**, 879–85.

Rosa, M. E. and Fortes, M. A. (1991) *J. Mat. Sci.*, **26**, 341–8.

Zimmerman, M. H. and Brown, C. L. (1971) *Trees Structure and Function*, p. 88. Springer, Berlin.

Chapter 13

Sources, suppliers and property data

13.1   Introduction and synopsis

Manufacturers of foams produce data-sheets, listing their properties. We have assembled a database of available foams and their suppliers, and illustrate it here. The foams and their suppliers are cataloged in the Appendix 13A, as Tables 13.A1 and 13.A2. The first lists foams by chemistry and trade name, attaching a manufacturer or supplier code to each. The second relates this code to a company, an address, and, where possible, a telephone and fax number. The contents of the tables are based on information obtained from suppliers in 1995. Products, of course, evolve and develop, and new manufacturers and materials appear, so a completely up-to-date compilation is not possible. But this catalogue gives a starting point.

Data-plots are used to illustrate the range of foams properties. Two case studies illustrate methods of selecting foams for specific applications.

13.2   The compilation of materials and suppliers

There are three main difficulties in locating a given foam for a given engineering application. First, there are almost no standards, either national or international.[†] Second, foams and cellular solids are marketed under weird trade

---

[†]Some structural foams now meet approved standards on certain properties – notably fire resistance and durability – set by organisations such as the American Bureau of Shipping, Lloyd's Register of Shipping, Norske Veritas, Registro Italiano, Navale, and Bureau Veritas.

468

names ('Neopolen', 'Cellobond') which give little guidance in identifying either their chemistry or their supplier. And third, while some large-volume foams are widely marketed by well-known multi-nationals, many are the products of small, specialized producers, not otherwise known in the market place.

Sources for data for foams – other than the suppliers' data-sheets – are very limited. Texts and handbooks such as Saechtling (1983), Seymour (1987), The Handbook of Industrial Materials (1992) and the Encyclopedia of Polymer Science and Engineering (1985) contain scattered information. Two software packages – Plascams (1995) and the Cambridge Materials Selector (CMS, 1995) contain data for cellular solids. But in the end it is the manufacturer to whom one has to turn: the data-sheets for their products are the most reliable source of information.

This is the source from which the data, illustrated below using the CMS (1995) software, was drawn. The database includes data for polymeric, metallic and ceramic foams, and for natural cellular solids such as balsa wood. Suppliers for honeycombs are listed, but honeycombs do not appear on the figures because they are not normally used in isolation, but as cores for sandwich structures.

13.3 Property ranges for available cellular materials

Cellular solids are commonly used in five broad classes of application. First there are the purely *mechanical applications*, most notably as core materials for sandwich structures. Second, there are uses which aim at *impact mitigation*: packaging, energy absorption and shock protection. Third, there are applications which exploit the low-thermal conductivity of foams: *thermal insulation* for housing, for refrigeration and for high-temperature equipment. Fourth, there are applications which exploit the low densities of foams: *floatation* and buoyancy-aids. Finally, there are applications which utilize the open porosity and large internal surface area of foams for *filtration*, as permeable *membranes* and as *carriers of catalysts*.

The figures which follow illustrate some of the properties of commercially available foams of relevance to these applications. All the data refer to currently-available commercial foams and highly porous solids. Each is labelled with a shorthand description of the material, followed in brackets by the density in Mg/m^3; thus LDPE (0.018) means 'low-density polyethylene foam with a density of $0.018 \, Mg/m^{3}$'. The other short names which appear in the figures are explained in Table 13.1. The data envelopes give an idea of the range of properties exhibited by each foam. Part of a typical record used to create figures is shown in Table 13.2.

Sandwich cores require materials which are stiff and strong (to keep the faces of the sandwich apart) but which are also as light as possible. Figure 13.1 shows

Table 13.1   Short names, material class and long name of foams

Short name	Class[†]	Long name
ABS	P	acrylonitrile butadiene styrene
Alumina	C	alumina
Al/SiC(p)	M	*Duralcan* Al/Si-C Alloy
Balsa	N	balsa wood
Brick	C	low density fire brick
CORD	C	cordierite
Cork	N	cork
EPS	P	expanded polystyrene
FSZ	C	fully stabilised zirconia
Glass	C	glass
HDPE	P	high density polyethylene
LDPE	P	low density polyethylene
MEL	P	melamine
Mullite	C	mullite
PC	P	polycarbonate
PE	P	polyethylene
PHEN	P	phenolic
PMA	P	polymethylmethacrylate
PMACR	P	polymethacrylimid
PB	P	polypropylene
PS	P	polystyrene
PSL	P	polyethersulphone
PSZ	C	partly stabilized zirconia
PU	P	polyurethane
PVC	P	polyvinylchloride
SiC	C	silicon carbide
UF	P	ureaformaldehyde
YZA	C	yttria zirconia alumina
ZTA	C	zirconia toughened alumina

†P = polymer foam; C = ceramic foam; M = metal foam; N = natural cellular solid

Young's modulus E plotted against density, ρ. The diagonal line roughly separates rigid from elastomeric foams; materials which are light and stiff lie in the upper left side of the diagram. The range of moduli is enormous: the most flexible foams are more than 10^6 times less stiff than the most rigid ones – a range far greater than that spanned by fully dense solids (roughly 1 to 1000 GPa – a

Table 13.2 Part of a datafile for a low-density polyethylene foam

Name			
Short name	LDPE(0.018)		
Composition	(CH2)n		

General properties

Cells count	0.1	–	1	/mm^3
Density	0.016	–	0.02	Mg/m^3
Relative density	0.017	–	0.022	
Energy content	120	–	160	MJ/kg
Oxygen index	16	–	17	%
Price	1.4	–	1.5	£/kg
Water absorption	0.4	–	0.5	%

Mechanical properties

Bulk modulus	2.50E-4	–	3.00E-4	GPa
Compressive strength	0.01	–	0.015	MPa
Compr. stress @ 25% strain	0.031	–	0.035	MPa
Compr. stress @ 50% strain	0.09	–	0.1	MPa
Densification strain	0.94	–	0.96	
Ductility	0.9	–	1	
Elastic limit	0.01	–	0.015	MPa
Endurance limit	0.15	–	0.2	MPa
Flexural modulus	2.50E-4	–	3.00E-4	GPa
Fracture toughness	0.005	–	0.01	MPa.m$^{1/2}$
Hardness	0.01	–	0.015	MPa
Loss coefficient	0.1	–	0.2	
Modulus of rupture	0.01	–	0.015	MPa
Poisson's ratio	0.07	–	0.1	
Shear modulus	1.00E-4	–	2.00E-4	GPa
Tensile strength	0.26	–	0.3	MPa
Young's modulus	2.50E-4	–	3.00E-4	GPa

Thermal properties

Glass temperature	160	–	200	K
Maximum service temperature	360	–	370	K
Melting point	385	–	390	K
Minimum service temperature	200	–	210	K
Specific heat	1.95E+3	–	2.10E+3	J/kg.K
Thermal conductivity	0.036	–	0.038	W/m.K
Thermal expansion	190E-6	–	220E-6	/K

Table 13.2   *(cont.)*

Electrical properties

Breakdown potential	4.0E+6	–	6.0E+6	V/m
Dielectric constant	1.05	–	1.1	
Resistivity	1.00E+13	–	1.00E+15	ohm.m
Power factor	1.00E-4	–	2.00E-4	

Typical uses: Packaging, buoyancy, insulation, cushioning, sleeping mats, automative, furnishing

References: Datasheets: Zotefoams 001, Zotefoams 002,BASF 001; BASF 002; BASF 003

Tradenames: PLASTAZOTE®, EVAZOTE®, SUPAZOTE®, NEOPOLEN®

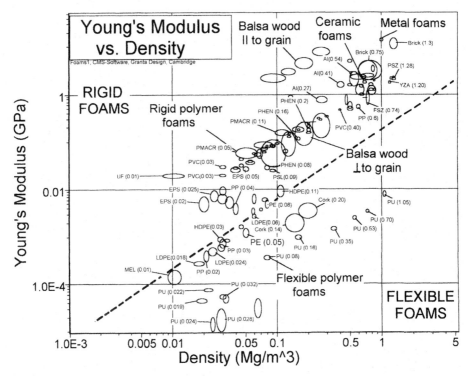

Figure 13.1   Young's modulus plotted against density for commercially-available foams. Rigid foams lie above the diagonal line; elastomeric (or 'flexible') foams lie below.

range of 10^3). Figure 13.2 shows the second important group of properties: the compressive strength σ_c (at 5% strain) plotted against density. Strengths, like moduli, span an enormous range (a factor of 10^5). These large ranges, shown by other foam properties also, offer the designer wide scope for matching material to design requirements. In selecting materials for sandwich cores, Fig. 13.3 is perhaps more helpful still. It shows the specific strength, σ_c/ρ plotted against the specific modulus, E/ρ. Materials near the top right have characteristics which make them suitable for the cores of sandwich structures. At the extreme top right lie 'end grain' balsa woods: they are exceptionally well suited to this application, and are marketed specifically for it ('end grain' means that the longitudinal properties are exploited).

Impact mitigation depends on the ability to absorb energy. Figure 13.4 shows the maximum compressive strain, ε_D, plotted against another measure of the compressive strength: it is $\sigma_{c(0.25)}$, the plateau stress at 25% strain level. The ability of a foam to absorb energy is measured approximately by the product $\varepsilon_D \cdot \sigma_{c(0.25)}$. It is shown as a family of diagonal contours. In general, the application determines the desired value of this, and of the plateau stress, because, in

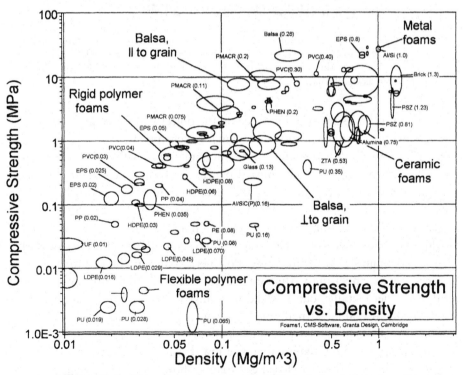

Figure 13.2   The compressive strength at 5% strain (or plateau strength) plotted against density for commercially-available foams. Metal foams are the strongest but do not have the highest specific strength (see Fig. 13.3).

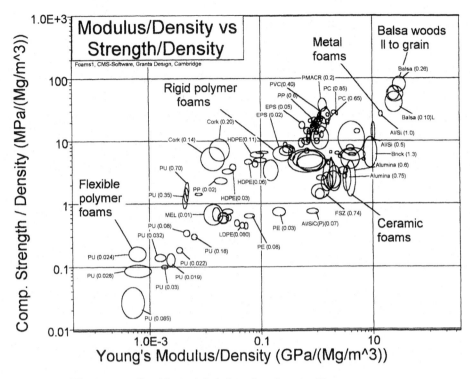

Figure 13.3 Specific modulus plotted against specific strength for commercially-available foams. End-grain balsa has the best properties, but only when loaded parallel to the grain – it is widely used as sandwich cores. Metal foams have promise as an alternative.

packaging, this limits the reaction-force on the object within the package. The figure allows the rational selection of a foam for such applications.

Most *thermal insulation* applications require materials with as low a thermal conductivity as possible, but they must also meet constraints on strength and service temperature. Figure 13.5 shows the thermal conductivity plotted against compressive strength at 5% strain. Both conductivity and strength fall with decreasing density, the conductivity reaching a minimum at $\lambda = 0.025$ W/m.K, roughly the conductivity of still air. Figure 13.6 shows the conductivity plotted against maximum service temperature. Polymeric foams are limited to applications near room temperature. Aluminium foams are useful to slightly higher temperatures; other metallic foams extend this range considerably, but are not yet commercially available. At really high temperatures, glass and ceramic foams become attractive, though their fragility demands special attention.

Buoyancy applications require foams of low density with closed cells. Figure 13.7 isolates information for closed-cell foams, showing metals, polymers and natural materials, such as balsa wood and cork. Polymers offer the lowest densities, and are the common choice for flotation. Among natural materials,

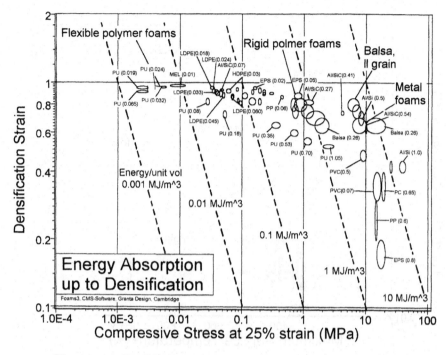

Figure 13.4 Densification strain ϵ_D plotted against plateau stress $\sigma_{c(0.25)}$ (the compressive strength at 25% strain) for commercially-available foams. The contours show energy absorption $\epsilon_D \cdot \sigma_{c(0.25)}$. per unit volume.

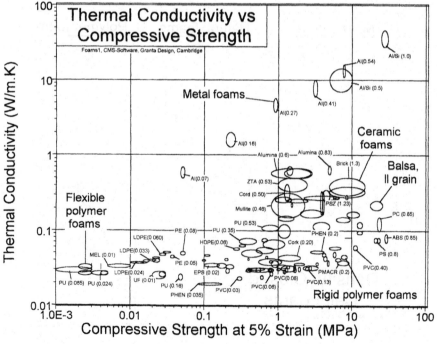

Figure 13.5 Thermal conductivity plotted against compressive strength for commercially-available foams. Strength and low conductivity are required when insulation is used to give structural rigidity, as in the walls of refrigerators.

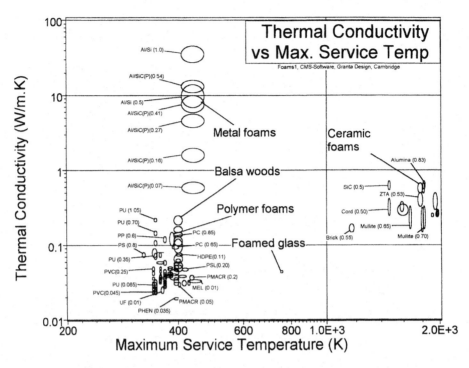

Figure 13.6 Thermal conductivity plotted against maximum service temperature for commercially-available foams. The figure allows selection of foams for high-temperature applications.

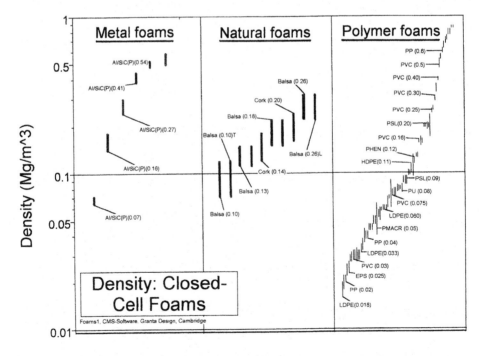

Figure 13.7 Buoyancy requires low-density, closed-cell foams. The densities of commercially-available closed-cell foams are shown here.

low-density balsa is good. It is interesting to note that foamed aluminium offers good buoyancy properties, combined with good durability.

Filtration and *catalysis* requires open-cell foams. The cell sizes of open-cell foams are shown in Fig. 13.8. Ceramic foams offer a range of between 0.1 and 3 mm; they find application in filtration of liquid metals, and for gas and liquid-phase catalysis at elevated temperatures. Most polymer foams have a similar range of cell sizes; though new techniques allow fabrication of micro-cellular foams with cells as small as 1 μm in size. They are used as dust-filters in air conditioners, air-cooled electronics and face-masks.

13.4   Case studies

The two case studies of this section illustrate how a database for foam properties can be used for selecting materials. The underlying methodology for selection, used here, is described in more detail in Ashby (1992).

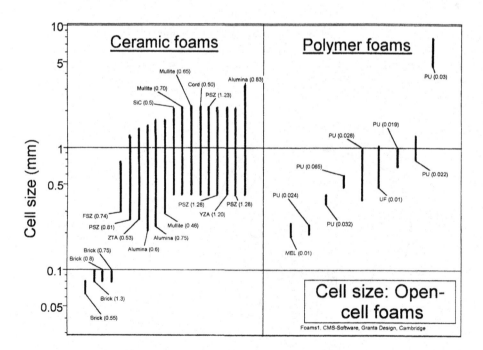

Figure 13.8   Filtration requires open-cell foams, of controlled cell size; catalysis depends on a high internal surface area. The cell sizes of commercially-available open-cell foams are shown here.

(a) Foam selection for a cycle helmet

Even in a small country such as England, cycling fatalities exceed 200 per year. An increasing number of cyclists wear helmets, giving a significant level of head-protection in an accident. The major impact-absorbing element of the helmet is a foamed polymer liner, commonly made of expanded polystyrene (EPS). Polymer foams are chosen because they are easily fabricated and because, unlike honeycombs, their ability to absorb energy is omni-directional. The helmet designer empirically selects the density and thickness to meet standard impact-tests which are at constant velocity (5 m/s) onto rigid anvils (Mills and Gilchrist, 1991). Could some of this empiricism be replaced by a more rational selection procedure?

Figure 13.9 is a schematic of a helmet. The foam thickness is limited by practicalities and (to some extent) styling; all helmets have almost the same liner-thickness (20 mm). The best choice of foam is then that which absorbs the most energy/unit volume, while limiting the load on the head to a less-than-damaging level (Table 13.3).

Table 13.3 Design requirement for cycle helmet

Function	Protective cycle helmet
Objective	Maximise energy absorption/unit volume
Constraints	Load on skull < damage-load

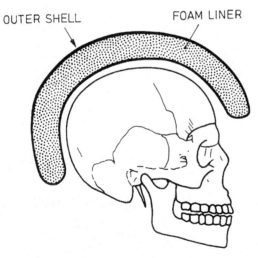

OUTER SHELL   FOAM LINER

Figure 13.9 A cycle helmet. The liner must absorb as much energy per unit volume as possible without allowing the force on the skull to reach a damaging level.

The helmet liner performs two-impact-mitigating functions. First, it redistributes a localised external force over a larger area, reducing the local stress on the skull. And second, it sets an upper limit, determined by the plateau-stress of the foam, to the magnitude of this distributed force. The key step in selecting a material for the liner is that of establishing the acceptable maximum value for this distributed force. The arguments here can become very sophisticated, but the underlying reasonings are these. The maximum tolerable deceleration, a, of the human head is 300g, provided it is applied for milliseconds only; more causes irreversible injury. The mass m of a head is approximately 3 kg, so the maximum allowable force, from Newton's law, is

$$F = ma \approx 9\,kN.$$

As the foam crushes against the obstacle (on the outside) and the skull (on the inside) it beds-down, distributing the load over a projected area A of order $10^{-2}\,m^2$. To prevent F rising above 9 kN, the foam must crush with a plateau-stress of

$$\sigma_c^* = \frac{F}{A} \approx 0.9\,MPa.$$

Figure 13.4 allows selection of foams for energy absorption. The foam which absorbs the most energy per unit volume while maintaining a plateau-stress σ_c^* of less than 0.9 MPa can be read from the figure: it is expanded polystyrene with a density of $0.05\,Mg/m^3$ (EPS(0.05)), which can absorb about $0.8\,MJ/m^3$. Most current cycle helmets are made of EPS, but they vary considerably in density, and thus weight – there is a tendency to select a low-density foam because it makes the helmet lighter. The figure shows that alternative selections with the same plateau-stress – polyurethane with a density of $0.53\,Mg/m^3$ (PU(0.53)), for instance – absorb much less energy.

The value of the diagram is the ease with which a first selection can be made, giving a short-list of viable candidates. Had the maximum permissible stress been 0.03 MPa, for instance, the best choice among commercially available foams would be the low-density polyethylenes; had it been 10 MPa, then Al-Si metal foams or end-grain balsa would become the best choices, absorbing almost $10\,MJ/m^3$.

(b) Energy versus cost: materials for insulation for refrigerators

The objective, in insulating a refrigerator (Fig. 13.10), is to minimize the energy lost from it, and thus the running cost over the design life. But the insulation itself has a capital cost associated with it, and it has a finite life. The most economical choice of material for insulation is that which minimizes the total. There is at least one constraint: an upper limit on the thickness of the insulation (Table 13.4).

Table 13.4   Design requirement for refrigerator insulation

Function	Thermal insulation
Objective	Minimize life-cost (material plus energy costs)
Constraints	Thickness $\leq x_{max}$

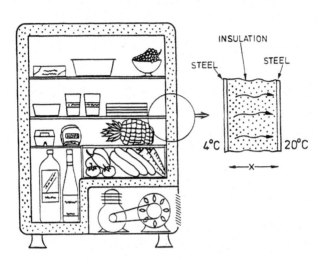

INSULATION

STEEL   STEEL

$4°C$ $20°C$

←—x—→

Figure 13.10   Insulation for refrigerators. The objectives are to minimize heat loss from the interior and to minimize the cost of the insulation itself.

The total cost per unit area of refrigerator wall, over the design life t_l, is the sum of two terms: the cost of the insulation, and the cost of the energy lost by heat transfer through the wall:

$$C_{tot} = x\rho^* C_m + \frac{\lambda \Delta T}{x} t_l C_E. \qquad (13.1)$$

Here x is the wall thickness, C_m is the cost/kg of the insulation, ρ^* is its density, C_E ($/J) is the cost of energy, λ is the thermal conductivity of the material, and ΔT is the temperature difference between the inside and the outside of the insulation layer. The equation contains two groupings of material properties:

$$M_1 = \frac{1}{\rho C_m} \quad \text{and} \quad M_2 = \frac{1}{\lambda}. \qquad (13.2)$$

We rewrite the equation in the form:

$$\frac{C_{tot}}{x} = \frac{1}{M_1} + \left[\frac{\Delta T}{x^2} t_l C_E\right] \frac{1}{M_2}. \qquad (13.3)$$

Everything in the equation is specified except the material groups M_1 and M_2.

Figure 13.11 shows M_2 plotted against M_1. Examining Eqn. (13.3) we see that the two contributions to $\frac{C_{tot}}{x}$ are equal when

$$M_2 = \left[\frac{\Delta T}{x^2} t_l C_E\right] M_1. \tag{13.4}$$

This describes a linear relationship between M_2 and M_1 which appears on Fig. 13.11 as a family of straight parallel lines, corresponding to differing values of the *coupling constant*, the quantity $\left[\frac{\Delta T}{x^2} t_l C_E\right]$. Two are shown, corresponding to two different values of the design life t_l: 10 years (typical of a household refrigerator) and 1 month (a hypothetical disposable cold-transport container, perhaps); the other variables are set at $\Delta T = 20°C$, $x = 10\,mm$ and $C_E = \$0.01/$ MJ. Centred around each line is a set of curved contours. They are plots of Eqn. (13.3). Moving upwards and to the right along a coupling line, the values of the cost contours decrease. For a design life of 10 years, the most economical choice minimizing the sum of the insulation and energy costs, is a phenolic foam with a density of 0.035 Mg/m^3 (PHEN(0.035)). But for a design life of 1 month, it is a low-density polystyrene foam such as the expanded polystyrene EPS (0.02) or the polypropylene foam PP(0.02)).

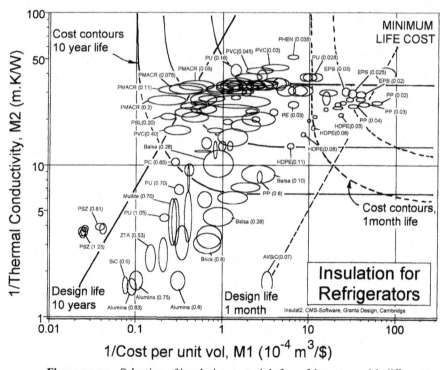

Figure 13.11   Selection of insulating materials for refrigerators with different design lives. The full lines show the coupling constant for two values of the design life. The contours show constant values of the life-cost; their values are relative.

13.5   **Conclusions**

Sources of information for the properties of polymer, metallic, ceramic and natural foams can be found in data-sheets issued by manufacturers and suppliers, and in occasional compilations. The Appendix to this chapter lists suppliers and their products, both by material and by company name. The data they contain is always incomplete, listing properties relevant to the applications for which the foam is normally used, but omitting others deemed by the manufacturer to be of secondary importance. Nonetheless, the data-sheets are an important source, and (until such information becomes available on WWW) the only one which is completely up-to-date.

Among compilations, the two largest are those contained in PLASCAMS (1995) and CMS (1995), the second of which was used to create the figures shown in this chapter. They provide a valuable resource, because they allow a comparison of foams and provide tools for selection exercises of the sort illustrated by the case studies of Section 13.5 – something the suppliers' data-sheets do not.

Appendix 13A: Commercially-available foams and their suppliers

Table 13A1 Materials and suppliers

Foam type & trade name	Supplier code
(a) Polymeric foams	
ABS foams	
BGA®	Borg-Warner 001
BGB®	Borg-Warner 001
BJF®	Borg-Warner 001
BIA®	Borg-Warner 001
LASTILAC®	CCC 001
Epoxy foams	
SYNCORE®	Hysol 001/002
Melamine elastomeric foam	
BASOTECT®	BASF 001/002/003
Phenolic foams	
KOOLPHEN K®	Kooltherm 001
CELLOBOND K®	BP 001
Polycarbonate foams	
LEXAN®	GE 001
POLYCARBFILL®	DSM 001
Polyester foams: see Polyurethane foams	
Polyethersulphone foams	
VICTREX®	Aerex 001
Polyethylene foams	
PLASTAZOTE®	Zotefoams 001/002
EVAZOTE®	Zotefoams 001/002
SUPAZOTE®	Zotefoams 001/002
NEOPOLEN®	BASF 001/002/003
Polymethacrylimid foams	
ROHACEL®	Roehm 001
Polypropylene foams	
NEOPOLEN P®	BASF 001/002/003
VESTOLEN P®	Huls 001
PROFIL®	DSM 001

Table 13.A1   *(cont.)*

Foam type & trade name	Supplier code
Polystyrene foam	
General	Dow 001
LACQURENE®	Elf 001/002/003
EPS®	BASF 001/002/003
STYRODUR®	BASF 001/002/003
STYROPOR®	BASF 001/002/003
VESTYRON®	Huls 001
STYRAFIL®	DSM 001
Polyurethane foams: flexible	
HYPERLAST®	Kemira 001; MacPherson 001
MICROVON®	Dunlop 001
FILTERCREST®	Crest-Foam 001: Scotfoam 001
SORANE 500®	Avalon 001
LAST-A-FOAM®	GPM 001
FOAMEX®	Foamex 001
Polyurethane foams: rigid	
HYPERLAST®	Kemira 001: MacPherson 001
BAYDUR®	Bayer 001
PROPOCON®	Lankro 001
LAST-A-FOAM®	GPM 001
Polyvinylchloride foams	
DIVINYCELL®	DIAB 001/002/003
KLEGECELL®	PELF 001
TERMINO®	Scot Bader 001
PLASTICELL®	Permali 001
VESTOLIT®	Huls 001
Urea formaldehyde foams	
POLLOPAS®	Huls 001
BEETLE UREA®	BIP 001
PERSTOP UREA®	Perstop-Ferguson 001

Table 13.A1 *(cont.)*

Foam type & trade name	Supplier code
(b) Natural foams	
Balsa	
FLEXICORE®	Flexicore 002
CONTOURKORE®	Baltek 001/002
PRO-BALSA®	DIAB 001/002/003
Cork	Amorim 001/002
(c) Ceramic foams	
Ceramic foams	
DUOCEL®	ERG 001
SELEE®	Selee 001
RETICEL®	Hi-Tech 001
RETICEL®	AM 001
Glass foams	
FOAMGLAS®	Pittsburg-Corning 001
(d) Metal foams	
Metal foams	
DUOCEL®	ERG 001
DURALCAN®	Alcan 001
ASTROMET 100®/200®/300®	AM 001
ALULIGHT®	Mepura 001
(e) Honeycombs	
Aluminium honeycombs	
AEROWEB® 3003	Ciba 001
AEROWEB® 5052	Ciba 001
DURACORE® 5052	Cyanamid 001/002/003
DURACORE® 5056	Cyanamid 001/002/003
Non-metallic honeycombs	
AEROWEB® G	Ciba 001
AEROWEB® A1	Ciba 001

Table 13.A2 Suppliers and addresses

Supplier code	Reference
Aerex 001	Aerex Limited Speciality Foams, CH-5643 Sins, SWITZERLAND Tel: (41)-42-66-00-66 Fax: (41)-42-66-17-07
Alcan 001	Alcan International Ltd. Box 8400, Kingston Ontario, K7L 5L9, CANADA Tel: 613-541-2400 Fax: 613-541-2134
AM 001	Astro Met Inc, 9974 Springfield Pike, Springfield, Ohio 45215, USA Tel: 513-772-1242 Fax: 513-772-9080
Amorim 001	Corticeira Amorim, LDA PO Box 1 – Mozelos 4539 Lourosa Codex, PORTUGAL Tel: 764-2395
Amorim 002	Amorim Cork, PO Box 1 – Santa Maria de Lamas 4538 Lourosa Codex, PORTUGAL Tel: 764-2013
Avalon 001	Avalon Chemical Co. Hitchen Lane, Shepton Mallet Somerset BA4 5TZ, UK Tel: 01749-343061 Fax: 01749-346283
Baltek 001	Baltek Corporation 10 Fairway Court, PO Box 195, North Vale, NJ 07647, USA Tel: 201-767-1400

Table 13.A2   *(cont.)*

Supplier code	Reference
Baltek 002	Baltek S.A., 61 rue de la Fontaine, 75016, Paris, FRANCE Tel: 33-1-647-5850
Baltek 003	Baltek Plc. Green Dragon House, 64, High Street Croydon CRO 9XN, UK Tel: 0181-688-4398 Fax: 0181-688-5740
BASF 001	BASF Aktiengesellchaft, D-6700 Ludwigshafen, GERMANY
BASF 002	BASF Corporation Engineering Plastics Bridgeport NJ 08014-0405, USA
BASF 003	BASF plc, PO Box 4, Earl Road, Cheadle Hume, Cheadle, Cheshire SK8 6QG, UK Tel: 0161-485-6222 Fax: 0161-488-4258
Bayer 001	Bayer House Strawberry Hill, Newbury, Berks RG13 1JA, UK Tel: 01635-39000 Fax: 01635-39513
BIP 001	BIP Chemicals, Popes Lane, Oldbury Warley, W. Midlands B69 4PD, UK Tel: 0121-552-1551 Fax: 0121-552-4267
BORG-WARNER 001	Borg-Warner Chemicals Inc., International Centre, Parkersburg, WV26102, USA Tel: 305-424-5411

Table 13.A2   *(cont.)*

Supplier code	Reference
BP 001	BP Chemicals Belgrave House, 76 Buckingham Palace Road, London SW1W 0SU, UK Tel: 0171-581-1388 Fax: 0171-581-6411
CCC 001	Central Chemicals Co Ltd, Unit, 7 Fallings Park Ind. Estate, Park Lane, Wolverhampton, W. Midlands, WV10 9QA, UK Tel: 01902-727544 Fax: 01902-864260
Ciba 001	Ciba-Geigy Plastics, Duxford, Cambridge CB2 4QD, UK Tel: 01233-833141
Crest-Foam 001	Crest-Foam Corp., 100 Carol Place, Moonachie, NJ 07074, USA Tel: 201-807-0809 Fax: 201-907-1113
Cyanamid 001	American Cyanamid Company, Industrial Chemicals Division, Wyane, New Jersey 07470, USA Tel: 201-831-2000
Cyanamid 002	ADCO Industries, Suite 1505 Fourth National Bank Building, 19 West Sixth Street, Tulsa, Oklahoma 74119, USA Tel: 918-587-6189
Cyanamid 003	Cyanamid B.V. Post Office Box 1523, 3000 BM Rotterdam, THE NETHERLANDS Tel: 010-116340

Table 13.A2 *(cont.)*

Supplier code	Reference
DIAB 001	Diab-Barracuda Inc., 1100 Avenue S, Grande Prairie, Texas 75050, USA Tel: 214-641-3014 Fax: 214-641-6677
DIAB 002	Diab-Barracuda Limited, 1 Eastville Close, Gloucester GL4 7SJ, UK Tel: 01452-501-860
DIAB 003	DIAB-Polimex France Centre d'Activites Saint Roch 61, rue Coquillet 452000 Montargis, FRANCE Tel: 38-93-80-20 Fax: 38-93-80-29
Dow 001	Dow Chemical Co 2040 Dow Center Midland, Michigan 48674, USA Tel: 1-800-441-4369
Dow 002	Dow Chemical Styrofoam Applications, PO Box 515-3825 Columbus Road SW, Granville, OH 43023, USA
Dow 003	Dow Europe SA Bachtobelstrasses 3 CH-8810 Horgen, SWITZERLAND Tel: 155-0407
Dow 004	Dow Information Centre, PO Box 12121 1100 AC Amsterdam Zuidoost, THE NETHERLANDS Fax: 31-20-69-418

Table 13.A2   *(cont.)*

Supplier code	Reference
DSM 001	DSM UK Ltd., Kingfisher House, Redditch, Worcestershire, B97 4EX, UK Tel: 01527-68254 Fax: 01527-68949
Dunlop 001	Dunlop, Ltd. GRG Division, Cambridge Street 1, Manchester M30 0BH, UK Tel: 0161-236-2131 Fax: 0161-236-1599
Dupont 001	E.I du Pont de Nemours & Co. 1007 Market Street Wilmington Delaware 19898, USA Tel: (302)-774-1000
Dupont 002	Du Pont (UK) Limited Maylands Avenue, Hemel Hempstead Hertfordshire HP2 7DP, UK Tel: (01442) 218-500 Fax: (01422) 249-463
Dupon 003	Du Pont de Nemours International SA Speciality Polymers 2, Chemin du Pavillon, PO Box 50 CH-1218 Le Grand-Saconnex/GE SWITZERLAND Tel: (22) -717-51-11 Fax: (22)-717-51-09
ELF 001	Elf Actochem North America Inc., 3 Parkway, Philadelphia, PA 19102, USA Tel: 215-587-7000 Fax: 215-587-7930

Table 13.A2 *(cont.)*

Supplier code	Reference
ELF 002	ELF Actochem Agence, Lyon, Chemin de la Lône, BP 40, 69310 Pierre-Benit, FRANCE Tel: 1-72-39-65-00 Fax: 1-72-39-69-73
ELF 003	Elf Actochem (UK) Ltd., Globe House, Bayley Street, Stalybridge, Cheshire SK15 1PU, UK Tel: 0161-338-4411 Fax: 0161-303-1908
ERG 001	ERG Materials Division, 900 Stanford Avenue, Oakland, CA 94608, USA Tel: 510-658-9785
Flexicore 001	Flexicore UK Limited, Earls Colne Industrial Park, Earls Court, Colchester, Essex CO6 2NS, UK Tel: 017875-3502 Fax: 017875-4330
Foamex 001	Foamex Ltd., Technical Products Group, 1500 East Second Street, Eddystone, PA 19022, USA Tel: 1-800-767-4997 Fax: 215-876-2341
GE 001	GE Plastics Ltd. Old Hall Road, Sale Cheshire M33 2HF, UK Tel: 0161-905-5000 Fax: 0161-905-5119

Table 13.A2   *(cont.)*

Supplier code	Reference
GPM 001	General Plastics Manufacturing Co. 4910 Burlington Way, Tacoma, Washington 98409, USA Tel: 206-473-5000 Fax: 206-473-5104
Helf 001	Helf S.P.A., Via Frigielica 2, 35139 Padova, ITALY Tel: 049-664-855
Hi-Tech 001	Hi-Tech Ceramics Inc. PO Box 1105, Alfred, NY 14802, USA Tel: 607-587-9146 Fax: 607-587-8770
Huls 001	Huls UK, Featherstone Road, Wolverton Mill South, Milton Keynes MK12 5TB, UK Tel: 01908-226444 Fax: 01908-224950
Hysol 001	Hysol Grafil Company, PO Box 312, 2850 Willow Pass Road, Pittsburgh, CA 94565, USA Tel: 415-938-5533
Hysol 002	Hysol Grafil Limited, PO Box 16, 345 Foles Hill Road, Coventry CV6 5AE, U.K. Tel: 01203-688-771
ICI 001	ICI Polyurethanes, Everslaan 45-B-3078, Kortenberg, BELGIUM

Table 13A2 *(cont.)*

Supplier code	Reference
Kemira 001	Kemira Polymers, Station Road, Birchvale, Stockport, Cheshire SK12 5BR, UK Tel: 01663-746518 Fax: 01663-746605
Kooltherm 001	Kooltherm, PO Box 3, Charlestown, Glossop, Derbyshire SK13 8LE, UK Tel: 01457-861611 Fax: 01457-852319
Lankro 001	Lankro Chemicals Urethane Div. PO Box 1, Silk Street, Eccles, Manchester M60 1PD, UK Tel: 0161-789-7300 Fax: 0161-788-7886
MacPherson 001	MacPherson Polymers, Station Road, Birchvale, Stockport, Cheshire SK12 5BR, UK Tel: 01663-46518 Fax: 01663-46605
Mepura 001	Metallpulvergesellschaft m.b.H Ranshofen A-5282 Braunau-Ranshofen AUSTRIA Tel: +43-7722-2216 Fax: +43-7722-68154
Mobay 001	Mobay Corporation, Polyurethane Division, Mobay Road, Pittsburgh, PA 15205-9741, USA

Table 13.A2   *(cont.)*

Supplier code	Reference
Perstorp Ferguson 001	Perstorp Ferguson, Aycliffe Industrial Estate, Newton Aycliffe, Country Durham DL5 6EF, UK Tel: 01325-300666 Fax: 01325-300385
Pittsburgh-Corning 001	Pittsburgh Corning 800 Presque Isle Drive Pittsburgh PA 15239, USA Tel: 412-327-6100
Roehm 001	Roehm Limited, Plastics Division, 18–19 Bermondsey Trading Estate, Rotherhithe New Road, London SE16 3LL Tel: 0181-237-2236 Fax: 0181-232-0604
Scot Bader 001	Scot Bader Company Limited, Polyester Division, Wollaston, Wellingborough, Northants NN9 7RL Tel: 01933-663-100
Scotfoam 001	Scotfoam Corp., 1500E, 2nd Street, Eddystone, PA 19013, USA Tel: 215-876-2551 (1-800-222-8470)
Selee 001	Selee Corporation, 700 Shepherd Street Hendersonville, NC 28792, USA Tel: 704-693-0256 Fax: 704-935-0861

Table 13.A2 *(cont.)*

Supplier code	Reference
Versar 001	Versar Manufacturing Inc., 14120-A Sullyfield Circle, Chantilly, VA 22012, USA
Zotefoams 001	Zotefoams Limited, 675 Mitcham Road, Croydon, Surrey CR9 3AL, UK Tel: 0181-684-3622 Fax: 0181-684-7571
Zotefoams 002	Zotefoams Inc., 12 Airport Road, Hackettstown, NJ 07840, USA Tel: 908-850-7294 Fax: 908-850-7216

References

The references for the data-sheets from which the data are drawn are listed in the Appendix.

Ashby, M. F. (1992) *Materials Selection in Mechanical Design*, Pergamon Press, Oxford, UK.

CMS (1995): Cambridge Materials Selector, Granta Design Limited, 20 Trumpington Street, Cambridge CB2 1QA, U.K.

Encyclopedia of Polymer Science and Engineering (1985), Volume 3, 2nd edition, section C, Wiley, N.Y., U.S.A.

Handbook of Industrial Materials (1992), 2nd edition, pp. 537, Elsevier Advanced Technology, Elsevier, Oxford, U.K.

Mills, N. J. and Gilchrist, A. (1991) The effectiveness of foams in bicycle and motorcycle helmets, *Accid. Anal. and Prev.*, **23**, 153–63.

Plascams (1995), Version 6, Plastics Computer-Aided Materials Selector, RAPRA Technology Limited, Shawbury, Shrewsbury, Shropshire, SY4 4NR, U.K.

Saechtling (1983) *International Plastics Handbook*, editor: Dr. Hans Jurgen, Saechtling, MacMillan Publishing Co (English edition), London, U.K.

Seymour, R. P. (1987) *Polymers for Engineering Applications*, ASM International, Metals Park, Ohio 44037, U.S.A.

Appendix

The linear-elasticity of anisotropic cellular solids

The formal description of elastic anisotropy

For some purposes the linear-elastic response of a foam can be thought of as roughly isotropic: its moduli are the same for all directions of loading. Then it is completely characterized by just two moduli (any two of Young's modulus, E^*, the shear modulus, G^*, the bulk modulus, K^*, and Poisson's ratio, v^*, Chapter 5). But most man-made foams are *anisotropic* (Chapter 6): the Young's modulus in the rise direction is often twice as great as that in the other two perpendicular directions and the shear moduli and Poisson's ratios, too, depend on the direction of loading. Natural cellular solids are more anisotropic: the moduli of wood can differ by a factor of 10 along the grain and across it. And honeycombs are more anisotropic still, with moduli normal to the plane of the honeycomb which can be hundreds of times greater than those in-plane. As the symmetry of the material decreases, more moduli are required to describe the elastic response completely. It is helpful to know how many are needed, what they describe, and how they relate to each other.

Consider the cube of solid shown in Fig. A1. It is subjected to normal stresses σ_1, σ_2 and σ_3, and to shear stresses σ_{12}, σ_{23} and σ_{31}. The strains are given by Hooke's law. At its most general, it takes the form (Hearmon, 1961)[†]:

[†]The equations describing linear elasticity are valid for fully dense solids, honeycombs and foams. Asterisks are included only when the equation is specifically for a cellular solid.

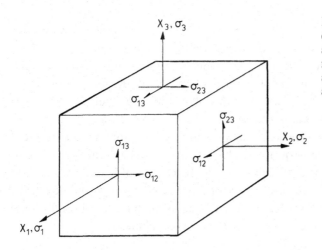

$$\epsilon_1 = S_{11}\sigma_1 + S_{12}\sigma_2 + S_{13}\sigma_3 + S_{14}\sigma_4 + S_{15}\sigma_5 + S_{16}\sigma_6$$
$$\epsilon_2 = S_{21}\sigma_1 + S_{22}\sigma_2 + S_{23}\sigma_3 + S_{24}\sigma_4 + S_{25}\sigma_5 + S_{26}\sigma_6$$
$$\epsilon_3 = S_{31}\sigma_1 + S_{32}\sigma_2 + S_{33}\sigma_3 + S_{34}\sigma_4 + S_{35}\sigma_5 + S_{36}\sigma_6$$
$$\epsilon_4 = S_{41}\sigma_1 + S_{42}\sigma_2 + S_{43}\sigma_3 + S_{44}\sigma_4 + S_{45}\sigma_5 + S_{46}\sigma_6$$
$$\epsilon_5 = S_{51}\sigma_1 + S_{52}\sigma_2 + S_{53}\sigma_3 + S_{54}\sigma_4 + S_{55}\sigma_5 + S_{56}\sigma_6$$
$$\epsilon_6 = S_{61}\sigma_1 + S_{62}\sigma_2 + S_{63}\sigma_3 + S_{64}\sigma_4 + S_{65}\sigma_5 + S_{66}\sigma_6$$

where ϵ_1, ϵ_2 and ϵ_3 are the extensional strains, $\epsilon_4 = \gamma_{23}$, $\epsilon_5 = \gamma_{31}$ and $\epsilon_6 = \gamma_{12}$ are the shear strains, and $\sigma_4 = \sigma_{23}$, $\sigma_5 = \sigma_{31}$ and $\sigma_6 = \sigma_{12}$ are the shear stresses. There is more than one convention for the ordering of the subscripts in this equation; we have used that of Nye (1957) throughout. Using the condensed notation the set of six equations can be written:

$$\epsilon_i = \sum_{i=1}^{6} S_{ij}\sigma_j. \tag{A1}$$

The coefficients S_{ij} (that is, S_{12}, S_{23}, etc.) are the components of the *compliance matrix*. At first sight there are 36 of them, but even for the most asymmetric of materials this is not so. Considerations of strain energy show that the compliance matrix is symmetrical (meaning that $S_{12} = S_{21}$, etc.), reducing the number to 21. This is the number that is needed to describe the most general anisotropic material.

Anisotropy in cellular solids reflects the anisotropy in their structure and that in the cell-wall material itself. If there is nothing directional about either of these – that is, if the cells have equiaxed shapes and the cell walls have constant thickness and uniform properties – then the stiffness and strength are not directional either. Man-made foams are anisotropic because of the way they are made: the foaming process gives flattened or elongated cells, or cell walls which are thicker in one direction than another. Natural cellular solids (like wood,

Chapter 10) are anisotropic because of the way they grow: almost always their cells have an elongated shape and the cell walls themselves are anisotropic in their properties. Honeycombs are anisotropic because of the two-dimensional arrangement of the cells: in the plane of the honeycomb the moduli are determined by cell-wall bending; normal to this plane they are determined by their axial extension or compression. But despite the differing reasons for anisotropy, almost all cellular solids have *orthotropic* symmetry, meaning that the structure has three perpendicular mirror planes. This symmetry reduces further the number of moduli (Voigt, 1910; Hearmon, 1961). If the axes of loading are aligned with the normals of these planes (which form an orthogonal coordinate set: the X_1, X_2 and X_3 axes) the S matrix for an orthotropic material, either fully dense or cellular, simplifies to a set of nine independent compliances:

$$S_{ij} = \begin{bmatrix} S_{11} & S_{12} & S_{13} & - & - & - \\ S_{12} & S_{22} & S_{23} & - & - & - \\ S_{13} & S_{23} & S_{33} & - & - & - \\ - & - & - & S_{44} & - & - \\ - & - & - & - & S_{55} & - \\ - & - & - & - & - & S_{66} \end{bmatrix}. \tag{A2}$$

Of the materials discussed in this book, honeycombs and woods have the least symmetry; then all nine constants are needed for a full description. The constants are directly related to the conventional engineering moduli. The three Young's moduli for loading in the X_1, X_2 and X_3 directions are:

$$E_1 = \frac{1}{S_{11}}, \quad E_2 = \frac{1}{S_{22}}, \quad E_3 = \frac{1}{S_{33}}. \tag{A3}$$

The three shear moduli are:

$$G_{23} = \frac{1}{S_{44}}, \quad G_{31} = \frac{1}{S_{55}}, \quad G_{12} = \frac{1}{S_{66}}. \tag{A4}$$

Poisson's ratio v_{ij} is defined as the negative of the strain in the j direction divided by the strain in the i direction, for nomal loading in the i direction ($v_{ij} = -\epsilon_j/\epsilon_i$), so that:

$$v_{12} = -\frac{S_{21}}{S_{11}} \quad v_{13} = -\frac{S_{31}}{S_{11}} \quad v_{23} = -\frac{S_{32}}{S_{22}}$$
$$v_{21} = -\frac{S_{12}}{S_{22}} \quad v_{31} = -\frac{S_{13}}{S_{33}} \quad v_{32} = -\frac{S_{23}}{S_{33}}. \tag{A5}$$

But only three of these are independent. Substitution from Eqn. (A3), and remembering that $S_{12} = S_{21}$, gives the *reciprocal relations*:

$$\frac{v_{12}}{E_1} = \frac{v_{21}}{E_2} = -S_{12}; \quad \frac{v_{13}}{E_1} = \frac{v_{31}}{E_3} = -S_{13}; \quad \frac{v_{23}}{E_2} = \frac{v_{32}}{E_3} = -S_{23} \tag{A6}$$

and the S matrix becomes:

$$S_{ij} = \begin{bmatrix} \dfrac{1}{E_1} & -\dfrac{v_{21}}{E_2} & -\dfrac{v_{31}}{E_3} & - & - & - \\[2mm] -\dfrac{v_{12}}{E_1} & \dfrac{1}{E_2} & -\dfrac{v_{32}}{E_3} & - & - & - \\[2mm] -\dfrac{v_{13}}{E_1} & -\dfrac{v_{23}}{E_2} & \dfrac{1}{E_3} & - & - & - \\[2mm] - & - & - & \dfrac{1}{G_{23}} & - & - \\[2mm] - & - & - & - & \dfrac{1}{G_{13}} & - \\[2mm] - & - & - & - & - & \dfrac{1}{G_{12}} \end{bmatrix}.$$ (A7)

Expanding, and using the reciprocal relations (Eqn. (A6)) gives the stress–strain relations for an orthotropic material in a particularly simple form:

$$\epsilon_1 = \frac{1}{E_1}(\sigma_1 - v_{12}\sigma_2 - v_{13}\sigma_3)$$

$$\epsilon_2 = \frac{1}{E_2}(\sigma_2 - v_{23}\sigma_3 - v_{21}\sigma_1)$$

$$\epsilon_3 = \frac{1}{E_3}(\sigma_3 - v_{31}\sigma_1 - v_{32}\sigma_2)$$ (A8)

$$\gamma_{23} = \frac{\sigma_{23}}{G_{23}}, \quad \gamma_{13} = \frac{\sigma_{13}}{G_{13}}, \quad \gamma_{12} = \frac{\sigma_{12}}{G_{12}}.$$

It is also useful to examine the moduli of a material with circular symmetry; that is, one that is isotropic in one plane, but has different moduli normal to that plane. Many man-made foams, and some natural ones (like cork, Chapter 12) are like this; and wood is sometimes approximately so. For a material with circular symmetry in the $X_1 X_2$ plane the compliance matrix reduces to five independent components:

$$S_{ij} = \begin{bmatrix} S_{11} & S_{12} & S_{13} & - & - & - \\ S_{12} & S_{11} & S_{13} & - & - & - \\ S_{13} & S_{13} & S_{33} & - & - & - \\ - & - & - & S_{44} & - & - \\ - & - & - & - & S_{44} & - \\ - & - & - & - & - & 2(S_{11} - S_{12}) \end{bmatrix}.$$ (A9)

In terms of the engineering moduli,

$$E_1 = E_2$$
$$G_{23} = G_{31}$$
$$v_{31} = v_{32}$$ (A10)
$$G_{12} = \frac{E_1}{2(1 - v_{12})}$$

and the reciprocal relations (Eqn. (A6)) still hold.

Isotropy, of course, requires that

$$E_1 = E_2 = E_3 = E$$
$$G_{23} = G_{31} = G_{12} = G$$

$$(\text{AII})$$

and

$$v_{12} = v_{21} = v_{13} = v_{31} = v_{23} = v_{32} = v.$$

The compliance matrix then becomes:

$$S_{ij} = \begin{bmatrix} \dfrac{1}{E} & -\dfrac{v}{E} & -\dfrac{v}{E} & - & - & - \\ -\dfrac{v}{E} & \dfrac{1}{E} & -\dfrac{v}{E} & - & - & - \\ -\dfrac{v}{E} & -\dfrac{v}{E} & \dfrac{1}{E} & - & - & - \\ - & - & - & \dfrac{1}{G} & - & - \\ - & - & - & - & \dfrac{1}{G} & - \\ - & - & - & - & - & \dfrac{1}{G} \end{bmatrix}.$$

$$(\text{AI2})$$

The wood literature uses a different terminology. Wood is generally approximated as an orthotropic material, and analysed in the rectangular coordinate system shown in Fig. A2. The X_1, X_2 and X_3 coordinates are replaced by a coincident set called R, T, and A. The *axial* or A-axis (sometimes called the *longitudinal* axis, and replacing our X_3) lies parallel to the axis of the tree, and thus to the grain. The *radial* or R-axis (replacing X_1) lies along a radius of the trunk, and is parallel to the rays in the wood. The *tangential* or T-axis (replacing X_2) lies paral-

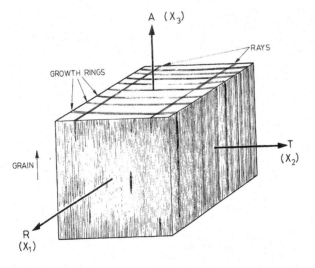

Figure A2 The coordinate system for wood, showing the relationship between the X_1, X_2 and X_3 coordinates and the R, T and A coordinates.

lel to the growth rings which, in a section cut from the trunk, are assumed to be planar. There are then three Young's moduli:

$$E_R^* = \frac{1}{S_{11}^*} \quad E_T^* = \frac{1}{S_{22}^*} \quad E_A^* = \frac{1}{S_{33}^*}. \tag{A13}$$

There are three shear moduli:

$$G_{TA}^* = \frac{1}{S_{44}^*} \quad G_{AR}^* = \frac{1}{S_{55}^*} \quad G_{RT}^* = \frac{1}{S_{66}^*} \tag{A14}$$

(where G_{TR} means a shear stress applied in the RT plane). And from the reciprocal relations (Eqns. (A6)) there are three independent Poisson's ratios:

$$\frac{v_{RT}^*}{E_R^*} = \frac{v_{TR}^*}{E_T^*} = -S_{12}^*; \quad \frac{v_{AR}^*}{E_A^*} = \frac{v_{RA}^*}{E_R^*} = -S_{13}^*; \quad \frac{v_{AT}^*}{E_A^*} = \frac{v_{TA}^*}{E_T^*} = -S_{23}^* \tag{A15}$$

where v_{AT} means: (contraction in T-direction)/(extension in A-direction), for tension in the A-direction, etc. Data for these nine constants are tabulated for a number of woods (Price, 1928; Hearmon, 1948; Dinwoodie, 1981; Bodig and Jayne, 1982). Using them, the strains caused by a general stress state can be calculated by using the expansion of Eqn. (A8) with terms defined by Eqns. (A13) to (A15):

$$\epsilon_R = \frac{1}{E_R^*}(\sigma_R - v_{RT}^*\sigma_T - v_{RA}^*\sigma_A)$$

$$\epsilon_T = \frac{1}{E_T^*}(\sigma_T - v_{TR}^*\sigma_R - v_{TA}^*\sigma_A)$$

$$\epsilon_A = \frac{1}{E_A^*}(\sigma_A - v_{AR}^*\sigma_R - v_{AT}^*\sigma_T) \tag{A16}$$

$$\gamma_{TA} = \frac{\sigma_{TA}}{G_{TA}^*}; \quad \gamma_{RA} = \frac{\sigma_{RA}}{G_{RA}^*}; \quad \gamma_{RT} = \frac{\sigma_{RT}}{G_{RT}^*}.$$

Measurements on many types of wood show that it often has roughly circular symmetry in the RT plane. Then wood has five independent compliances; they are (Eqn. (A10)):

$$E_R^* = E_T^*$$
$$G_{AR}^* = G_{AT}^*$$
$$v_{AR}^* = v_{AT}^* \tag{A17}$$
$$G_{RT}^* = \frac{E_R^*}{2(1 - v_{RT}^*)}$$

and the reciprocal relations (Eqns. (A15)) apply.

References

Bodig, J. and Jayne, B. A. (1982) *Mechanics of Wood and Wood Composites*. Van Nostrand Reinhold, New York.

Dinwoodie, J. M. (1981) *Timber, Its Nature and Behaviour*. Van Nostrand Reinhold, New York.

Hearmon, R. F. S. (1948) Special Report on Forest Products Research No. 7. Her Majesty's Stationery Office, London.

Hearmon, R. F. S. (1961) *Introduction to Applied Anisotropic Elasticity*. Oxford University Press, Oxford.

Nye, J. F. (1957) *Physical Properties of Crystals*. Oxford University Press, Oxford.

Price, A. T. (1928) *Phil. Trans. Roy. Soc.*, **228A**, 1.

Voigt, W. (1910) *Lehrbuch der Kristallphysik*. Trubner, Leipzig.

Index

Page numbers in *italics* refer to major entries.

503

Printed in the United States
By Bookmasters